W9-CGL-836

FUNDAMENTALS OF FLIGHT

FUNDAMENTALS OF FLIGHT

Second Edition

RICHARD S. SHEVELL

Stanford University

PRENTICE HALL, Englewood Cliffs, New Jersey 07632

Library of Congress Cataloging-in-Publication Data

Shevell, Richard Shepard.
Fundamentals of Flight / Richard S. Shevell.–2nd ed.
p. cm.
Includes bibliographies and index.
ISBN 0-13-339060-8
1. Aerodynamics. 2. Airplanes–Design and construction.
I. Title.
TL570.S462 1989
629.132'3–dc19 88-7279 CIP

Editorial/production supervision and
interior design: Gertrude Szyferblatt
Cover design: Wanda Lubelska Design
Manufacturing buyer: Mary Noonan

A Division of Simon & Schuster
Englewood Cliffs, New Jersey 07632

Printed in the United States of America

10 9 8 7 6 5 4 3

ISBN 0-13-339060-8

Prentice-Hall International (UK) Limited, *London*
Prentice-Hall of Australia Pty. Limited, *Sydney*
Prentice-Hall Canada Inc., *Toronto*
Prentice-Hall Hispanoamericana, S.A., *Mexico*
Prentice-Hall of India Private Limited, *New Delhi*
Prentice-Hall of Japan, Inc., *Tokyo*
Simon & Schuster Asia Pte. Ltd., *Singapore*
Editora Prentice-Hall do Brasil, Ltda., *Rio de Janeiro*

To Lorraine

CONTENTS

PREFACE

This second edition of *Fundamentals of Flight* follows the general outline of the first edition but includes many enhancements and some totally new subjects. A new chapter on hypersonic flow is added. Substantial sections are added to previous chapters to cover the subjects of winglets, effects of sweepback on lift curve slope, deep stall, overall propulsion efficiency, ramjets, and elliptical satellite orbits. In addition, dozens of smaller additions improve the depth and breadth of the coverage of many subjects.

OBJECTIVES

The main objective of the book is to provide a clear introductory understanding of the science and engineering of heavier-than-air flight vehicles. The intent is to give the reader a thorough understanding of the basics of fluid mechanics, as required for aeronautics, the production of lift and drag, the importance and effects of viscosity and compressibility, methods of estimating performance, the elements of stability, and the impact of airplane design characteristics on performance and stability. Lift, drag dependent on lift, and other phenomena, such as downwash, ground effect, and wake vortex turbulence, can be understood properly only by studying the theory of circulation. Therefore, the theory of circulation about an airfoil is discussed extensively. The book includes a history of aeronautical development, emphasizes applied aerodynamics, and provides insight into the technologies of propulsion and structures. Since my background includes many years in the aircraft industry, as well as almost two decades of university teaching, I have tried to explain, in appropriate parts of the text, the major parameters that affect the design integration of an airplane. Wing design and propulsion integration are examples of this. Drag prediction and critical Mach number estimation methods are examples of the use of extensive applied aerodynamic experience adapted to an introductory text.

The elements of rocket propulsion, rocket trajectories, and orbital performance are also presented.

A secondary objective is to show the relation between indispensable and beautiful pure theory, and the empirical modifications to theory that so often are necessary to fine-tune the theoretical results to agree with real life. To obtain solutions to theoretical formulations of physical phenomena, it is usually necessary to make some simplifications. The empiricism corrects for these simplifications. Sometimes, relatively simple methods give excellent answers and illustrate the essential relationships governing important physical phenomena. Such methods have been included in this book not only to permit estimation of aerodynamic characteristics, such as induced drag, or performance capabilities, such as takeoff distance, but also to demonstrate how these important characteristics are affected by changes in aircraft design parameters.

AUDIENCE

The book is primarily intended for a one-semester undergraduate course. It can be used in a one-quarter course, although some subjects may have to be given light treatment. The book will serve well as a first introduction to the essential elements of aeronautics for students planning a career in aeronautical engineering. It should provide good preparation for in-depth courses in theoretical aerodynamics, flight mechanics, propulsion, and structures. The book also is very suitable for students in other engineering disciplines who wish to acquire breadth in their technical knowledge or are just interested in aeronautics for its own sake. Students in nontechnical fields such as business or the social sciences may acquire a sound understanding of aircraft to pursue an avocational or secondary technical interest. I have found among students a wide range of reasons for pursuing such a course; interested students vary from advanced freshmen to graduate students. The desirable prerequisites are elementary physics and an understanding of calculus fundamentals. Calculus is used in many of the derivations, although the methods and most of the concepts can be understood without it.

I also hope that practicing aeronautical engineers will find the book a useful reference for the review of flight fundamentals with which they do not normally work. Obviously, an introductory text cannot pretend to educate a specialist in his or her own field.

UNITS

The subject of units is a difficult one. United States policy is to convert to Systeme International (SI) units. However, the aircraft industry has generally disregarded this policy in its internal work. Most teaching is in SI units, but U.S. aircraft data are almost always presented in English units. The result is that a student must become familiar with both the SI and the English engineering systems of units, an unfortunate complication. This book is generally written using *SI units with English units following in parentheses*. Example problems alternate between the systems. The performance section is primarily in English units because that is the way all U.S. aircraft performance data are given. Few American aeronautical engineers have any feeling for specific range in kilograms of fuel per kilometer or engine power in kilowatts. Perhaps this unfortunate necessity to cope with two sets of units will disappear someday, but it will take decades, not years.

ACKNOWLEDGMENTS

I am indebted to many colleagues and students who have reviewed all or portions of the manuscript for this book and have made many valuable suggestions. I owe particular thanks to Professors Donald Baganoff, Arthur Bryson, I-Dee Chang, Nicholas Hoff, and Ilan Kroo of Stanford University, Professor William Sears of the University of Arizona, Professor John Bertin of the University of Texas, and Dr. John Scull of the Jet Propulsion Laboratory. Mr. Ron E. G. Davies and the Smithsonian Air and Space Museum have provided many of the historical photographs. I am grateful to Maribel Calderon, whose skillful typing and retyping of the original manuscript contributed so much. Finally, I would like to acknowledge the indirect contribution of the late Dr. Clark Millikan, former chairman of the aeronautics department at the California Institute of Technology, under whom I studied many years ago. Dr. Millikan was a great teacher, whose clear thinking and articulate presentations have always been an inspiration.

Richard S. Shevell

Conversion Factors Between SI Units and English Units

To obtain numerical solutions to some of the homework problems or to many of the applications in industry, it is necessary to be able to convert English-unit values for various parameters to the equivalent SI-unit values, and vice versa. A table of conversion factors appropriate to the problems discussed in this text is presented here.

Mass: 1.00 kg = 0.06853 slug

1.00 slug = 14.592 kg

At the surface of the earth, an object with a mass of 1.00 kg weighs 9.8 N or 2.205 lb, and an object with a mass of 1.00 slug weighs 32.17 lb or 143.1 N.

Length: 1.00 m = 3.2808 ft

1.00 ft = 0.3048 m = 30.48 cm

Force: 1.00 N = 0.2248 lb

1.00 lb = 4.4482 N

Temperature: 1.00 K = 1.8°R °R = °F + 460

1.00°R = 0.5556 K K = °C + 273

Pressure: 1.00 N/m^2 = 1.4504 × 10^{-4} lb/in.^{2} = 2.0886 × 10^{-2} lb/ft^2

1.00 lb/in.^{2} = 6.8947 × 10^3 N/m^2

1.00 lb/ft^2 = 47.88 N/m^2

Velocity: 1.00 m/s = 3.2808 ft/s = 2.2369 mi/h

1.00 ft/s = 0.6818 mi/h = 0.3048 m/s

Density: 1.00 kg/m^3 = 1.9404 × 10^{-3} slug/ft^3

1.00 slug/ft^3 = 515.36 kg/m^3

Viscosity:

$$1.00\frac{\text{kg}}{\text{m}\cdot\text{s}} = 2.0886 \times 10^{-2}\frac{\text{lb}\cdot\text{s}}{\text{ft}^2}$$

$$1.00\frac{\text{lb}\cdot\text{s}}{\text{ft}^2} = 47.879\frac{\text{kg}}{\text{m}\cdot\text{s}}$$

Specific heat:

$$1.00\frac{\text{N}\cdot\text{m}}{\text{kg}\cdot\text{K}} = 1.00\frac{\text{J}}{\text{kg}\cdot\text{K}} = 2.3928 \times 10^{-4}\frac{\text{Btu}}{\text{lb}_\text{m}\cdot{}^\circ\text{R}}$$

$$= 5.9895\frac{\text{ft}\cdot\text{lb}_\text{f}}{\text{slug}\cdot{}^\circ\text{R}}$$

$$1.00\frac{\text{ft-lb}}{\text{slug}^\circ\text{R}} = 1.6728 \times 10^{-1}\frac{\text{N}\cdot\text{m}}{\text{kg}\cdot\text{K}} = 1.6728 \times 10^{-1}\frac{\text{J}}{\text{kg}\cdot\text{K}}$$

Frequently Used Equivalents

1 bhp = 550 ft-lb/s = 33,000 ft-lb/min
1 knot (i.e., 1 nautical mile per hour) = 1.152 statute miles per hour
1 knot (nautical mile per hour) = 1.69 ft/s
1 statute mile per hour = 0.868 knot (nautical miles per hour)
1 statute mile per hour = 1.467 ft/s
1 ft/s = 0.682 statute mile per hour
1 ft/s = 0.592 knot (nautical miles per hour)
1 kilometer = 0.621 statute mile
1 kilometer = 0.539 nautical mile
1 statute mile = 1.609 kilometers
1 nautical mile = 1.854 kilometers
1 radian = 57.3 degrees

Note that the preceding values are *equivalents*. The conversion factors are the reciprocals. For example, to convert a speed in knots to a speed in feet per second,

$$V_{\text{ft/sec}} = V_{\text{knots}} \times 1.69$$

Frequently Used Constants

γ = 1.4 (air)

gas constant, R (air) = 287.05 newton-meters/kilogram K
= 1718 ft-lb/slug °R

Specific heat, c_p (air) = 1004.7 newton-meters/kilogram K (joules/kg K)
= 6006 ft-lb/slug °R

Gravitational constant at sea level, g_0 = 9.8 m/s^2
= 32.17 ft/s^2

Radius of the earth, r_0 = 6.378×10^6 m
= 20.92×10^6 ft

NOMENCLATURE

English Symbols

a	acceleration; speed of sound; temperature lapse rate in the atmosphere $= dT/dh$; wing lift curve slope; fractional increase in velocity through a propeller; semimajor axis of ellipse
a.c.	aerodynamic center of the wing
a_0	two-dimensional section lift curve slope
A	area of a propeller or actuator disk
AR	aspect ratio $= b^2/S$
b	wing span; fractional increase in slipstream velocity far behind a propeller; semiminor axis of ellipse
bhp	engine brake horsepower
BSFC	brake specific fuel consumption, same as c for shaft engines
c	chord length; specific fuel consumption; linear eccentricity of ellipse
$\bar{c}$	mean aerodynamic chord or m.a.c.
CAS	calibrated airspeed
c_d	two-dimensional or section drag coefficient
C_D	drag coefficient based on the reference area for the vehicle $= D/qS$
C_{D_i}	induced drag coefficient $= D_i/qS$
C_{D_P}	parasite drag coefficient $= D_P/qS$
ΔC_{D_C}	incremental drag coefficient due to compressibility
c_e	effective exhaust velocity of a rocket motor
C_f	total smooth flat plate skin friction drag coefficient, the average drag coefficient over a surface from the leading edge to the trailing edge
C_{fl}	local smooth flat plate skin friction drag coefficient at a point l units from the leading edge $= \tau/q$
C_F	force coefficient
c_l	two-dimensional or section lift coefficient
$c_{l_{\max}}$	section maximum lift coefficient

C_L	lift coefficient
$C_{L_{\text{MAX}}}$	wing or airplane maximum lift coefficient
C_M	moment coefficient
c_p	specific heat at constant pressure
C_P	pressure coefficient $= \Delta p/q$; propeller power coefficient
C_R	root chord of an airfoil surface
C_T	tip chord of an airfoil surface
c_v	specific heat at constant volume
d	diameter
d_{LO}	distance to lift off
D	total drag; propeller diameter
D_i	induced drag
D_p	parasite drag
e	airplane efficiency factor, the total correction factor applied to the equation for induced drag to account for deviations from ideal elliptic lift distribution and for parasite drag variations with C_L; eccentricity of ellipse
E	total aircraft energy
EAS	equivalent airspeed
f	equivalent parasite drag area $= C_{D_p} S_{\text{REF}}$
F	force; thrust; F factor correction $= V_E/V_{\text{CAL}}$; focus of ellipse
F_G	gross thrust of a jet engine
F_N	net thrust of a jet engine
g	gravitational constant, the acceleration due to gravity
g_0	gravitational constant at sea level
h	height or altitude
h_a	absolute altitude from the center of the earth
h_b	rocket altitude at the end of fuel burn
h_e	specific energy of a vehicle
h_f	heat energy available per unit weight of fuel
I_{sp}	specific impulse or specific thrust
I_t	total impulse
J	propeller advance ratio $= V/nD$
K	form factor for surface or bodies
K_L	induced drag correction due to ground effect
l	distance from the leading edge of a surface
l_H	distance from the center of gravity to the aerodynamic center of the horizontal tail
l'_H	distance from the aerodynamic center of the wing to the aerodynamic center of the horizontal tail
l_V	distance from the center of gravity to the aerodynamic center of the vertical tail
L	lift; overall characteristic length of a wing or body
L/D	ratio of lift to drag
L'	lift per unit span
m	mass
m.a.c.	mean aerodynamic chord or $\bar{c}$
M	mass; pitching moment; Mach number; mass flow through a propeller or actuator disc per unit time
$\overline{M}$	molecular weight of a gas
M_{cc}	crest critical Mach number, the freestream Mach number at which the local Mach number perpendicular to the isobar at the crest becomes 1.0

M_{DIV}	drag divergence Mach number, the freestream Mach number at which the drag starts to rise abruptly
M_1	rocket mass without propellant
mi/lb	specific range, miles flown per pound of fuel consumed
n	propeller revolutions per second; load factor
p	pressure
q	dynamic pressure $= \rho V^2/2$
r	radius; radius from the center of the earth
r_a	radius from the center of the earth at apogee
r_0	radius of the earth
r_p	radius from the center of the earth at perigee
R	gas constant $= p/\rho T$; radius of a turn; mass ratio of a rocket; range
$\overline{R}$	universal gas constant
RC	rate of climb
RN	Reynolds number based on the characteristic length $L = \rho VL/\mu$
RN_l	local Reynolds number based on the distance l from the leading edge
s	induced drag correction factor due to fuselage interference
S	area, either cross section, surface or wing platform as the context requires
S_π	frontal or cross section area for bodies
S_{REF}	reference area upon which the coefficients are based; S_{REF} is usually the wing planform area
S_{WET}	wetted area
t	time; thickness
t_b	burn-out time of a rocket
t_e	endurance
t/c	airfoil or wing thickness ratio
t_R	airfoil thickness at the root of a wing
t_T	airfoil thickness at the tip of a wing
T	time when referring to the dimension of time; temperature; thrust
thp	thrust horsepower
thp_{avail}	thrust horsepower available
thp_{req}	thrust horsepower required for level flight
thp_{excess}	excess thrust horsepower available for climbing or acceleration
u	induced drag correction factor for deviation from an elliptic lift distribution
U_g	vertical gust velocity
V	velocity
V_a	velocity at apogee
V_b	burn-out velocity when all rocket fuel is consumed
V_{CAL}	calibrated airspeed
V_e	exit velocity of a jet; escape velocity of a rocket
V_E	equivalent airspeed
V_{IND}	indicated airspeed
V_{LO}	lift-off speed
V_o	undisturbed freestream velocity
V_p	velocity at perigee
V_S	speed at the stall; satellite or orbital velocity
v	local velocity
w	downwash velocity
W	weight

W_F	fuel flow per unit time; fuel used (in energy analysis)
W_f	final cruise weight
W_i	initial cruise weight
x	distance from the wing aerodynamic center to the center of gravity, positive with the center of gravity aft

Greek Symbols

α	angle of attack of the reference line with respect to the freestream flow. The reference line is usually the chordline for an airfoil and the fuselage reference line for an airplane.
α_a	absolute angle of attack, i.e., the geometric angle of attack above the angle for zero lift $= \alpha - \alpha_{L=0}$
α_e	effective angle of attack $= \alpha_a - \epsilon$
$\alpha_{L=0}$	angle of attack for zero lift
β	oblique Mach wave angle = 90 degrees − sweepback angle of wave; angle of sideslip
δ	flow deflection angle in supersonic flow; boundary layer thickness
ϵ	downwash angle
γ	ratio of the specific heat at constant pressure c_p to the specific heat at constant volume c_v; flight path angle
Γ	circulation, the strength of a circulatory or vortex flow; dihedral angle
Λ	angle of sweepback
μ	coefficient of viscosity; braking coefficient of friction
μ'	rolling coefficient of friction
ν	kinematic viscosity $= \mu/\rho$
η	empirical factor applied to theoretical two-dimensional airfoil lift curve slope; propeller efficiency
η_H	tail efficiency
η_{overall}	overall propulsion efficiency
η_{prop}	propulsive efficiency
η_{th}	thermal efficiency
ρ	mass density
σ	density ratio, the ratio of the air density to the density at sea level in the standard atmosphere; taper ratio = tip chord/root chord
τ	shear stress; orbital period
ϕ	angle of bank

Subscripts

ac	aerodynamic center
C	compressible
cg	center of gravity
CR	critical
e	exit
H	horizontal tail
INC	incompressible
l	lower surface

n	neutral point
o	freestream conditions: This subscript is used to avoid any ambiguity. When the quantity, such as V, p, T, etc., clearly refers to the freestream, the subscript may be omitted.
s	sea level standard atmosphere conditions
t	test section
T	total, stagnation or reservoir conditions
u	upper surface
w	wall
W	wing

Superscripts

*	designates channel throat where $M = 1.0$

1

A BRIEF HISTORY OF AERONAUTICS

Man's urge to fly must have originated in prehistoric times. The beauty, grace, and freedom of soaring birds has always drawn our admiration and envy. The circling of a hawk in a rising column of air is a beautiful sight as the bird maneuvers in carefully controlled turns to stay within the updraft (Figure 1.1). The freedom to move in any direction over all obstacles is a capability that all of us would enjoy.

ORNITHOPTERS

Early concepts of flight sought to imitate birds by the use of flapping wings. A flying machine utilizing flapping wings is called an *ornithopter*. The famous Greek legend about Daedalus and his son, Icarus, told of Daedalus fashioning wings of feathers and wax to escape from the labyrinth of Cretan King Minos. Icarus, a head-strong youth, ignored his father's advice to avoid flying too close to the sun. The heat of the sun's rays melted the wax and the wings disintegrated, plunging Icarus into the sea (Figure 1.2). Leonardo da Vinci (1452–1519) created dozens of sketches of flying machines, mostly based on the flapping-wing principle. Detailed drawings showed the levers and pulleys through which the pilot would flap the wings. All this was doomed to failure because the remarkable physiological capabilities of birds can never be matched by human beings. The energy output per unit weight of birds is indicated by the 800 heart beats per minute of a sparrow or the respiration rate of 400 per minute exhibited by a pigeon in flight.

Figure 1.1 Soaring hawk. (Courtesy of K. W. Gardiner.)

Figure 1.2 Icarus. (Courtesy of National Air and Space Museum, Smithsonian Institution.)

LIGHTER-THAN-AIR CRAFT

The idea of filling a closed container with a substance that normally rises through the atmosphere was proposed as early as the thirteenth century. Roger Bacon proposed filling a thin metal globe with ethereal air or liquid fire. The exact description and source of ethereal air or liquid fire was omitted. Another suggestion was the use of water vapor, which would rise as does the dew from the grass in early morning. Following the invention of the air pump in 1650, the concept of evacuating a thin sphere was proposed. The sphere would then float in the atmosphere. The problem was to make the evacuated sphere light enough and yet strong enough to avoid collapsing under the pressure of the outside air. Also, there might have been difficulty getting down, an issue that does not seem to have been considered at that time.

Although history records various other attempts at lighter-than-air devices, the Montgolfier brothers, in France, are generally conceded to have built the first practical balloons. The Montgolfiers were paper mill owners with a scientific interest. In 1782, observing sparks and smoke rising in a fireplace, they sought to capture the gas produced by the fire and contain it in a bag. Starting with a small bag made of silk, they burned paper beneath an opening in the bottom and saw the bag swell and then rise to the ceiling of the room. They repeated this outdoors with bags made of paper lined with silk. They were using a fire produced by burning chopped wood and straw and thought they had discovered a new gas with remarkable properties. This they called Montgolfier gas. The mysterious gas was really nothing but hot air which, because of the elevated temperature, has a reduced density. In 1783, the Montgolfiers carried out a public demonstration of a larger balloon with a circumference of more than 100 ft (Figure 1.3). A height of about 6000 ft was attained. This event attracted the attention of the French Academy of Sciences, which authorized a young physicist, J. A. C. Charles, to perform further research. Charles soon realized that the "Montgolfier gas" was not nearly as effective as hydrogen in providing lift. He, therefore, began the development of hydrogen balloons with basic features that differ little from modern balloons. Hydrogen and hot-air balloons were developed in parallel with dozens of flights in the decade following the Montgolfiers' first flight. Accidents were few, the most dramatic being the result of an attempt by Jean-François Pilatre de Rozier, a physicist who had been the first man to fly in a Montgolfier balloon, to combine the lifting efficiency of hydrogen and the lift control of the burners on hot-air balloons. The explosive death of Pilatre in 1785 cooled the enthusiasm for hot-air balloons. A few years later the French revolution set back the level of all ballooning.

Although balloon enthusiasts continued their activities in Europe and America, the balloon was basically handicapped by a total lack of directional control. Tethered balloons were used for observation in the American civil war. In 1870–1871, the French used balloons for observation and transportation during the siege of Paris in the Franco-Prussian war. But the use of lighter-than-air craft was severely limited until power plants could be developed with a ratio of power to weight that permitted efficient propulsion. In the mid-nineteenth century, several models of people-carrying elongated balloons were built that began to combine the concepts of streamlined shape to reduce drag and engine-driven propellers to provide propulsion. Both steam and electric engines were tried but were too heavy to produce more than marginal vehicles. The first successful combination of the increasingly efficient internal combustion

Figure 1.3 Montgolfier balloon. (Courtesy of National Air and Space Museum, Smithsonian Institution.)

engine and the streamlined balloon was built by a rich Brazilian named Alberto Santos-Dumont who was living in Paris. This courageous and persistent man designed, built, and flew more than a dozen small airships (Figure 1.4). He won public acclaim and aroused public interest in flight by winning a 100,000-franc prize for being the first to fly a round trip from Saint-Cloud to the Eiffel Tower in less than 30 minutes. Santos-Dumont later successfully turned his attention to heavier-than-air craft.

Figure 1.4 Santos-Dumont airship. (Courtesy of National Air and Space Museum, Smithsonian Institution.)

Figure 1.5 Dirigible Akron. (Courtesy of Goodyear Aerospace.)

The final development of lighter-than-air machines was led by Count Ferdinand von Zeppelin, whose name is synonomous with large rigid *dirigibles*. The term "dirigible" really means controllable. "Rigid" means that there is a structure that gives the airship its shape even when there is no gas pressure. In a rigid airship the gas is contained in large bags inside the rigid framework. Zeppelin started building rigid airships in 1900. Germany used some of these in World War I. In the early 1930s, the German Graf Zeppelin was the most famous airship and made several flights to the United States, in addition to running a scheduled service to South America. The larger Hindenburg was equally successful until destroyed by fire in 1937 while landing at Lakehurst, New Jersey, at the conclusion of a trans-Atlantic flight. These airships used flammable hydrogen for lift. United States dirigibles used helium, less efficient but nonflammable. However, the large American naval rigid airships, the Akron (Figure 1.5) and the Macon, built by the Goodyear Zeppelin Co., were lost in bad-weather mishaps in 1933 and 1935. These large machines, almost 800 ft long, were not built for violent weather, but their operator, the U.S. Navy, seemed to ignore these limitations. The lighter-than-air era ended with the loss of these three airships. In the mid-1970s, a brief sentimental attempt to revive the huge and dramatic lighter-than-air ship as a partial solution to the energy shortage was not successful. Low-speed, low-operating-cruise altitudes, and fundamentally high costs make lighter-than-air craft economically undesirable and operationally undependable. They are at least two to three times more expensive per cargo ton-mile than comparable airplanes that offer six times the speed.

HEAVIER-THAN-AIR CRAFT

Sir George Cayley

The development of the airplane has its beginnings in the realization that human beings should copy the soaring bird and not the flapping one. The first design concept that utilized a fixed wing for lift and different surfaces for control and propulsion was created by Sir George Cayley (1773–1857). An English nobleman, Cayley was a well-educated person with a deep interest in science. In 1796, he experimented with model helicopters. Three years later, when he was only 26 years of age, he inscribed on a silver disk a drawing of an airplane (Figure 1.6). The design showed a fixed wing for lift, horizontal and vertical tail surfaces for stability and control, and paddles for propulsion. On the other side of the disk, he engraved a force diagram that divided the resultant force on the wing into lift and drag components. In 1804, he built a whirling-arm test rig. Airfoils were carried at the end of the arm, and the forces on the airfoils were measured. This was really the first aerodynamic test of the type run in wind tunnels, although wind tunnels blow air over the models instead of whirling the models through the air.

Cayley extended his aeronautical research to kites based on his glider design. He hinged the aft part of the model to vary the tail incidence angle, a great step forward. He discovered that setting the wings with a slight dihedral angle to each other provided lateral stability. In 1809, Cayley published the results of his work in Nicholson's *Journal of Natural Philosophy*. He defined the flying problem as being the requirement "to make a surface support a given weight by the application of power to the resistance of air." Cayley really understood many of the fundamentals of flight. He seems to have turned to other interests for about 30 years, but in the period from 1848 to 1854 he revived his aeronautical work. In 1849, he built a glider that carried a 10-year-old boy to a height of several meters. In 1853, George Cayley built and flew a person-carrying glider. It is said this glider carried Cayley's coachman several hundred yards across a shallow valley. The coachman immediately resigned to avoid having to repeat the experience.

Although a few enthusiasts tried to carry on Cayley's work, his results seem to have been largely forgotten until the 1940s. Renewed interest in Cayley's work led to

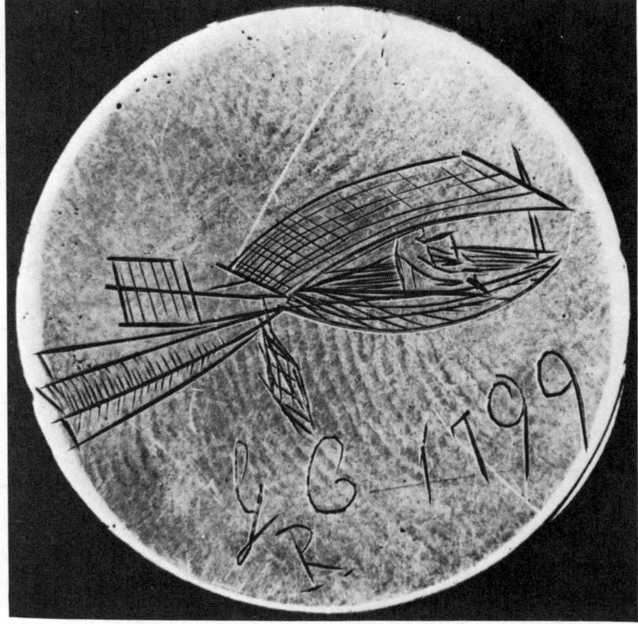

Figure 1.6 Sir George Cayley's person carrier. (Courtesy of Science Museum, London.)

several publications, among them a book published in 1962 by the British aviation historian Charles H. Gibbs-Smith, entitled *Sir George Cayley's Aeronautics*. It appears that later aviation pioneers developed their own ideas without much knowledge of Cayley's conclusions. A student of Cayley's works would have learned most of the concepts necessary for flight. Powered flight would not have been possible in Cayley's time in spite of his clear understanding because no power plant existed with a sufficiently high ratio of power to weight. In fact, this problem remained unsolved for person-carrying airplanes until the Wright brothers built the Wright Kitty Hawk Flyer in 1903.

Although many efforts to fly continued in the next 40 years with attempts to use steam engines, none resulted in a controllable machine capable of sustained flight. Usually, the flights were so short that even controllability was not demonstrated. The next important step forward was the design and repeated flights of controllable gliders by Otto Lilienthal, whose methodical engineering and flight studies of gliders provided much of the inspiration and data for the developments that led to successful powered flight.

Otto Lilienthal

Otto Lilienthal was the first of the great pioneers to fly heavier-than-air craft in a consistent and frequent pattern. Although there were theoreticians studying fluid flow and finding that according to existing theory neither lift nor drag could exist, and there were experimentalists studying the factors influencing lift and drag, few of these investigators actually flew. Born in 1848, Lilienthal graduated from the Berlin Technical Academy as a mechanical engineer in 1870. He was always interested in flight but did not turn seriously to aeronautics until the late 1880s. He published a book *Bird Flight as the Basis of Aviation* in 1889. He studied both the structure and the aerodynamics of bird wings. His book contained aerodynamic data on wings that were later used, and corrected, by the Wright brothers. He felt the need to fly himself in order to learn and stated: "The only way that leads us to a quick development in human flight is a systematic and energetic practice in actual flying experiments." Lilienthal put this into practice by making over 2000 glides between 1891, when he built his first successful glider, and 1896, when a gust caused his glider to stall and crash. Lilienthal suffered fatal injuries.

Lilienthal used birdlike wing planforms. The gliders had wings with cambered airfoil sections and vertical and horizontal tail planes (Figure 1.7). Control was obtained by shifting weight just as in modern hang gliders. Lilienthal recognized lift and control as two major requirements for flight. At his death, he was preparing to attempt powered flight, although he does not seem to have discovered the concept of the propeller. Instead, he was experimenting with powered flapping wing tips. Photos of his flights as well as his technical papers and aerodynamic data were an inspiration to those who pursued the enticing world of flight after him.

Among those who followed Lilienthal's work was Octave Chanute, a French-born naturalized American citizen living in Chicago. A well-established civil engineer, Chanute became interested in flight about 1875 and began to collect and correlate all the aeronautical information he could find. In 1894, he published a book called *Progress in Flying Machines,* the first factual and trustworthy history of human

Figure 1.7 Lilienthal glider. (Courtesy of National Air and Space Museum, Smithsonian Institution.)

attempts to fly. In addition, Chanute began to design gliders in 1896 using his bridge-building knowledge to improve the wing structural strength. Chanute sought designs with built-in stability. He designed and built many gliders with a rear tail, his best being a biplane. No movable control surfaces were utilized, control still being obtained by using body movements to change the center of gravity position (Figure 1.8). Chanute's greatest contribution was the collection and dissemination of knowledge. After the Wright brothers read Chanute's book, they began a correspondence that blossomed into a firm friendship that gave great encouragement and assistance to the Wrights during the exciting and frustrating years in pursuit of powered flight.

Orville and Wilbur Wright

Wilbur Wright was born in 1867, four years earlier than his brother Orville. They were brought up in Dayton, Ohio. The brothers' first acquaintance with flying is said to have been playing with a toy helicopter in 1878. Their first business venture was publishing a neighborhood newspaper. Following the invention of the bicycle, they opened a shop where they engaged in the manufacture and repair of bicycles.

About 1894 or 1895 they learned of the gliding activities of Lilienthal. Obviously attracted by the possibilities of flight, they wrote to the Smithsonian Institution seeking more information. The reply included reports on the work of Lilienthal and other experimenters. The Smithsonian emphasized the value of Chanute's writings.

The Wrights methodically studied these references. They had observed that birds seemed to twist their wings to maintain lateral balance and were critical of Lilienthal's control by pure weight shifting. This led to the concept of wing warping and started the Wright's preoccupation with airplane control, a subject that had been given relatively little attention by earlier experimenters. The Wrights built and flew a box

Figure 1.8 Octave Chanute's glider. (Courtesy of National Air and Space Museum, Smithsonian Institution.)

kite controlled by warping the wing tips through four control strings. From the knowledge thus gained, they built their first glider, a biplane with wing warping capability and a forward control surface for pitch control. Inquiries to the Weather Bureau seeking a suitable flight test area with hills, high prevailing winds, and soft ground for landing resulted in the selection of Kitty Hawk, North Carolina, as their initial test site.

The first tests of the glider in 1900 were flown mostly using the glider as a kite. Although a lot was learned, the tests were not too successful. The Wrights built a second glider with a larger span, 22 ft, and returned to Kitty Hawk in 1901. Results were much improved. Several hundred glides were made, some up to 370 ft in length. The Wrights had expected better performance, however, and decided that the basic airfoil data they were using, mostly from Lilienthal, was probably to blame. Upon their return to Dayton, they began a systematic investigation of the properties of different airfoil shapes. At first they used a rig on a bicycle to compare the forces on two airfoils. Later they built a wind tunnel, 6 ft long with a test section 16 in. square. They tested over 200 model wings of different camber, thickness ratio, aspect ratio, and wing tip shape. Their results were consistent but differed from previous data, so they decided to rely only on their own results.

From these aerodynamic studies came the 1902 glider, a machine with a 32-ft span, a 5-ft chord, and a forward elevator farther ahead to give a better moment arm

(Figure 1.9). In the 1901 tests they had encountered adverse yaw due to the wing warping, although it was not recognized as such for some time. Adverse yaw is a yawing or turning moment in a direction opposite to that naturally resulting from banking the airplane. It results from increased drag on the side of the wing providing the additional lift. The need for a yaw control and directional stability eventually became clear, so an aft-controllable vertical surface was added. The 1902 glider was the first aircraft with three-axis control—pitch, roll, and yaw—a great landmark in the development of aircraft. In 1902, a total of over 800 flights were made at Kitty Hawk with this very successful glider. Distances up to 600 ft and more, in a strong headwind, and durations up to a minute were attained. The great problem of control had been conquered.

Filled with confidence that they knew how to fly, the Wrights then turned their attention to power. Unable to find an engine manufacturer to meet their specification of developing 8 hp with an engine weight less than 200 lb, they designed and built their own engine. In this they were greatly aided by their bicycle mechanic, Charlie Taylor. Taylor's engine actually developed about 12 hp, but almost 3 hp was consumed by the chain-drive system that transmitted the power to the two pusher propellers.

The design and construction of the propellers were equally important challenges. Little knowledge existed on this subject, but the Wrights seem to have worked out a generally correct theory of propeller operation. Using the results of their airfoil experiments, they made surprisingly good estimates of propeller efficiency.

A new airframe was built for the 1903 *Kitty Hawk Flyer* (Figure 1.10). The Wrights returned to Kitty Hawk for their first attempt at powered flight. After a failure due to overcontrolling the craft, which responded differently from their gliders, the Wrights finally demonstrated true powered flight on December 17, 1903. Alternating as pilots, they made four flights, the longest being over a ground distance of 852 ft against a 24-mph headwind. The equivalent still-air distance was over half a mile. These were indeed the first powered flights of a heavier-than-air machine that met the criteria of rising from level ground, flying forward under control without loss of speed,

Figure 1.9 Wright Brothers 1902 glider flight. (Courtesy of National Air and Space Museum, Smithsonian Institution.)

Figure 1.10 The Wright Brothers *Kitty Hawk Flyer* (December 17, 1903). (Courtesy of National Air and Space Museum, Smithsonian Institution.)

and landing without damage at a point as high as the takeoff point. Human beings had finally really flown.

Following the fourth flight, a strong gust tipped over the aircraft. It was badly damaged and the *Kitty Hawk Flyer* never flew again. The Wrights set about building a new aircraft. The 1904 *Flyer* was quite similar in design but had a more powerful engine. It was flown from a pasture near Dayton, Ohio. Because of the higher altitude and warm weather, which reduced engine power, the Wrights developed a catapult to help launch the machine. They made over 100 flights, some as long as 5 minutes. In 1905, they built an improved airplane and flew 24 miles with a duration of 38 minutes. They offered to demonstrate their machine to the War Department, which refused to believe that flight was possible and rejected their invitation. In 1908, the Wrights gave very successful demonstrations in Europe, where they were received enthusiastically by the press and even by royalty. Not until 1908 did the Wrights get to demonstrate their capabilities to the American military. In 1909, the first Wright airplane was delivered to the Army. This airplane was conceptually very similar to the original *Kitty Hawk Flyer*. However, about this time the first versions with the elevators in back were built.

The Wrights have sometimes been pictured as persistent, lucky bicycle mechanics. In truth, they were fine engineers. They lacked formal training in mathematics and science, but they possessed logical, curious, orderly minds that are basic to good engineering. The Wrights were methodical students, read everything available, and organized and interpreted the available data and the results they generated themselves. They ran their own extensive wind tunnel tests. The recognition that birds warp their wings for lateral control led to appreciating the importance of control in human flight. Therefore, they built the wing warp system, a movable elevator, and later a rudder. After achieving control in their gliders, they concentrated on a light internal combustion engine and on propeller design. They did a real engineering job and knew what they were trying to do at each stage. Their success was not luck but the result of methodical, rational, persistent progress.

AFTER THE WRIGHT BROTHERS

Following the Wrights' success, a flurry of aeronautical activity took place in Europe. Notable among these was the work of Santos-Dumont, whose 1907 airplane was less advanced, however, than the Wrights' 1903 aircraft. In France, Louis Bleriot built a historic monoplane in which he flew across the English Channel in 1909 (Figure 1.11). This flight really showed the potential of the airplane for transportation. The British felt less safe on their island and began to ponder the need for defenses other than their powerful navy. Bleriot gave the initiative in airplane development to the French, a lead they held for some years. In the United States, Glenn Curtiss designed and built a series of successful aircraft, notable among which were marine aircraft, both on floats and with flying boat hulls.

World War I gave a tremendous impetus to aircraft development. Airplanes were at first considered useful only for reconnaissance. In a few short years, 1914–1918, aircraft were first armed with pistols to stop enemy observation planes, then fitted with machine guns, after means were developed to avoid shooting off the propellers, and finally equipped with bombs. Airplanes were designed specifically to be reconnaissance airplanes, fighters, or bombers (Figure 1.12).

Figure 1.11 The monoplane in which Bleriot crossed the English Channel in 1909. (Courtesy of National Air and Space Museum, Smithsonian Institution.)

Figure 1.12 The French Spad, a leading World War I fighter. (Courtesy of National Air and Space Museum, Smithsonian Institution.)

AFTER WORLD WAR I

The end of World War I naturally terminated the high level of military aircraft design and construction. Many companies disappeared. Barnstorming with surplus World War I airplanes was the major aviation activity. Nevertheless, many individuals dedicated to improving the art of flight persisted in building new and often, but not always, improved airplanes. Very few of these aircraft were built in significant quantities, but some of them made their mark in attaining new levels of performance. Noteworthy was the U.S. Navy Curtiss NC-4, a flying boat powered by four 400-hp, 12-cylinder liquid-cooled Liberty engines (Figure 1.13). In 1919, one of these airplanes, out of an original group of three, flew from Long Island to Lisbon, Portugal, in 13 days. The longest hop was from Newfoundland to the Azores, 2250 km (1400 statute miles). The other two NC-4's made forced landings at sea and abandoned the flight. A few weeks later, Alcock and Brown flew a converted British Vickers Vimy bomber with two Eagle engines on the first nonstop transatlantic flight from Newfoundland to Ireland. In 1924, two U.S. Air Service Douglas World Cruiser single-engine biplanes completed a round the world flight in 175 days. Two other starters were lost.

In 1927, these pioneering long-distance flights were climaxed by the 34-hour nonstop flight by Charles Lindbergh from New York to Paris in a Ryan single-engine monoplane, *Spirit of St. Louis* (Figure 1.14). This flight fired the world's imagination

Figure 1.13 Curtiss NC-4 flying boat. (Courtesy of National Air and Space Museum, Smithsonian Institution.)

Figure 1.14 The *Spirit of St. Louis*. (Courtesy of National Air and Space Museum, Smithsonian Institution.)

as to the potential of aircraft and stimulated a great growth in aircraft development. In the same period, many pilots tried to be the first to fly various transoceanic routes successfully. Many of these attempts ended tragically. The stories of the trailblazing pilots of the late 1920s are fascinating from both the human and technical points of view. It must be remembered that radio navigation aids were nonexistent. The magnetic compass, human judgment, and luck were the sole means of navigation.

THE BEGINNINGS OF TRANSPORT AIRCRAFT

The 1920s saw the beginnings of commercial transport aviation. At first, the European countries were the leaders. In 1920, the British Handley-Page Company built a twin-engine 12-passenger transport. In 1922, a revised version, the W8b, powered by Rolls-Royce 360-hp engines, entered service on a London–Paris–Brussels route (Figure 1.15). In 1926 the Armstrong-Whitworth Argosy, a 20-passenger plane, started operating between London and various European cities. The French and Germans also built many different types of transports, the most notable being the Junkers aircraft. Junkers introduced all-metal construction as early as 1915. The Junkers F-13, a four-passenger single-engine monoplane, was an outstanding success with 322 units being built. The airplane operated on airlines throughout Europe, South America, and Africa until well into the 1930s. The F-13 had a cantilever wing, a remarkable achievement for its day. In 1932, the Junkers Ju52, a trimotor carrying up to 17 passengers, started a successful career (Figure 1.16). Over 200 aircraft were sold and operated until the late 1940s on airlines throughout Europe. In 1924 in Holland, Anthony Fokker started a successful series of transport aircraft, topped in 1928 by the Fokker FVIIb-3m, a 10-passenger aircraft powered by three Wright Whirlwind 300-hp engines. This aircraft was sold worldwide.

In the United States, air transport was slower in developing during the 1920s. Nevertheless, many companies built transport aircraft for a limited market. Among the early aircraft was the Fairchild FC-2W, a single-engine four-passenger monoplane. The first successful American multiengine transport was the Ford 4AT Trimotor (Figure 1.17). Flight-tested in 1926, this aircraft featured an all-metal structure cov-

Figure 1.15 The Handley-Page W8b 12-passenger transport entered service in 1922. (Courtesy of British Aerospace.)

Figure 1.16 The Junkers Ju52, a 17-passenger trimotor transport, was introduced in 1932. (Courtesy of Flight International.)

Figure 1.17 Ford trimotor, 1927. (Courtesy of Gordon S. Williams.)

ered with corrugated sheet aluminum similar to the Junkers construction. Almost all American airlines used the Ford Trimotor, as did many airlines in Canada, Mexico, South and Central America, and Europe. In 1928, Boeing introduced the Model 80A, an 18-passenger biplane powered by three Pratt & Whitney Hornet 525-hp engines. This was the last gasp of the biplane with fixed landing gear, and only about 16 were built. In 1931, the Stinson Aircraft Co. developed a 10-passenger high-wing monoplane transport powered by three Lycoming 215-hp engines. This airplane was quite successful for its day. In 1933, the Curtiss T-32 Condor, a second-generation twin-engine biplane transport, made its appearance, and 45 were sold to Eastern Air Transport and American Airways. This airplane is interesting because its genesis was in the older biplane design, but it featured retractable landing gear and NACA cowls around the engines, a new development that greatly reduced drag (Figure 1.18). Its life was short, however, because a much advanced design appeared at almost the same

Figure 1.18 Curtiss T-32 Condor, 1933. (Courtesy of American Airlines.)

time and immediately rendered obsolete all other transports. This was the Boeing 247, a 10-passenger, all-metal twin-engine airplane with retractable landing gear and a cantilever wing (Figure 1.19). Powered by two Pratt & Whitney Wasp 550-hp engines, the Boeing 247 cruised at a speed of 155 mph. Most previous aircraft flew at 110 to 125 mph, although the Curtiss Condor achieved 145 mph. The clean design of the Boeing compared to its predecessors reflected the reduced drag and resulting greater efficiency.

Because of its corporate relationship, Boeing committed its entire production of the 247 to United Airlines. This turned out to be a devastating mistake because TWA, badly outdistanced by United, prepared a specification for a new and further improved airplane. Douglas Aircraft Co. responded with the DC-1, the production version of which was the DC-2. In speed (170 mph), comfort, and economy, this aircraft so surpassed the 247 that the 247 production line was soon terminated. Even United Airlines was forced to buy the DC-3, the final and large-scale-production version. The DC-3 was a twin-engine monoplane with cantilever wings, wing flaps, and retractable landing gear (Figure 1.20). The structure utilized stress-carrying skin, which saved considerable structural weight. Designed for 21 passengers to be carried much more comfortably than ever before, it was powered by either Pratt & Whitney or Wright engines of about 1200 hp and cruised at 180 mph. Over 800 DC-3's were sold to almost every airline in the world before the United States entered the war in 1941. Including

Figure 1.19 Boeing 247D transport, 1934. (Courtesy of R. S. Shevell.)

Figure 1.20 Douglas DC-3, 1936. (Courtesy of Air Canada.)

the massive wartime production, almost 11,000 DC-3's were built. Even in 1987, 52 years after its first flight, DC-3's can be seen in service at many airports throughout the world. The DC-3 has been active for over 60% of the time since the Wrights first flew!

During the 1930s some dramatic flying boats were built. These include the Sikorsky S-42 and the Martin M-130 China Clipper (Figure 1.21). Built for Pan American Airways, these aircraft carried 32 to 48 passengers and were monuments to the belief that overwater aircraft should be able to safely land on water. Eventually, the reliability of engines changed the philosophy to designing so that one would not have to land on water. The extra weight and drag of the flying boat hull could then be avoided. Nevertheless, these beautiful aircraft pioneered the overwater routes and developed the basis of long-distance overwater air transportation.

Except for the Boeing Stratoliner, the first pressurized transport airplane, no aircraft challenged the DC-3 until after World War II. The four-engine Stratoliner carried 33 passengers and started service in 1940. The war limited its production, and after the war more efficient aircraft appeared. Starting in 1942, the four-engine Douglas DC-4, a 44-passenger airplane, originally designed for the airlines, was produced for the Air Force as the C-54 passenger/cargo plane. Production totaled 1300

Figure 1.21 Martin China Clipper, 1935. (Courtesy of Pan American World Airways.)

aircraft. The airlines were well populated immediately after the war with surplus C-54's refurbished as DC-4's. The DC-4 was powered by 1450-hp Pratt & Whitney R-2000 engines and cruised at 230 mph (Figure 1.22).

Before turning to the postwar period, we should mention that the advances in transport aircraft were paralleled or led by racing and military aircraft. During the 1920s and 1930s, a series of air races stimulated the development of engine and airframe technology. The Schneider, Thompson, and Bendix Trophy races had the greatest impact. The Schneider race was for seaplanes. Among the winners were the Curtiss R3C-2, flown by Jimmy Doolittle in 1925, and the great British Supermarine racing aircraft, whose designer Reginald Mitchell was also responsible for the Spitfire fighter (a Supermarine S.6B is shown in Figure 1.23). The Thompson Trophy races produced famous airplanes such as the Wedell-Williams racer flown by Roscoe Turner in one of his victories (Figure 1.24). For more discussion of these exciting times, the reader is referred to the references at the end of this chapter.

Figure 1.22 Douglas DC-4. (Courtesy of Pan American World Airways.)

Figure 1.23 Supermarine S.6B racer, 1931. (Courtesy of National Air and Space Museum, Smithsonian Institution.)

Figure 1.24 Wedell-Williams racer, 1932. (Courtesy of National Air and Space Museum, Smithsonian Institution.)

Among American military aircraft, airplanes such as the 1932 Boeing F4B-4 and P-26A (Figure 1.25), the Navy Vought Corsair, the 1935 Martin B-10 bomber, and the Grumman F3F and F4F fighters earned a place in history. During World War II, a whole series of fighters and bombers was developed, such as the North American Mustang, the Lockheed P-38 (Figure 1.26), the Boeing B-17 and B-29 (Figures 1.27 and 1.28), the Douglas A-20 and A-26, and the Convair B-24. The British aircraft industry produced the outstanding Supermarine Spitfire and Hawker Hurricane fighters and the Avro Lancaster bombers. These aircraft were produced by the thousands and did much to turn the tide for the Allies. The German Stuka dive bomber first wreaked havoc on the allied forces and showed the awesome power of the airplane.

Figure 1.25 Boeing P-26A fighter, 1934. (Courtesy of Boeing Commercial Airplane Company.)

Figure 1.26 Lockheed P-38 fighter, 1940. (Courtesy of Lockheed-California Company).

Figure 1.27 Boeing B-17 flying fortress. (Courtesy of Boeing Commercial Airplane Company.)

Figure 1.28 Boeing B-29 super-fortress. (Courtesy of Boeing Commercial Airplane Company.)

AFTER WORLD WAR II

Following the end of World War II, technological advances developed for combat were applied to commercial aircraft. In addition, developments that never got into wartime production appeared in military aircraft. The first new airliners were the Lockheed Constellation (Figure 1.29) and the Douglas DC-6 introduced in 1946.

The Constellation and DC-6 represented giant steps forward in air transportation. Powered by four engines, Pratt & Whitney R-2800 engines for the DC-6 and Wright R-3350 engines for the Constellation, these 52-passenger aircraft increased

Figure 1.29 Lockheed Constellation. (Courtesy of British Airways.)

airline cruising speeds from the DC-4 speed of 230 mph to 310 to 320 mph. Pressurized cabins improved comfort both because of the benefits of increased pressure in the cabin and because the higher cruise altitudes permitted flight above and around much of the turbulent air. Operating costs were significantly reduced. The DC-6 was the more economical, a characteristic that kept it in operation long after it was replaced in first-line service. Shortly after the DC-6 and the Constellation entered service, the Convair 240 and the Martin 404 brought the same technology to smaller, short-range aircraft. The Convair was the more successful, and turboprop conversions of that airplane are still in service on short, low-density routes in 1987.

The DC-6's and Constellations were followed in the early 1950s by growth versions such as the DC-6B (Figure 1.30), DC-7, DC-7C, and Super Constellation. These aircraft were lengthened to increase payload and improve economy. The DC-7 and the Super Constellation were also equipped with Wright R-3350 turbocompound engines, reciprocating engines that used the energy in the exhaust gases to drive turbines geared to the engine driveshaft. This recovered much of the energy usually lost in the exhaust gases. The greater power and efficiency of these engines gave transcontinental range and higher speeds to these airplanes. They became the first-line transport aircraft until they were replaced in that position by the Boeing 707 and Douglas DC-8 jet transports starting in 1959. Although the compound engine provided more power and fuel efficiency, its complexity decreased reliability and raised costs. The DC-7 and the Super Constellation disappeared from service after the entry of the jets much more rapidly than the earlier but more efficient DC-6 and DC-6B.

Another prominent airplane of the postwar period was the Boeing 377 Stratocruiser. Developed from the B-29 bomber, the Stratocruiser configuration originally appeared as a military cargo airplane, the C-97. It first flew in late 1944 and was very successful in the military cargo role, a total of 888 airplanes being built. In 1947, Boeing first flew a commercial version. This two-deck airplane was unusually spacious and its downstairs bar was much appreciated by its passengers. Powered by four 3500-bhp Pratt & Whitney R-4360 Wasp Major radial engines with 28 cylinders, this airplane had competitive speed and comfort. However, its economics were poor and as a result only 55 commercial versions of the Stratocruiser were built.

Figure 1.30 Douglas DC-6B. (Courtesy of Northwest Airlines.)

Postwar military developments were focused on the aeronautical gas turbine engine. The gas turbine engine had been conceptually discussed as early as the beginning of the nineteenth century, but always in the context of providing power to a shaft. In modern aircraft, we call this kind of engine application a turboprop. A turboprop engine is a gas turbine that drives a propeller. The propeller itself was a serious airplane speed limitation, however. The high Mach number of the propeller tips at high airplane forward speeds causes a serious loss of propulsive efficiency. This results from the fact that the drag of airfoils rises sharply as the speed approaches the speed of sound. In 1928, a young British air cadet, Frank Whittle, wrote a college thesis on a means of overcoming this speed limitation. He proposed to use a gas turbine to produce a high-speed aft-moving flow of exhaust gases that would produce thrust without the use of a propeller. This proposal was amazingly farsighted, since in 1928 maximum racing airplane speeds were under 350 mph, and designers of practical military aircraft were struggling to reach 200 mph. Neither of these speeds was limited by propeller-tip velocities.

Little attention was paid to Whittle's work, although he patented his ideas in 1930. It was not until 1936 that a company was organized to put his concepts into practice. In April 1937, the first Whittle turbojet was run on a test stand. The British government gave full support to the development in March 1940, resulting in the flight of a single-engine prototype jet fighter, the Gloster E.28/39, in May 1941. This prototype evolved into the twin-engine Gloster Meteor fighter, which saw active service starting in the summer of 1944 (Figure 1.31).

The first American jet fighter was the Bell Airacomet XP-59A equipped with a Whittle engine. The first practical U.S. jet was the Lockheed F-80 Shooting Star, which was produced with a General Electric engine based on the Whittle design (Figure 1.32). The F-80 became operational in 1945.

The Germans had worked with the jet engine almost concurrently with the British effort, and the Messerschmitt Me 262 twin-jet fighter played an active role

Figure 1.31 Gloster meteor fighter. (Courtesy of National Air and Space Museum, Smithsonian Institution.)

Figure 1.32 Lockheed F-80 shooting star. (Courtesy of National Air and Space Museum, Smithsonian Institution.)

starting in the summer of 1944. Actually, the first jet-powered airplane was the German Heinkel He 178 flown on August 27, 1939, with an engine developed by Pabst von Ohain.

World War II was followed by a period of fantastic technological progress in military aircraft. The jet engine was rapidly improved to increase thrust and reduce fuel consumption. Wing sweepback was employed to ameliorate the adverse effects of high Mach number (the ratio of airplane speed to the speed of sound). Experimental research aircraft probed the high subsonic and finally the supersonic flight regimes. These research aircraft were powered by rocket engines and carried aloft by a "mother plane" such as a B-29 bomber. The Bell X-1 flown by Chuck Yeager exceeded the speed of sound in 1947 (Figure 1.33). The Douglas D-558-II piloted by Bill Bridgeman attained a Mach number of 1.89 in 1951. In December 1953, Yeager flew 1650 mph, a Mach number of 2.5, in the Bell X-1A. In 1962 the North American X-15 rocket-powered airplane attained a Mach number of 6.7 and an altitude of 108 km (67 mi) (Figure 1.34). It remained the fastest airplane ever built until the advent of the Space Shuttle in 1981.

Concurrent with probing the limits of technology both in flight and in wind tunnels, operational fighters and bombers were being designed using the fruits of this research. Among them were the Republic F-84 and the North American F-86 fighters, first flown in the 1946–1947 period (Figures 1.35 and 1.36). The Boeing

Figure 1.33 Bell X-1 experimental rocket plane. (Courtesy of Bell Aerospace Textron, Buffalo, N.Y. 14240.)

Figure 1.34 North American X-15 rocket plane. (Courtesy of Rockwell International.)

Figure 1.35 Republic F-84. (Courtesy of Fairchild Industries.)

Figure 1.36 North American F-86. (Courtesy of Rockwell International.)

B-47 jet bomber powered by six GE J-47 turbojet engines was the standard U.S. Air Force bomber from 1950 to about 1957 when it was replaced by the Boeing B-52 (Figures 1.37 and 1.38). The B-52, an eight-engine giant using Pratt & Whitney J-57 engines, is still the U.S. long-range bomber.

Equally impressive for longevity are the Navy Douglas A-4 and the Navy/Air Force McDonnell F-4 fighters (Figures 1.39 and 1.40). These aircraft entered service in 1956 and 1961, respectively, and are still important weapons systems more than 25 years later. Of course, an aircraft type that remains useful for so long a period changes with time. Engines, electronics, weight-carrying ability, and performance are

Figure 1.37 Boeing B-47 bomber. (Courtesy of Boeing Commercial Airplane Company.)

Figure 1.38 Boeing B-52 bomber. (Courtesy of Boeing Commercial Airplane Company.)

Figure 1.39 Douglas A-4 attack airplane. (Courtesy of McDonnell Douglas Corporation.)

Figure 1.40 McDonnell F-4 fighter. (Courtesy of McDonnell Douglas Corporation.)

continually improved so that a model built 10 years after the first version has much greater capability. Nevertheless, the basic designs are unchanged.

In the 1970s, a new generation of combat aircraft was developed. Powered by afterburning turbofans, the twin-engine Grumman F-14A Navy fleet defense fighter, the twin-engine McDonnell-Douglas F-15 air superiority fighter, and the General Dynamics F-16 fighter have supersonic capability to Mach 2 to 2.5 (Figures 1.41 and 1.42). They are equipped with highly sophisticated target acquisition, guidance, aiming, and defensive electronics, the cost of which comprises a large percentage of the total airplane price. A new supersonic bomber, the Rockwell B-1, planned for the 1970s, was test flown but not put into production because of debate about its usefulness and cost. In the early 1980s, the B-1 program was resurrected and an updated version, the B-1B (Figure 1.43), was produced as the first-line United States bomber.

Figure 1.41 McDonnell-Douglas F-15 fighter. (Courtesy of McDonnell Douglas Corporation.)

Figure 1.42 General Dynamics F-16 fighter. (Courtesy of General Dynamics.)

Figure 1.43 Rockwell B-1B bomber.

THE COMING OF THE JET TRANSPORTS

Shortly after the introduction of jets into military aircraft, designers of commercial airplanes began exploring the technical and economic characteristics of jet transports. They found that the fear that jet aircraft efficiency demanded flying very close to a prescribed flight path was unfounded. Although jet aircraft must fly high and fast, moderate deviations from the optimum altitude had only slight effects on fuel consumption. Early beliefs that jets were useful only for very long ranges were also proved wrong. However, in the late 1940s, studies also showed that jet operating costs were considerably greater than for existing transports powered by reciprocating piston engines. Furthermore, jet aircraft ranges with existing takeoff runway lengths were limited to about 750 mi, roughly the distance from New York to Chicago. Another major concern was the magnitude of cabin pressurization necessary at jet cruising altitudes of 30,000 to 40,000 ft. Previous aircraft seldom flew above 25,000 ft. The higher pressure difference between the inside and the outside of the jet transport fuselage posed a serious danger. A fuselage structural crack could lead to an explosive decompression.

Although American companies were worrying about these problems, the British de Havilland Company, aided by an order from British Overseas Airways Corporation (BOAC), proceeded with the design of the world's first jet transport, the Comet I (Figure 1.44). Powered by four de Havilland Ghost turbojets, this 36-passenger airplane first flew in 1949 and entered service in 1952. The Comet quickly demonstrated both the speed and the comfort of quiet, vibrationless jet flight. Although the economics were poor, the airplane's attractiveness resulted in quite a few orders. Unfortunately, three accidents between May 1953 and April 1954 brought the Comet's career to an abrupt end. The first accident was due to taking off at too low a speed, without the safety margin later found necessary for jet aircraft. The last two crashes were due to fuselage explosions caused by fatigue failures at the corner of a window. Although the Comet was a failure, it showed the advantage of jet flight and from its troubles came knowledge that avoided similar aerodynamic and structural problems in later jet aircraft. In 1958, a redesigned larger Comet 4 entered service, but it was outclassed by the Boeing 707 and Douglas DC-8, which appeared shortly thereafter.

Figure 1.44 DeHavilland Comet, 1952. (Courtesy of British Aerospace.)

While the British were building the Comet, American companies were conducting intensive aerodynamic and structural studies of jet transports. In 1952, these studies, the experience gained in building the B-47, and advances in engine efficiency and reliability provided the confidence for Boeing to begin building a prototype jet transport powered by four Pratt & Whitney J-57 engines. This forerunner of the 707 transport first flew in 1954. Its success stimulated interest by the world's airlines in seriously considering moving into the jet era. In 1955, Douglas also became convinced that the improved engines available would make jet airplanes both safe and economically feasible. It therefore joined Boeing in offering commercial jet transports in 1955. As a result, the Boeing 707 entered commercial service in 1958 followed by the Douglas DC-8 in 1959 (Figures 1.45 and 1.46). These 135-passenger (mixed-class) aircraft soon populated the airlines of the world. They brought a new level of speed, about 545 mph (877 km/h or 473 knots), more than 50% faster than the previous fastest aircraft, the DC-7. Comfort was greatly improved due to very large reductions in noise, vibration, and ride roughness. The ride roughness improvement resulted from the higher wing loadings of the jets, the higher, smoother cruise altitudes, and the ability to fly around turbulent regions because of the great speed and range. Furthermore, these jet transports proved to be even more economical than the predicted 20% direct operating-cost improvement relative to the DC-6B. Shortly after these aircraft entered service, improved Pratt & Whitney JT3D turbofan engines became available. These low-bypass-ratio turbofans reduced fuel consumption about 15% and reduced noise, which was becoming a serious community problem. Production of the 707 and the DC-8 was switched to the new engine, and many of the earlier aircraft were retrofitted with the JT3D.

Figure 1.45 Boeing 707-320. (Courtesy of Boeing Commercial Airplane Company.)

Figure 1.46 DC-8-63, a 1966 stretched version. (Courtesy of McDonnell Douglas Corporation.)

While the jets were being developed, there were those who still thought that turboprops were more practical, especially for short and medium range. This led to the development of such turboprop aircraft as the Lockheed L-188 Electra and the Vickers Viscount (Figures 1.47 and 1.48). These attracted a substantial market for a few years but were superseded when smaller, shorter-range jets became available in the mid-1960s.

Air travel was made so much more pleasant, useful, and economical by the jets that an enormous growth in air travel occurred in the 1960s. This, in turn, stimulated the development of smaller jet transports to bring jet travel advantages to the shorter-range and less densely traveled routes. This trend had started with the development of the French Sud-Aviation Caravelle in 1959. The Caravelle pioneered the fashion of mounting engines on the rear of the fuselage. The U.S. responses to the shorter-range need were the Boeing 727 in 1964 (Figure 1.49), the Douglas DC-9 in 1966 (Figure 1.50), and the Boeing 737 in 1968. All these airplanes spawned stretched fuselage versions to increase capacity in the later models. These smaller U.S. transports were all powered by Pratt & Whitney JT8D low-bypass-ratio turbofan engines, while the earlier Caravelle used the Rolls-Royce Avon.

Figure 1.47 Lockheed Electra. (Courtesy of Lockheed-California Company.)

Figure 1.48 Vickers Viscount. (Courtesy of British Aerospace.)

Figure 1.49 Boeing 727-200, 135 passengers. (Courtesy of Boeing Commercial Airplane Company.)

Figure 1.50 Douglas DC-9-30, 95 passengers. (Courtesy of Air Canada.)

The growth in airline traffic in the 1960s produced a requirement for much larger aircraft. The number of flights required to carry the increasing numbers of passengers was leading to serious congestion both along the airways and on the airports. In 1966, Douglas produced the stretched DC-8-61, -62, -63 series, the largest of which carried 50% more passengers than the original DC-8. This 200-passenger (mixed-class) airplane resulted in another large improvement in operating cost per seat-mile. It is interesting to note that the DC-8-61 and -63 airplanes are so efficient that they were retrofitted with high-bypass-ratio GE CFM-56 engines in the early 1980s when they were up to 16 years old. They began a new life that will probably last well into the 1990s.

In the mid-1960s, Boeing was unable to stretch the 707 to compete with the new DC-8's. Therefore, Boeing took a giant step forward and started design on the 747, a 365-passenger long-range airplane (Figure 1.51). The 747 was designed partly from the technology Boeing developed in competing for the C-5 Air Force cargo airplane in a contest won by Lockheed. The 747 was built around four 43,000-lb Pratt & Whitney JT9D high-bypass-ratio engines, which offered large improvements in fuel efficiency and noise. Because of the large passenger capacity, two cabin aisles were used, a feature that became the trademark of airplanes called wide-body aircraft. Actually, the wide body simply results from the large passenger capacity and a reasonable fuselage fineness ratio. The important design features are really large size and the high-bypass-ratio engines. The 747 entered service in 1970, followed in 1971 by the Lockheed L-1011 and the Douglas DC-10 wide-body trijets (Figure 1.52). Powered by three Rolls-Royce RB-211 and General Electric CF6-6 engines, re-

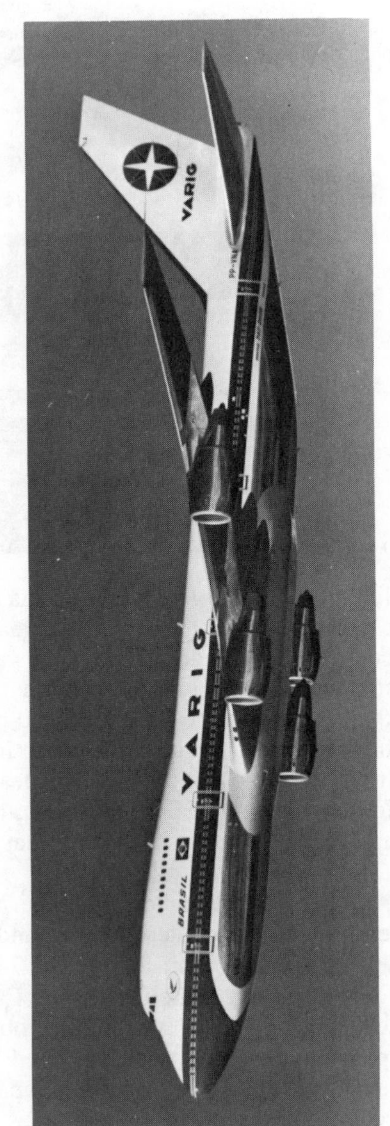

Figure 1.51 Boeing 747-200, 365 passengers. (Courtesy of Boeing Commercial Airplane Company.)

Figure 1.52 Douglas DC-10-30, 270 passengers. (Courtesy of McDonnell Douglas Corporation.)

spectively, these aircraft brought the technology of the high-bypass-ratio engine to aircraft about 25% to 30% smaller than the 747, with more flexibility for shorter-range operation. All three of these airplanes have been built in various versions to accommodate cargo and to adapt to shorter- or longer-range requirements. As with all transport airplanes, the number of passengers varies with the airline's choice of interior arrangement. The numbers of passengers cited in this chapter are based on normal airline mixed-class interiors with about 15% first-class seats. Maximum coach seating is 15% to 40% greater.

Another airplane in the same technology class is the A-300B, a twin-engine 230-passenger airplane with most versions powered by the General Electric CF6-50C 51,000-lb engine. Produced by Airbus Industrie, the airplane is built as a combined effort of France, Germany, and Great Britain. It serves the need for an airplane that is smaller and has shorter range than the wide-body trijets. After a slow start, it is carving out a substantial niche in the world market. In the United States, Boeing moved to compete with the A-300B by offering the 757 and 767 aircraft (Figure 1.53). Both of these twin-engine designs entered service in the early 1980s. The 757 is a single-aisle, 186-passenger airplane with a range of 3500 nautical miles, while the two-aisle 767 will carry 210 passengers for about the same range. Later versions of the 767 have

Figure 1.53 Boeing 767, 1982. (Courtesy of Boeing Commercial Airplane Company.)

greater range or larger capacity. There are modest aerodynamic and structural improvements, but these aircraft served mostly to bring high-bypass-ratio engine technology and modern guidance and control systems, which permit two-person flight crews, to this size of airplane.

The late 1970s and the 1980s brought important derivative aircraft. A stretched version of the McDonnell-Douglas DC-9, called the MD-80 (Figure 1.54), became very popular. Most "stretched" aircraft were developed by lengthening the fuselage, raising the maximum weight, and using a more powerful, uprated version of the original engine. Wing tip extensions to increase the span were often involved. The MD-80 has a large fuselage extension, along with a 20% wing area increase obtained from 5-ft spanwise plugs added at the wing root, plus 2-ft wing tip span extensions. Similar extensions were applied to the tail surfaces. The engines are Pratt & Whitney JT8D-209 or -217 engines, major modifications to the original engines, which include larger fans and provide a large thrust increase and significantly less noise and fuel consumption. Passenger capacity of about 145 is double that of the original 1966 version of the DC-9. Boeing achieved similar success with the 737-300, featuring a more modest fuselage extension, a minor wing area change, but a completely new engine installation using a down-rated engine developed from the basic General Electric CF6 engine that powers the DC-10. Further fuselage length changes and more advanced engine installations were being offered or proposed for both the 737 and the MD-80. The 747-400 aircraft featured a stretched upper deck, which increased the passenger capacity, and new Pratt & Whitney R-4000 engines with about 10% less fuel consumption. A stretched version of the DC-10 with new engines and a two-person computerized cockpit has been committed to production and is to be called the MD-11.

Although much of the activity was concerned with derivatives, Airbus was building an advanced 150-passenger airplane, the A-320, for service in the late 1980s and studying the possibility of entering the market with larger, longer-range aircraft. The most interesting expectation was the return of the propeller in the exotic form of

Figure 1.54 McDonnell Douglas MD-80. (Courtesy of McDonnell Douglas Corporation.)

Figure 1.55 Concept of the MD-80 adapted to the prop-fan.

the 8 to 12 wide blade propellers called prop-fans, unducted fans or ultra-high-bypass-ratio fans (see Chapter 17). These very thin, highly loaded blades were swept back at the tips to reduce high Mach number tip effects and offered fuel savings of 20% to 25% compared to turbofans with the same technology level. Technical problems remaining to be solved were primarily interior noise, gear box reliability, if a gear box is used, and structural integrity of the blades. Boeing was estimating the introduction into service of a moderate-range, 150-passenger airplane with prop-fan engines about 1992. McDonnell Douglas was studying prop-fans applied to the MD-80 (Figure 1.55), as well as a new airplane. The actual introduction of these all-new power systems, which require much larger turbo-shaft engines than currently available, will be strongly influenced by the price of fuel. If fuel costs are low, even the large fuel savings may not pay for the high initial costs and risks of new engines, propellers, and possibly completely new aircraft compared to the A320 or derivative MD-80's and 737's with turbofans. But high fuel costs will encourage the appearance of the prop-fan. The airlines and the aircraft manufacturers must project fuel costs many years ahead of the time when the aircraft will first enter service. The time from decision to design and build to first commercial use is at least 3.5 years, and the midlife of the aircraft is perhaps 10 years later than that. This is truly a difficult problem since we do not seem able to predict fuel costs even 6 months ahead.

GENERAL AVIATION

Although this historical review emphasizes military and transport aircraft because these types of airplanes led aeronautical technological development, general aviation has always played an important role. In aerodynamics and overall configuration, however, small private aircraft have not changed much in the last 40 to 50 years (Figures 1.56 and 1.57). The two-place Piper Cub of the 1930s was being built until the early 1980s. A 1981 Cessna 152 differs from its 40-year-old predecessor mainly in the use of a tricycle landing gear. Of course, engine powers have increased, radio and navigation aids have shown the benefits of advanced avionics and micro-processors, and retractable landing gears are quite common among the more expensive small aircraft.

Figure 1.56 1946 Cessna 140. (Courtesy of Cessna Aircraft Company.)

Figure 1.57 1981 Cessna 152. (Courtesy of Cessna Aircraft Company.)

Larger general aviation aircraft have absorbed many of the advanced features of transports. Turboprop, turbojet, and turbofan power plants and pressurized cabins are widely used for executive aircraft (Figures 1.58 and 1.59). In fact, some of the latest executive turbofan aircraft use advanced airfoils not yet applied to transports. Some of these applications are less than successful because only sophisticated engineering departments are usually equipped to introduce new technology. Nevertheless, some of the executive jets are highly advanced machines.

Two unique executive aircraft under development are the Beech Starship and the Piaggio Avanti. Both of these aircraft feature two turboprop engines placed on the aft portion of the wing. The driving design requirement is to move the propellers behind the passenger compartment for noise reasons. This requires an aft wing location and seriously reduces the moment arm of a conventional aft horizontal tail. The result could be high tail loads and resulting trim drag. To alleviate this problem, the Starship (Figure 1.60) eliminates the aft tail and controls the airplane with a canard (i.e., a forward tail). The Avanti retains a small aft tail but adds a canard. The Starship is built

Figure 1.58 Beechcraft King Air F90. (Courtesy of Beech Aircraft Corporation.)

Figure 1.59 Cessna Citation III. (Courtesy of Cessna Aircraft Company.)

Figure 1.60 Beech Starship. (Courtesy of Beech Aircraft Corporation.)

primarily of graphite-epoxy composite material, which has a large weight saving potential. Both aircraft are attempts by general aviation manufacturers to use advanced technology to improve performance and comfort.

A PERSPECTIVE ON TRANSPORT AIRCRAFT DEVELOPMENT

This review has mentioned only a few of the hundreds of airplane types developed in the last 50 years. Although we have discussed only representative successful airplanes at each stage, and have thereby omitted important airplane types, many designs never reached the flight stage. For many more, one airplane was more than enough to saturate the market. To be successful, an airplane had to have either better performance or lower operating costs or be of a more suitable size than its predecessor. In transport aircraft, the rapid progression of technical advances usually provided all three of these.

The history of transport aircraft is summarized in Figures 1.61 to 1.64. Figure 1.61 shows that speed increased fivefold from 1928 to 1959. The beginning of the jet age in 1959, however, established speed levels that have shown little change for the past 30 years. We have known how to increase speed above the typical airline cruise Mach numbers of 0.80 to 0.84 for decades, but we have not known how to increase these speeds substantially while simultaneously complying with the other major aircraft trend, ever-decreasing cost.

Figure 1.62 shows the relative direct operating cost per seat-mile from the DC-3 to the present 747 and the DC-10. The original 707/DC-8 turbofans are arbitrarily assigned a relative cost value of 1.0. This chart has been constructed on an approximate constant-dollar basis by using the cost ratios between one aircraft and the one preceding it, as determined at the time of the aircraft's development. The smaller aircraft, such as the DC-9, show costs for their own shorter ranges that are similar to the larger aircraft with the same technology at the same short range. This occurs only because the smaller aircraft are designed for the shorter range. Normally, increasing size for a given design range decreases seat-mile cost up to 350- to 400-passenger

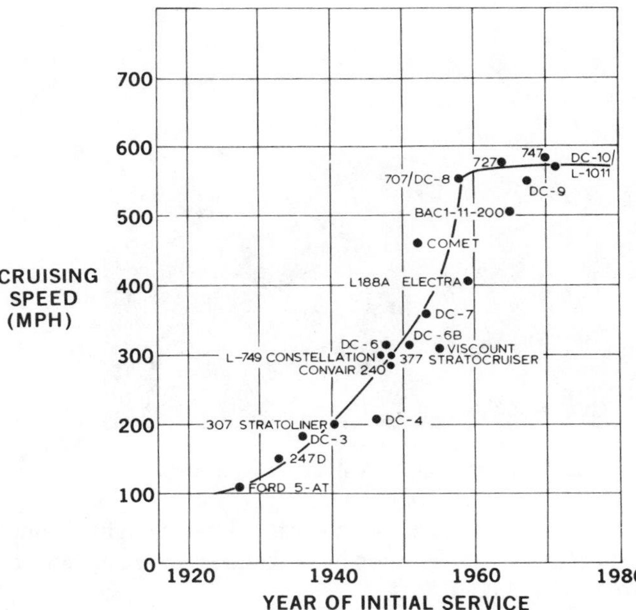

Figure 1.61 Speed history of transport aircraft. From "Technological Development of Transport Aircraft, Past and Future," by R. S. Shevell. Reprinted from *Journal of Aircraft,* Vol. 17, No. 2, Feb. 1980.

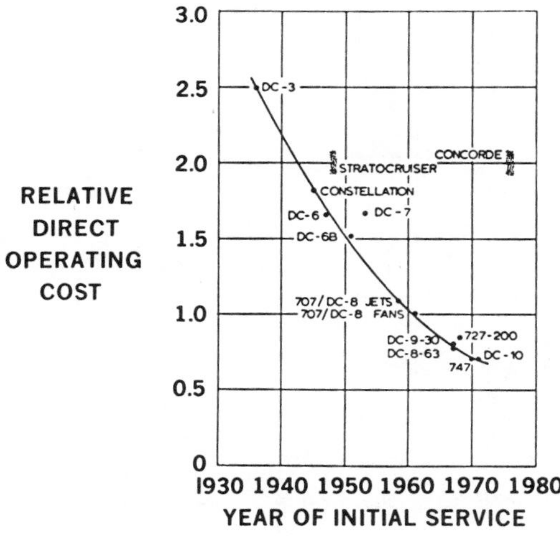

Figure 1.62 Relative direct operating costs. From "Technological Development of Transport Aircraft, Past and Future," by R. S. Shevell. Reprinted from *Journal of Aircraft,* Vol. 17, No. 2, Feb. 1980.

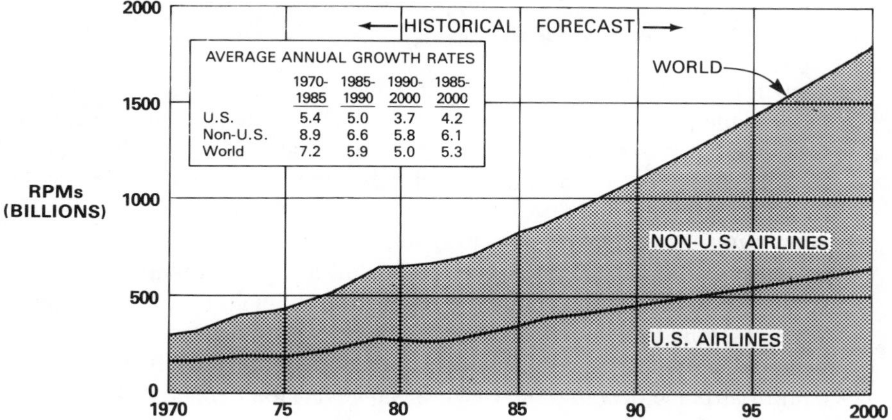

AVERAGE ANNUAL GROWTH RATES

	1970-1985	1985-1990	1990-2000	1985-2000
U.S.	5.4	5.0	3.7	4.2
Non-U.S.	8.9	6.6	5.8	6.1
World	7.2	5.9	5.0	5.3

Figure 1.63 World revenue passenger-miles. (Courtesy of Boeing Commercial Airplane Company.)

capacities, after which the trend flattens out. Increasing range decreases operating cost up to about 3000 mi.

Figure 1.62 shows that operating costs have almost always improved with each successive airplane generation. The only exceptions were the Boeing Stratocruiser, the DC-7, the Comet, and the Concorde. The message from 50 years of aircraft history is clear. Successful transport aircraft have almost always had equal or lower costs compared with their predecessor, while offering service improvement in speed, range, comfort, or combinations of these. The exceptions have offered a unique service at a moderately higher (10% to 15%) direct operating cost, and their success was limited. Those with even higher costs were economic failures.

The rapid improvement in speed, the main commodity offered by aircraft, together with comparable gains in range, comfort, and operating cost, led to the huge

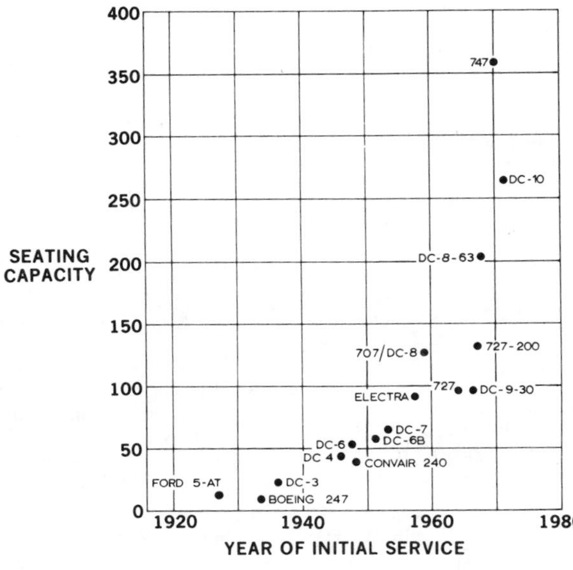

Figure 1.64 Growth of airplane passenger capacity. From "Technological Development of Transport Aircraft, Past and Future," by R. S. Shevell. Reprinted from *Journal of Aircraft,* Vol. 17, No. 2, Feb. 1980.

growth in travel shown in Figure 1.63 and in aircraft capacity shown in Figure 1.64.

Table 1.1 shows the chief technological developments that enabled each generation of aircraft to reign supreme, although usually for only a brief period. The table lists items that are inventions, or perhaps just the practical implementation of earlier inventions. Not listed specifically but continuing throughout the entire period have been significant improvements in airfoil design, flap systems, structural materials, and increasingly sophisticated methods of detail design and manufacture. The enormous contributions of avionics to aerial navigation, both en route and in the terminal area, were essential to the growth and safety of the air transport system. Great strides have been made in the electrical, hydraulic, pneumatic, and mechanical systems. From the overall airplane design point of view, however, the items listed in Table 1.1 were the most significant in making each succeeding design possible. Rapid technological progress imposed technological functional obsolescence on each succeeding transport aircraft generation within two to seven years.

In 1958, that rate of functional obsolescence changed. The 707/DC-8 jet transports boosted speed to the threshold of the transonic region, greatly reduced operating cost, practically eliminated vibration, reduced internal noise almost as much as one could want, and essentially eliminated ride roughness as a significant problem. One knew that these aircraft would be a hard act to follow!

Twenty-five years later, most of these aircraft were still in service, approaching obsolescence only because of the huge fuel price increases that started in 1973 and the noise regulations instituted in the late 1960s. Although the wide-body aircraft brought together much improved high-bypass-ratio turbofan propulsion, many modest improvements in aerodynamic components such as airfoils, flaps and slats, and structural gains in construction and material properties, their functional gains were primarily in large size. Of course, the new engines and acoustically treated nacelles achieved large

TABLE 1.1 TECHNOLOGICAL ADVANCES IN TRANSPORT AIRCRAFT (FROM REF. 1.7)

Approx. year	Aircraft	Multi-engine	Engine cowl	Flaps	Canti-lever wing	Load carry-ing skin	Retract-able landing gear	Engine super-chargers	Cabin super-chargers	Com-pound engine	Turbine engines: Turbo-prop	Turbine engines: Turbo-jet	Swept-back wings	Rip-stop struc-ture	Turbo-fan engines	Lead-ing edge slats	High-bypass-ratio turbofan
1927	Ford Trimotor	√	—	—	—	—	—	—	—	—	—	—	—	—	—	—	—
1933	Boeing 247	√	1/2	—	√	1/2	√	—	—	—	—	—	—	—	—	—	—
1936	DC-3	√	√	√	√	√	√	—	—	—	—	—	—	—	—	—	—
1939	Stratoliner	√	√	√	√	√	√	√	√	—	—	—	—	—	—	—	—
1946	DC-4	√	√	√	√	√	√	√	—	—	—	—	—	—	—	—	—
1947	Constellation/ DC-6/240/202/ Stratocruiser	√	√	√	√	√	√	√	√	—	—	—	—	—	—	—	—
1952	DC-7/1049	√	√	√	√	√	√	√	√	√	—	—	—	—	—	—	—
1955	Viscount/ Electra	√	—	√	√	√	√	—	√	—	√	—	—	—	—	—	—
1952	Comet	√	—	√	√	√	√	—	√	—	—	√	1/2	—	—	—	—
1958	707/DC-8 Turbojet	√	—	√	√	√	√	—	√	—	—	√	√	√	—	—	—
1961	707/DC-8 Turbofan	√	—	√	√	√	√	—	√	—	—	—	√	√	√	—	—
1965	727/DC-9-30/ 737	√	—	√	√	√	√	—	√	—	—	—	√	√	√	√	—
1970	747/DC-10/1011	√	—	√	√	√	√	—	√	—	—	—	√	√	—	√	√

reductions in community noise. The 747/L-1011/DC-10 aircraft also continued the progress toward lower operating costs. Nevertheless, the old milestones of progress, such as speed and comfort, have not been significantly improved since 1960.

SUPERSONIC TRANSPORTS

We cannot complete this historical review without mentioning the great technical achievement of the British/French supersonic transport, the Concorde (Figure 1.65). Prototype construction began in 1965 after years of intensive study. First flight of the experimental aircraft occurred in 1969, followed by the first flight of a production version in 1973. This airplane, which flies at a Mach number of 2, proved to be a great technical success but an economic disaster. The engineering problems of supersonic flight are difficult to solve. Although many military aircraft fly supersonically, few maintain those speeds for prolonged periods of time. The Concorde cruises steadily at Mach 2. Concorde's failure lies in its very high operating costs. Even without the fuel price rises of the 1970s and beyond, the seat-mile cost of the Concorde would have been much higher than that of the 747. Because of its high fuel consumption, however, the Concorde is more sensitive to the cost of fuel. Figure 1.62 shows the relative Concorde costs based on 1976 fuel costs. With 1981 fuel costs, Concorde direct operating costs were probably about four times as high as for the subsonic wide-body aircraft. The result is a relatively low total patronage even though the airplane flies at a loss. Because of these economic difficulties, only 16 airplanes were built before production was discontinued in 1979. Occasional reports about the Concorde becoming profitable are based on zero depreciation costs since the British government has written off the capital costs of the airplane.

With the technologies of the 1980s, a much more efficient supersonic transport could be built. The supersonic Boeing 2707, canceled in 1971 due to a variety of economic, environmental, and political causes, would have been at least one-third less costly per seat-mile than the Concorde. Current technology would further reduce costs substantially. The costs would still be considerably greater than for subsonic aircraft, however. Whether the performance gains would create a large enough market to make a second-generation SST economically feasible remains unclear. Furthermore, the capital requirements are very large, perhaps $10 billion. This would require government financing, throwing the project into the political arena. Even the improved SST

Figure 1.65 Concorde supersonic transport. (Courtesy of British Airways.)

would consume twice the fuel per seat-mile of the subsonic transports. With economic and energy-saving problems dominating the world scene, it will probably be 10 to 20 years before a second-generation SST can be developed.

THE LEADERSHIP OF MILITARY AVIATION

The history of commercial aviation shows enormous technological growth. It is important to realize, however, that while commercial aircraft builders and operators were responsible for the progress of air transportation, most of the important technical developments were derived from military aircraft. For example, improved propulsion was often the most important factor in the development of more efficient transports with better performance. Designing, testing, and producing a new engine is very expensive. Proven reliability can be assured only after an extensive operating history has been accumulated. A commercial airline could not usually afford either the financial risk or the safety hazard of starting to operate a new airplane with a new engine. The engine manufacturer could not usually afford the financial burden of running extensive proving trials involving thousands of hours of engine operating time. Therefore, military aircraft, at the forefront of technology, served the function of operating new power plants. When all the "bugs" had been found and eliminated, the commercial airplane designers would design an airplane around the proven engine, or some modest modification of it. Similarly, wing sweepback appeared first in fighters. Each speed increment was explored first by military aircraft. Advanced composite materials and computerized active controls are being used more adventurously in military aircraft.

There are some exceptions to the dependence of commercial technology on military progress. When the commercial industry developed a large enough market, starting in the 1960s, the engine industry was able to develop engines of particular sizes and improved technology specifically for transport airplanes. The basic technology had still been evolved in military projects, however.

Figure 1.66 shows the speed history of military and research aircraft. Comparison with the transport speed history in Figure 1.61 shows that military aircraft have always pioneered the frontiers of speed. Obviously, commercial aircraft designers placed safety first. Then economics dominated the design decisions.

CONCLUDING REMARKS

It is always instructive when flying in a modern transport airplane, which may be carrying up to 400 people at 885 km/h (550 mph) over ranges up to 10,000 km (6200 statute miles), to think back about 80 years to two young men struggling to achieve powered controlled flight. Whereas we dine and watch a motion picture, the Wrights were lying on their stomachs teaching themselves how to control their strange vehicle, just one step beyond a kite. A little over 50 years ago, a brave young pilot may have traveled the same path across the ocean as our airliner, but he was alone with a compass and one engine, fighting to keep awake for 34 hours. About 25 years ago, you could have made the same trip as a passenger, but with much noise and vibration.

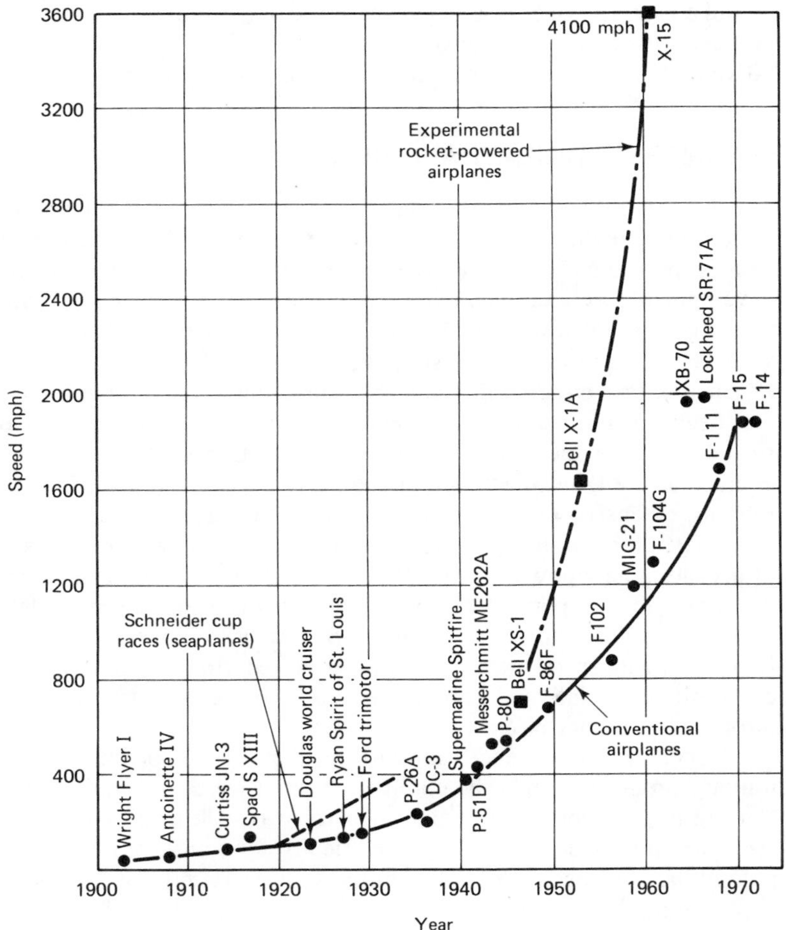

Figure 1.66 Speed history of military and research aircraft. (From Ref. 1.10.)

The flight would have taken twice as long and a rough ride was a fair possibility. Truly a fantastic history of progress!

Our musing about the history of flight brings forth many questions. How can an airplane carry 400 people into the air? What is the nature of the air through which we fly, and how can it sustain the airplane? Exactly what process produces the required lift? How much drag is associated with flight and how can we predict it without testing the actual airplane? How can we produce the thrust required to balance the drag? How can these huge machines be controlled so accurately? Going beyond aircraft, how can a satellite stay continuously in orbit? The purpose of this book is to provide a sound engineering understanding of these and many other aspects of flight and flight vehicle design. First, we turn to a brief discussion of the components of an airplane.

PROBLEMS

1.1. People responsible for technological advances range from highly educated to uneducated (technically speaking) individuals with great interest, drive, and enthusiasm. Of course, most people fall between these extremes. Discuss the kind of people who contributed to important steps in the development of our ability to fly. Give examples. After the pioneering efforts (i.e., after about 1930), what were the primary drives leading to aircraft advances?

1.2. The cruising speeds of widely used transport aircraft increased rapidly from 1930 to about 1960 and then leveled off. Why?

1.3. What was the role of military aviation in the progress of aeronautics?

REFERENCES

1.1. Garber, Paul E., "The Wright Brothers Contributions to Airplane Design," *Proceedings of the Diamond Jubilee of Flight,* Dayton–Cincinnati section, AIAA, and the Air Force Museum, Dayton, Ohio, December 1978.

1.2. Combs, Harry, *Kill Devil Hill,* Houghton Mifflin, Boston, 1979.

1.3. Taylor, John W. R., et al., *The Lore of Flight,* Crescent Books, New York, 1976.

1.4. Josephy, Alvin M., Jr., *American Heritage History of Flight,* American Heritage, New York, 1962.

1.5. Angelucci, E., and P. Matricardi, *World Aircraft, 1918 to 1935,* Rand McNally, Chicago, 1979.

1.6. Francillon, Rene J., *McDonnell Douglas Aircraft since 1920,* Putnam, London, 1979.

1.7. Shevell, Richard S., "Technological Development of Transport Aircraft—Past and Future," *Journal of Aircraft,* Vol. 17, Feb. 1980, pp. 67–80.

1.8. Steiner, Jack E., "Transcontinental Rapid Transit—The 367–80 and a Transport Revolution," *Proceedings of the Diamond Jubilee of Flight,* Dayton–Cincinnati section, AIAA, and the Air Force Museum, Dayton, Ohio, December 1978.

1.9. Davies, Ronald E. G., *Airlines of the United States since 1914,* Putnam, London, 1972.

1.10. Anderson, John D., Jr., *Introduction to Flight,* McGraw-Hill, New York, 1978.

2

THE ANATOMY OF THE AIRPLANE

Before beginning our discussion of the nature of the atmosphere and the physics of aircraft flight, it is useful to review the names of the various components of airplanes. The body of the airplane, which carries the crew and payload, such as passengers or cargo, is called the *fuselage*. The surface providing the lift is a *wing*. The stabilizing surfaces at the rear of the airplane are the horizontal and vertical *tails*. Streamline enclosures carrying the power plants are *nacelles*. The nacelles are often supported by struts known as *pylons*.

These major components are often divided into smaller elements. For example, the forward, usually fixed, part of the horizontal tail is the *horizontal stabilizer*. Attached to it is the movable control surface called the *elevator*. Changing the elevator deflection changes the lift on the horizontal tail and thereby controls the angle of attack and lift of the wing. Similarly, the vertical tail is divided into the fixed *vertical stabilizer,* or *fin,* and the *rudder*. Control surfaces on the outer part of the wing are the *ailerons,* which operate differentially to raise the lift on one side of the wing and lower it on the other. The result is to roll the airplane about its longitudinal axis and produce a turn. On high-speed aircraft a small inboard aileron is also used. *Spoilers* on the wing disturb the flow and reduce the lift to get more load on the wheels during a braked ground run, to lower a wing and help the rolling process if operated on one side only, and to create drag and increase the rate of descent at the end of a flight. Movable surfaces on the rear of the inboard part of the wing are deflected to increase lift during takeoff and landing. These are called *wing flaps*. Movable surfaces on the leading edges of the wing also serve to permit higher lift during takeoff and landing.

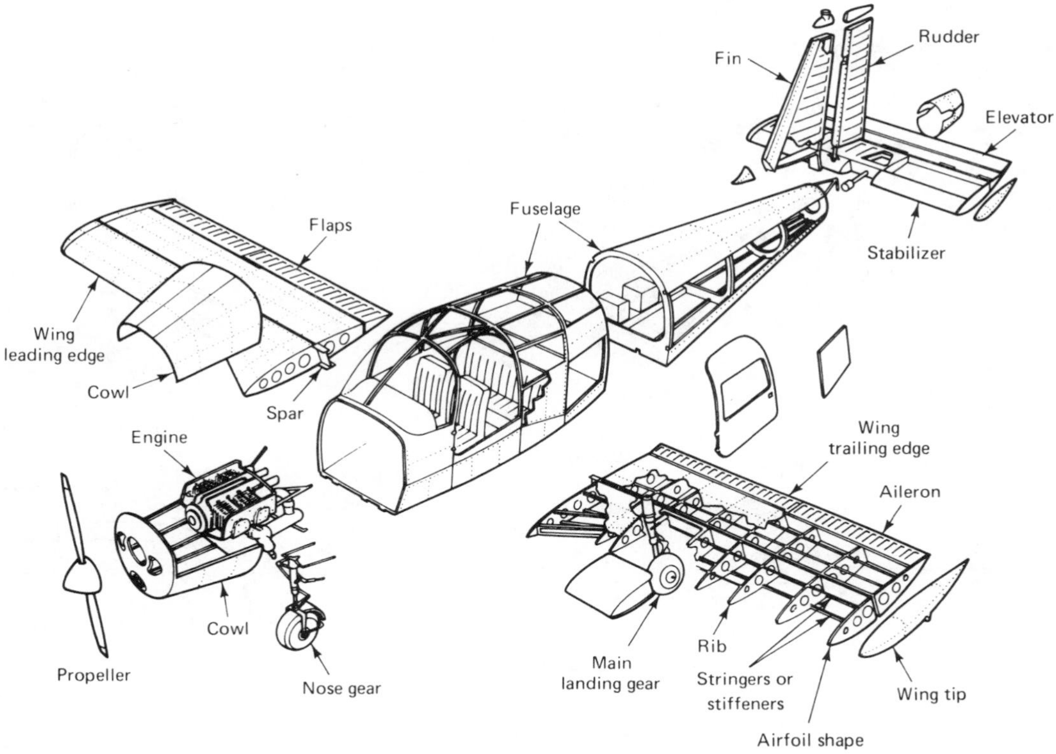

Figure 2.1 Airplane components: small private airplane. (Adapted with permission of A. B. Nordbok, Gothenburg, Sweden.)

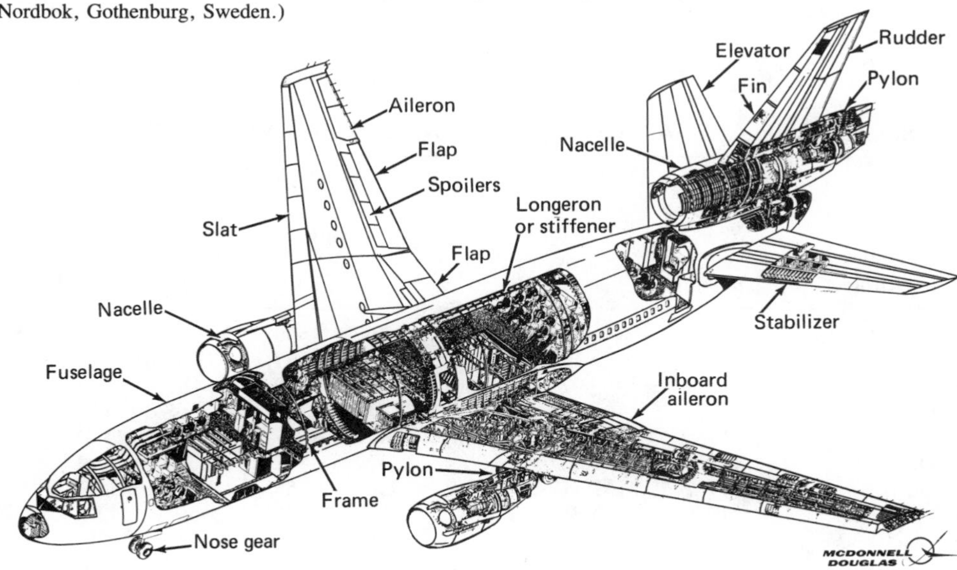

Figure 2.2 Airplane components: transport airplane. (Courtesy of McDonnell Douglas Corporation.)

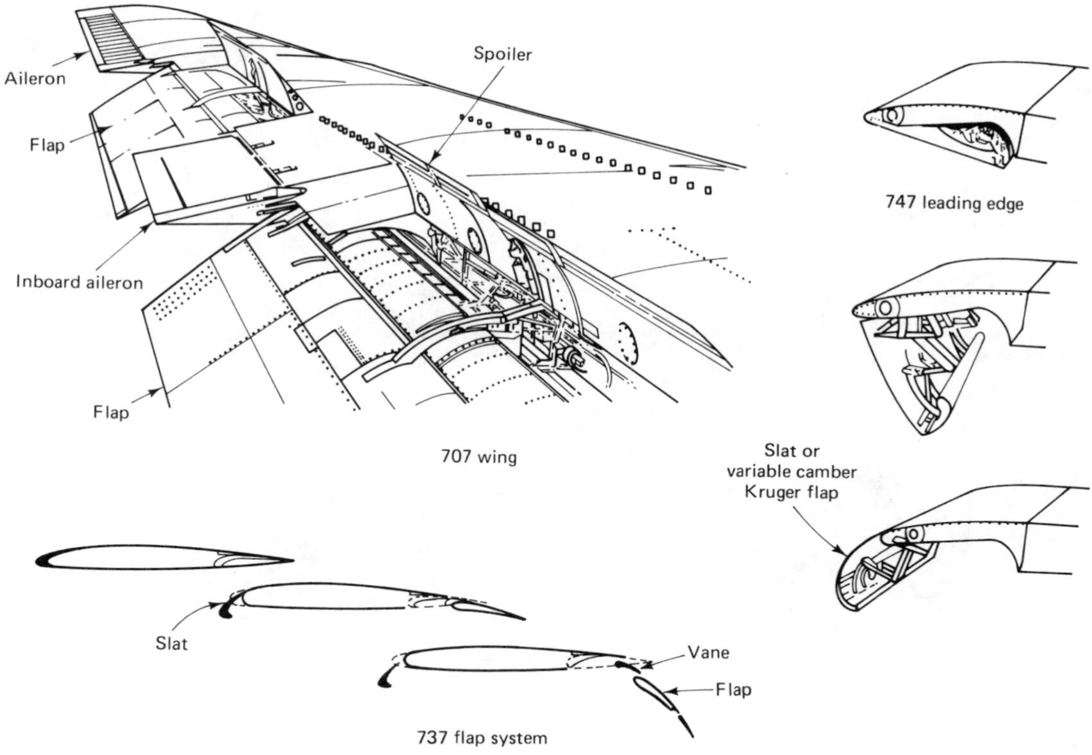

Figure 2.3 Wing controls and high-lift system. (Adapted with permission of A. B. Nordbok, Gothenburg, Sweden.)

TABLE 2.1 AIRPLANE TAKEOFF WEIGHTS, WING AREAS, WING LOADINGS, AND SPEEDS

Airplane	Year of intro-duction	Takeoff weight, kg (lb)	Wing area, m^2 (ft^2)	Wing loading: $\frac{\text{weight}}{\text{area}}$, $\frac{kg}{m^2}\left(\frac{lb}{ft^2}\right)$	Approx. cruising speed, km/h (mph)
Wright Flyer	1903	340 (750)	46.5 (500)	7.3 (1.5)	56 (35)
Cessna 150		726 (1,600)	14.6 (157)	49.8 (10.2)	188 (117)
Piper Cherokee		975 (2,150)	14.9 (160)	65.4 (13.4)	214 (133)
Ford Trimotor	1927	4,990 (11,000)	72.9 (785)	68.4 (14.0)	177 (110)
DC-3	1935	11,430 (25,200)	91.7 (987)	124 (25.3)	290 (180)
DC-6	1947	47,627 (105,000)	136 (1,462)	352 (72)	507 (315)
DC-8-50	1959	147,417 (325,000)	256 (2,758)	576 (118)	875 (544)
DC-9-30	1966	54,884 (121,000)	92.9 (1,000)	591 (121)	845 (525)
DC-10-30	1971	256,278 (565,000)	336 (3,620)	762 (156)	908 (564)
747-B	1970	362,872 (800,000)	511 (5,500)	708 (145)	917 (570)

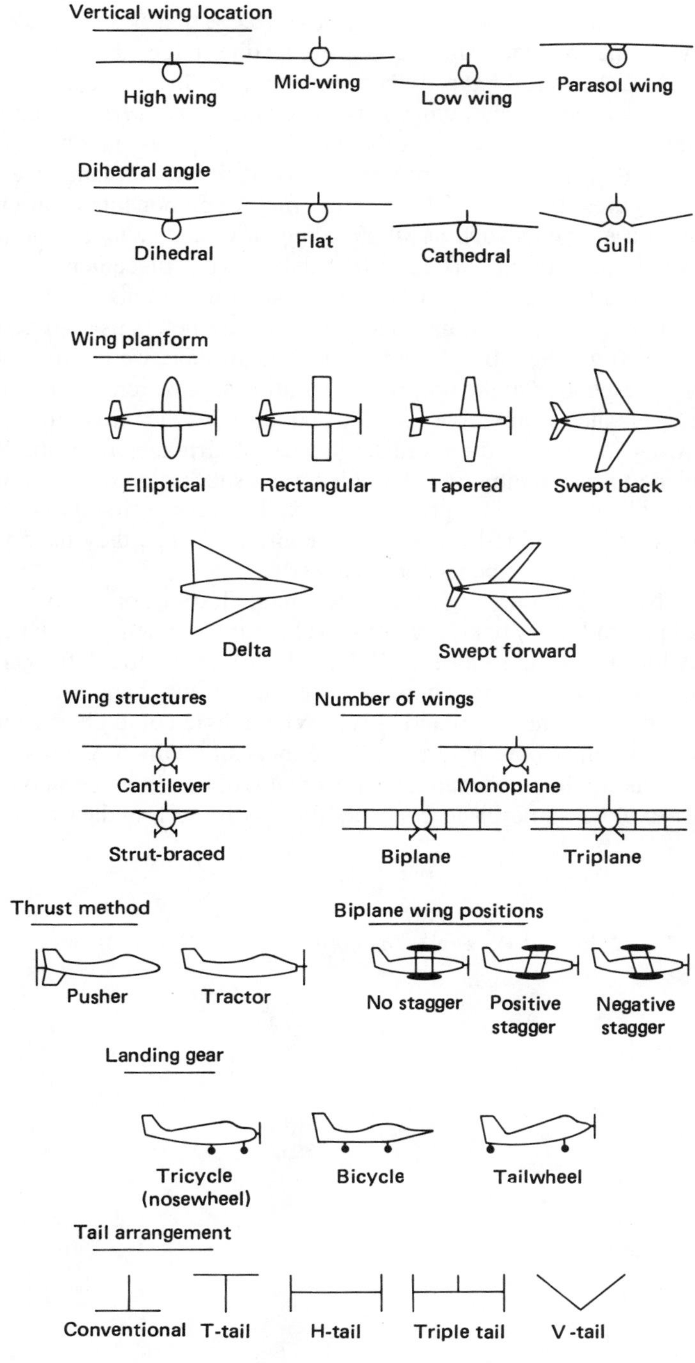

Figure 2.4 Wing, tail, and landing gear configurations.

They are known as *leading edge flaps*, except that if a carefully designed gap exists between the extended surface and the leading edge, the device is a *slat*.

Not all aircraft have all these components. In a few designs the horizontal control surface is ahead of the wing. It is then called a *canard*. In such a case the canard is destabilizing and the wing is the stabilizer. Engines may be installed in the body or wing so that no nacelles are needed. Very simple airplanes may not use flaps.

Figures 2.1 to 2.3 illustrate the major components of aircraft, as well as some structural elements such as *spars, wing ribs,* and *stringers* or *stiffeners*. Figure 2.4 shows the definitions of terms that define wing placement, wing shape, number of wings, and tail surface and landing gear arrangements.

One of the most important technical parameters used to describe airplanes is the *wing loading,* the ratio of the fully loaded airplane weight to the wing planform area. Some perspective of the progress in aviation and the remarkable lifting ability of wings can be gleaned from Table 2.1. This table shows the maximum weight, the wing area, the wing loading, and the cruising speed of airplanes from the Wright brothers' first airplane to the Boeing 747. Included are a succession of transport aircraft, as well as two small contemporary private aircraft. The year of first introduction is shown for the transports but omitted for the private aircraft, since they have been built with small continuing improvements for decades.

Note that the largest intercontinental transports have wing loadings up to 762 kg/m^2 (156 lb/ft^2). Ponder this value for a moment. The largest standard suitcase, fully loaded, weighs close to 17.7 kg (39 lb). Imagine a 1-ft^2 piece of metal with such a suitcase placed on it. Then add a second, a third, and a fourth suitcase. Then step back and issue the command "Fly!" What magic could obey your command? Yet that is what airplanes are able to do. The means by which wings can produce such great lift is among the important questions that will be answered in our studies. But first we must explore some basic concepts that form the foundation of aerodynamic theory.

REFERENCE

2.1. Taylor, John W. R., et al., *The Lore of Flight,* Crescent Books, New York, 1976.

3

THE NATURE OF AERODYNAMIC FORCES: DIMENSIONAL ANALYSIS

The basis of flight is found in the study of *aerodynamics*. Aerodynamics is a branch of the science of fluid mechanics. A *fluid* is a substance, such as a liquid or a gas, that changes shape readily and continuously when acted on by forces. *Fluid mechanics* is the science of how the fluid qualities respond to such forces and what forces the fluid exerts on solids in contact with the fluid.

Many of the fundamental laws of fluid mechanics apply to both liquids and gases. *Liquids* are distinguished from gases by the fact that liquids are nearly incompressible. Unlike a gas, the volume of a given mass of the liquid remains almost the same when a pressure is applied to the fluid. Our interest is centered on gases since the atmosphere in which aircraft operate is a gas we know as air. Composed mostly of nitrogen (78%) and oxygen (21%), air is a viscous, compressible fluid, but, as we shall see, it is often convenient and reasonably accurate to assume it to be an inviscid, incompressible gas.

A *gas* consists of a large number of molecules in random motion, each molecule having a particular velocity, position, and energy, varying because of collisions between molecules. The force per unit area created on a surface by the time rate of change of momentum of the rebounding molecules is called the *pressure*. As long as the molecules are sufficiently far apart so that the intermolecular magnetic forces are negligible, the gas acts as a continuous material in which the properties are determined by a statistical average of the particle effects. Such a gas is called a *perfect* or *ideal gas*.

AERODYNAMIC FORCES

Forces exerted by a fluid moving past a body such as a sphere, a wing, or a streamlined body such as an airplane fuselage fall into general classes: *normal* or *pressure forces* and *tangential* or *shearing forces*. Before exploring the nature of these types of forces, it is useful to note that the force due to a relative velocity between a fluid and a body depends only on that relative velocity. Whether the body is moving through a fluid at rest as in Figure 3.1a or the fluid is moving past a body at rest as in Figure 3.1b does not influence the resultant forces on the body. It will be readily seen that this relativity principle is the reason why wind tunnels, in which air is blown past an airplane model held firmly in place by appropriate struts, can be used to determine the forces that will occur on the airplane when it is flown through the atmosphere.

Consider a small section ΔS of the surface of a body immersed in a moving fluid as shown in Figure 3.2. The figure represents a plane perpendicular to the surface. The vector ΔF represents the force that the fluid exerts on this section, which is so small that it may be considered to be flat. The force ΔF may be resolved into two components normal and tangential to the surface element ΔS. These components are indicated by ΔF_n and ΔF_t. If these force components are divided by the area of the surface

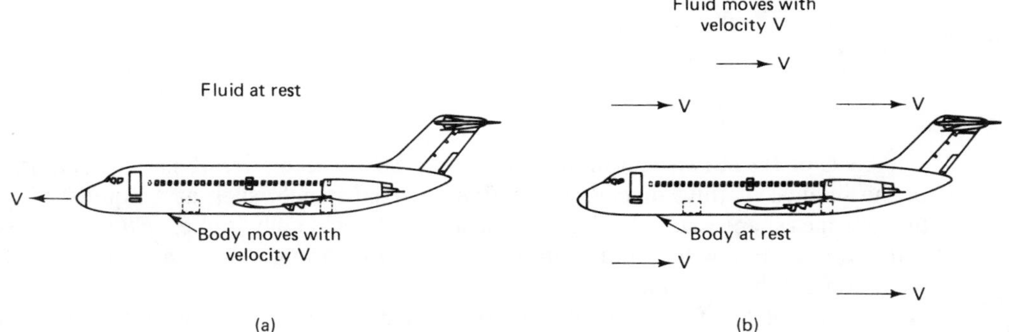

Figure 3.1 Relative motion of a body and a fluid: (a) Body moves with velocity V; (b) body at rest. (Courtesy of McDonnell Douglas Corporation.)

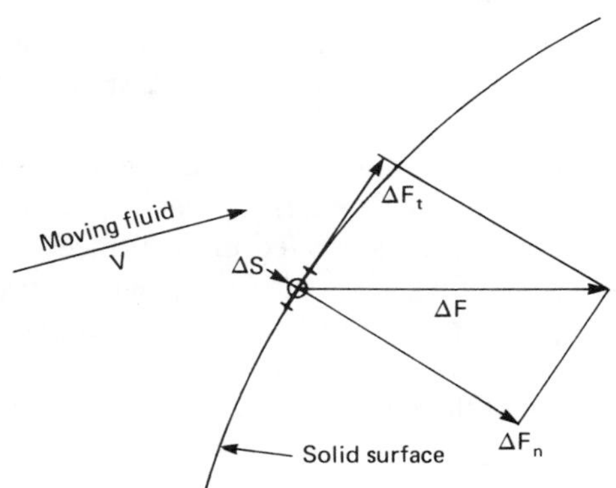

Figure 3.2 Force components on a surface element.

element ΔS, quantities are obtained that have the nature of stresses, or forces per unit area. These stresses are the normal stress or pressure acting on the surface, p, and the shear stress τ.

The pressure obviously results from the gas pressure, which can only act normal to the surface. The effect of the sum of the molecular collisions of the gas cannot create a force parallel to a smooth surface. The tangential force does not depend on pressure but on the viscosity of the fluid. *Viscosity* is a property of a fluid through which momentum is transferred between parallel layers of fluid aligned with the flow direction but moving with different velocities. The viscous force per unit area is the *shear stress*.

The shear stress can be best understood by reference to Figure 3.3, which depicts the velocity profile in the boundary layer of a laminar, viscous flow over a surface. The layer of fluid molecules adjacent to a solid surface sticks to the surface and thus has zero velocity. The next fluid layer has a small velocity, and each succeeding layer has a greater speed until, at the outer edge of what is defined as the *boundary layer*, the velocity becomes that of the undisturbed freestream. The value of the shear stress was first given by Newton as

$$\tau = \mu\left(\frac{dv}{dy}\right)$$

where

τ = shear stress in the fluid at the point y_1 at which dv/dy is measured

μ = constant of proportionality known as the coefficient of viscosity

$\dfrac{dv}{dy}$ = velocity gradient

The *coefficient of viscosity* is the physical constant that determines the nature of any fluid with regard to frictional phenomena. It is larger for sticky fluids such as oil and small for water or air.

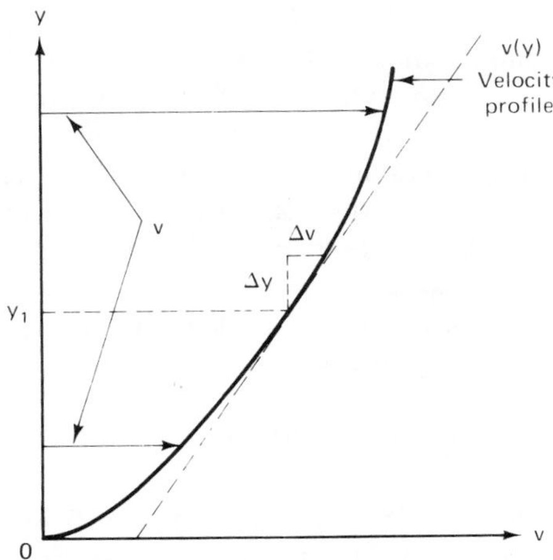

Figure 3.3 Notation for the boundary layer velocity profile.

The total pressure force on the body can be found by adding, vectorially, the normal forces over the entire body. Similarly, the tangential or shear stresses can be integrated over the body to determine the total "friction" force.

PARAMETERS AFFECTING AERODYNAMIC FORCES: DIMENSIONAL ANALYSIS

The objective of science is to obtain knowledge about physical relationships involved in a process. The related but somewhat different goal of engineering is to develop quantitative methods for the prediction of the effects of the applicable parameters on the output quantities of interest. In the case of an aerodynamic body, we need to know what characteristics of the fluid and the body influence the forces on the body. As a start, it is clear that the *drag force,* which retards the motion of the body through the air, and the *lift force,* which is a measure of the usefulness of the device, are the primary forces of interest. We shall see that there are several ways to determine the essential relationships upon which the lift and drag forces depend. First we shall examine a simple approach known as *dimensional analysis*.

Consider a body moving in a given direction with uniform velocity through a homogeneous fluid that extends to infinity. We must first list the factors that may influence the force F on the body. Gravity is not involved because the effects of gravity on the forces exerted by the fluid are buoyancy forces, which have the same value whether the body is moving or at rest. The velocity V of the body through the fluid must be significant. The properties of the fluid such as the density ρ and the viscosity μ are primary quantities influencing the interaction between the fluid and the body. Another fluid characteristic that may be important is elasticity, which can be determined by the density and the speed of sound a. Finally, the body shape and size are important. The body size can be represented by a characteristic dimension l, such as the diameter for a sphere or the length for a streamlined shape such as a fuselage or airfoil. We can summarize these thoughts by writing the equation

$$F = F(\rho, \mu, V, l, a, \text{shape}) \tag{3.1}$$

The relation between F and the independent parameters ρ, μ, V, l, and a can be made considerably more precise by applying the principle of dimensional homogeneity. This principle is based on the *pi theorem* (Refs. 3.1 and 3.2) and essentially states that an algebraic equation expressing a relation among physical quantities can be meaningful only if all terms in the equation have the same dimensions. More colloquially, one cannot add apples and oranges. This statement leads to a simple method of determining the combinations in which the variables may appear in an equation. It is in fact a means of finding the allowable combinations of exponents. In a general form, equation 3.1 may be expressed as

$$F = \sum_i C_i \rho^a \mu^b V^c l^d a^e \tag{3.2}$$

where the exponents $a \ldots e$ can have any value and the coefficients C_i are dimensionless numbers that must introduce the effect of body shape. The symbol Σ means that all terms for any number of values of the index i are to be added so that the first terms of the equation would be

$$F = C_1\rho^a\mu^bV^cl^da^e + C_2\rho^a\mu^bV^cl^da^e + C_3\rho^a\mu^bV^cl^da^e + \cdots \tag{3.3}$$

The *principle of homogeneity* requires that each term in the series have the same dimensions. Therefore, defining Dim F to mean the dimensions of the term F, we get

$$\text{Dim } F = \text{Dim}(\rho^a\mu^bV^cl^dq^e) \tag{3.4}$$

The only dimensions that appear in mechanics are mass M, length L, and time T. The dimensions of the quantities in equation 3.3 are familiar, except, perhaps, for μ. For example,

$$F = Ma \qquad \text{so that} \qquad \text{Dim } F = \frac{ML}{T^2}$$

and

$$\rho = \frac{\text{mass}}{\text{volume}} \qquad \text{so that} \qquad \text{Dim } \rho = \frac{M}{L^3}$$

Since $\mu = \tau/(dv/dy)$,

$$\begin{aligned}\text{Dim } \mu &= \text{Dim}\left[\frac{\text{force/area}}{\text{velocity/length}}\right] \\ &= \frac{(ML/T^2)/L^2}{(L/T)/L} \\ &= \frac{M}{LT}\end{aligned}$$

Before proceeding with our example of dimensional analysis, it is useful to make an assumption that is acceptable for a wide range of fluid mechanics problems. This is the assumption that the fluid is inelastic or incompressible. Liquids such as water are so nearly incompressible that the compressibility effects are negligible. Even for gases such as air, this assumption introduces negligible errors when the velocity is less than about 40% the speed of sound. We shall see later that at higher speeds compressibility effects must be accounted for to obtain good accuracy.

The effect of neglecting compressibility is to remove the speed of sound a as a parameter. Equation 3.4 then becomes

$$\text{Dim } F = \text{Dim}(\rho^a\mu^bV^cl^d) \tag{3.5}$$

By substituting in equation 3.5 the dimensions of the various parameters determined above, we obtain

$$\frac{ML}{T^2} = \frac{M^a}{L^{3a}}\frac{M^b}{L^bT^b}\frac{L^c}{T^c}L^d$$

and, by collecting terms,

$$\frac{ML}{T^2} = \frac{M^{a+b}L^{c+d-3a-b}}{T^{b+c}}$$

Since the dimensions on the two sides of the equation must be equal, the following equations in the exponents must be true:

$$a + b = 1\text{: } c + d - 3a - b = 1\text{: } b + c = 2$$

Here we have three equations with four unknowns. By solving for a, c, and d in terms of b, we can obtain

$$a = 1 - b, \qquad c = 2 - b, \qquad d = 2 - b$$

Substituting these exponents in the typical series term shown in equation 3.2 gives

$$\rho^a \mu^b V^c l^d = \rho^{1-b}\mu^b V^{2-b} l^{2-b} = \rho V^2 l^2 \left(\frac{\mu}{\rho V l}\right)^b$$

The entire general series expression for the force F is then

$$F = \rho V^2 l^2 \sum_i C_i \left(\frac{\mu}{\rho V l}\right)^{b_i} \tag{3.6}$$

Thus we have shown that the force on a body of a given shape depends on the density of the fluid ρ, the square of the velocity of the fluid past the body, the square of a characteristic length, and some function of the ratio $\mu/\rho V l$. The l^2 term represents a characteristic area usually taken as the frontal area for bodies and the wing planform area for wings and complete airplanes. The inverse of the ratio $\mu/\rho V l$ is known as the *Reynolds number* (RN), in honor of its discoverer, Osborne Reynolds.

$$\text{RN} = \frac{\rho V l}{\mu} \tag{3.7}$$

We shall show later that the Reynolds number represents the ratio of the kinetic or inertia forces in the fluid to the viscous forces. High Reynolds numbers indicate that the viscous forces are relatively low and in many problems can be neglected.

The series function of Reynolds number in equation 3.6 is also a primary function of body shape. The variation with Reynolds number is often very gradual over most of the flight regime so that this function, $\Sigma_i C_i (\mu/\rho V l)^{b_i}$, is often considered to be a constant for a particular configuration and attitude. It is convenient to define this constant, or more correctly this function of Reynolds number, as half the force coefficient C_F. Thus,

$$\sum_i C_i \left(\frac{\mu}{\rho V l}\right)^{b_i} = \frac{1}{2} C_F$$

and

$$F = \frac{1}{2}C_F\rho V^2 S$$
$$= C_F\frac{\rho}{2}V^2 S \tag{3.8}$$

where

S = wing planform area for wings and complete airplanes
$S = S_\pi$ = frontal area for bodies

The reason for the "$\frac{1}{2}$" is that the term $\frac{1}{2}(\rho V^2)$ is a very important quantity known as the *dynamic pressure*. The dynamic pressure is equal to the kinetic energy of a unit volume of air. Although we have shown the dependency of aerodynamic forces on density and the square of the velocity from dimensional considerations only, we will find later that one-dimensional fluid mechanics theory leads to the same result. Furthermore, the same expression can be derived from potential theory. Although potential theory is beyond the scope of this book, its results will be discussed extensively in later chapters. A fourth proof of the significance of $\frac{1}{2}(\rho V^2)$ can be shown experimentally in pressurized wind tunnels in which both the air density and the velocity can be varied and the effect on the aerodynamic forces measured.

When the force being considered is drag, a force in the direction of motion of the vehicle taken positive in the downstream direction, the force coefficient is called the *drag coefficient* C_D. When the force is a lift, perpendicular to the direction of motion, the coefficient is called the *lift coefficient* C_L.

In addition to these force equations, it is often important to analyze the moments acting on a wing or body. Since a moment = force × length, it is clear that the moment equation analogous to equation 3.6 requires an additional length on the right side. This length is usually chosen as the length appearing in the definition of Reynolds number. The length is body length for bodies and the wing mean aerodynamic chord $\bar{c}$ for wings and complete airplanes. $\bar{c}$ is a mean effective chord and is defined in equation 11.1. Summarizing these primary airplane coefficients, we have

$$C_L = \frac{\text{lift}}{\frac{1}{2}\rho V^2 S} = \text{lift coefficient}$$
$$C_D = \frac{\text{drag}}{\frac{1}{2}\rho V^2 S} = \text{drag coefficient} \tag{3.9}$$
$$C_M = \frac{\text{pitching moment}}{\frac{1}{2}\rho V^2 S\bar{c}} = \text{pitching moment coefficient}$$

It is important to realize that the term *shape* includes the attitude or angle of attack α of the body with respect to the fluid flow direction. For a given body, the attitude is generally described by the angle between the direction of motion V and a reference

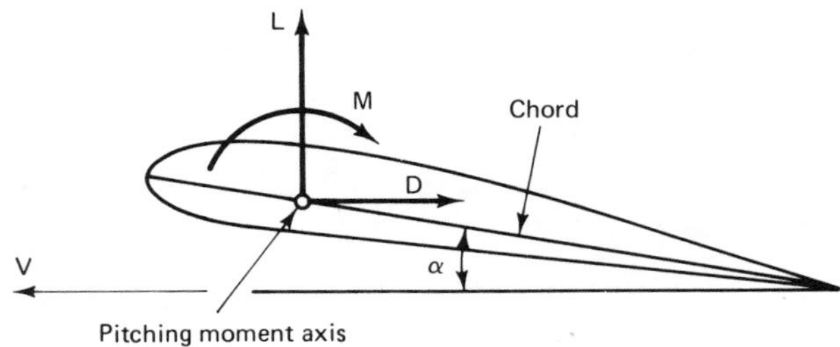

Figure 3.4 Airfoil force and moment definitions.

axis in the body, such as the fuselage floor or the wing chord line. We have now reduced the problem of determining lift, drag, and moment to the determination of dimensionless coefficients, which are functions of Reynolds number.

Figure 3.4 summarizes the definitions of lift, drag, moment, and angle of attack. We should emphasize again that in incompressible flow all these coefficients are constant for a specific configuration (i.e., body shape and attitude) and for a given Reynolds number. We have suggested that for many flight conditions, cruise flight for example, the effects of Reynolds number changes are small and may be neglected. This is valid provided that the wing and body shapes and attitudes are such that the air flows smoothly over the surfaces without separation until the trailing edge is reached. It is also limited to flight regimes in which changes in Reynolds numbers from one part of the flight to another are relatively small, perhaps plus or minus 20%, and an average value of Reynolds number is used in the analysis. At the extremes of the flight regime, such as at high angles of attack near the stall, where the flow separates from the wing, modest changes in Reynolds number may completely change the type of flow and the resulting force and moment coefficients. Even with smooth, attached flow, large Reynolds number changes have an important effect on the drag coefficient. The effects of Reynolds number on airplane force and moment coefficients are major topics in aerodynamics. They will be explored in considerable detail in appropriate chapters of this book.

To obtain experimental information about the forces on an airplane, tests of small models are conducted in wind tunnels. Obviously, if the full-scale flow is to be properly represented, the model must be geometrically similar to the real airplane. Since exact reproduction of the flow on the model requires the same Reynolds number in the wind tunnel as occurs in flight, Reynolds number is called a *similarity parameter*. Although we have restricted our discussion thus far to an incompressible fluid, it must be noted that in flight faster than about 40% to 50% of the speed of sound, air must be considered to be a compressible fluid. Then another similarity parameter, the Mach number, must be matched also. The *Mach number* is the ratio of the speed of the body through the fluid to the speed of sound.

PROBLEMS

3.1. The gross weight of a two-place Piper Cherokee is 2000 lb and its wing area is 160 ft^2. What is its wing loading?

3.2. A Boeing 747B has a wing loading of 123 lb/ft^2 at takeoff and a wing area of 5500 ft^2. What is its takeoff weight?

3.3. A 747 with a gross (total) weight of 650,000 lb is cruising at 35,000 ft at a speed of 465 knots. The 747 has a wing area of 5500 ft^2. The density at 35,000 ft is 0.0007382 slug/ft^3. What is the value of the cruise lift coefficient, C_L, and the wing loading? (When using equations 3.9, V must be in feet per second in the English system.)

3.4. If the 747 in Problem 3.3 has a drag of 40,000 lb, what is the drag coefficient?

3.5. A DC-9 is carrying 100 passengers at 30,000 ft at a C_L of 0.35. The density at 30,000 ft is 0.0008907 slug/ft^3. The DC-9 wing area is 1000 ft^2 and its weight is 100,000 lb. What is the cruise speed in feet per second? in miles per hour? in knots?

3.6. A Cessna 150 is cruising at 115 mph at 7000 ft. The airplane weighs 1500 lb and has a wing area of 157 ft^2. What is the lift coefficient? If the drag coefficient is 0.0300, what is the drag in pounds? The air density, ρ, at 7000 ft is 0.001927 slug/ft^3.

REFERENCES

3.1. Millikan, Clark B., *Aerodynamics of the Airplane,* Wiley, New York, 1941.

3.2. Kuethe, Arnold M., and Chow, Chuen-Yen, *Foundations of Aerodynamics,* Wiley, New York, 1976.

3.3. Bertin, John J., and Smith, Michael L., *Aerodynamics for Engineers,* Prentice-Hall, Englewood Cliffs, N.J., 1979.

4

THEORY AND EXPERIMENT: WIND TUNNELS

All engineering is a blend of theory and empiricism. The exact mix depends on the type of engineering and the particular problem. In nineteenth-century aeronautics there were important, and mathematically correct, theories that never gave the right answers. The "right" answers were the results of tests. It later turned out that the defective theories left out an important element of the flow problem, viscosity. Often the changes in the theory to include the essential contribution of viscosity were relatively simple, but the resultant answers were very different. At the appropriate time, we shall discuss the way that viscosity was brought into the equations for lift and drag and why it affected some quantities much more than others. We shall also learn that almost all analytical methods in aeronautics are themselves a blend of beautiful theory and empirical fine tuning to account for quantities either omitted from the theory altogether or too laborious to take into account with precision.

Almost all theories must be simplified to allow mathematical solutions to be obtained. The validity of the basic theory and the simplifications must always be confirmed by test. In aeronautics, a flight test is ideal, but it is enormously expensive to engineer and build a full-scale airplane to find out if the design is correct. We must have substantial experimental verification of the safety and efficiency of the design before we fly. Furthermore, if we wish to test any variations of a design, for example different wing planforms and airfoil shapes, it is almost impossible to do such tests full scale. In the uncertain atmosphere, it is difficult to find perfectly smooth air to allow ideal steady-state flight conditions to be set up for a test. For all these reasons, wind tunnels were invented.

You will recall that we have stated that the forces on an airplane moving through the air at a particular speed are the same as the forces on a fixed airplane with the air moving past it at the same speed. This principle led to the invention of the wind tunnel, basically a duct or pipe through which air is blown or drawn. In the tunnel a precise scale model of an airplane is mounted on some sort of strut, and the loads transmitted from the model to the strut are measured by balances. The balances can be scales or strain gauges.

The loads measured for any model configuration will obviously vary if the model size, the speed of the air, or the air density is changed. How can these tests be interpreted in terms of the full-scale aircraft we plan to build and fly at different altitudes, which lead to different densities, and at different speeds? The answer is found in the equations for force we have just derived. To obtain meaningful values for the lift, for example, we divide the measured lift in pounds or newtons by one-half the density, the square of the velocity, and the reference wing area. From equation 3.9, we can see that the result is the lift coefficient. This lift coefficient, found at any normal flight attitude in the wind tunnel, can be expected to apply to the full-scale airplane in flight at the same attitude, except for effects of differing Reynolds numbers, which are often quite small. A similar statement can be made about the drag coefficient, except that the effect of varying Reynolds number is much greater and must be accounted for.

To minimize the calculated corrections from the wind tunnel test Reynolds number to the full-scale Reynolds number, it is desirable to test at a Reynolds number as close to full scale as possible. Therefore, a major goal in wind tunnel design is to obtain values of the test Reynolds numbers, based on the model wing mean aerodynamic chord, that are as large as possible. Recalling that Reynolds number is $\rho VL/\mu$, where L is a characteristic length, it is clear that the usual small model scale produces a small L and a Reynolds number much lower than associated with the real airplane. Because Mach number is even more important, at least at high Mach numbers above about $M = 0.6$, the wind tunnel must match the full-scale Mach number and then do its best with respect to Reynolds number. The wind-tunnel-test Reynolds number can be increased by having the largest possible wind tunnel test section to accommodate the largest possible model and by pressurizing the wind tunnel to raise the air density. Therefore, we find very large wind tunnels and highly pressurized wind tunnels. Because of the huge cost in structure and power, we do not have very large, highly pressurized tunnels.

The cross section of the wind tunnel through which the air flows can vary from a few inches in diameter to over a hundred feet. Although wind tunnels had been built 30 years earlier, the Wright Brothers were the first to plan and carry out a large systematic series of airfoil wind tunnel tests. Their tunnel, built in 1901, was 6 ft long and had a 16-in. square cross section. It is illustrated in Figure 4.1. The flow was produced by a two-bladed fan powered by a gasoline engine. An excellent example of a low-speed wind tunnel is the California Institute of Technology (Cal Tech) wind tunnel shown in Figure 4.2. This tunnel, built in the early 1930s and still in full-time use, has a 10-ft-diameter circular test section in which the model is placed and a maximum speed of about 200 mph. To reduce the energy loss from the friction between the air and the walls and to permit more efficient propellers or fans, the

Figure 4.1 The Wright Brothers' wind tunnel. (Courtesy of National Air and Space Museum, Smithsonian Institution.)

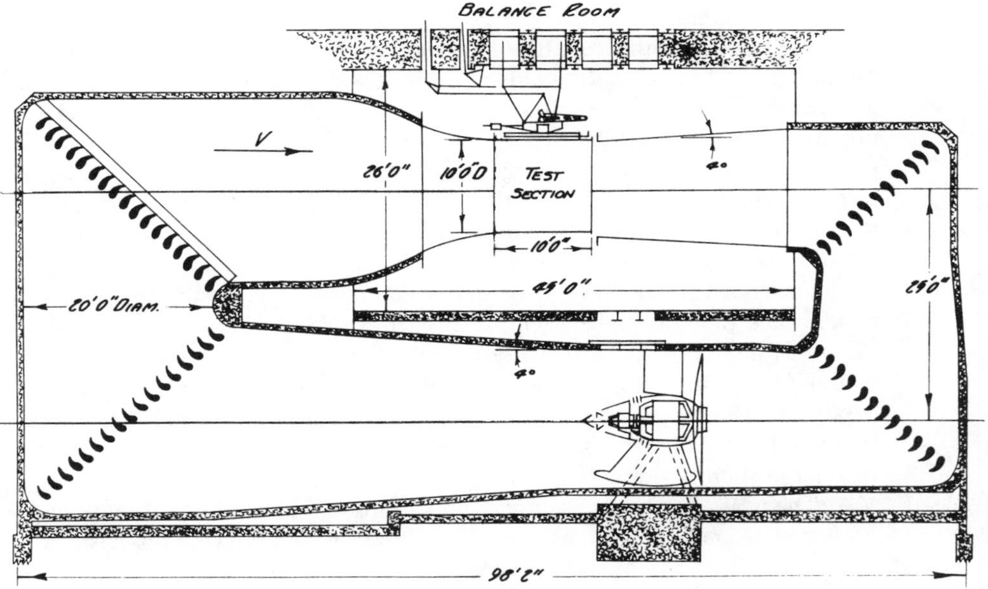

Figure 4.2 Vertical section through 10-ft wind tunnel at the California Institute of Technology. (Courtesy of the Guggenheim Aeronautical Laboratory, California Institute of Technology.)

highest velocity occurs only in the test section. The velocity is reduced elsewhere by increasing the size of the wind tunnel cross sections. The Cal Tech tunnel has a maximum diameter of 20 ft ahead of the test section. The reason for the larger section causing lower air velocities will become clear in our study of fluid dynamics in Chapter 6. The Cal Tech tunnel is a closed-return type, recirculating the air to save power, whereas the Wright Brothers tunnel was an open-return tunnel dumping the air into whatever space exists beyond it. Although tunnels have varying diameters along the circuit, they are usually designated by the size of the test section (e.g., the Cal Tech 10-ft tunnel).

The largest wind tunnel in the world is the 40- by 80-ft test section wind tunnel built in 1944 at the NASA Ames Research Center at Moffett Field, California. The low-speed cross section of this tunnel is 132 by 176 ft. The test section maximum velocity is 265 mph. Full-scale small aircraft or very large models of large aircraft can be placed in the test section. A 1981 modification of this tunnel added an even larger, but lower-speed, test section 80 by 120 ft in size. Its purpose is the testing of V/STOL aircraft.

High-speed wind tunnels require very high powers and specialized test section design. Such tunnels can be high-speed subsonic, transonic, supersonic (greater than the speed of sound), or hypersonic (greater than five times the speed of sound). As noted earlier, some tunnels are pressurized to several times normal sea-level atmospheric pressure to increase the density to compensate for the small scale of the models. One of the finest pressurized high-subsonic-speed tunnels is illustrated in Figure 4.3. It is the 11-ft pressurized wind tunnel at the Ames Research Center, an important tool in the development of most U.S. high-speed transport aircraft.

Figure 4.4 shows the operating curves for the Ames Research Center 11-ft tunnel. The Reynolds number per foot of characteristic length of the model is shown plotted against Mach number for constant values of stagnation pressure and test section dynamic pressure. Also shown are the operating limits determined by the tunnel structural strength or the available power. In practice, an additional limit may be imposed by the strength of the model. For large-span, thin wings, the wings may not be able to withstand the bending loads imposed at high angles of attack by the dynamic pressures associated with high Mach numbers and high Reynolds numbers.

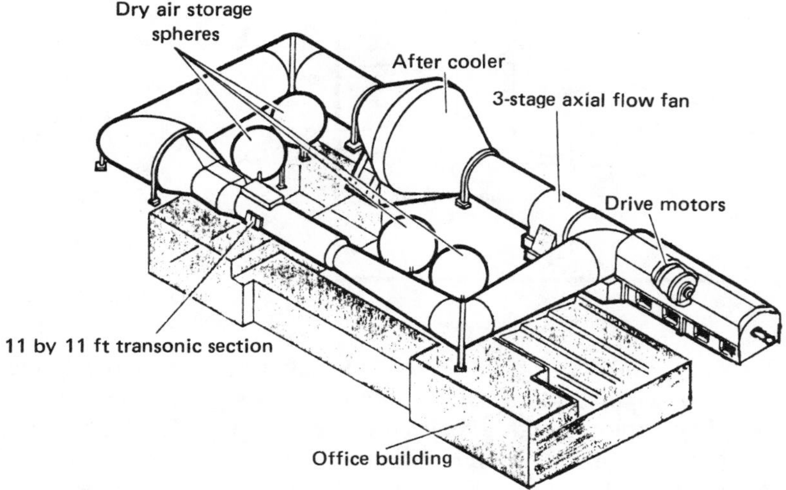

Figure 4.3 NASA-Ames Research Center 11-ft transonic variable density wind tunnel. (Courtesy of NASA.)

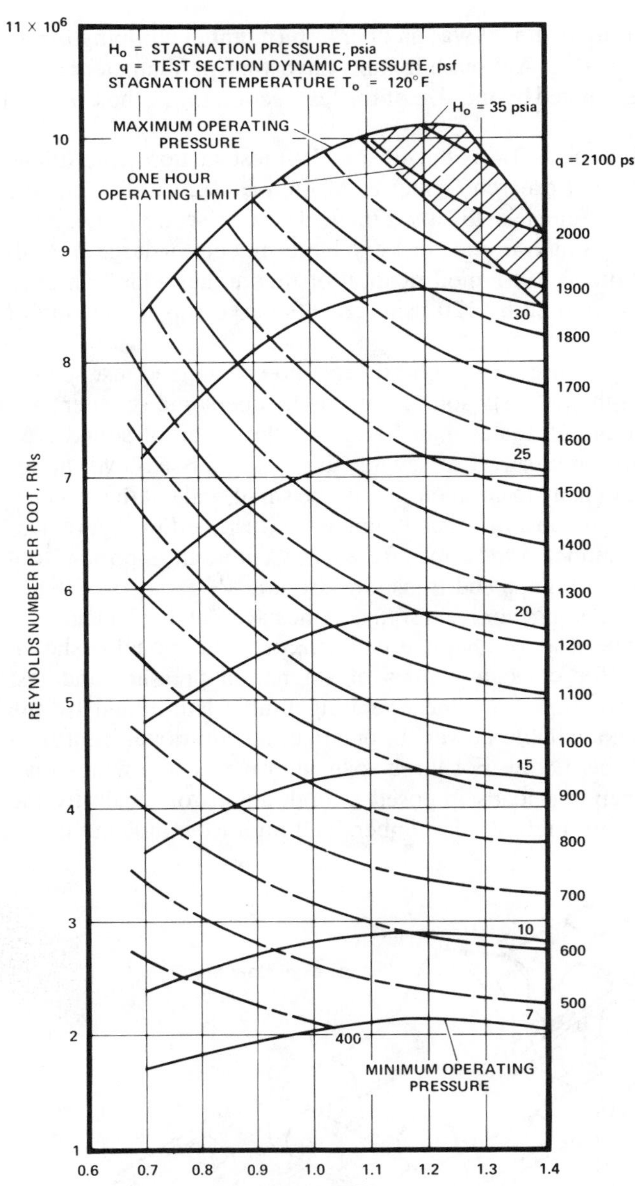

Figure 4.4 Operating curves for the NASA-Ames Research Center 11-ft transonic wind tunnel (Courtesy of NASA.)

Tests at the highest Reynolds numbers may then cover only a limited angle of attack range, which may still satisfy the normal cruising conditions. Higher angles of attack would be explored with a reduced tunnel pressure. The highest attainable Reynolds number for a typical transport airplane model in the ARC 11-ft tunnel is about 6.5 to 7 million based on the wing chord. Full-scale Reynolds numbers are 20 to 60 million.

To avoid the power requirements of continuous-running, closed-circuit, pressurized high-speed tunnels, some tunnels operate for only a short period on the order of

15 to 30 seconds by exhausting the air from a high-pressure tank through an open-circuit tunnel. Data can be obtained over a wide range of angles of attack in a single brief run in these intermittent-flow or blowdown tunnels with the aid of automatic pitch attitude change and recording equipment. The tank is then pumped up for the next run, a process taking 15 to 60 minutes.

A relatively new wind tunnel development is the use of a very low temperature gas as the fluid in a highly pressurized wind tunnel. The low temperature lowers the speed of sound so that a desired Mach number can be obtained at a lower speed. The density is greatly increased by the low temperature, as well as by the high pressure. The result is a high Reynolds number obtained with lower power requirements. The NASA National Transonic Facility is a cryogenic wind tunnel utilizing nitrogen as the fluid, with a test section temperature as low as 115 K (207°R). The test section is 2.5 m square (8.2 ft square). Reynolds numbers as high as 80 million can be obtained based on the wing chord of an airplane model. The Mach numbers may be varied from 0.2 to 1.2. Maximum power required is 120,000 bhp.

Figure 4.5 shows a wind tunnel model mounted in a low-speed tunnel. The portions of the struts attached to the model and exposed to the airflow are very thin to minimize the corrections that must be made to the measured model forces. The large strut fairings are bolted to the tunnel floor and shield most of the actual struts from the airflow.

Figure 4.5 A DC-9-50 wind tunnel model in the test section of the McDonnell Douglas 8- by 12-ft wind tunnel. (Courtesy of McDonnell Douglas Corporation.)

5

THE ATMOSPHERE

Since aircraft operate in the earth's atmosphere and their lift and drag characteristics depend on the properties of that atmosphere, density and viscosity, it is essential to be able to define those properties. Even spacecraft pass through the atmosphere during launch and, when applicable, during return to earth. Density and viscosity are functions of altitude. More precisely, density ρ varies with pressure p and temperature T, while viscosity μ varies with temperature only. Since meteorological conditions are constantly varying, there is no steady-state "normal" atmosphere, but to establish a representative set of conditions, a standard atmosphere has been defined. The atmosphere is defined from the equation of state of a perfect gas,

$$p = \rho RT \qquad \text{equation of state} \tag{5.1}$$

where

p = pressure [newtons/square meter (N/m^2) or lb/ft^2]
ρ = density [kilograms/cubic meter (kg/m^3) or $slugs/ft^3$]
T = absolute temperature [Kelvin (K) or degrees Rankine (°R)]
R = gas constant [287.05 newton meters/kilogram Kelvin (n-m/kg K) or 1718 ft-lb/slug °R for air]

and the hydrostatic equation,

$$dp = -\rho g\, dh \qquad \text{hydrostatic equation} \tag{5.2}$$

where

g = gravitational constant = 9.8 m/s^2 or 32.17 ft/s^2 at sea level
h = altitude above sea level (meters or feet)

Note that all these units are given first in SI (Système International) units, followed by English engineering units.

To derive the hydrostatic equation, consider the force balance on an element of fluid at rest (Figure 5.1). We assume the element has rectangular sides. The top and bottom surfaces have sides of unit length; that is, the sides may be assumed to have a length of 1. The height of the element is an infinitesimally small height dh. On the bottom surface, the pressure p acts to produce an upward force of $p(1)(1)$ exerted on the fluid element. The top surface is higher in altitude by the distance dh. Because pressure varies with altitude, the downward pressure on the top surface will differ from the pressure on the bottom surface by the infinitesimally small value dp. Therefore, on the top surface, the downward pressure force on the fluid element will be $(p + dp)(1)(1)$. The volume of the fluid element is $(1)(1)\,dh = dh$, and since ρ is the mass per unit volume, the mass of the fluid element is $\rho(1)(1)\,dh = \rho\,dh$. If the local acceleration of gravity is g, the weight of the fluid element is $\rho g\,dh$. The forces acting on the fluid element, pressure forces on the top and bottom and the weight, must balance because the fluid element is stationary. Hence

$$p = (p + dp) + \rho g\,dh$$

Thus

$$dp = -\rho g\,dh$$

This equation is the hydrostatic equation and applies to any fluid (i.e., liquids as well as gases). It is a differential equation; that is, it relates an infinitesimally small change in altitude dh to a corresponding infinitesimally small change in pressure dp.

Before integrating the hydrostatic equation to find a useful expression for the variation of pressure with altitude, $p = p(h)$, it is helpful to make the assumption that

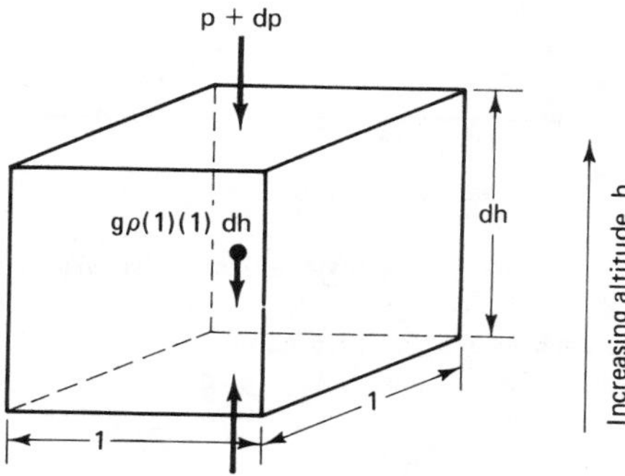

Figure 5.1 Force diagram for the hydrostatic equation.

g is constant through the atmosphere, equal to its value at sea level, g_0.* Hence we can write

$$dp = -\rho g_0 \, dh \tag{5.3}$$

Only one more relationship is needed to define the standard atmosphere: a defined variation of T with altitude. This variation is based on experimental evidence and is shown in Figure 5.2 for the U.S. standard atmosphere. From sea level to 11 km (36,150 ft) in the gradient region called the *troposphere,* the temperature decreases linearly. Above 11 km in the isothermal region known as the *stratosphere,* the temperature is constant. At 25.1 km (82,300 ft), a temperature gradient again occurs, but here the temperature increases to 47 km (154,000 ft). The temperature variations are shown in Figure 5.2 only to 100,000 ft, since the regions of interest are generally below 15 km (49,200 ft) for conventional aircraft and below 22 km (72,160 ft) even for extremely high altitude aircraft. Given $T = T(h)$ as defined by Figure 5.2, then

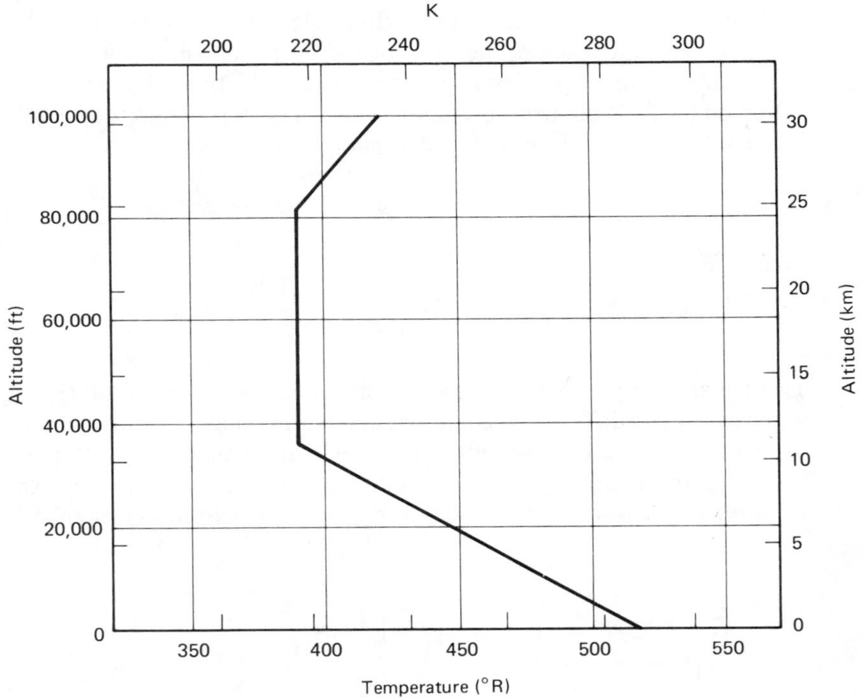

Figure 5.2 Temperature variation with altitude in the U.S. standard atmosphere.

*Actually, g varies inversely as the square of the distance from the center of the earth in accordance with Newton's law of gravitation. If r_0 is the radius of the earth and h is the geometric altitude,

$$g = g_0 \left(\frac{r_0}{r_0 + h} \right)^2$$

p and ρ can be found for any altitude from the hydrostatic equation and the equation of state. Consider first the hydrostatic equation

$$dp = -\rho g_0 \, dh$$

Divide by the equation of state, $p = \rho RT$:

$$\frac{dp}{p} = -\frac{\rho g_0 \, dh}{\rho RT} = -\left(\frac{g_0}{RT}\right) dh \tag{5.4}$$

We must now introduce the variation of T with altitude h. Consider first the *isothermal* (constant temperature) layers of the standard atmosphere, as given by the vertical line in Figure 5.2. Let the temperature, pressure, and density at a point in the isothermal layer be T_1, p_1, and ρ_1, respectively. The altitude at this point is h_1. Now consider a higher point in the isothermal layer, where the altitude is h and the pressure is p. The relationship between the pressures p and p_1 can be obtained by integrating equation 5.4 between h_1 and h.

$$\int_{p_1}^{p} \frac{dp}{p} = -\frac{g_0}{RT} \int_{h_1}^{h} dh \tag{5.5}$$

Note that g_0, R, and T are constants that can be taken outside the integral. Performing the integration in equation 5.5, we obtain

$$\ln \frac{p}{p_1} = -\frac{g_0}{RT}(h - h_1)$$

or

$$\boxed{\frac{p}{p_1} = e^{-(g_0/RT)(h-h_1)}} \quad \text{isothermal region} \tag{5.6}$$

Relating pressure and density from the equation of state yields

$$\frac{p}{p_1} = \frac{\rho T}{\rho_1 T_1} = \frac{\rho}{\rho_1}$$

Thus

$$\boxed{\frac{\rho}{\rho_1} = e^{-(g_0/RT)(h-h_1)}} \quad \text{isothermal region} \tag{5.7}$$

Equations 5.6 and 5.7 give the variation of p and ρ versus geometric altitude for the isothermal layers of the standard atmosphere.

Turning to the *gradient* layers in which the temperature varies linearly with altitude,

$$T = T_1 + a(h - h_1) \tag{5.8}$$

where $a = dT/dh$ and is the change in temperature for each meter (or foot) increase in altitude as defined in Figure 5.2. Here a is a constant for each gradient layer and is called the *lapse rate*; h_1 and T_1 are the altitude and corresponding temperature at some point in the gradient layer, and T and h represent some higher altitude. Since $a = dT/dh$,

$$dh = \frac{1}{a} dT \tag{5.9}$$

Substitute equation 5.9 into 5.4:

$$\frac{dp}{p} = -\left(\frac{g_0}{aR}\right)\frac{dT}{T} \tag{5.10}$$

Integrating between h_1 and some higher altitude h, equation 5.10 yields

$$\int_{p_1}^{p} \frac{dp}{p} = -\frac{g_0}{aR}\int_{T_1}^{T} \frac{dT}{T}$$

$$\ln\frac{p}{p_1} = -\frac{g_0}{aR}\ln\frac{T}{T_1}$$

Thus

$$\boxed{\frac{p}{p_1} = \left(\frac{T}{T_1}\right)^{-g_0/aR}} \quad \text{gradient region} \tag{5.11}$$

To find the density relationship, we recall from the equation of state that

$$\frac{p}{p_1} = \frac{\rho T}{\rho_1 T_1}$$

Hence equation 5.11 becomes

$$\frac{\rho T}{\rho_1 T_1} = \left(\frac{T}{T_1}\right)^{-g_0/aR}$$

$$\frac{\rho}{\rho_1} = \left(\frac{T}{T_1}\right)^{(-g_0/aR)-1}$$

or

$$\boxed{\frac{\rho}{\rho_1} = \left(\frac{T}{T_1}\right)^{-[(g_0/aR)+1]}} \quad \text{gradient region} \tag{5.12}$$

To apply equations 5.11 and 5.12, we need to know the lapse rate a in the equation for T. Figure 5.2 shows that from sea level to 36,150 ft (11,019 m)

$$a = -0.0065 \text{ K per meter}$$
$$a = -0.00356°\text{R per foot}$$

Equation 5.8 gives $T = T(h)$ for the gradient layers. When it is substituted into equation 5.11, we obtain $p = p(h)$; similarly, from equation 5.12, we obtain $\rho = \rho(h)$.

We now have the tools required to calculate the standard atmosphere. Referring to the temperature variation in Figure 5.2, we start at sea level ($h = 0$), where standard sea-level values of pressure, density, and temperature, p_s, ρ_s, and T_s, respectively, are

$$p_s = 1.01325 \times 10^5 \text{ N/m}^2 = 2116.2 \text{ lb/ft}^2$$
$$\rho_s = 1.2250 \text{ kg/m}^3 = 0.002377 \text{ slug/ft}^3$$
$$T_s = 288.16 \text{ K} = 518.69°\text{R}$$

The density ratio ρ/ρ_s is used so frequently that it is given a special symbol, σ.

The sea-level values are the reference values for the first gradient region. We use equation 5.8 to obtain values of T as a function of h until $T = 216.66$ K (389.99°R), which occurs at $h = 11{,}019$ m (36,150 ft). With these values of T, use equations 5.11 and 5.12 to obtain the corresponding values of p and ρ in the first gradient layer. Next, starting at $h = 11{,}019$ m as the base of the first isothermal region (see Figure 5.2), use equations 5.6 and 5.7 to determine values of p and ρ versus h, until $h = 25.1$ km, which is the starting point of the next gradient region. With this procedure, using Figure 5.2 and equations 5.6, 5.7, 5.8, 5.11, and 5.12, a table of the characteristics of the standard atmosphere can be constructed.

Example 5.1

Calculate the values of pressure, pressure ratio, density, density ratio, and temperature for the standard atmosphere at an altitude of 8000 m. Show the results in SI units.

The standard sea-level values are pressure = 101,325 N/m^2, density = 1.2250 kg/m^3, and temperature = 288.16 K. The temperature lapse rate $a = -0.0065$ K/m.

Solution: At 8000 m, $T = 288.16 - (8000)(0.0065) = 236.16$ K. In the gradient region, from equation 5.11,

$$\frac{p}{p_1} = \left(\frac{T}{T_1}\right)^{-g_0/aR}$$

Taking as the initial conditions the sea-level values,

$$\text{pressure ratio} = \frac{p}{101{,}325} = \left(\frac{T}{288.16}\right)^{-g_0/aR}$$

$$= \left(\frac{236.16}{288.16}\right)^{-9.8/(-0.0065)(287.05)}$$

$$= \underline{0.3516}$$

and

$$p = 101{,}325(0.3516) = \underline{35{,}625 \text{ N/m}^2}$$

The density is found from the equation of state of a gas:

$$\rho = \frac{p}{RT} = \frac{35{,}625}{(287.05)(236.16)} = \underline{0.52552\ \text{kg/m}^3}$$

$$\sigma = \frac{\rho}{\rho_s} = \frac{0.52554}{1.2250} = \underline{0.4290}$$

Example 5.2

Calculate the values of pressure, pressure ratio, density, density ratio, and temperature for the standard atmosphere at an altitude of 25,000 ft. Show results in English units.

The standard sea-level values are pressure = 2116 lb/ft², density = 0.002377 slug/ft³, and temperature = 519°R (59°F). The temperature lapse rate $a = -0.00356$°R/ft.

Solution: At 25,000 ft, $T = 519 - (25{,}000)(0.00356) = 430$°R. In the gradient region, from equation 5.11,

$$\frac{p}{p_1} = \left(\frac{T}{T_1}\right)^{-g_0/aR}$$

Taking as the initial conditions the sea-level values,

$$\text{pressure ratio} = \frac{p}{2116} = \left(\frac{T}{519}\right)^{-g_0/aR}$$

$$= \left(\frac{430}{519}\right)^{-32.2/(-0.00356)(1718)}$$

$$= \underline{0.3714}$$

and

$$p = 2116(0.3714) = \underline{785.6\ \text{lb/ft}^2}$$

The density is found from the equation of state of a gas:

$$\rho = \frac{p}{RT} = \frac{785.6}{(1718)(430)} = \underline{0.001063\ \text{slug/ft}^3}$$

$$\sigma = \frac{\rho}{\rho_s} = \frac{0.001063}{0.002377} = \underline{0.4472}$$

Example 5.3

Air flowing at high speed in the working section of a wind tunnel has pressure and temperature values equal to 0.6 atm at sea level and −40°C, respectively. Calculate, in English engineering units:

(a) Air density.
(b) Density ratio.
(c) Specific volume.

Solution:

(a) Pressure = $0.6 \times 2116 = 1269.6$ lb/ft^2. Temperature = $(-40 \times \frac{9}{5}) + 32 =$ -40°F, or, since we must work with absolute temperature,

$$T = -40 + 460 = 420°\text{R}$$

From the equation of state,

$$\rho = \frac{p}{RT} = \frac{1269.6}{(1718)(420)} = \underline{0.001760 \text{ slug/ft}^3}$$

(b) $\sigma = \dfrac{0.001760}{\rho_{\text{sea level std. day}}} = \dfrac{0.001760}{0.002377} = \underline{0.740.}$

(c) Specific volume $= \dfrac{1}{\rho} = \underline{568.18 \text{ ft}^3\text{/slug.}}$

Example 5.4

Repeat Example 5.3 using SI units.

Solution:

(a) $p = 0.6 \times 101{,}325 = 60{,}795$ N/m^2
$T = 273 - 40 = 233$ K (273 K corresponds to 0°C)

$$\rho = \frac{60{,}795}{(287.05)(233)} = 0.90898 \text{ kg/m}^3$$

(b) $\sigma = \dfrac{0.90898}{1.2250} = 0.742.$

(c) Specific volume $= \dfrac{1}{\rho} = 1.100$ m^3/kg.

PROPERTIES OF THE U.S. STANDARD ATMOSPHERE

The properties of the U.S. standard atmosphere are given in the tables in Appendix A. Both SI and English engineering units are shown for altitude, temperature, pressure, density, speed of sound, and kinematic viscosity.

Figure 5.3 shows the variation of density ratio ρ/ρ_s, pressure ratio p/p_s, and kinematic viscosity ratio ν/ν_s with altitude for the standard atmosphere. The subscript $(\cdot)_s$ designates sea-level standard conditions. Kinematic viscosity is μ/ρ and is convenient in calculating Reynolds number, which then becomes

$$\text{RN} = \frac{Vl}{\nu} \tag{5.13}$$

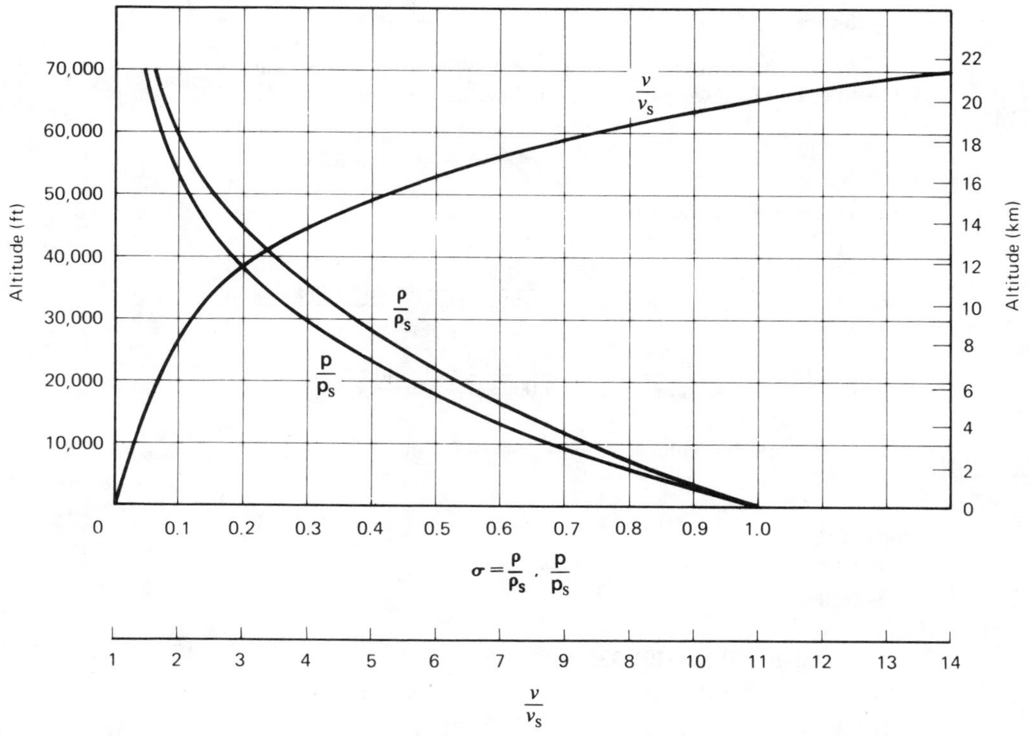

Figure 5.3 Variation of density, pressure, and kinematic viscosity ratios for the U.S. standard atmosphere.

DEFINITIONS OF ALTITUDE

In aerospace terms, there are several definitions of altitude. The first and most obvious is *geometric altitude h,* the physical altitude in meters or feet above sea level. The second is *density altitude* h_d, the geometric altitude on a standard day at which the density would be equal to the actual air density experienced by the vehicle. Since the forces acting on a wing or body are a direct function of density, the behavior of an airplane depends only on density altitude (although the power or thrust available from the engine depends also on pressure and temperature). The third altitude is *pressure altitude* h_p, the geometric altitude on a standard day for which the pressure is equal to the existing atmospheric (or ambient) pressure. Altimeters are pressure instruments and are therefore calibrated to read the pressure altitude. Pressure altitude, density altitude, and temperature are related through the equation of state $p = \rho RT$, as shown in Figure 5.4.

There is one other altitude of interest to orbital or space flights, although it is of little importance to airborne vehicles. The *absolute altitude* h_a is the distance from the center of the earth. If r_0 is the radius of the earth, then $h_a = h + r_0$. The absolute altitude is important for space flight because the local acceleration of gravity g varies

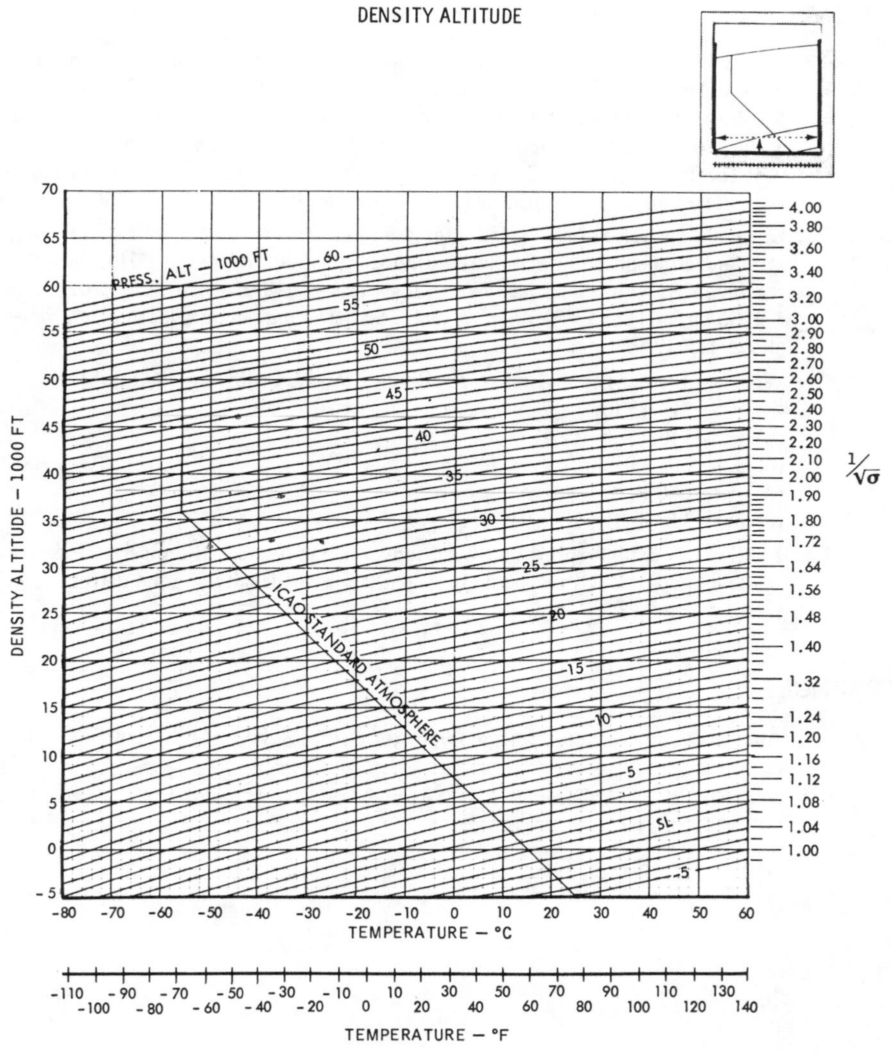

Figure 5.4 Density altitude versus pressure altitude and temperature. (From USAF Series T.O. 1F-5a-1, *Flight Manual.*)

with h_a. From Newton's law of gravitation, g varies inversely as the square of the distance from the center of the earth. Letting g_0 be the gravitational acceleration at sea level, the local gravitational acceleration g at a given absolute altitude h_a is

$$g = g_0\left(\frac{r_0}{h_a}\right)^2 = g_0\left(\frac{r_0}{r_0 + h}\right)^2 \tag{5.14}$$

The variation of g with altitude must be taken into account when dealing with complete mathematical models of the atmosphere.

PROBLEMS

5.1. Calculate the values of pressure, pressure ratio, density, density ratio, and temperature for a nonstandard atmosphere at geometric altitudes of 15,000, 25,000, 36,150, 45,000, and 55,000 ft. Show results in English units. The atmosphere is nonstandard in that temperatures at all altitudes are 20°F above standard for that altitude. The sea-level values are pressure = 2116 lb/ft^2 and temperature = 519°R + 20°R. Find the sea-level density from the equation of state. The temperature lapse rate $a = -3.56$°R/1000 ft; above 36,150 ft, $a = 0$.

5.2. Calculate the values of pressure, pressure ratio, density, density ratio, and temperature for the standard atmosphere at geometric altitudes of 5000, 10,000, 11,019, and 15,000 m. Show results in SI units. The sea-level values are pressure = 1.01325×10^5 N/m^2, density = 1.2250 kg/m^3, and temperature = 288.16 K. The temperature lapse rate $a = -0.0065$ K/m; above 11,019 m, $a = 0$.

5.3. What are the density and pressure altitudes for the following conditions?

(a) $p = 719.15$ lb/ft^2; $T = 422.4$°R.
(b) $p = 719.15$ lb/ft^2; $T = 442.4$°R.
(c) $p = 719.15$ lb/ft^2; $T = 400$°R.
(d) $p = 600$ lb/ft^2; $T = -30$°F.
(e) $p = 390$ lb/ft^2; $T = -50$°F.

5.4. Air flowing at high speed in a wind tunnel has pressure and temperature equal to 0.8 atm at sea level and −25°C, respectively. What is the air density and the density ratio? What is the specific volume?

REFERENCES

5.1. Anderson, John D., Jr., *Introduction to Flight,* McGraw-Hill, New York, 1978, especially pp. 54–60.

5.2. Minzner, R. A., Champion, K. S. W., and Pond, H. L., *The ARDC Model Atmosphere,* AF CRC-TR-50-267, 1959.

6

INCOMPRESSIBLE ONE-DIMENSIONAL FLOW

We have seen that aerodynamic forces can be divided into shear forces tangential to the surface of a body and pressure forces acting normal to the surface. Shear forces can result only from viscosity, whereas normal forces arise from changes in the fluid pressure as the fluid moves past the body. It is now appropriate to study these pressure changes. It will be seen that for simple but very useful flows the applicable equations can be derived from a few fundamental laws of physics.

First, we must consider the characteristics of the fluid, *air,* with which we are concerned. Air is a viscous compressible fluid. The viscosity is small, however, and the compressibility effects on the flow are usually negligible at speeds below about 40% of the speed of sound. The compressibility effects in which we are interested are changes in density resulting from the pressure changes in the fluid. It is therefore convenient to develop the pressure–velocity relationships for a fluid defined as an incompressible perfect fluid, where "perfect" means having zero viscosity. We shall see later that viscosity is, in fact, enormously important, but that its effects are usually confined to a thin layer of fluid around the body. The flow outside this layer acts as though the fluid were inviscid.

THE CONTINUITY EQUATION: BERNOULLI'S EQUATION

We can begin our study by visualizing a flow through a channel of varying cross sections. The particular channel with a constriction shown in Figure 6.1 is known as a *venturi* and is an important element of internal combustion engine carburetors and

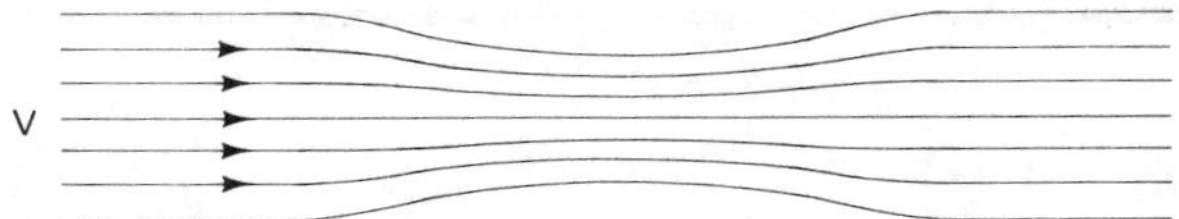

Figure 6.1 Streamlines for flow through a channel with constriction (Venturi).

other devices involved in the measurement of flow. We shall simplify this flow by defining it as *two dimensional,* which means that the velocity at every point has a direction lying in a plane parallel to a fixed reference plane such as that of the paper. Furthermore, the flow pattern is identical in all planes parallel to the reference plane. In our analysis of channels, we also assume that the flow velocity components perpendicular to the axis of the channel are sufficiently small compared to the flow velocities parallel to the channel axis so that they can be neglected. Thus we are really studying a *one-dimensional* flow.

The flow through the venturi in Figure 6.1 is indicated by a family of lines called *streamlines*. If every particle of fluid were to be made visible by injecting smoke into the air, we could follow the particle paths through the channel. The paths so determined are streamlines. A more specific definition of a streamline is a path whose tangent at every point is in the direction of the velocity vector at that point. An important property of streamlines is that they cannot have any end unless fluid is created or destroyed. Streamlines must extend to infinity or form closed paths.

The streamlines at the left side of the diagram of the channel in Figure 6.1 are drawn with equal spacing. Assuming that the velocity is uniform over the entering cross section, the flow quantity between neighboring streamlines is the same. Since by definition there is no flow across the streamlines, the same quantity of fluid flows between adjoining streamlines along their entire length. Clearly, the streamlines in the constriction are closer together than in the wider sections. Consider a volume element bounded by two neighboring streamlines in the contracting part of the channel (Figure 6.2). The element is also bounded by two planes, parallel to the flow and unit distance apart, and by two faces perpendicular to the flow placed anywhere in the contracting portion of the channel. The left face of the element has an area S_1 and a velocity V_1 across it, while the area and velocity of the right face are S_2 and V_2, respectively.

We must now introduce one of the basic principles of Newtonian mechanics, the conservation of mass, which states that mass can be neither created nor destroyed. If

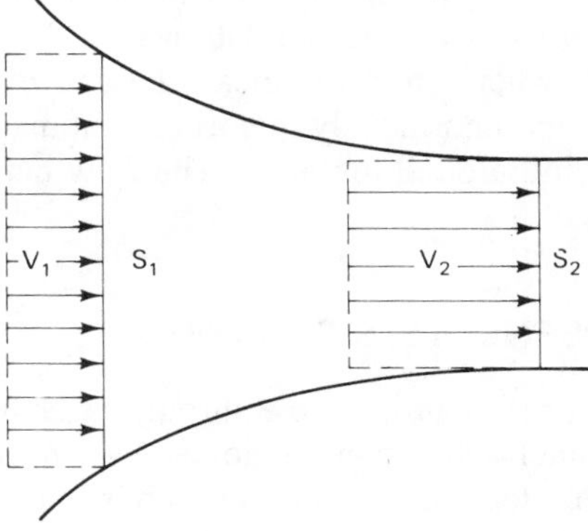

Figure 6.2 Flow between converging streamlines.

no fluid is to be created or destroyed in the fluid element, the mass of fluid entering the element across the face S_1 in the unit time must equal the mass of fluid leaving the element across face S_2. The volume of fluid crossing the face S_1 per second is the cross-sectional area times the velocity or $S_1 \times V_1$. The mass of fluid passing through the section is the density times the volume, or $\rho_1 \times S_1 \times V_1$. Similarly, the mass leaving the element is $\rho_2 \times S_2 \times V_2$. Since the two masses are equal,

$$\rho_1 S_1 V_1 = \rho_2 S_2 V_2 \qquad \text{continuity equation} \tag{6.1}$$

This is the continuity equation, one of the fundamental physical laws that underlie the solution of fluid dynamic problems. Since for this chapter we are assuming an incompressible fluid with constant density, $\rho_1 = \rho_2$, and the continuity equation becomes

$$S_1 V_1 = S_2 V_2 \qquad \text{incompressible continuity equation} \tag{6.2}$$

Equation 6.2 states that the product of S times V is a constant along any pair of streamlines, where S is the spacing between streamlines. Thus the velocity V is inversely proportional to the spacing between the streamlines. A narrowing of the space between streamlines requires an increase in velocity, and a wider spacing slows the flow. A three-dimensional region enclosed by streamlines is called a *streamtube,* and the flow through sections perpendicular to the flow direction is essentially the same as the two-dimensional flow we have described.

The continuity equation can be used to deduce the general nature of more complex flows than a simple channel. Figure 6.3 illustrates the flow around a circular cylinder, the flow far from the cylinder being unaffected by its presence. This is a two-dimensional flow since the flow in all planes parallel to the paper is assumed to be identical, and there is no velocity component parallel to the axis of the cylinder. The streamlines, which are straight far ahead of the body, must curve as they approach the body to avoid the obstruction. Far above and below the cylinder the flow would be unaffected by the body. The streamlines around the cylinder are shown and have the nature of the venturi flow, except that the area is constricted by an obstruction in the middle of the flow instead of a narrowing of the walls of a channel. Since constriction of the streamlines is associated with high velocity, it can be seen that the velocity is increased over the top and bottom of the cylinder and reduced ahead and behind it. The general velocity pattern over many simple bodies can be determined in this way, but the force on the body depends not on velocity but on pressure. The relationship between velocity and pressure is found most simply in the Bernoulli equation, named after Daniel Bernoulli (1700–1782), a Swiss mathematician. In spite of its name, the Bernoulli equation was actually derived by Leonhard Euler (1707–1783), also a Swiss mathematician.

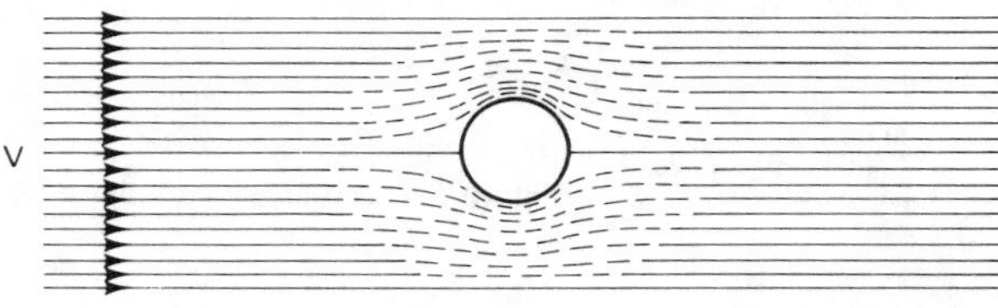

Figure 6.3 Streamline picture for flow around a circular cylinder.

To derive the Bernoulli equation, we consider the forces acting on a very small element of volume through which a perfect incompressible fluid is flowing with *steady* two-dimensional motion. The term "steady" means that the flow at any point is unaffected by time. The element is bounded on its sides by streamlines, which are everywhere tangent to the velocity (Figure 6.4). Streamlines are curved in general, but the size of the element is so small that these sides may be considered to be straight. The element has unit thickness perpendicular to the plane of flow (i.e., the plane of the paper). The left and right ends of the element are perpendicular to the direction of flow and are a small distance dx apart.

The fluid entering the volume element at the left face has the velocity V and the pressure p. The left face has an area s. The fluid leaving the element a small distance dx downstream will have a different velocity and pressure since conditions are changing along the streamtube. If the rate of velocity and pressure change are dV/dx and dp/dx, respectively, the values at the right face will be $V + dV$ and $p + dp$. The average pressure over the sides will be the average between the two faces, or $p + dp/2$.

It is important to clarify the nature of the pressure to which we are referring. This pressure is called the *static pressure*. It is the pressure that would be measured at a small orifice in a plane parallel to the flow so that no effect of air velocity is felt. Static pressure is caused by the random motion of gas molecules. It is equal to the time rate of change of momentum of the particles rebounding against a surface moving with the fluid.

Returning now to our discussion of the forces acting on the fluid element in Figure 6.4, we recall that Newton's second law applied to fluid mechanics states: the net force acting on a fluid particle is equal to the time rate of change of the linear momentum of that particle. This is just a statement of the familiar equation $F = ma$ or, alternatively, $F = m(dV/dt)$. The forces on the element arise entirely from the pressures since a perfect fluid is assumed so that the viscosity $\mu = 0$. The momentum can appear only in the net momentum flowing into and out of the element boundaries. Since the sides have been chosen along streamlines, flow can cross only the left and right ends of the element.

The mass entering from the left per unit time is equal to ρsV, and the momentum is ρsVV. The mass leaving across the right face must equal that entering the element because of the conservation of mass, but the velocity will differ by the increment dV. Thus the momentum leaving is $\rho sV(V + dV)$. The excess of momentum is then

$$\rho sV(V + dV) - \rho sV^2 = \rho sV\,dV \tag{6.3}$$

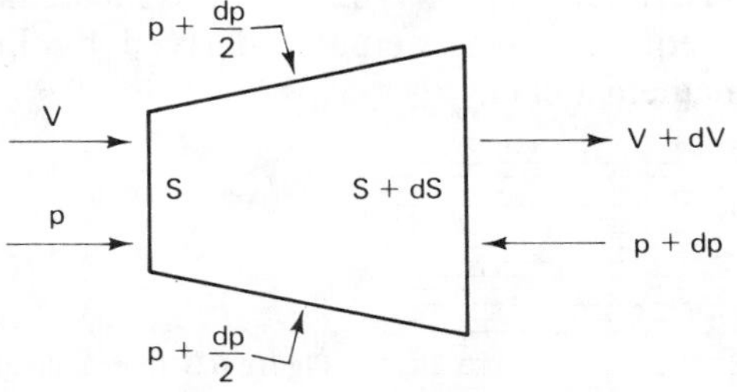

Figure 6.4 Infinitesimal volume element in the flow.

This change in momentum per unit time must equal the net force. Taking the force as positive to the right, the force on the left face is ps, while the force on the right face is $-(p + dp)(s + ds) = -ps - p\,ds - s\,dp - dp\,ds$. As the element becomes vanishingly small, the product of the differentials, $dp\,ds$, becomes negligibly small and may be dropped. The resultant force from the ends is then

$$-p\,ds - s\,dp$$

The net force from the sides can be seen from Figure 6.4 to be $(p + dp/2)ds = p\,ds + (dp\,ds)/2$. The second term is again a product of differentials and may be neglected. The resultant force from the sides is, therefore,

$$p\,ds$$

and the total force is

$$-p\,ds - s\,dp + p\,ds = -s\,dp \tag{6.4}$$

From Newton's second law, we must equate equations 6.3 and 6.4:

$$-s\,dp = \rho s V\,dV$$

so that

$$dp = -\rho V\,dV \qquad \text{Euler's equation} \tag{6.5}$$

The Euler or momentum equation is a differential relationship between pressure and velocity. To obtain a direct relationship, we can integrate from one point on a streamline designated by $(\cdot)_1$ to another designated by $(\cdot)_2$. Then, assuming that ρ = constant,

$$\int_{p_1}^{p_2} dp = \int_{V_1}^{V_2} -\rho V\,dV$$

$$p_2 - p_1 = \frac{\rho}{2}V_1^2 - \frac{\rho}{2}V_2^2 \tag{6.6}$$

$$\boxed{p_2 + \frac{\rho}{2}V_2^2 = p_1 + \frac{\rho}{2}V_1^2 = B} \qquad \text{Bernoulli's equation} \tag{6.7}$$

Bernoulli's equation applies only to an incompressible, inviscid fluid. Nevertheless, it is a very useful tool, the first of many we shall develop, and certainly one of the most remarkable because of the wealth of information it yields and its simplicity. The equation tells us that, as the flow progresses from one point to another, an increase in speed will be accompanied by a decrease in pressure. This follows because the sum of the static pressure p, a form of potential energy, and the term $(\rho/2)V^2$, the kinetic energy of a unit volume of fluid usually called the dynamic pressure, is a constant along a streamline unless heat is added to the flow. The constant B represents the total energy in a unit volume of fluid and is best known as the *total* or *stagnation pressure*, p_T. The name "total pressure" arises because it is the sum of the static and dynamic pressures, while the stagnation designation arises from the fact that, when the velocity is reduced to zero (i.e., stagnation), the static pressure is equal to the total pressure.

In a flowing stream of fluid with constant pressure and velocity such as that approaching a flying vehicle before the influence of the vehicle is felt, the total pressure will clearly be uniform throughout the flow. The Bernoulli equation states that this total pressure will remain unchanged along streamlines around the vehicle. Furthermore, since the total pressure is uniform throughout the flow, the Bernoulli equation can be applied from any point in the fluid to any other point. This will be true unless some external force is applied to the fluid. A propeller or jet engine would be typical external forces. The Bernoulli equation cannot be applied in studying flow through a propeller, although it can be applied in the region ahead of the propeller and then, as a separate calculation, in the region behind a propeller. We shall study this problem in Chapter 17.

We can immediately apply the Bernoulli equation to learn more about the flow over the circular cylinder in Figure 6.3. We had noted that the velocity was highest over the top and bottom of the cylinder and lowest at the front and rear. Now we can deduce the relative pressures since from Bernoulli's equation the high velocities will cause low pressures on the top and bottom, and, conversely, the pressures will be highest at the front and rear of the body. This is indeed confirmed by experiment except at the rear, where viscosity effects, so far ignored, become predominant. This is discussed in detail later.

Let us assume that p_0 and V_0 are the pressure and velocity, respectively, in the freestream atmosphere and that p_2 and V_2 are the pressure and velocity at some point on the surface of a body or a wing. Then, to relate the freestream conditions to those on the airplane, we can write Bernoulli's equation as

$$p_2 - p_0 = \frac{\rho}{2}(V_0^2 - V_2^2) = \frac{\rho V_0^2}{2}\left[1 - \left(\frac{V_2}{V_0}\right)^2\right] \tag{6.8}$$

Thus the change in pressure from the freestream, or ambient, air pressure to a point on the surface of a wing or body depends on the air density, the square of the freestream velocity, and the ratio of the local velocity on the wing to the freestream velocity. The velocity ratio V_2/V_0 is nondimensional and is a function only of the shape of the wing and its angle of attack. (At higher speeds at which compressibility cannot be neglected, the Mach number, the ratio of the freestream velocity to the speed of sound, must also be considered.) Since the force F on the wing is simply the integral of the pressures over the wing surface area, it has now been shown from fundamental physical principles that $F = C_F(\rho V^2/2)S$, where C_F is a coefficient dependent on the body shape and attitude. This confirms the results of dimensional analysis.

We have now introduced three important physical relationships involving the primary fluid parameters p, ρ, T, and V. It is helpful to remember a basic distinction between the equation of state, $p = \rho RT$, which applies at any point in the fluid, and the Bernoulli (energy) equation and the continuity equation, both of which relate quantities at a point to the comparable quantities at another point along a fluid streamline.

APPLICATIONS OF BERNOULLI'S EQUATION: AIRSPEED INDICATORS AND WIND TUNNELS

Bernoulli's equation is the basis of all airspeed indicators. Writing the equation as

$$p_0 + \frac{\rho}{2}V_0^2 = p_T \tag{6.9}$$

where $(\cdot)_0$ refers to undisturbed freestream flow and p_T is the total pressure, or the static pressure when the velocity is zero, it is seen that the dynamic pressure can be calculated if the freestream and total pressures can be measured. An airspeed system is designed to do exactly that.

Figure 6.5 is a schematic drawing of a *pitot tube,* a device for measuring total pressure. A pitot tube is essentially a small pipe with one end open and facing into the oncoming fluid stream. The other end is connected by a tube to a pressure-measuring instrument such as a manometer, an aneroid gauge, or an electrical pressure gauge. The flow around the pitot tube is that of a perfect fluid that far upstream has a uniform velocity and pressure so that the Bernoulli constant B (i.e., the total pressure) is constant throughout the fluid. The flow divides at the front of the pitot tube, half of the flow going above the tube and half below it. The streamline intersecting the tube at its center must come to a complete stop. This point is called a *stagnation point.* Since the pitot tube and the lines to the pressure gauge are a closed system, there is no flow through the pitot tube in a steady-state condition. The pressure at the gauge is the same as the pressure at the stagnation point at the front of the pitot tube.

Figure 6.6 is a drawing of a pitot-static tube, a device that adds to a pitot tube the capability of measuring static pressure. This is accomplished by placing the forward-facing pitot orifice in a body specially designed so that the static pressure at an orifice further aft on the body is the same as the static pressure in the freestream far from the body. A separate pressure line leads from the static orifice to one or more pressure gauges. The ambient static pressure can now be used for two purposes. First, it can be used directly in an altimeter to provide altitude information. Second, from equation 6.9, the difference between the total pressure and the static pressure is the dynamic pressure. If the ambient air density is known, the velocity of the air past the body can be found. Thus

$$\text{dynamic pressure} = \frac{\rho}{2}V_0^2 = p_T - p_0 \tag{6.10}$$

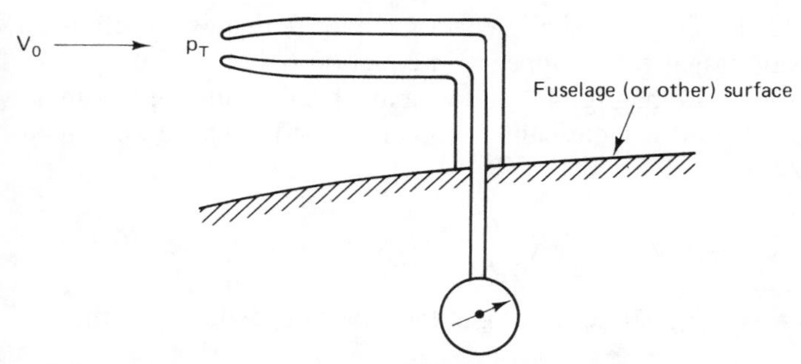

Figure 6.5 Schematic drawing of a pitot tube.

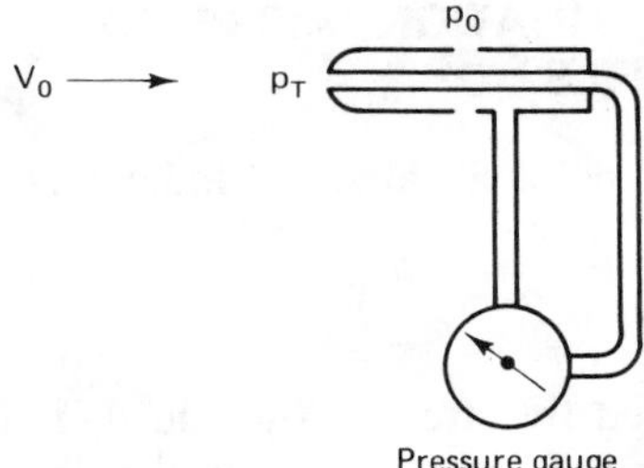

Figure 6.6 Sketch of a pitot-static probe that can be used to measure the airspeed.

and

$$\boxed{V_0 = \sqrt{\frac{2(p_T - p_0)}{\rho}}} \tag{6.11}$$

Note that a pitot-static tube will read the dynamic pressure of the freestream only if it is placed at a location at which there is no influence on the static pressure from a body or surface. If the static orifice has been incorrectly placed so that the pressure at that point differs from the ambient freestream pressure, the measured airspeed will still be the correct speed past the orifice, but it will not be the speed of the aircraft through the air.

The dynamic pressure is of such great importance that it is often designated by a special symbol, q. The device used to determine the dynamic pressure, which is really the difference between two measured pressures, is a gauge that measures only that pressure difference. Calibrating the airspeed indicator in knots or miles per hour from equation 6.11, so that a measured pressure difference is presented on the dial in velocity units, requires knowing the air density, which varies with altitude and temperature. This problem is solved by calibrating all airspeed indicators assuming that the density is the sea-level standard-day value.

The speed read on a perfect airspeed indicator with zero instrument error using a static source that records true ambient air pressure is called the calibrated airspeed, V_{CAL}:

$$\boxed{V_{\text{CAL}} \text{ (calibrated airspeed)} = \sqrt{\frac{2\,(p_T - p_0)}{\rho_s}}} \tag{6.12}$$

where ρ_s = sea-level standard density. We shall see in Chapter 7 that when compressibility is considered an additional factor appears in equation 6.12.

When compressibility effects are negligible, and we have been assuming this in our discussion, the calibrated airspeed is identical to another defined airspeed known as the equivalent airspeed, V_E.

$$V_{\text{CAL}_{\text{(incompressible)}}} = V_E = V_0\sqrt{\frac{\rho}{\rho_s}} = V_0\sqrt{\sigma} \tag{6.13}$$

where σ is the density ratio $(= \rho/\rho_s)$. Note that V_0 is the true airspeed, sometimes called V_{true}.

The equivalent airspeed is the speed at sea level, or more specifically with sea-level standard density, for which the dynamic pressure is the same as at the actual speed and density conditions. Then, by definition,

$$\frac{\rho_s V_E^2}{2} = \frac{\rho V_0^2}{2}$$

and

$$V_E^2 = V_0^2 \frac{\rho}{\rho_s}$$

$$\boxed{V_E = V_0 \sqrt{\frac{\rho}{\rho_s}}} \quad \text{and} \quad \boxed{V_0 = V_E \sqrt{\frac{\rho_s}{\rho}}}$$

In high-speed aircraft, V_{CAL} differs from V_E by a compressibility correction discussed later.

On many aircraft the pitot tube and the static orifice are not combined in a single device. The pitot tube is then mounted on the wing or on the top or side of the fuselage nose. The static orifice is located at a point on the fuselage established by both calculation and experiment as a place where the local pressure is very close to the ambient pressure over the entire flight range of angle of attack and Mach number (Figure 6.7a). In recent years the pitot-static tube has returned as the preferred approach on sophisticated aircraft (Figure 6.7b). This arises from the use of microprocessors to correct static readings to true ambient pressure. Great accuracy is not required in the pressure reading since the air data computer receives pressure, angle of attack, and Mach number information and uses a preprogrammed calibration to provide corrected airspeed and altitude outputs to the pilot's instruments.

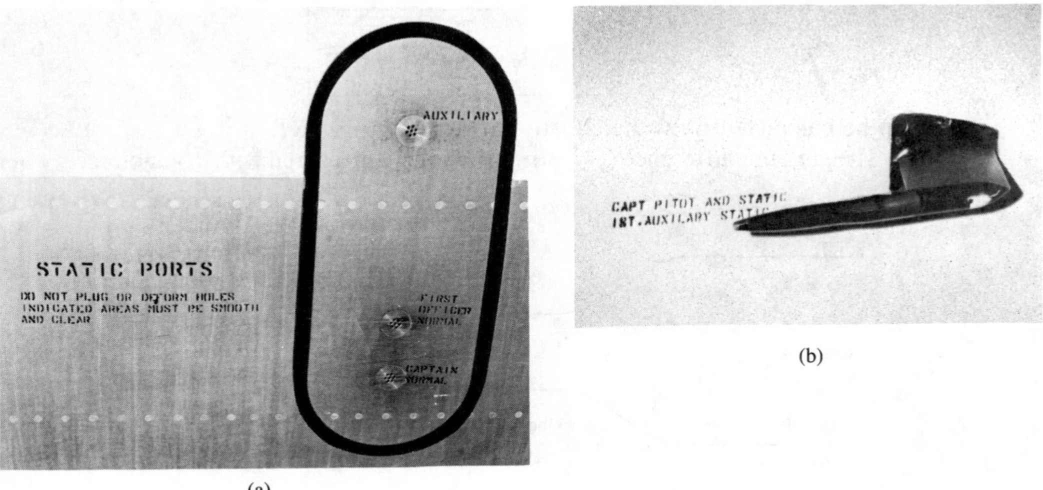

(a)

(b)

Figure 6.7 Airspeed and altimeter sensors: (a) 727 static orifice located on fuselage; (b) 747 pitot-static probe. (Courtesy of United Airlines.)

Another and related application of the Bernoulli equation is measuring the airspeed in a low-speed (< 200 mph) wind tunnel (Figure 6.8). Here, instead of using the difference between the static and total pressure at a point, the static pressures at two points along the tunnel wall are determined. Then, from the Bernoulli equation,

$$p_1 + \frac{\rho V_1^2}{2} = p_2 + \frac{\rho V_2^2}{2}$$

$$p_1 - p_2 = \frac{\rho V_2^2}{2} - \frac{\rho V_1^2}{2} \tag{6.14}$$

$$= \frac{\rho V_2^2}{2}\left[1 - \left(\frac{V_1}{V_2}\right)^2\right]$$

But, from the continuity equation for incompressible fluids (since $\rho_1 = \rho_2$),

$$S_1 V_1 = S_2 V_2$$

$$\frac{V_1}{V_2} = \frac{S_2}{S_1}$$

Therefore,

$$p_1 - p_2 = \frac{\rho V_2^2}{2}\left[1 - \left(\frac{S_2}{S_1}\right)^2\right] \tag{6.15}$$

The dynamic pressure in the working section is

$$\frac{\rho V_2^2}{2} = \frac{p_1 - p_2}{1 - (S_2/S_1)^2}$$

and

$$\boxed{V_2 = \sqrt{\frac{2(p_1 - p_2)}{\rho[1 - (S_2/S_1)^2]}}} \tag{6.16}$$

ρ can be calculated from the equation of state, $\rho = p/RT$.

The pressure difference $(p_1 - p_2)$ is typically measured with a manometer, where

$$p_1 - p_2 = w\Delta h \tag{6.17}$$

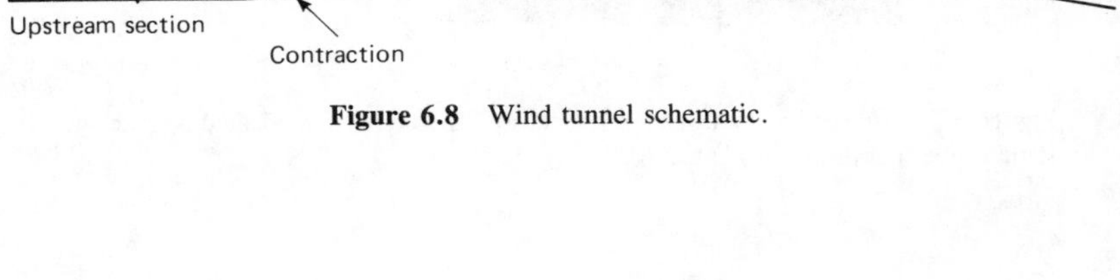

Figure 6.8 Wind tunnel schematic.

where

w = weight per unit volume of manometer fluid
Δh = height of manometer fluid column

In high-speed wind tunnels the same general procedure is used except that the equations must include the effects of compressibility (i.e., $\rho_1 \neq \rho_2$). This is discussed later.

Example 6.1

Visualize a sea-level, low-speed wind tunnel of circular cross section with a diameter upstream of the contraction of 20 ft and a test section diameter of 10 ft. The test section is vented to atmosphere. If the working section velocity is 180 mph, find:

(a) Upsteam section velocity.
(b) Upstream pressure.
(c) Height of a mercury column being used to regulate tunnel speed by measuring the difference between the upsteam and working section pressures. (Mercury weighs 0.49 lb/in.^{3}.)

Solution: First draw the wind tunnel cross section. List the given information on the sketch. Since the test section is vented to the local atmosphere, which is defined as sea level, the working section pressure is 1 atm or 2116 lb/ft^2.

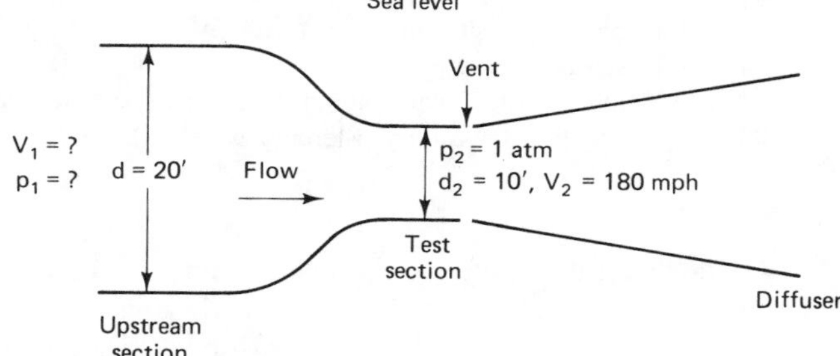

(a) From the continuity equation,

$$\frac{V_2}{V_1} = \frac{S_1}{S_2} = \frac{(\pi/4)d_1^2}{(\pi/4)d_2^2} = \frac{20^2}{10^2} = 4$$

Then

$$V_1 = \frac{V_2}{4} = \frac{180}{4} = 45 \text{ mph} = \underline{66.00 \text{ ft/s}}$$

(b) From the Bernoulli equation,

$$p_1 + \frac{\rho}{2}V_1^2 = p_2 + \frac{\rho}{2}V_2^2$$

$$p_1 = p_2 + \frac{\rho}{2}(V_2^2 - V_1^2)$$

$$= 2116 + \frac{0.002377}{2}\left[\left(180 \times \frac{88}{60}\right)^2 - 66.00^2\right)\right]$$

Note the need to change the 180-mph speed to feet per second.

$$= 2116 + \frac{0.002377}{2}(65{,}340) = 2116 + 77.7$$

$$= \underline{2193.7\ \text{lb/ft}^2} \qquad \text{upstream pressure}$$

(c) $p_1 - p_2 = w\Delta h$, from equation 6.17.

$$\text{height of column, } \Delta h = \frac{p_1 - p_2}{w} = \frac{2193.7 - 2116}{0.49 \times (12)^3} = \frac{77.7}{0.49 \times 1728}$$

changes weight to lb/ft³ from lb/in.³

$$= \underline{0.0918\ \text{ft}} = \underline{1.101\ \text{in.}}$$

Example 6.2

A wind tunnel, located at a pressure altitude of 500 m, has a circular test section with a 3.0-m diameter. The airspeed is 80 m/s in the test section, which is vented to the ambient atmosphere. The airspeed in the larger-diameter section just upstream of the contraction is 16 m/s. Determine:

(a) Upstream diameter.
(b) Dynamic pressure in the test section.
(c) Upstream pressure.
(d) Height of a mercury column measuring the difference between the upstream and test section pressures. (Mercury weighs 0.133 N/cm³.)

Solution:

(a) From the continuity equation, defining d_2, V_2 as conditions at the test section,

$$\frac{V_2}{V_1} = \frac{S_1}{S_2} = \frac{(\pi/4)d_1^2}{(\pi/4)d_2^2}$$

Then

$$\frac{80}{16} = \frac{d_1^2}{3^2}$$

and

$$d_1 = \sqrt{\frac{80}{16} \times 9} = \underline{6.71\ \text{m}}$$

(b) At an altitude of 500 m, interpolating from Table A.1, we obtain

$$\rho = 1.1674\ \text{kg/m}^3, \qquad p = 95{,}472\ \text{N/m}^2$$

$$\text{dynamic pressure} = \frac{\rho V^2}{2} = \frac{(1.1674)(80)^2}{2} = \underline{3735.7\ \text{N/m}^2}$$

(c) $p_1 = p_2 + \frac{\rho}{2}(V_2^2 - V_1^2)$

$= 95{,}472 + \frac{1.1674}{2}(80^2 - 16^2)$

$= \underline{99{,}058\ \text{N/m}^2}$ upstream pressure

(d) $p_1 - p_2 = w\Delta h$, from equation 6.17.

$$\text{height of column, } \Delta h = \frac{p_1 - p_2}{w} = \frac{99{,}058 - 95{,}472}{0.133 \times (100)^3} = \frac{3586}{133{,}000}$$

↑ changes weight to N/m³ from N/cm³

$$= \underline{0.0270\ \text{m}} = \underline{2.70\ \text{cm}}$$

Example 6.3

An airfoil is moving through the air at 220 mph at sea level. The atmospheric pressure is 2116 lb/ft² and the temperature is 80°F. At a point on the airfoil upper surface the local velocity is 260 mph.

(a) Determine the pressure at the upper surface point.
(b) If the pressure in part (a) is the average upper surface pressure, how much lift per square foot (referred to ambient pressure) is being provided by the upper surface?
(c) If the average speed on the lower surface is 200 mph, what is the pressure and the average lift per square foot on the lower surface?
(d) What is the total lift per square foot of wing area?

Solution:

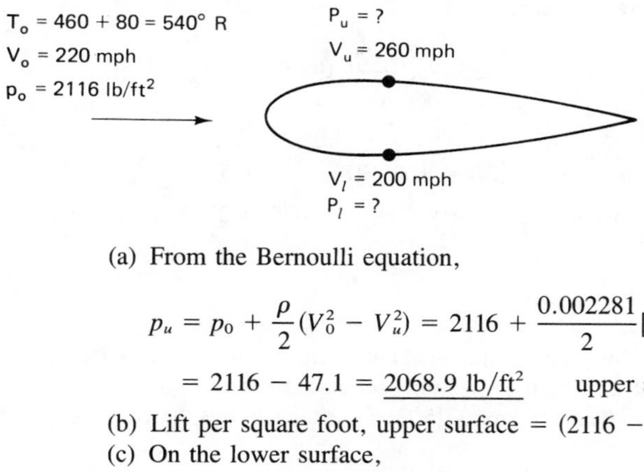

$T_o = 460 + 80 = 540°\ \text{R}$
$V_o = 220$ mph
$p_o = 2116\ \text{lb/ft}^2$

$P_u = ?$
$V_u = 260$ mph

$V_l = 200$ mph
$P_l = ?$

$220\ \text{mph} = 220 \times \frac{88}{60} = 322.67\ \text{ft/s}$

$260\ \text{mph} = 381.33\ \text{ft/s}$

$200\ \text{mph} = 293.33\ \text{ft/s}$

$\rho_o = \frac{p_o}{RT_o} = \frac{2116}{(1718)(540)} = 0.002281$

(a) From the Bernoulli equation,

$$p_u = p_o + \frac{\rho}{2}(V_o^2 - V_u^2) = 2116 + \frac{0.002281}{2}[(322.67)^2 - (381.33)^2]$$

$$= 2116 - 47.1 = \underline{2068.9\ \text{lb/ft}^2} \quad \text{upper surface pressure}$$

(b) Lift per square foot, upper surface $= (2116 - 2068.9) = \underline{47.1\ \text{lb/ft}^2}$.
(c) On the lower surface,

$$p_l = 2116 + \frac{0.002281}{2}[(322.67)^2 - (293.33)^2]$$

$$= 2116 + 20.61 = \underline{2136.61\ \text{lb/ft}^2}$$

$$\text{lower surface lift} = 2136.61 - 2116 = \underline{20.61\ \text{lb/ft}^2}$$

(d) Total lift per square foot $= 47.1 + 20.61 = 67.71\ \text{lb/ft}^2$.

Example 6.4

An airfoil is moving through the air at 355 km/h at sea level. The atmospheric pressure is 101,325 N/m^2 and temperature is 27°C. At a point on the airfoil upper surface the local velocity is 420 km/h.

(a) Determine the pressure at the upper surface point.
(b) If the pressure in part (a) is the average upper surface pressure, how much lift per square meter (referred to ambient pressure) is being provided by the upper surface?
(c) If the average speed on the lower surface is 320 km/h, what is the pressure and the average lift per square meter on the lower surface?
(d) What is the total lift per square meter of wing area?

Solution: In the SI system, temperature must be expressed in Kelvin and velocities in m/s.

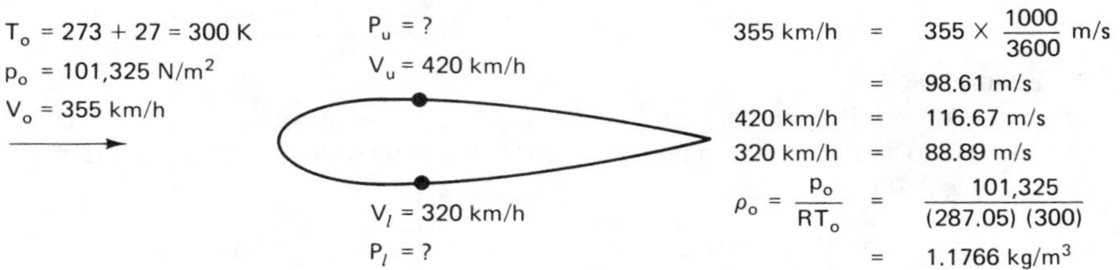

(a) From the Bernoulli equation,

$$p_u = p_0 + \frac{\rho}{2}(V_0^2 - V_u^2) = 101{,}325 + \frac{1.1766}{2}[(98.61)^2 - (116.67)^2]$$

$$= 101{,}325 - 2287 = \underline{99{,}038 \text{ N/m}^2} \quad \text{upper surface pressure}$$

(b) Lift per square meter, upper surface = 101,325 − 99,038 = $\underline{2287 \text{ N/m}^2}$.

(c) On the lower surface,

$$p_l = 101{,}325 + \frac{1.1766}{2}[(98.61)^2 - (88.89)^2]$$

$$= 101{,}325 + 1072 = \underline{102{,}397 \text{ N/m}^2}$$

$$\text{lower surface lift} = 102{,}397 - 101{,}325 = \underline{1072 \text{ N/m}^2}$$

(d) Total lift per square meter = 2287 + 1072 = 3359 N/m^2.

PROBLEMS

6.1. Consider a sea-level, low-speed wind tunnel of circular cross section with a diameter upstream of the contraction of 22 ft and a working section diameter of 10 ft. The working section is vented to atmosphere. Temperature is 15°F above standard. If the working section velocity is 210 mph, what is:

(a) the upstream (22 ft section) velocity?
(b) the upstream pressure?
(c) the height of a mercury column being used to regulate the tunnel speed by measuring the pressure difference between the upstream and working section pressure? Mercury weighs 0.49 lb/in.^{3}.

6.2. An airfoil is moving through the air at 180 mph at 5000 ft pressure altitude. The temperature is 80°F. At a point on the airfoil upper surface the local velocity is 215 mph. What is the pressure at that point? If this is the average pressure on the surface, how much lift per square foot is being provided by the top surface? If the average speed on the lower surface is 160 mph, what is the pressure and average lift per square foot on the lower surface? What is the total lift per square foot of wing area?

6.3. A wind tunnel has a rectangular test section 7 ft × 10 ft in size. The upstream section has a cross-section area that is 5 times the test section area. If the pressure difference between the two sections is 100 lb/ft^2, what is the test section velocity? The test section is vented to atmosphere. The temperature is standard for the 1000 ft pressure altitude of the tunnel.

6.4. An airfoil is moving through the air at 360 km/h at 3000 m pressure altitude. The temperature is 7°C. At a point on the airfoil upper surface the local velocity is 432 km/h. What is the pressure at that point? If this is the average pressure on the surface, how much lift per square meter is being provided by the top surface? Use SI units. If the average speed on the lower surface is 330 km/h, what is the pressure and average lift per square meter on the lower surface? What is the total lift per square meter of wing area?

6.5. An airfoil is moving through the air at 300 mph at 15,000 ft pressure altitude. The temperature is 45°F. At a point on the airfoil upper surface the local velocity is 360 mph. What is the pressure at that point? If this is the average pressure on the surface, how much lift per square foot is being provided by the top surface? If the average speed on the lower surface is 280 mph, what is the pressure and average lift per square foot on the lower surface? What is the total lift per square foot of wing area?

REFERENCE

6.1. Millikan, Clark B., *Aerodynamics of the Airplane,* Wiley, New York, 1941.

7 ONE-DIMENSIONAL FLOW IN A COMPRESSIBLE FLUID

COMPRESSIBLE FLOWS

All matter is compressible to some extent; that is, the volume of an element of matter will decrease under pressure, thereby increasing the density. For solids, the decrease is insignificantly small. For liquids the change in volume is also very small, so that for water, for example, the density ρ is essentially constant. Pressure changes in gases do produce significant volume and density changes, although if the percentage change in pressure is small, the effect of the density change on the forces involved may be very small. This applies to most flows over wings and bodies at speeds below approximately 40% of the speed of sound.

We can now characterize two classes of aerodynamic flows: incompressible flow and compressible flow.

Incompressible flow. An incompressible flow is one in which the density of the fluid elements can be assumed to be constant. For this case, the continuity equation can be written, since $\rho_1 = \rho_2$,

$$S_1V_1 = S_2V_2$$

Incompressible flow is an idealization that never actually occurs in nature. However, as noted above, for those flows where the actual variation of ρ is negligibly small, it is convenient, to simplify our analysis, to make the assumption that ρ is constant. Engineers and scientists must frequently make idealized assumptions about real physical systems in order to obtain useful mathematical descriptions of such systems. Of course, it is essential to limit the use of such idealizations to those problems where the assumptions are realistic. The assumption of incompressible flow is an excellent approximation for the flow of liquids, such as water or oil, and for the low-speed flow of air, where $V < 200$ to 300 mph. Since early aircraft did not fly faster than this, incompressible theory was used almost exclusively until the 1940s, when fighters reached into the compressible flow regions. Transport aircraft were affected only slightly by compressibility until the 1950s.

Compressible flow. A compressible flow is one in which the density of fluid elements must be considered variable from point to point along the streamlines. In such cases, the equations must treat ρ as a variable. To some extent, this is the situation with all modern transport and combat aircraft, causing small to large changes in aerodynamic forces depending on the speed of flight and the size of the velocity increments induced by the airplane bodies and surfaces. The latter is a function of the design and the flight regime (e.g., high or low angle of attack).

ONE-DIMENSIONAL COMPRESSIBLE FLOW CONCEPTS

Since most current aircraft operate at speeds at which compressibility effects are significant, we must now develop the equations that relate p, ρ, and V at various points along a streamtube without the simplifying assumption that ρ = constant. Later we shall find that these equations can be used to solve many practical aircraft problems. As noted earlier, the impact of compressibility will be shown to be a function of the Mach number M, the ratio of the air velocity to the speed of sound in air. In discussing flow along a streamtube, we can neglect the small velocity components perpendicular to the main flow direction. Such a flow is a one-dimensional flow.

In high-speed flows, the kinetic energy of the fluid elements is large. When high-speed flows are slowed down, the reduction in kinetic energy appears as a significant increase in pressure and temperature. To study these changes, the fundamental energy relationships in a gas must be understood. These relationships are the foundations of the science of thermodynamics. For our purposes, we shall simply introduce the essential equations, the derivation of which can be found in texts on thermodynamics or on the flow of compressible fluids. The derivations are based on the *first law of thermodynamics,* a statement of the law of conservation of energy that specifies that the heat energy and mechanical energy are equivalent and interchangeable. An alternative statement of the first law is that the change in the internal energy of a gas is equal to the sum of the heat added to and the work done on the system.

Before approaching these thermodynamic equations, we must introduce the concept of isentropic flow. Consider these definitions.

1. *Adiabatic process:* a process in which no heat is added or taken away, $\delta q = 0$ (where δ implies an increment, q = heat).
2. *Reversible process:* a process in which no frictional or other dissipative effects occur.
3. *Isentropic process:* a process that is both adiabatic and reversible.

Most aerodynamic flows are adiabatic, and outside the thin "boundary" layer of air next to a surface, they are usually also essentially frictionless and thus isentropic. The exceptions are flows through shock waves.

The relation between pressure p and density ρ at two different points along a streamline for an isentropic change of state is

$$\boxed{\frac{p_2}{p_1} = \left(\frac{\rho_2}{\rho_1}\right)^{\gamma}} \quad \text{isentropic gas law} \tag{7.1}$$

where γ is the ratio of the specific heat at constant pressure c_p to the specific heat at constant volume c_v. $\gamma = 1.4$ for air. Other forms of this equation can be easily derived by substituting the equation of state, $\rho = p/RT$, in equation 7.1. Then

$$\frac{p_2}{p_1} = \left(\frac{p_2/RT_2}{p_1/RT_1}\right)^{\gamma}$$

$$\left(\frac{p_2}{p_1}\right)^{1-\gamma} = \left(\frac{T_1}{T_2}\right)^{\gamma} = \left(\frac{T_2}{T_1}\right)^{-\gamma}$$

and

$$\boxed{\frac{p_2}{p_1} = \left(\frac{T_2}{T_1}\right)^{\gamma/(\gamma-1)}} \tag{7.2}$$

Also,

$$\frac{p_2}{p_1} = \left(\frac{\rho_2}{\rho_1}\right)^{\gamma} = \left(\frac{T_2}{T_1}\right)^{\gamma/(\gamma-1)}$$

so that

$$\boxed{\frac{\rho_2}{\rho_1} = \left(\frac{T_2}{T_1}\right)^{1/(\gamma-1)}} \tag{7.3}$$

Thus the relationships among p, ρ, and T at various points in the fluid are established.

We need one other important equation, the energy equation relating temperature and velocity at points along a streamline in adiabatic flow:

$$c_pT + \frac{V^2}{2} = \text{constant} = c_pT_T$$

or

$$\boxed{c_pT_1 + \frac{V_1^2}{2} = c_pT_2 + \frac{V_2^2}{2}} \tag{7.4}$$

where

c_p = specific heat at constant pressure
= 1004.7 N-m/kg K (J/kg K) for air
= 6006 ft-lb/slug °R

Equations 7.1 to 7.4 are very powerful tools in solving flow problems.

We return now to the Euler or momentum equation, treating ρ as a variable. From equation 6.5,

$$dp = -\rho V\,dV$$

Letting $(\cdot)_T$ denote the stagnation conditions at $V = 0$, recall that, from equation 7.1,

$$\rho = \rho_T\left(\frac{p}{p_T}\right)^{1/\gamma}$$

Substituting this in the Euler equation, we obtain

$$dp = -\rho_T\left(\frac{p}{p_T}\right)^{1/\gamma} V\,dV$$

$$dp\left(\frac{p}{p_T}\right)^{-1/\gamma}\frac{1}{\rho_T} + V\,dV = 0$$

Integrating from the stagnation point to a point along the streamline yields

$$\int_{p_T}^{p}\left(\frac{p}{p_T}\right)^{-1/\gamma}\frac{1}{\rho_T}dp + \int_0^V V\,dV = 0 \tag{7.5}$$

$$\boxed{\frac{\gamma}{\gamma-1}\frac{p_T}{\rho_T}\left(\frac{p}{p_T}\right)^{(\gamma-1)/\gamma} + \frac{V^2}{2} = \frac{\gamma}{\gamma-1}\frac{p_T}{\rho_T}} \quad \text{compressible Bernoulli equation} \tag{7.6}$$

The details of the mathematical manipulations involved in the last step are shown in Appendix B.

A somewhat similar equation can be obtained by substituting $T = p/\rho R$ from the equation of state 5.1 into equation 7.4 and using the relationship $c_p = \gamma R/(\gamma - 1)$ (equation 7.17):

$$\boxed{\frac{\gamma}{\gamma-1}\frac{p}{\rho} + \frac{V^2}{2} = \frac{\gamma}{\gamma-1}\frac{p_T}{\rho_T}} \tag{7.7}$$

In equations 7.6 and 7.7, the stagnation pressure p_T is sometimes called the reservoir pressure for the flow, as it corresponds to the pressure in a reservoir out of which the flow would issue and accelerate isentropically to speed V and pressure p. p_T corresponds to the total pressure in the incompressible fluid Bernoulli equation. The dynamic pressure $q = \frac{1}{2}\rho V^2$ has lost some of its physical significance since it is no longer the difference between the local freestream pressure p and the total pressure p_T. However, q can be related to $(p_T - p)$ by the application of a function of Mach number, which we shall soon discuss.

Equations 7.6 and 7.7 differ in that equation 7.7 is derived from the energy equation, valid for adiabatic flows, and the equation of state. Thus equation 7.7 is valid for any adiabatic flow. Equation 7.6, on the other hand, was derived using the isentropic gas equation 7.1 and, therefore, is valid only in isentropic flow.

THE SPEED OF SOUND

If a disturbance, such as a sudden increase of pressure, occurs at some point of an incompressible fluid, the disturbance is transmitted instantaneously to all parts of the fluid. In a compressible fluid, the disturbance travels through the fluid in the form of a pressure wave at a definite velocity, which is in fact the velocity of sound in the fluid. When this disturbance is caused by an approaching wing or body and moves much faster than the speed of the airplane, the signal that the airplane is approaching is transmitted throughout the fluid, allowing the fluid to accommodate gradually to the oncoming flow changes almost as if the fluid were incompressible. If the airplane speed approaches the speed of the disturbance, the flow adjustment occurs more abruptly. In the extreme in which the airplane moves faster than the speed of the sound, *supersonic flight,* the fluid is unaware of the approaching airplane until it impacts. In this case the fluid adjustment is essentially instantaneous, resulting in a thin sheet or disturbance wave known as a *shock wave*. In passing through the shock wave, the flow experiences sudden changes in velocity, pressure, density, and temperature. The changes are those required to permit the air to flow tangent to the surface of the wing or body.

Let us consider a weak disturbance or sound wave traveling through the fluid in a channel. Imagine that we are moving with the sound wave. We then observe a steady flow to the right in the channel in Figure 7.1 at velocity u and pressure and density p and ρ. In passing through the wave, the velocity, pressure, and density are slightly changed by the amounts du, dp, and $d\rho$, respectively. From the continuity equation,

$$\rho S u = (\rho + d\rho)(S + dS)(u + du) \tag{7.8}$$

Since the channel has constant area across the wave, $dS = 0$.

$$\rho u = (\rho + d\rho)(u + du) = \rho u + \rho\, du + u\, d\rho + d\rho\, du$$

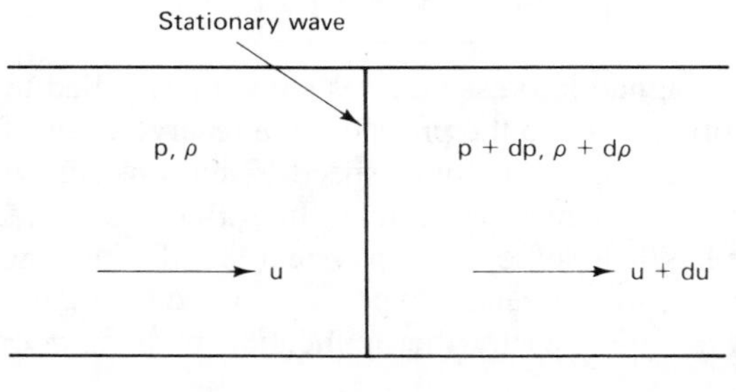

Figure 7.1 Weak stationary wave.

Since we are dealing with a weak disturbance, $d\rho$ and du are small (i.e., $d\rho/\rho$ and du/u are much less than 1.0). Thus the product of two small quantities is very small compared to the other terms and can be neglected. Then

$$u = -\rho\frac{du}{d\rho} \tag{7.9}$$

From the Euler (momentum) equation,

$$dp = -\rho u\, du$$

and

$$du = -\frac{dp}{\rho u}$$

Substituting in equation 7.9 yields

$$u = +\frac{\rho}{d\rho}\frac{dp}{\rho u} = \frac{dp}{d\rho}\frac{1}{u}$$

and

$$\boxed{u^2 = \frac{dp}{d\rho}}$$

where u is the speed of the sound wave.

The changes in pressure, temperature, and velocity across a weak sound wave are so small that there is no exchange of heat between adjacent fluid elements, and the effect of friction is negligible. Therefore, the flow through a sound wave is isentropic, and the isentropic gas law applies. Thus

$$\frac{p_2}{p_1} = \left(\frac{\rho_2}{\rho_1}\right)^\gamma$$

$$\frac{p_2}{(\rho_2)^\gamma} = \frac{p_1}{(\rho_1)^\gamma} = \text{constant} = C \tag{7.10}$$

Then

$$\frac{dp}{d\rho} = \frac{d(C\rho^\gamma)}{d\rho} = \gamma C\rho^{\gamma-1} \tag{7.11}$$

But, from equation 7.10,

$$C = \frac{p}{\rho^\gamma}$$

Substituting, we obtain

$$\frac{dp}{d\rho} = \frac{\gamma p}{\rho^\gamma}\rho^{\gamma-1} = \frac{\gamma p}{\rho} \tag{7.12}$$

Thus the square of the velocity of the sound wave is

$$u^2 = \frac{dp}{d\rho} = \frac{\gamma p}{\rho}$$

The speed of sound is often called a, so

$$a^2 = \frac{dp}{d\rho} = \frac{\gamma p}{\rho}, \qquad a = \sqrt{\frac{\gamma p}{\rho}} \tag{7.13}$$

Furthermore, from the equation of state $p = \rho RT$,

$$\frac{p}{\rho} = RT$$

Substituting yields

$$\boxed{a^2 = \gamma RT, \qquad a = \sqrt{\gamma RT}} \qquad \text{speed of sound} \tag{7.14}$$

Thus the speed of sound in a perfect gas depends only on the square root of the temperature of the gas.

At sea level on a standard day, $T = 288.15$ K. As noted earlier, $\gamma = 1.4$ and R in SI units is 287.05 N-m/kg K. Then

$$a = \sqrt{(1.4)(287.05)(288.15)} = 340.29 \text{ m/s}$$

In English engineering units, $T = 518.69°$R, $R = 1718$ ft-lb/(slug °R), and

$$a = \sqrt{(1.4)(1718)(518.69)} = 1116.94 \text{ ft/s}$$

Note that the speed-of-sound equation was derived using the continuity and momentum (Euler) equations together with the isentropic and state gas equations, a typical example of extracting important results from fundamental theorems.

Many aerodynamic phenomena, such as lift and drag, are influenced by the ratio of the air velocity to the speed of sound. This nondimensional ratio is called the *Mach number, M:*

$$M = \frac{V}{a}$$

The Mach number is named after Ernst Mach (1838–1916), an Austrian physicist and philosopher.

COMPRESSIBLE FLOW EQUATIONS IN A VARIABLE-AREA STREAMTUBE

We can now derive several important and useful one-dimensional compressible flow equations.

Recall the energy equation along a streamline,

$$c_pT_1 + \frac{V_1^2}{2} = c_pT_2 + \frac{V_2^2}{2}$$

Consider a stagnation point in a compressible flow where the flow is isentropically compressed to $V = 0$. Then the energy equation becomes

$$c_p T_1 + \frac{V_1^2}{2} = c_p T_T \tag{7.15}$$

where $(\cdot)_T$ signifies stagnation ($V = 0$) conditions. Dividing by $c_p T_1$, we obtain

$$\frac{T_T}{T_1} = 1 + \frac{V_1^2}{2c_p T_1} \tag{7.16}$$

Now, it is shown in fundamental thermodynamics that

$$c_p = \frac{\gamma R}{\gamma - 1} \tag{7.17}$$

Substituting in equation 7.16, we get

$$\frac{T_T}{T_1} = 1 + \frac{V_1^2}{2[\gamma R/(\gamma - 1)]T_1} = 1 + \frac{\gamma - 1}{2}\frac{V_1^2}{\gamma R T_1}$$

But

$$a_1^2 = \gamma R T_1$$

Then

$$\frac{T_T}{T_1} = 1 + \frac{\gamma - 1}{2}\frac{V_1^2}{a_1^2}$$

$$\boxed{\frac{T_T}{T_1} = 1 + \frac{\gamma - 1}{2}M_1^2} \tag{7.18}$$

Remembering the isentropic laws, we obtain

$$\frac{p_T}{p_1} = \left(\frac{\rho_T}{\rho_1}\right)^{\gamma} = \left(\frac{T_T}{T_1}\right)^{\gamma/(\gamma-1)}$$

We can substitute

$$\frac{T_T}{T_1} = \left(\frac{p_T}{p_1}\right)^{(\gamma-1)/\gamma} = \left(\frac{\rho_T}{\rho_1}\right)^{\gamma-1}$$

in equation 7.18 and obtain

$$\boxed{\frac{p_T}{p_1} = \left(1 + \frac{\gamma - 1}{2}M_1^2\right)^{\gamma/(\gamma-1)}} \tag{7.19}$$

and

$$\boxed{\frac{\rho_T}{\rho_1} = \left(1 + \frac{\gamma - 1}{2}M_1^2\right)^{1/(\gamma-1)}} \tag{7.20}$$

Equations 7.18, 7.19, and 7.20 are highly useful relations for compressible isentropic flow. They describe flow along a streamtube, but also fit the flow in a duct such as a wind tunnel or a rocket nozzle. If the tunnel is a blowdown type with the flow issuing from a large pressurized reservoir, T_T, p_T, and ρ_T represent the conditions at $M = 0$ in the reservoir. If we know M_1 and T_1, p_1 or ρ_1 at some point along the channel, we can calculate the "total" values or equivalent reservoir values, T_T, p_T, and ρ_T. If we then measure T_1, p_1, or ρ_1 elsewhere along the streamtube, we can calculate the local Mach number, M_1. The ratios T_T/T_1, p_T/p_1, and ρ_T/ρ_1 are dependent only on the Mach number at the point $(\cdot)_1$. To calculate Mach number M_1, we can solve equation 7.19 for M_1:

$$\boxed{M_1 = \sqrt{\frac{2}{\gamma - 1}\left[\left(\frac{p_T}{p_1}\right)^{(\gamma-1)/\gamma} - 1\right]}} \tag{7.21}$$

Thus the Mach number is a direct function of the ratio of total to static pressure. An instrument recording this ratio and properly calibrated in accordance with equation 7.21 is called a *Machmeter*. Remember that the system used to determine equivalent airspeed measures the difference between these pressures rather than the ratio.

Example 7.1

A DC-10 transport is flying at a pressure altitude of 35,000 ft. A pitot tube measures a pressure of 799 lb/ft^2. We wish to calculate the Mach number at which the airplane is flying. If the ambient air temperature is 10°F above standard, determine the true airspeed and the equivalent airspeed.

Solution: From the standard atmosphere tables, at a pressure altitude of 35,000 ft,

$$p = 499.34 \text{ lb/ft}^2$$

$$T_{\text{standard}} = 394°\text{R}$$

$$T_{\text{ambient}} = 394 + 10 = 404°\text{R}$$

From equation 7.21,

$$M_1 = \sqrt{\frac{2}{\gamma - 1}\left[\left(\frac{p_T}{p_1}\right)^{(\gamma-1)/\gamma} - 1\right]} = \sqrt{\frac{2}{1.4 - 1}\left[\left(\frac{799}{499.34}\right)^{(1.4-1)/1.4} - 1\right]} = \underline{0.85}$$

$$a = \sqrt{\gamma R T} = \sqrt{(1.4)(1718)(404)} = 985.7 \text{ ft/s}$$

$$V_{\text{true}} = M_1 a = 0.85 \times 985.7 = 837.9 \text{ ft/s}$$

$$= \underline{570.8 \text{ mph}}$$

(Note that $V_{\text{mph}} = V_{\text{ft/s}} \times 60/88$.) Also,

$$V_E = V_{\text{true}}\sqrt{\sigma} = V_{\text{true}}\sqrt{\frac{\rho}{\rho_s}}$$

From the equation of state,

$$\rho = \frac{p}{RT} = \frac{499.34}{(1718)(404)} = 0.000718 \text{ slug/ft}^3$$

$$\rho_s = 0.002377 \text{ slug/ft}^3$$

Then

$$V_E = 570.8\sqrt{\frac{0.000718}{0.002378}} = 570.8\sqrt{0.302} = \underline{313.7 \text{ mph}}$$

Now suppose that a pressure is read on the upper surface of the wing and is found to be 440 lb/ft^2. What is the local Mach number, that is, the Mach number at that point on the wing?

We again use equation 7.21 with p_T, the total pressure read by the pitot tube, unchanged, but the static pressure is 440 lb/ft^2. Then

$$M_1 = \sqrt{\frac{2}{1.4-1}\left[\left(\frac{799}{440}\right)^{(1.4-1)/1.4} - 1\right]} = 0.964$$

The local Mach number is considerably higher than the Mach number of the airplane through the air.

To obtain the local temperature, we would first have to find the total temperature. We have found that in the freestream $M_1 = 0.85$ and $T_1 = 404°\text{R}$; then, from equation 7.18,

$$T_T = T_1\left(1 + \frac{\gamma - 1}{2}M_1^2\right) = 404\left[1 + \frac{1.4-1}{2}(0.85)^2\right]$$
$$= 462.4°\text{R}$$

At the point on the wing, $M_1 = 0.964$. Then the temperature at that point is

$$T_1 = \frac{T_T}{1 + [(\gamma - 1)/2]M_1^2} = \frac{462.4}{1 + [(1.4-1)/2](0.964)^2} = 389.9°\text{R}$$

Thus the air cools as it accelerates. Incidentally, this is the temperature just outside the boundary layer, the region of air slowed by friction. So far we have been ignoring friction.

In this section we have used $(\cdot)_1$ to designate a point of interest in the flow. As long as we remember the point to which we are referring, it is convenient to drop this subscript from here on or to use a more descriptive subscript.

Example 7.2

This problem is identical to Example 6.3 except that the *compressible* fluid equations are to be used.

An airfoil is moving through the air at 220 mph at sea level. The atmospheric pressure is 2116 lb/ft^2 and the temperature is 80°F. At a point on the airfoil upper surface the local velocity is 260 mph.

(a) Determine the pressure at the upper surface point.
(b) If the pressure in part (a) is the average upper surface pressure, how much lift per square foot (referred to ambient pressure) is being provided by the upper surface?
(c) If the average speed on the lower surface is 200 mph, what is the pressure and average lift per square foot on the lower surface?
(d) What is the total lift per square foot of wing area?

Solution: This problem can be solved using the compressible form of Bernoulli equation 7.6 after determining the stagnation conditions from equations 7.19 and 7.20. Thus

$$p_T = p_0\left(1 + \frac{\gamma - 1}{2}M_0^2\right)^{\gamma/(\gamma-1)} = 2116(1 + 0.2M_0^2)^{3.5}$$

The ambient speed of sound

$$a = \sqrt{\gamma RT} = \sqrt{1.4(1718)(540)} = 1139.65 \text{ ft/s}$$

so

$$M_0 = \frac{220 \times 88/60}{1139.65} = 0.2831$$

Then

$$p_T = 2116[1 + (0.2)(0.2831)^2]^{3.5} = (2116)(1.0572)$$
$$= 2237.11 \text{ lb/ft}^2$$

Also,

$$\rho_T = \rho_0[1 + (0.2)(0.2831)^2]^{1/(\gamma-1)}$$

Since

$$\rho_0 = \frac{p_0}{RT_0} = \frac{2116}{(1718)(540)} = 0.002281$$
$$\rho_T = (0.002281)(1.01603)^{2.5} = 0.002374 \text{ slug/ft}^3$$

From the compressible Bernoulli equation,

$$\frac{\gamma}{\gamma - 1}\frac{p_T}{\rho_T}\left(\frac{p}{p_T}\right)^{(\gamma-1)/\gamma} + \frac{V^2}{2} = \frac{\gamma}{\gamma - 1}\frac{p_T}{\rho_T}$$

(a) For the upper surface,

$$3.5\left(\frac{2237.11}{0.002374}\right)\left(\frac{p_u}{2237.11}\right)^{0.2857} + \frac{(381.33)^2}{2} = \frac{(3.5)(2237.11)}{0.002374}$$
$$364{,}141 p_u^{0.2857} + 72{,}706.3 = 3{,}298{,}180$$
$$p_u^{0.2857} = 8.8578$$
$$p_u = \underline{2069.22 \text{ lb/ft}^2}$$

(b) The upper surface lift is (2116 − 2069.22) = $\underline{46.78 \text{ lb/ft}^2}$.

(c) On the lower surface, using the compressible form of the Bernoulli equation, we obtain

$$\frac{(3.5)(2237.11)}{0.002374}\left(\frac{p_l}{2237.11}\right)^{0.2857} + \frac{(293.33)^2}{2} = \frac{(3.5)(2237.11)}{0.002374}$$
$$364{,}141 p_l^{0.2857} + 43{,}021.2 = 3{,}298{,}180$$
$$p_l^{0.2857} = 8.9393$$
$$p_l = \underline{2135.79 \text{ lb/ft}^2}$$
$$\text{lower surface lift} = 2135.79 - 2116 = \underline{19.79 \text{ lb/ft}^2}$$

(d) Total lift = 46.78 + 19.79 = $\underline{66.57 \text{ lb/ft}^2}$. This answer is very close to the incompressible result and shows that at a low Mach number, 0.2831 in this example, the simple incompressible equation serves very well. As the Mach number increases, the more complicated compressible equation yields results increasingly different from the incompressible answer. Above about $M = 0.6$, the difference becomes very significant.

Example 7.3

A DC-10 is cruising at its assigned altitude at a Mach number of 0.85. The outside air temperature is 232 K. At a given point on the upper surface of the wing the pressure is measured at 20,100 N/m^2. The temperature at this point is 221 K. How much lift per square meter (referred to ambient pressure) is provided by the upper surface at this point? What is the assigned pressure altitude? What is the density altitude? What is the true speed of the airplane? (Use SI units, but also give the upper surface lift in lb/ft^2.)

Solution: Since the cruise Mach number is 0.85, the compressible equations must be used.

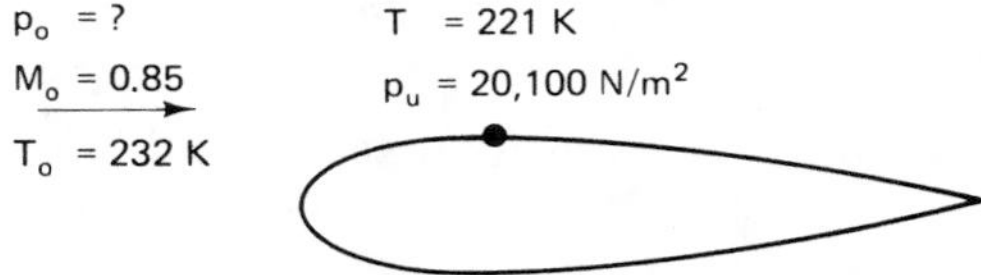

We are given the temperature at two points along a streamline, and the pressure at one of them. We must find the ambient pressure p_0. From equation 7.2,

$$\frac{p_0}{p_u} = \left(\frac{T_0}{T_u}\right)^{\gamma/(\gamma-1)} = \left(\frac{232}{221}\right)^{1.4/(1.4-1)} = 1.1853$$

$$p_0 = (1.1853)(20{,}100) = 23{,}825 \text{ N/m}^2$$

The upper surface lift is the difference between the ambient and upper surface pressures.

$$\text{lift on upper surface} = 23{,}825 - 20{,}100 = \underline{3725 \text{ N/m}^2}$$

$$= 3725 \times \underset{(\text{lb/N})}{0.2248} \times \underset{(\text{m}^2/\text{ft}^2)}{\frac{1}{10.76}} = \underline{77.82 \text{ lb/ft}^2}$$

Since the ambient pressure is 23,825 N/m^2, the pressure altitude is found from Table A.1 by interpolation to be <u>10,692 m</u>, or 35,080 ft.

From the equation of state, $p = \rho RT$, the density is

$$\rho_0 = \frac{p_0}{RT_0} = \frac{23{,}825}{(287.05)(232)} = 0.3578 \text{ kg/m}^3$$

From Table A.1, the density altitude is <u>11,128 m</u> or 36,510 ft.

$$\text{speed of sound } a = \sqrt{\gamma RT} = \sqrt{(1.4)(287.05)(232)} = 305.34 \text{ m/s}$$

$$\text{airplane true speed} = Ma = (0.85)(305.34) = \underline{259.54 \text{ m/s}}$$

APPLICATION TO AIRCRAFT AIRSPEED MEASUREMENT

The one-dimensional compressible equations are useful in improving the accuracy of airspeed systems of high-speed airplanes. It is interesting to compare the compressible equation for total pressure (measured by a pitot tube) with the incompressible approximation of Bernoulli.

From the incompressible Bernoulli equation,

$$p_T = p + \frac{\rho}{2}V^2$$

Now

$$a^2 = \frac{\gamma p}{\rho} \qquad \text{so} \qquad \rho = \frac{\gamma p}{a^2}$$

Substituting yields

$$\boxed{p_T = p + \frac{\gamma p}{2a^2}V^2 = p + \frac{\gamma}{2}pM^2} \tag{7.22}$$

This introduces an alternative, very useful form for dynamic pressure, q:

$$q = \frac{\rho}{2}V^2 = \frac{\gamma}{2}pM^2 \tag{7.23}$$

In a compressible fluid, from equation 7.19,

$$p_T = p\left(1 + \frac{\gamma - 1}{2}M^2\right)^{\gamma/(\gamma-1)}$$

For values of M less than $\sqrt{2/(\gamma - 1)} = 2.24$, the right-hand side can be expanded in a power series so that

$$p_T = p + \frac{\gamma}{2}pM^2\left[1 + \frac{M^2}{4} + \frac{M^4}{12}\left(1 - \frac{\gamma}{2}\right) + \frac{M^6}{48}\left(1 - \frac{\gamma}{2}\right)\left(\frac{3}{2} - \gamma\right) + \cdots\right]$$

For air, $\gamma = 1.4$ and

$$\boxed{p_T = p + \frac{\gamma}{2}pM^2\left(1 + \frac{M^2}{4} + \frac{M^4}{40} + \frac{M^6}{1600} + \cdots\right)} \tag{7.24}$$

For $M < 1.0$, the error in the expansion with three terms is less than 0.1%.

Comparing with equation 7.22, we find that the compressible equation for the total pressure differs from the incompressible approximation by the function of M in parentheses. The dynamic pressure becomes

$$\frac{\gamma}{2}pM^2 = \frac{\rho}{2}V^2 = \frac{p_T - p}{1 + M^2/4 + M^4/40 + M^6/1600 + \cdots}$$

and

$$V_E = \sqrt{\frac{2}{\rho_s}(p_T - p)\frac{1}{1 + M^2/4 + M^4/40 + M^6/1600}} \tag{7.25}$$

Equation 7.25 differs from equation 6.12 for incompressible flow in the Mach number terms. To find V_E from the measured values of p_T and p, the value of Mach number corresponding to V_E must be used in equation 7.25. This value of M will vary with pressure altitude.* To permit a standard calibration, all airspeed indicators are calibrated using values of M that correspond to each value of V_E at sea level. At other altitudes, there are different values of Mach number corresponding to each value of V_E. The speed read by a perfect airspeed system (i.e., zero instrument and static error) with the Mach number correction based on a standard-day sea-level relationship between V_E and M is the calibrated airspeed, V_{CAL}. At sea level, the Mach number correction is exact and

$$V_{\text{CAL}} = V_E$$

At other altitudes,

$$V_{\text{CAL}} = V_E + \Delta V_C$$

where ΔV_C is the difference between the true Mach number correction at the flight altitude and values based on sea level.

Another approach to the problem of measuring airspeed in a compressible flow is to solve for true speed in terms of the difference between the total pressure and the static pressure using equation 7.19. Thus

$$p_T - p = p\left(1 + \frac{\gamma - 1}{2} M^2\right)^{\gamma/(\gamma-1)} - p \tag{7.26}$$

Dividing by p yields

$$\frac{p_T - p}{p} = \left(1 + \frac{\gamma - 1}{2} M^2\right)^{\gamma/(\gamma-1)} - 1 \tag{7.27}$$

Solving for M, we obtain

$$M = \left(\frac{2}{\gamma - 1}\right)^{1/2}\left[\left(1 + \frac{p_T - p}{p}\right)^{(\gamma-1)/\gamma} - 1\right]^{1/2} \tag{7.28}$$

The true speed V is equal to Ma, where $a = \sqrt{\gamma p/\rho}$. Then

$$V = \left(\frac{2\gamma}{\gamma - 1}\right)^{1/2}\left(\frac{p}{\rho}\right)^{1/2}\left[\left(1 + \frac{p_T - p}{p}\right)^{(\gamma-1)/\gamma} - 1\right]^{1/2} \tag{7.29}$$

The equivalent airspeed equals $V\sqrt{\rho/\rho_s}$, so

$$V_E = \left(\frac{2\gamma}{\gamma - 1}\right)^{1/2}\left(\frac{p}{\rho_s}\right)^{1/2}\left[\left(1 + \frac{p_T - p}{p}\right)^{(\gamma-1)/\gamma} - 1\right]^{1/2} \tag{7.30}$$

This equation is another form of equation 7.25.

*Mach number is related to equivalent airspeed as follows:

$$M = \frac{V}{a} = \frac{V_E}{\sqrt{\rho/\rho_s}} \times \frac{1}{\sqrt{\gamma RT}} = \frac{V_E}{\sqrt{p/RT}} \times \frac{\sqrt{\rho_s}}{\sqrt{\gamma RT}} = \frac{V_E\sqrt{\rho_s}}{\sqrt{p}\,\sqrt{\gamma}}$$

$$= \frac{V_E\sqrt{0.002377}}{\sqrt{p}\,\sqrt{1.4}} = 0.04121\frac{V_E}{\sqrt{p}}$$

V_E is in ft/s; p in lb/ft^2.

The unknowns in equation 7.30 for equivalent airspeed are the static pressure and the difference between the total and static pressure. The latter is exactly what is read by the pitot-static system. As noted previously, it is standard practice to calibrate the indicator at sea level. This is accomplished by assuming sea-level standard pressure in the calibration. The speed given by such an indicator is the calibrated airspeed, V_{CAL}, and is defined as

$$V_{\text{CAL}} = \left(\frac{2\gamma}{\gamma - 1}\right)^{1/2}\left(\frac{p_s}{\rho_s}\right)^{1/2}\left[\left(1 + \frac{p_T - p}{p_s}\right)^{(\gamma-1)/\gamma} - 1\right]^{1/2} \tag{7.31}$$

To obtain the equivalent airspeed at altitudes other than sea level, a correction must be made. The correction is the ratio between equations 7.30 and 7.31. It is sometimes called the *F factor*. Thus

$$V_E = FV_{\text{CAL}} \tag{7.32}$$

where

$$F = \frac{\left(\dfrac{p}{\rho_s}\right)^{1/2}\left[\left(1 + \dfrac{p_T - p}{p}\right)^{(\gamma-1)/\gamma} - 1\right]^{1/2}}{\left(\dfrac{p_s}{\rho_s}\right)^{1/2}\left[\left(1 + \dfrac{p_T - p}{p_s}\right)^{(\gamma-1)/\gamma} - 1\right]^{1/2}} \tag{7.33}$$

The F factor is given in Table 7.1. From equation 7.33, it can be seen that, for a given value of the pressure difference $(p_T - p)$, F varies only with pressure, so it can be thought of as a pressure correction. However, the magnitude of F also varies strongly with the pressure difference measured by the instrument and that, in turn, is a function of speed. The derivation in equations 7.22 to 7.25 shows that the correction is, in fact, due to the difference between the compressibility effect at sea level, which is applied in calibrating the instrument, and the compressibility effect at other altitudes. Thus it is an altitude factor to be applied to the compressibility correction. For this reason, the correction is often called a compressibility correction to airspeed.

Figure 7.2 shows the compressibility correction given graphically as an increment rather than as a ratio like the F factor. It is, however, the same thing. Note that the correction is negligible for low-speed aircraft. All pilots of high-speed aircraft are provided with such charts or tables.

TABLE 7.1 *F* FACTOR TO BE APPLIED TO CALIBRATED AIRSPEED TO OBTAIN EQUIVALENT AIRSPEED

Pressure altitude (ft)	Calibrated airspeed (knots)							
	200	250	300	350	400	450	500	550
10,000	1.0	1.0	0.99	0.99	0.98	0.98	0.97	0.97
20,000	0.99	0.98	0.97	0.97	0.96	0.95	0.94	0.93
30,000	0.97	0.96	0.95	0.94	0.92	0.91	0.90	0.89
40,000	0.96	0.94	0.92	0.90	0.88	0.87	0.87	0.86
50,000	0.93	0.90	0.87	0.86	0.84	0.84	0.84	0.84

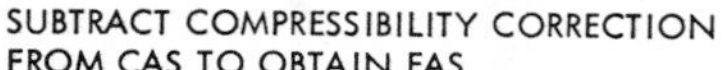

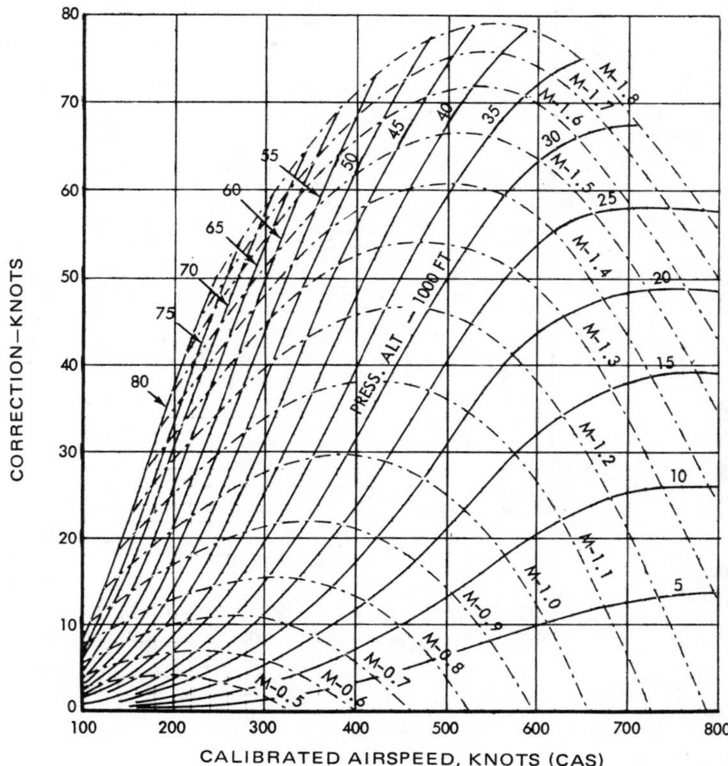

Figure 7.2 Compressibility correction to calibrated airspeed. (From USAF Series T.O. 1F-5a-1, *Flight Manual.*)

At supersonic speeds, a shock wave occurs ahead of the pitot tube. Since a total pressure loss occurs through a shock wave (see Chapter 12), equation 7.19 is replaced by a more complex equation called the Rayleigh pitot equation. The airspeed indicator calibration is adjusted accordingly.

Another defined aircraft speed is the indicated airspeed, V_{IND}. This is the calibrated airspeed plus any error due to an imperfect position of the static source. It is the instrument reading corrected only for instrument error,

$$V_{\text{IND}} = V_{\text{CAL}} + \Delta V_p \tag{7.34}$$

where ΔV_p is the static position error.

Since measurements of total pressure are used to determine calibrated airspeed, V_{CAL}, it is useful to evaluate the magnitude of the Mach number effects on the total pressure. This is done in Figure 7.3, where the ratio of the total pressure with compressibility from equation 7.19 to the total pressure neglecting compressibility from equation 7.22 is plotted versus M. The difference in total pressure due to compressibility is 0.4% at $M = 0.4$. Since the dynamic pressure q is 10% of the total pressure at $M = 0.4$, the error in q that would rise from neglecting the compressibility

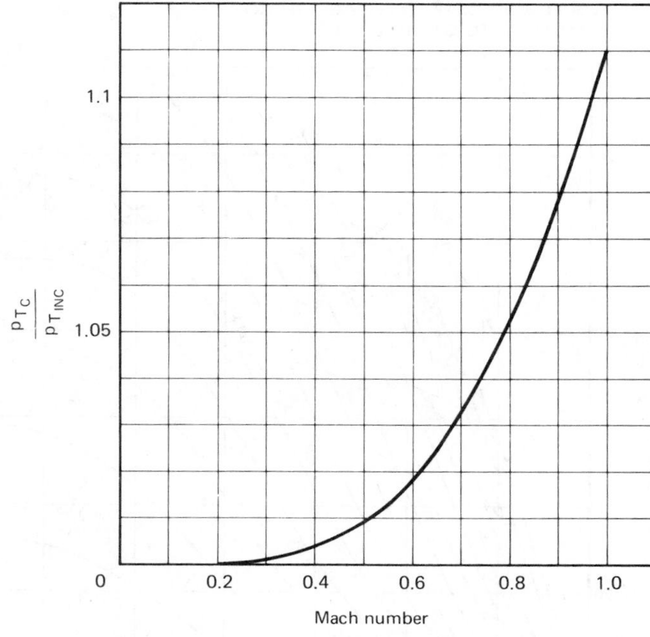

Figure 7.3 Ratio of total pressure with compressibility to calculated total pressure neglecting compressibility.

effect on p_T is 4%. This corresponds to a 2% speed error. At $M = 0.8$, the total pressure difference is 5.3%. Here the q is 29.4% of the total pressure. The q error is then $(0.053/0.294) \times 100$, or 18%, and the speed difference is approximately 9%, a very large error.

Note that the ratio of q to p_T is found by dividing equation 7.23 by equation 7.19, so that

$$\frac{q}{p_T} = \frac{(\gamma/2)pM^2}{p(1 + [(\gamma - 1)/2]M^2)^{\gamma/(\gamma-1)}} = \frac{0.7M^2}{(1 + 0.2M^2)^{3.5}} \tag{7.35}$$

APPLICATION TO CHANNELS AND WIND TUNNELS

Equations 7.18, 7.19, and 7.20 tell us about the variation of T, p, and ρ as the Mach number changes along a streamtube, but give no information about the corresponding area changes. An important relationship involving the cross-sectional area S can be derived as follows: From the continuity equation,

$$\rho SV = \text{constant}$$

or

$$\log \rho + \log S + \log V = \log (\text{constant}) \tag{7.36}$$

Differentiating, we obtain

$$\frac{d\rho}{\rho} + \frac{dS}{S} + \frac{dV}{V} = 0 \tag{7.37}$$

From the Euler momentum equation,

$$dp = -\rho V\,dV$$

or

$$\rho = -\frac{dp}{V\,dV}$$

Substituting in equation 7.37 yields

$$-\frac{d\rho V\,dV}{dp} + \frac{dS}{S} + \frac{dV}{V} = 0 \tag{7.38}$$

Since $dp/d\rho = a^2$, equation 7.38 becomes

$$-\frac{V\,dV}{a^2} + \frac{dS}{S} + \frac{dV}{V} = 0$$

and

$$\frac{dS}{S} = \frac{V\,dV}{a^2} - \frac{dV}{V} = \left(\frac{V^2}{a^2} - 1\right)\frac{dV}{V} \tag{7.39}$$

Since $V/a = M$, the Mach number,

$$\boxed{\frac{dS}{S} = (M^2 - 1)\frac{dV}{V}} \tag{7.40}$$

This area–velocity equation is of great significance. When the Mach number is less than 1, the velocity increases when the area decreases. As $M \to 0$, the magnitude of the percentage change in velocity becomes equal to the percentage change in area, but of course the sign is changed. This is the incompressible continuity result. As M approaches 1, the percentage change in velocity becomes large compared to the percentage change in area. For example, with $M = 0.9$,

$$\frac{dS}{S} = [(0.9)^2 - 1]\frac{dV}{V} = -0.19\left(\frac{dV}{V}\right)$$

$$\frac{dV}{V} = -5.26\left(\frac{dS}{S}\right)$$

Thus in a wind tunnel with $M = 0.9$ at the test section, a 1% reduction in cross-sectional area due to the blocking effect of a model will cause a 5% change in speed. To reduce such problems, models must be small in high-speed wind tunnels. A partial solution to this problem is to slot or perforate the test section walls, allowing some of the flow to pass through the tunnel walls into a surrounding chamber.

Returning to equation 7.40, as M becomes greater than 1, the sign of $(M^2 - 1)$ changes. At these speeds, the density decreases so rapidly for a given speed increase that the channel must actually expand as the speed rises. Thus subsonically ($M < 1.0$) the streamtube must contract as the speed increases, but supersonically ($M > 1.0$) the streamtube expands as the velocity increases. It follows that the flow pattern past a

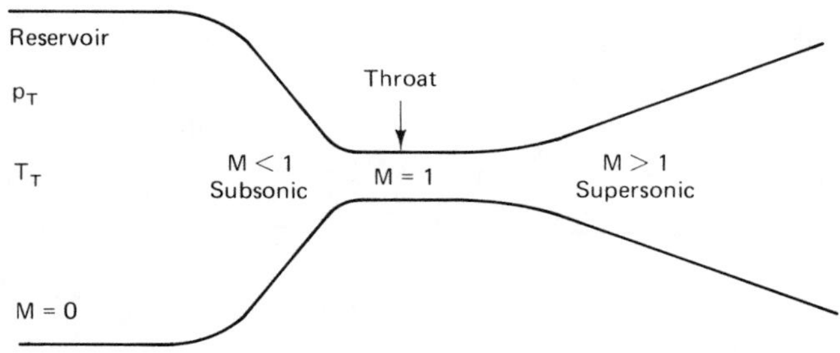

Figure 7.4 Converging-diverging duct.

body must change very considerably as the velocity approaches and then exceeds the velocity of sound.

The cross-sectional area of the streamtube, channel, or wind tunnel has a minimum value when $M = 1.0$. The minimum-area point in the channel is called a *throat*.

From the preceding discussion, it is clear that, to expand a gas to supersonic speeds starting with a gas under pressure in a reservoir, a duct of sufficiently converging–diverging shape must be used. Figure 7.4 shows such a duct, which is typical of supersonic wind tunnel nozzles and rocket engine nozzles. The flow through such nozzles is closely approximated by equations 7.18, 7.19, and 7.20, since the flow is nearly isentropic. Little or no heat is added or removed through the nozzle walls, and a vast core of the flow is virtually frictionless.

We still lack a quantitative relationship between area and Mach number. This can be constructed by some fairly involved algebraic manipulation of the compressible Bernoulli equation (7.6), the isentropic law $p/p_T = (\rho/\rho_T)^\gamma$, and the continuity equation. The result shows that

$$\boxed{\left(\frac{S}{S^*}\right)^2 = \frac{1}{M^2}\left[\frac{2}{\gamma + 1}\left(1 + \frac{\gamma - 1}{2}M^2\right)\right]^{(\gamma+1)/(\gamma-1)}} \tag{7.41}$$

where

S = channel area at the point where the Mach number is M

S^* = throat area where $M = 1.0$

A graph of this relation is shown in Figure 7.5. The curve verifies the preceding discussion, showing a minimum or throat area at $M = 1.0$. The corresponding area–pressure relationship is shown in Figure 7.6. The value of the pressure ratio p_T/p for which $M = 1$ is of particular significance since it defines the ratio of reservoir pressure to throat pressure required to cause $M = 1.0$ at the throat. This pressure ratio is 1.893. The inverse, p/p_T, is shown in Figure 7.6 and is 0.528 at the throat, where $M = 1.0$.

A summary of the one-dimensional flow equations is given in Appendix C.

Example 7.4

The air temperature and pressure in the reservoir of a supersonic blowdown wind tunnel are $T_T = 945$ K and $p_T = 8$ atm. The static temperatures at the throat and test section

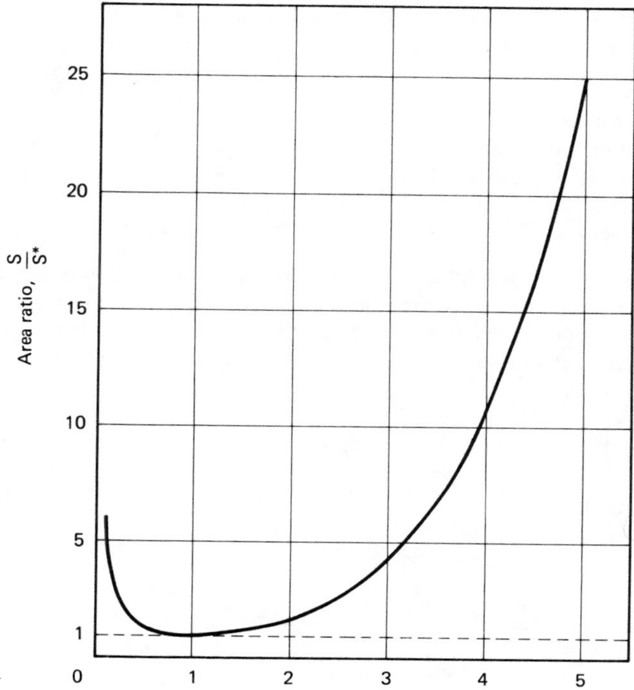

Figure 7.5 Mach number-area variation.

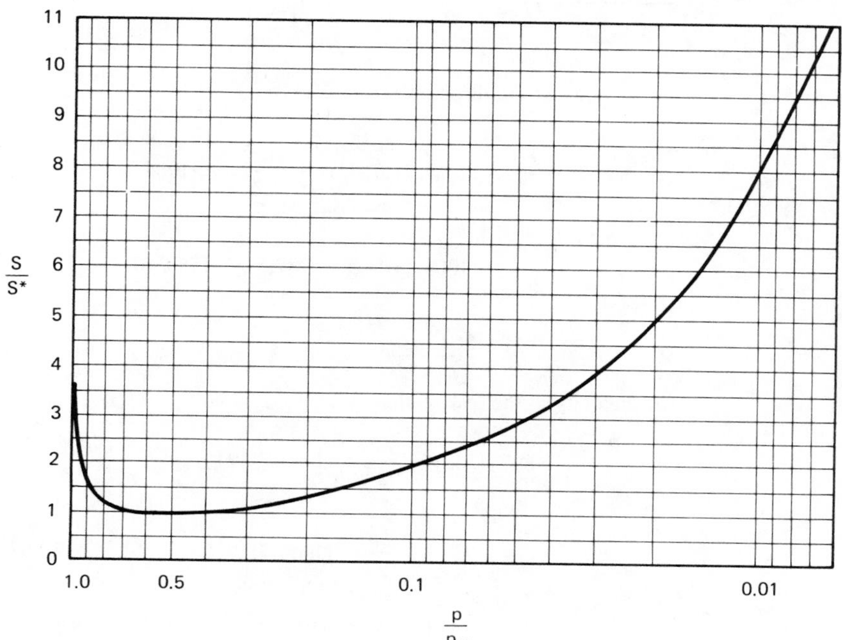

Figure 7.6 Pressure-area variation.

are $T^* = 787.5$ K and $T_t = 335$ K, respectively. The mass flow is 1.5 kg/s. For air, $C_p = 1004.7$ N-m/kg K. Calculate:

(a) Velocity at the throat, V^*.
(b) Velocity at the test section, V_t.
(c) Mach number at the throat, M^*.
(d) Mach number at the test section, M_t.
(e) Area of the throat, S^*.
(f) Area of the test section, S_t.

Solution: In all problems it is very helpful to draw the wind tunnel (or airfoil or whatever is indicated) to emphasize the given and desired quantities. From these, the appropriate equations can be selected. In this problem we are given the temperatures, including the stagnation temperature, and we desire the velocities. The energy equation 7.15 relates these quantities. Thus start with

$$c_p T + \frac{V^2}{2} = c_p T_T$$

where $c_p = 1004.7$ N-m/kg K.

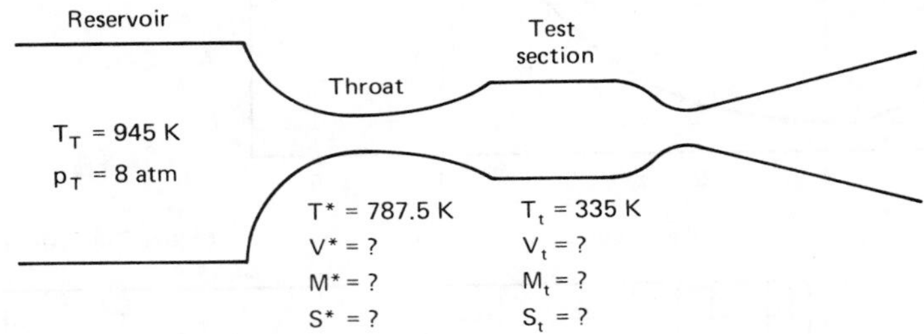

(a) Velocity at the throat, V^*:

$$c_p T + \frac{V^2}{2} = (1004.7)(787.5) + \frac{(V^*)^2}{2} = (1004.7)(945) = c_p T_T$$

$$\frac{(V^*)^2}{2} = (1004.7)(945 - 787.5) = 158{,}240$$

$$V^* = \underline{562 \text{ m/s}}$$

(b) Velocity at the test section, V_t:

$$(1004.7)(335) + \frac{V_t^2}{2} = (1004.7)(945)$$

$$\frac{V_t^2}{2} = (1004.7)(945 - 335) = 612{,}867$$

$$V_t = \underline{1107 \text{ m/s}}$$

(c) Mach number at the throat, M^*. From equation 7.18,

$$\frac{T_T}{T^*} = 1 + \left(\frac{\gamma - 1}{2}\right)M^{*2}$$

$$M^* = \sqrt{\frac{2}{\gamma - 1}\left(\frac{T_T}{T^*} - 1\right)} = \sqrt{5\left(\frac{945}{787.5} - 1\right)} = \underline{1.0}$$

This is, of course, the definition of the throat.

(d) Mach number at the test section:

$$M_t = \sqrt{5\left(\frac{945}{335} - 1\right)} = \underline{3.02}$$

(e) Area of the throat, S^*. The area obviously influences the mass flow, which has been given. These quantities are related by the continuity equation. Thus

$$\text{mass flow} = \rho^* S^* V^* = 1.5 \text{ kg/s}$$

But we do not know the density at the throat ρ^*. We do know the stagnation and throat temperatures and the stagnation pressure. Equation 7.2 relates temperature and pressure ratios and can be used to find the throat pressure. Then, since we already know the throat temperature, the equation of state, $p = \rho RT$, can provide the throat density.

$$\frac{p^*}{p_T} = \left(\frac{T^*}{T_T}\right)^{\gamma/(\gamma-1)} : p^* = p_T\left(\frac{T^*}{T_T}\right)^{\gamma/(\gamma-1)} = (8)(101{,}325)\left(\frac{787.5}{945}\right)^{3.5}$$

$$= 428{,}225 \text{ N/m}^2$$

Then, at the throat,

$$\rho^* = \frac{p^*}{RT^*} = \frac{428{,}225}{(287.05)(787.5)} = 1.894 \text{ kg/m}^3$$

and

$$\text{mass flow} = 1.5 \text{ kg/s} = (1.894)S^*(562)$$

$$S^* = \frac{1.5}{(1.894)(562)} = \underline{0.00141 \text{ m}^2}$$

$$= \underline{2.184 \text{ in.}^2}$$

(f) Area of the test section, S_t: Now that ρ^* is known, ρ_t can be obtained from equation 7.3. We could also use the method of part (e), but use of equation 7.3 saves a step. Thus

$$\frac{\rho_t}{\rho^*} = \left(\frac{T_t}{T^*}\right)^{1/(\gamma-1)} \quad \text{so} \quad \rho_t = \rho^*\left(\frac{T_t}{T^*}\right)^{1/(\gamma-1)} = 1.894\left(\frac{335}{787.5}\right)^{2.5}$$

$$= 0.2235 \text{ kg/m}^3$$

$$\text{mass flow} = \rho_t S_t V_t = (0.2235)S_t(1107) = 1.5$$

$$S_t = \frac{1.5}{(0.2235)(1107)} = \underline{0.00606 \text{ m}^2}$$

$$= \underline{9.39 \text{ in.}^2}$$

Example 7.5

A 747 is flying at 34,000 ft pressure altitude. Temperature is 15°C above standard and the cruise Mach number is 0.85.

(a) What is the equivalent airspeed?
(b) What is the total pressure at the pitot tube?
(c) What is the calibrated airspeed read by a defectively calibrated instrument where Mach number was ignored?
(d) What is the calibrated reading of an instrument correctly calibrated at sea level?
(e) What is the difference due to compressibility between the readings in parts (c) and (d) and the true equivalent airspeed?

Solution: $h_p = 34{,}000$ ft; $\quad p_0 = 523.47$ lb/ft^2

$$T_0 = 15^\circ\text{C} > \text{Std} = \left(15 \times \frac{9}{5}\right)^\circ\text{R} > \text{Std} = 27^\circ\text{R} > \text{Std}$$

$$= 27 + \underbrace{397.6}_{\text{Std. temp}} = 424.6^\circ\text{R}$$

$$\rho_0 = \frac{p_0}{RT_0} = \frac{523.47}{(1718)(424.6)} = 0.000718$$

$$a = \sqrt{\gamma R T_0} = \sqrt{1.4(1718)424.6} = 1010.6 \text{ ft/s}$$

$$V_0 = Ma = 0.85(1010.6) = 859.01 \text{ ft/s} = 585.69 \text{ stat. mph}$$

(a) $$V_E = V_0\sqrt{\sigma} = 859.01\sqrt{\frac{0.000718}{0.002377}} = 472.11 \text{ ft/s} = \underline{321.90 \text{ mph}}$$

(b) $$p_T = p_0\left(1 + \frac{\gamma - 1}{2}M_0^2\right)^{\gamma/(\gamma-1)} = 523.47[1 + 0.2(0.85)^2]^{3.5} = \underline{839.55 \text{ lb/ft}^2}$$

(c) $$V_{E_{M=0}} = \sqrt{\frac{2(p_T - p_0)}{\rho_s}} = \sqrt{\frac{2(839.55 - 523.47)}{0.002377}} = \underline{515.70 \text{ ft/s}}$$

or

$$\underline{351.62 \text{ mph}}$$

(d) This problem can be solved four ways.

1. With equation 7.25, using the Mach number associated with the calibrated speed at sea level: First guess at calibrated speed is the actual V_E. Then, at sea-level standard,

$$V_E = V_T \quad \text{and} \quad M = \frac{V_E}{1116.4} = \frac{472.11}{1116.4} = 0.423$$

From equation 7.25,

$$V_{\text{CAL}} = \sqrt{\frac{2(839.55 - 523.47)}{0.002377}\,\frac{1}{1 + \dfrac{(0.423)^2}{4} + \dfrac{(0.423)^4}{40}}}$$

$$= \sqrt{\frac{265948}{1 + 0.045 + 0.0008}} = 504.35 \text{ ft/s}$$

This differs from the first guess of 472.11 ft/s. Using the new value, $M = 504.35/1116.4 = 0.452$ and

$$V_{CAL} = \sqrt{\frac{265948}{1 + \frac{(0.452)^2}{4} + \frac{(0.452)^4}{40}}} = \underline{502.77 \text{ ft/s}} \quad \text{or} \quad \underline{342.79 \text{ mph}}$$

Close enough!

2. The second method is the use of equation 7.31:

$$V_{CAL} = \left(\frac{2\gamma}{\gamma - 1}\right)^{1/2}\left(\frac{p_s}{\rho_s}\right)^{1/2}\left[\left(1 + \frac{p_T - p_O}{p_s}\right)^{(\gamma-1)/\gamma} - 1\right]^{1/2}$$
$$= 2.646(943.55)(0.20143)$$
$$= 502.90 \text{ ft/s} \quad \text{or} \quad 342.88 \text{ mph}$$

Differences are due to rounding.

3. From Table 7.1 comes the third method. Guess V_{CAL} to be $V_E = 472.11$ ft/s or 279.36 knots. By interpolation at 34,000 ft, 279.36 knots, $F = 0.94$. Then

$$V_{CAL} = \frac{V_E}{F} = \frac{472.11}{0.94} = \underline{502.24 \text{ ft/s}} \quad \text{or} \quad \underline{342.44 \text{ mph}}$$

4. The fourth method is to use Figure 7.2: At $V_E = 279.36$ knots (assume this to be V_{CAL} to start) and 34,000 ft, $\Delta V_C = 15$ knots or 25.35 ft/s. Then $V_{CAL} = V_E + \Delta V_C = 472.11 + 25.35 = 497.46$ ft/s or 294.36 knots. Checking Figure 7.2 at 294.36 knots, $\Delta V_C = 18$ knots or 30.42 ft/s.

$$V_{CAL} = 472.11 + 30.42 = \underline{502.53 \text{ ft/s}} \quad \text{or} \quad \underline{342.63 \text{ mph}}$$

PROBLEMS

7.1. An airfoil is moving through the air at 540 mph at 15,000-ft pressure altitude. The temperature is 45°F. At a point on the airfoil upper surface, the local velocity is 630 mph. What is the pressure at that point? If this is the average pressure on the surface, how much lift per square foot is being provided by the top surface? If the average speed on the lower surface is 495 mph, what is the pressure and average lift per square foot on the lower surface? What is the total lift per square foot of wing area?

(a) Solve using *incompressible* equations.

(b) Solve using *compressible* equations.

(c) What are the percentage differences in the lift value determined by parts (a) and (b)?

(d) Find the local Mach number on the upper surface of the wing.

7.2. A Boeing 727 is cruising at its assigned altitude at a Mach number of 0.82. The outside air temperature is 227 K. At a given point on the upper surface of the wing, the pressure is measured at 19,000 N/m^2. The temperature at this point is 216 K. How much lift per square meter (referred to ambient pressure) is provided by the upper surface at this point? What is the assigned pressure altitude? What is the density altitude? What is the true speed of the airplane? (Use SI units, except also give the upper surface lift in lb/ft^2.)

7.3. The air temperature and pressure in the reservoir of a supersonic wind tunnel are $T_T = 500$ K and $P_T = 6$ atm. The static temperatures at the throat and test section are $T^* = 416.66$ K and $T_t = 136$ K, respectively. The mass flow is 2.82 kg/s. For air, $c_p = 1004.7$ N-m/kg K. Calculate:
(a) Velocity at the throat, V^*.
(b) Velocity at the test section, V_t.
(c) Area of the throat, S^*.
(d) Area of the test section, S_t.

7.4. A Boeing 737 is cruising at a pressure altitude of 30,000 ft at a Mach number of 0.78. Outside air temperature is −40°C. The airplane weight is 96,000 lb. The 737 has a wing area of 980 ft^2. If 85% of the total lift is developed by the upper surface, and if we assume the upper surface lift to be evenly distributed across the chord of the wing, what is the ratio of the upper surface local velocity to the freestream velocity?

7.5. A DC-10 is cruising at its assigned altitude at a Mach number of 0.85. The outside air temperature is 417°R. At a given point on the upper surface of the wing, the pressure is measured at 420 lb/ft^2. The temperature at this point is 397°R. How much lift per square foot (referred to ambient pressure) is provided by the upper surface at this point? What is the assigned pressure altitude? What is the density altitude? What is the true speed of the airplane?

7.6. The air temperature and pressure in the reservoir of a supersonic wind tunnel are $T_T = 1800$°R and $P_T = 6$ atm. The static temperatures at the throat and test section are $T^* = 1500$°R and $T_t = 490$°R, respectively. The weight flow is 4.0 lb/s. For air, $c_p = 6006$ ft-lb/slug °R. Calculate:
(a) Velocity at the throat, V^*.
(b) Velocity at the test section, V_t.
(c) Area of the throat, S^*.
(d) Area of the test section, S_t.

7.7. A DC-9 is cruising at its assigned altitude at a Mach number of 0.78. The outside air temperature is 232 K. At a given point on the upper surface of the wing, the pressure is measured at 20,100 N/m^2. The temperature at this point is 221 K. How much lift per square meter (referred to ambient pressure) is provided by the upper surface at this point? What is the assigned pressure altitude? What is the density altitude? What is the true speed of the airplane? (Use SI units, but also give the upper surface lift in lb/ft^2.)

7.8. The air temperature and pressure in the reservoir of a supersonic wind tunnel are $T_T = 1000$ K and $P_T = 6$ atm. The static temperatures at the throat and test section are $T^* = 833.33$ K and $T_t = 272$ K, respectively. The mass flow is 1.82 kg/s. For air, $C_p = 1004.7$ N-m/kg K. Calculate:
(a) Velocity at the throat, V^*.
(b) Velocity at the test section, V_t.
(c) Area of the throat, S^*.
(d) Area of the test section, S_t.

7.9. A Boeing 767 is flying at 37,000 ft pressure altitude, temperature is 10°C above standard, and the cruise Mach number is 0.80.
(a) What is the equivalent airspeed?
(b) What is the total pressure at the pitot tube?
(c) What is the calibrated airspeed read by a defectively calibrated instrument for which Mach number was ignored?
(d) What is the calibrated reading of an instrument correctly calibrated at sea level?
(e) What is the difference due to compressibility between the readings in parts (c) and (d) and the true equivalent airspeed?

7.10. A DC-10 is flying at 39,000 ft, temperature is 10°C below standard, and the cruise Mach number is 0.83.

(a) What is the equivalent airspeed?

(b) What is the total pressure at the pitot tube?

(c) What is the calibrated airspeed read by a defectively calibrated instrument where Mach number was ignored?

(d) What is the calibrated reading of an instrument correctly calibrated at sea level?

(e) What is the difference due to compressibility between the readings in parts (c) and (d) and the true equivalent airspeed?

REFERENCES

7.1. Liepmann, Hans Wolfgang, and Puckett, Allen E., *Aerodynamics of a Compressible Fluid,* Wiley, New York, 1947.

7.2. McCormick, Barnes W., *Aerodynamics, Aeronautics and Flight Mechanics,* Wiley, New York, 1979.

8

TWO-DIMENSIONAL FLOW: LIFT AND DRAG

LIMITATIONS OF THE ONE-DIMENSIONAL FLOW EQUATIONS: HYDRODYNAMIC THEORY

We have become familiar with a large number of fundamental equations describing (in modern parlance, modeling) the relationships between pressure, density, temperature, and velocity along a streamtube with a particular variation of cross-sectional area. We have noted that channels (such as carburetor venturis and wind tunnels) can be considered very nearly one dimensional, so these equations apply. From an intuitive feeling for the contraction and expansion of streamtubes in fluid flowing past a body or airfoil, we can sense where the pressures will be less than ambient and where they will be greater, as in, for example, Figure 6.3. But none of this defines quantitatively how the streamtube area changes as the air flows past a wing or body. Furthermore, the flow around a circular cylinder in an ideal fluid is clearly symmetrical front and rear (Figure 6.3). Thus the pressure at any vertical position on the cylinder is the same front and rear, so the net force in the (horizontal) direction of the flow must be zero. This force is the drag force, which experience shows is not zero.

Hydrodynamic theory holds the answer to the first problem, quantitatively specifying the details of the flow past a body so that the pressures at the surface can be determined. The basis of hydrodynamic theory is establishing two functions, the potential function ϕ (defined such that $\partial\phi/\partial x$ = velocity in the x-direction, $\partial\phi/\partial y$ = velocity in y-direction, etc.) and a stream function ψ, defining streamlines such that no fluid crosses a line of constant ψ. By applying the continuity and momentum concepts and an additional principle known as irrotationality, and by finding distributions of ϕ and ψ that satisfy these principles throughout the fluid and also the boundary conditions on the surface of the body (e.g., the velocity component perpendicular to the body surface equals zero), the complete flow field can be defined. Then the velocities and pressures can be determined on the surface of the body.

Hydrodynamic theory is beyond the limits of our interest here, but some of its results are very pertinent. The theory assumes a perfect fluid (i.e., a nonviscous or inviscid fluid). One of its results, first revealed by D'Alembert and known as *D'Alembert's paradox,* was that the resultant drag force on a body in a perfect fluid is zero. Also, there is no lift unless circulation is present. It turns out in the case of lift that, although the forces can be determined very closely by perfect (inviscid) fluid theory, the type of flow that must be assumed can exist only if viscosity is assumed to be present. As for drag, once the correct type of flow is assumed, perfect fluid theory shows the presence of a drag due to lift, but only for a wing of finite span. In addition, there are other elements of drag due to viscous skin friction and viscosity-caused pressure drag, the latter arising from deviations from ideal pressure distributions due to the presence of boundary layers. All these comprise the total drag. It is to the subjects of lift and drag that we now turn our attention.

The lift problem was solved by the concept of circulation, a circulatory type of flow that is the only way in which hydrodynamic theory could explain lift. Circulation is not only a theoretical idea but exists in physical reality. It explains many aerodynamic phenomena, as we shall learn in the next section.

THE THEORY OF LIFT: CIRCULATION

To begin our discussion of the creation of lift, it is necessary to study a very simple but vital flow, a steady two-dimensional flow in concentric circles in a perfect incompressible fluid. This flow is called *circulatory* or *vortex flow* and, when combined with a uniform linear flow, will be shown to produce lift. We are permitted to superimpose simple flow patterns to create more complex flows by the *principle of superposition,* which states that simple flows can be combined or added together provided that the addition is done vectorially. Since velocities are vectors with both magnitude and direction, the resultant velocity at any point in the fluid must be obtained by adding the individual flow velocities at that point vectorially, just as force vectors are summed to obtain a resultant force.

Figure 8.1 portrays the streamlines of a circular flow. The Bernoulli constant has the same value throughout the fluid. It is convenient to consider the flow to be occurring around the outside of a circular cylinder. Figure 8.2 illustrates a small fluid element at a distance r from the center of curvature. The element has a unit thickness

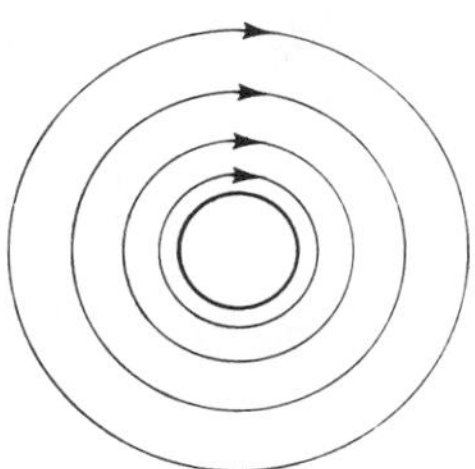

Figure 8.1 Streamlines for a concentric circular flow.

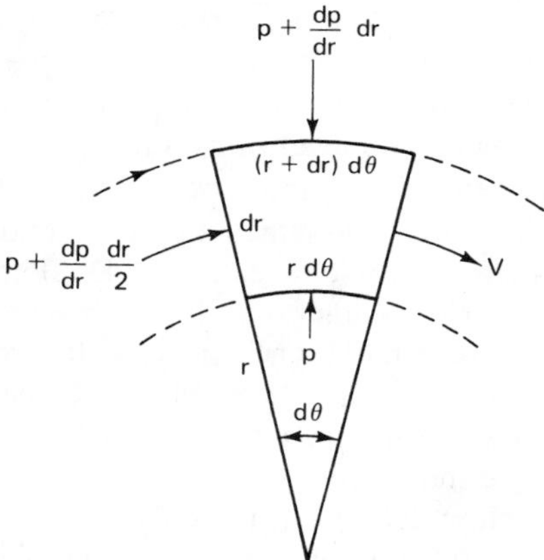

Figure 8.2 Volume element for a concentric circular flow.

in the direction perpendicular to the plane of flow, a width $r\,d\theta$ on its inner face and $(r + dr)\,d\theta$ on its outer face, and a length dr. The centrifugal forces acting on this element in the radially outward direction must be balanced by the pressure forces on the sides of the element. The centrifugal force is

$$\rho\frac{[r\,d\theta + (r + dr)\,d\theta]\,dr}{2}\frac{V^2}{r} = \rho\, r\, d\theta\, dr\frac{V^2}{r}$$

where the higher-order term has been dropped. The pressure forces, positive in the radially inward direction, are

$$-p r\, d\theta \qquad \text{on the inner face}$$

$$\left(p + \frac{dp}{dr}dr\right)(r + dr)\,d\theta \qquad \text{on the outer face}$$

$$-\left(p + \frac{dp}{dr}\frac{dr}{2}\right)[(r + dr)\,d\theta - r\,d\theta] \qquad \text{on the side faces}$$

Neglecting the higher-order terms, the total of the pressure forces is

$$r\,d\theta\,dp$$

Equating the pressure and the centrifugal forces, we obtain

$$r\,d\theta\,dp = \rho\, r\, d\theta\, dr\frac{V^2}{r}$$

and

$$dp = \rho V^2\frac{dr}{r}$$

From the Euler equation, $dp = -\rho V\,dV$. Substituting for dp yields

$$-\rho V\,dV = \rho V^2 \frac{dr}{r}$$

$$\frac{dV}{V} = -\frac{dr}{r}$$

Integrating, we obtain

$$\log V = -\log r + \text{constant}$$
$$\log(Vr) = \text{constant}$$
$$Vr = \text{constant}$$

and

$$\boxed{V = \frac{\text{constant}}{r} = \frac{A}{r}} \tag{8.1}$$

The equilibrium concentric flow of Figure 8.1 is shown by equation 8.1 to be one in which the velocity increases as the radius of curvature is reduced. At $r = 0$, V becomes infinite. The point $r = 0$ is a mathematical singular point. In hydrodynamics, it is called a *vortex*. Since infinite velocities cannot occur in nature, the flow described by equation 8.1 does not apply at small distances from the vortex. The rate of change of velocity radially becomes so large that the viscous effects become significant and the assumption of an inviscid flow breaks down. Thus the central core of the flow starts to rotate as a solid body, and vortex-type flow, in which velocity varies inversely as the radius, exists outside the core. The resultant velocity is shown schematically in Figure 8.3.

From Bernoulli's equation with constant total pressure, it is clear that, as the velocity increases, the pressure decreases, and vice versa. Figure 8.3 shows the pressure trend associated with a vortex flow. The pressure becomes very low near the center of a vortex flow. This is exactly the flow condition of a tornado, a waterspout, or the water flow around a bathtub drain. The low pressure is the force with which a tornado removes the roof from a house. The vortex flows are not mathematical devices but real phenomena that occurs frequently. We shall soon learn that every airplane is lifted by a vortex flow and trails vortices from its wing tips.

The strength of a circulatory flow is clearly some quantity that would define the velocity at any given distance from the center of rotation. From equation 8.1, this quantity must be related to the constant A. The strength of a circulatory flow is called the circulation Γ. It is defined mathematically as the line integral of $V \cos\theta$ around a given closed path or contour in the fluid (Figure 8.4). Thus

$$\Gamma = \oint V \cos\theta\,dl \tag{8.2}$$

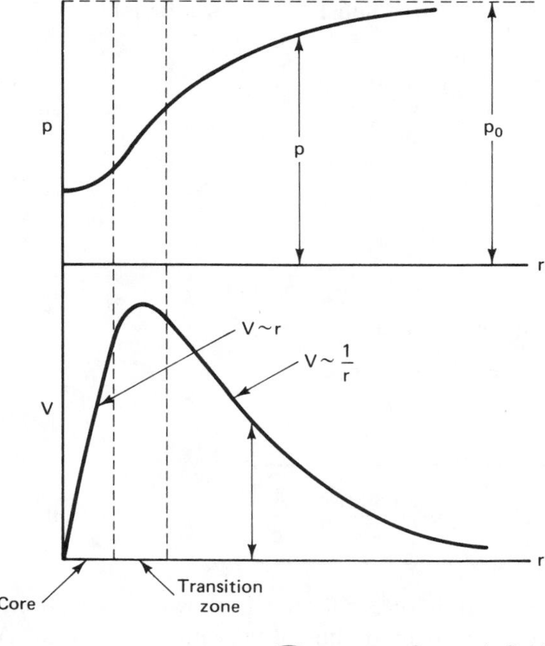

Figure 8.3 Pressure and velocity distribution in a vortex flow.

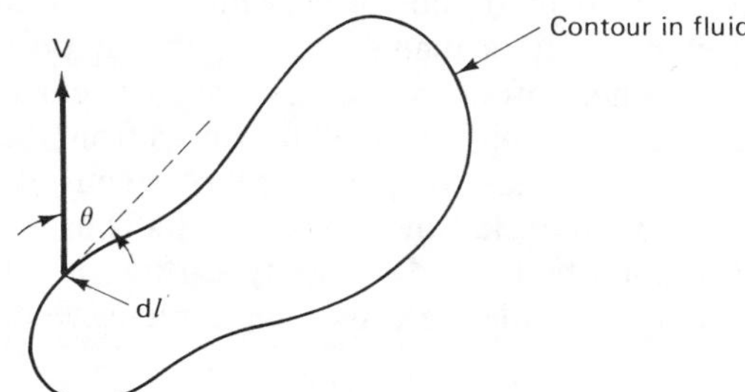

Figure 8.4 Contribution to circulation around a defined contour.

where

θ = angle between flow velocity V and the path at a point on the contour (Figure 8.4)

dl = element of length along the contour

The contribution to circulation of each small element of length around the contour is the product of the velocity component tangent to that small element times the length of the element. The total circulation is the sum of the contributions of all the elements on the closed contour. A simple case that clarifies this concept is the circular flow we have just studied. Around such a circular path the velocity is always tangent to the path (Figure 8.5), and its magnitude is given by equation 8.1.

$$V = \frac{A}{r}$$

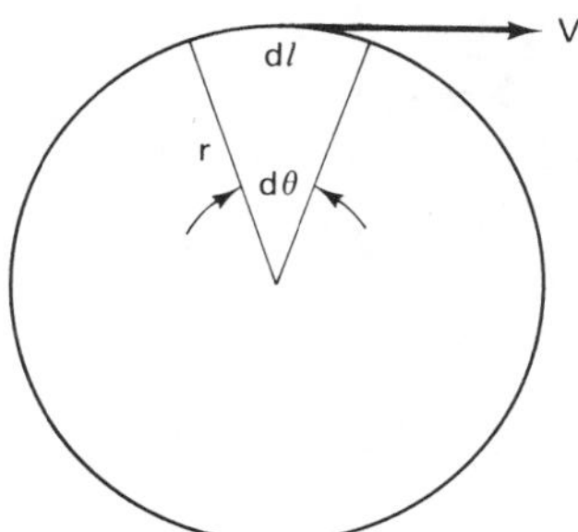

Figure 8.5 Definitions for circulation with concentric circle flow.

Since the velocity is tangent to the path, θ is 0 and $\cos\theta$ is 1.0. The element of length is $r\,d\theta$. Then equation 8.2 becomes

$$\Gamma = \oint V(1) r\,d\theta = \oint \frac{A}{r} r\,d\theta = \oint A\,d\theta$$

Integrating around the contour from any arbitrary angle θ_1 to $\theta_1 + 2\pi$ yields

$$\Gamma = 2\pi A \tag{8.3}$$

Thus

$$A = \frac{\Gamma}{2\pi}$$

Substituting in equation 8.1, we obtain

$$\boxed{V = \frac{\Gamma}{2\pi r}} \tag{8.4}$$

There are two important observations to be noted from equations 8.3 and 8.4. First, the circulation Γ is shown in equation 8.3 to be independent of the radius. The circulation is the same around any streamline in the circular flow. Second, for the circular-flow case, the circulation is equal to the product of the velocity and the length of the streamline at any selected distance from the vortex center. For this simple case, the integration was really unnecessary and the result is intuitively clear from the definition of circulation.

The theorems of classical hydrodynamics lead to generalizations of these ideas. Specifically, it can be shown that, in a perfect fluid with constant total pressure, the circulation is identical around every simple closed path enclosing a given set of vortices or solid bodies (Figure 8.6). Thus the path of integration does not have to follow a particular streamline but may have any shape, provided that the path does not enclose any body or vortex more than once. The expression "circulation around a body" has a well-defined unique value.

We can now show how the superposition of a vortex flow and a uniform linear flow can produce lift. Figure 8.7a shows a circular cylinder in a uniform linear flow

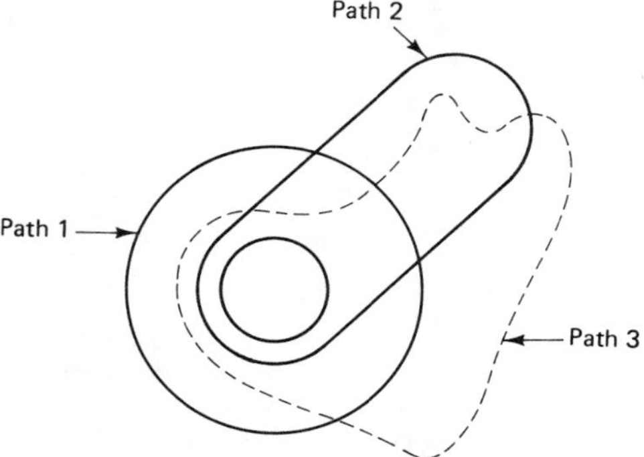

Figure 8.6 Alternative paths having the same circulation.

from the left. From the symmetry of the flow, the velocities on the top of the cylinder are clearly the same as the velocities at the corresponding points on the bottom. If we were to evaluate the line integral (equation 8.2) around the outer edge of the cylinder, taking the clockwise direction as positive, we would obtain zero circulation. The positive contributions obtained on the top would be exactly counteracted by the negative contributions to circulation on the bottom. Figure 8.7b shows a pure circulatory flow, while Figure 8.7c represents the superposition of the first two flows. The combined flow can be visualized by adding the velocity vectors of the uniform and circulatory flows. Thus some of the flow that passed under cylinder in the uniform case now goes over the top because of the vertical component of velocity added at the front of the cylinder by the circulation. Similarly, a downward component is added behind the cylinder. The general nature of the flow is confirmed by Figure 8.8, which shows a photograph of an actual flow around a rotating cylinder with the streamlines made visible by injecting fine particles into the flow. Above the cylinder the circulatory flow adds its velocity to the freestream velocity, whereas below the cylinder it subtracts. Therefore, the resultant velocity above the cylinder is higher than the freestream, whereas below the cylinder it is lower. From the Bernoulli equation, we know that higher velocity is associated with lower pressure, and the lower velocity increases the pressure. The sum of the forces in the vertical direction is upward. Lift has been created. The pressures are still symmetrical about the vertical axis of the cylinder, however, so there is no net force in the fore/aft direction. Thus the drag is zero.

Hydrodynamic theory shows that these results apply not only to circular cylinders but to cylinders of any shape, the most important of which is an airfoil shape. Thus, for a cylinder in steady two-dimensional flow in a perfect fluid, we have the following conclusions:

1. The drag is zero.
2. If there is no circulation, the lift is zero.
3. If circulation exists around a body in a uniform linear flow, lift is produced

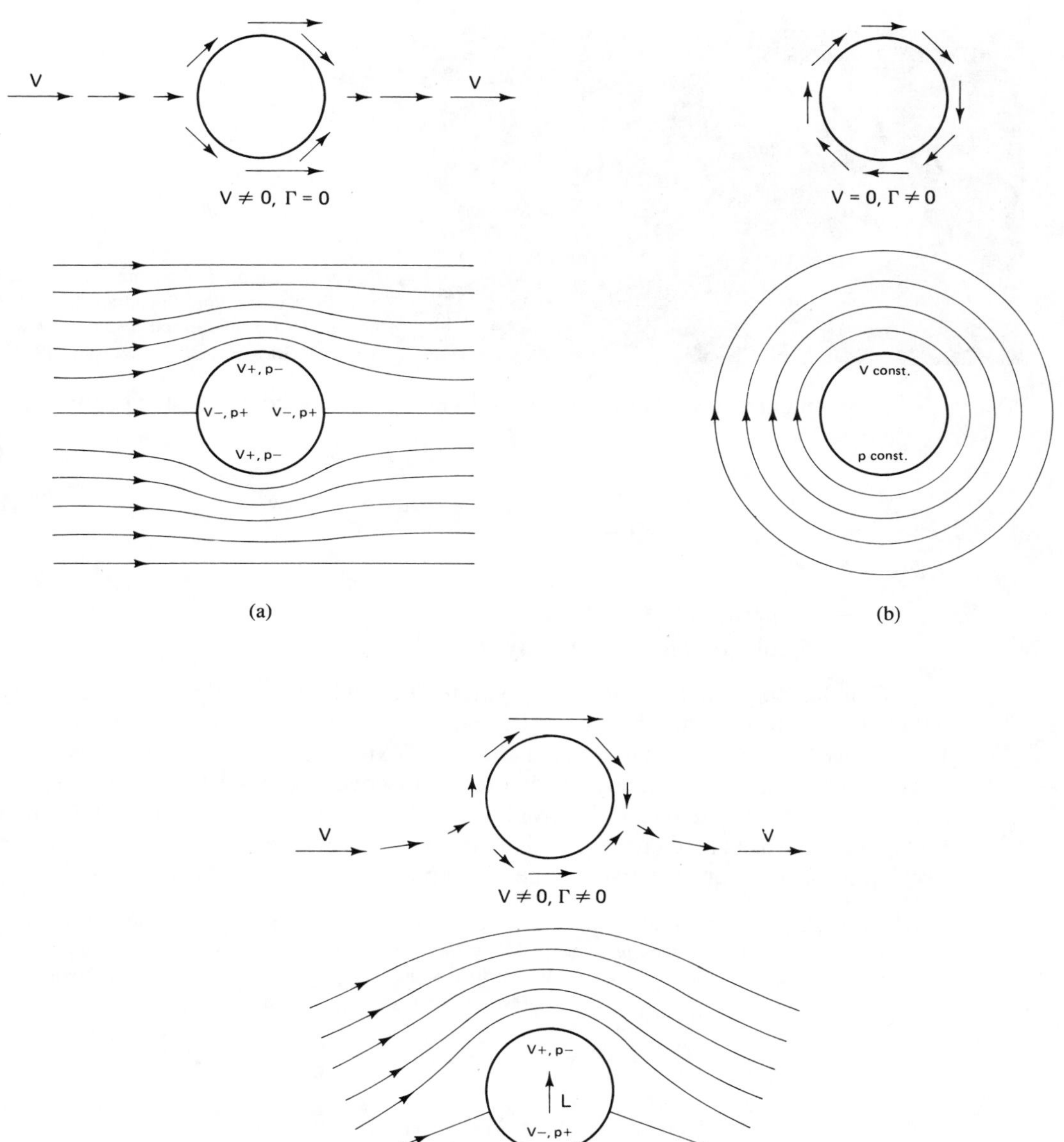

Figure 8.7 Perfect fluid flows around a circular cylinder.

Figure 8.8 Flow around circular cylinder with circulation. From Prandt and Tietjens, *Applied Hydro- and Aeromechanics*. New York: McGraw-Hill Book Company, 1934.

perpendicular to the direction of the freestream, and the magnitude of the lift is given by

$$\boxed{L' = \rho V \Gamma} \qquad \text{Kutta–Joukowski law*} \tag{8.5}$$

where

L' = lift per unit span
Γ = circulation around the body

Conclusions 1 and 2 together constitute D'Alembert's paradox. In effect, they state that, according to the mathematical theory of fluid flow, there is no lift or drag. Tests proved, however, that lift and drag did exist. The inability to resolve this discrepancy was used for many years after its discovery in 1744 to demonstrate the futility of solving practical fluid problems with theory. It was not until 1905 that Joukowski, a Russian scientist, and Kutta, a German mathematician, explained the development of lift quantitatively. The assumption required to successfully solve the

*A simple proof of the Kutta–Joukowski law can be given for a lifting flat plate airfoil, with chord c, assumed to have a constant increase in velocity, ΔV, on the upper surface and a constant decrease of the same magnitude on the lower surface. The subscripts u and l designate upper and lower surfaces. From Bernoulli's equation (6.7),

$$p_l = p_0 + \frac{\rho}{2}[V_0^2 - (V_0^2 - 2V_0\Delta V + \Delta V^2)] = p_0 + \frac{\rho}{2}[2V_0\Delta V - \Delta V^2]$$

$$p_u = p_0 - \frac{\rho}{2}[(V_0^2 + 2V_0\Delta V + \Delta V^2) - V_0^2] = p_0 - \frac{\rho}{2}[2V_0\Delta V + \Delta V^2]$$

Then

$$\text{lift per unit span} = (p_l - p_u)c = 2\rho V_0 \Delta V c$$

Also,

$$\text{circulation, } \Gamma = \oint \Delta V\, dx = 2\Delta V c$$

Substituting,

$$\text{lift per unit span} = \rho V_0 \Gamma$$

lift problem was first published by Kutta in 1902. This seemingly simple assumption was that the flow over both the upper and lower surfaces of an airfoil with a sharp trailing edge must leave the airfoil smoothly at the trailing edge.

The significance of the Kutta–Joukowski assumption is illustrated in Figure 8.9. Figure 8.9a shows schematically the streamlines over an airfoil with small and large angles of attack according to the results of hydrodynamic theory without circulation. On the lower surface, the fluid flows around the trailing edge to the stagnation point on the upper surface. Such a flow around the sharp, zero-radius trailing edge of an airfoil would require infinite velocities, in accordance with our prior discussion of circular flow. Since infinite velocities are impossible, this flow cannot exist, and the flow must leave the trailing edge smoothly from both top and bottom surfaces. To achieve this, the aft stagnation point must move aft to the trailing edge, as shown in Figure 8.9b. Kutta assumed that sufficient circulation was created around the airfoil to accomplish this. The requirement that the stagnation point be at the trailing edge is known as the *Kutta–Joukowski condition*. If the magnitude of the circulation required to satisfy the Kutta–Joukowski condition can be calculated, then the lift is found from equation 8.5. Although Kutta could not explain the physical process by which the circulation was developed, his assumption proved to be very accurate. We shall discuss the generation of circulation further after we study the effects of fluid friction in Chapter 10.

It is interesting that equation 8.5 also explains the curve ball of a baseball pitcher or the hook or slice that frequently haunts golfers. If the ball is given a spin around its vertical axis, friction starts a circulatory air flow around the ball. The resulting circulation produces a force in the lateral direction, curving the flight path.

Note that, in general, in a circulatory flow the air particles do not actually flow completely around the wing. Even though the circulation is positive, the freestream velocity is greater than the circulatory velocity in the opposite direction on the lower surface of the wing, so the net velocity is still aft.

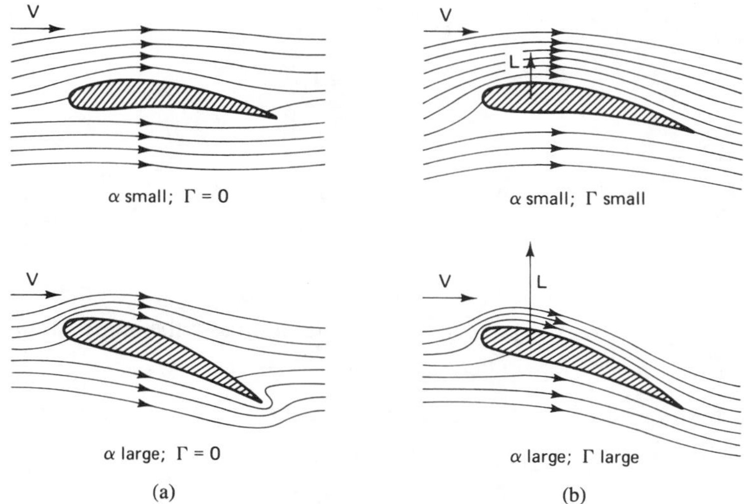

Figure 8.9 Flows around an airfoil without and with circulation.

We have now developed two equations for lift:

$$L' = \rho V \Gamma \tag{8.5}$$

and

$$L = C_L \frac{\rho}{2} V^2 S \tag{3.9}$$

where C_L is a constant for a given wing or body shape and attitude. If we are considering a section of wing with little or no spanwise variation in lift per unit span, the section may be assumed to be a two-dimensional section. The local or section lift coefficient for a section is usually designated by c_l, whereas C_L represents the average lift over an entire wing. The area of a section of unit span is simply the chord c. Thus equation 3.9 applied to a wing section of unit span becomes

$$L' = c_l \frac{\rho}{2} V^2 c \tag{8.6}$$

Equating equations 8.5 and 8.6 yields

$$\boxed{\Gamma = \frac{c_l V c}{2}} \tag{8.7}$$

We shall see in Chapter 9 that the circulation around the wing ends up in the fluid as trailing vortices. Equation 8.7 shows that large aircraft, with large wing chords and equipped with powerful wing flaps and slats that permit high lift coefficients on approach to landing, will produce a large value of circulation Γ. The trailing vortices will, therefore, be strong and dangerous to smaller aircraft that might get caught in the swirling flow.

Theory and experiment show that for a two-dimensional airfoil

$$\boxed{c_l = \frac{dc_l}{d\alpha} \alpha_a = a_0 \alpha_a} \tag{8.8}$$

where

$dc_l/d\alpha = a_0 =$ slope of the section lift curve, usually constant except at high angles of attack near the stall; the lift curve is the curve of c_l versus α

$\alpha_a =$ angle of attack measured from the angle of attack for zero lift, i.e., absolute angle of attack; the angle of attack for zero lift is called $\alpha_{L=0}$; then $\alpha_a = \alpha - \alpha_{L=0}$

Furthermore, hydrodynamic perfect fluid theory gives the amazing result that for thin airfoils

$$a_0 = 2\pi$$

where α_a is in radians.

Experimental results for airfoils deviate from this for two reasons. First, the perfect fluid thin airfoil theory applies to airfoils of zero thickness. Airfoil thickness is defined non-dimensionally as the ratio of the airfoil thickness to the chord. The thickness is often expressed as a percentage of the chord. Subsonic aircraft airfoil thickness is generally between 8% and 16%. The theoretical effect of thickness is to increase the lift curve slope by about 6% for 8% thick airfoils, 9% for 12% thick airfoils, and 12% for 16% thick airfoils.

The second reason for the lift curve slope deviation from 2π is that the boundary layer, the thin layer of slower-moving air next to the wing surface, changes the effective shape of the airfoil. We discuss the development of the boundary layer in Chapter 10. The boundary layer becomes thicker toward the rear of the airfoil, growing more rapidly on the upper surface. The difference between the upper and lower surface boundary layer thicknesses increases as the angle of attack is increased. This raises the effective trailing edge of the airfoil, reducing the effective angle of attack. The lift curve slope is therefore reduced. These two effects are largely counteracting so that experimental lift curve slopes show only a small variation with airfoil thickness with the average value being slightly less than 2π. The airfoil section lift curve slope is expressed as

$$\boxed{a_0 = 2\pi\eta} \qquad \text{per radian} \tag{8.9}$$

where η is an empirical factor accounting for the effects of thickness and viscosity, which inviscid thin airfoil theory ignores. The value of η is usually between 0.95 and 0.98 for Reynolds numbers above 3×10^6. Even small aircraft have cruise Reynolds numbers equal to or greater than 3×10^6.

If α is measured in degrees, then

$$a_0 = \frac{2\pi}{57.3}\eta = 0.110\eta \qquad \text{per degree}$$

AIRFOIL PRESSURE DISTRIBUTIONS

Although the detailed pressure distribution over an airfoil surface requires the results of potential theory, Figures 8.7 and 8.9 plus some thinking about the shape of an airfoil can yield considerable insight into this problem. From Figure 8.7 we noted that above a cylinder with clockwise circulation, in a linear flow from left to right, the velocity was increased and the pressure decreased with respect to the flow around a cylinder without circulation. The opposite is true on the lower surface. If we add these concepts to the effect of curvature on flow velocity as seen in equation 8.1, we can begin to draw a curve of pressure around the airfoil. Referring to Figure 8.9b, wherever the radius of curvature of the airfoil surface is low, the velocity will be high. Thus, above the stagnation point at the airfoil nose, the flow will accelerate rapidly to a maximum velocity, causing the minimum pressure on the airfoil. Then, as the radius of curvature increases across the middle and rear of the airfoil, the velocity will decrease and the pressure will increase until the trailing edge stagnation point is reached. On the lower surface, the velocity may always be below the freestream velocity at high angles of attack. At low angles of attack, the effect of thickness may cause some velocities above the freestream value even on the lower surface. In practice, the boundary layer growth along the airfoil keeps the trailing edge from being aerodynamically as sharp as it looks, so the pressure does not reach true stagnation.

Figure 8.10 shows typical pressure distributions for an airfoil. The curves are shown for various angles of attack and display the trends discussed previously. The pressures are plotted in the form of the pressure coefficient, a generalized non-dimensional quantity given by

$$\text{pressure coefficient } C_p = \frac{p_L - p_0}{\rho V_0^2/2} = \frac{\Delta p}{q} \tag{8.10}$$

Thus the pressure coefficient C_p is the difference between the local pressure p_L at a point on the airfoil and the freestream pressure p_0, normalized by dividing by the freestream dynamic pressure q. In incompressible flow, the pressure coefficient at the stagnation point at the nose is $+1.0$, the most positive possible value. Note that negative values of the pressure coefficient are plotted above the horizontal axis. Negative pressures with respect to freestream on the upper surface add to lift, while positive values on the lower surface also contribute lift. If there are negative pressures on the lower surface but not as negative as at the same chordwise position on the upper surface, lift is still contributed. Thus the area between the upper and lower surface pressure distributions is proportional to lift.

The lift on a wing can be calculated from pressure distributions available from theory or experiment by summing the lift contribution at each point along the chord. From Figure 8.11, on the surface element ds, with a unit span, the component of the force due to aerodynamic pressure that acts normal to ds, F_N, is

$$p\,ds$$

and the vertical or lift component is

$$p \cos\theta\, ds$$

where θ is the angle between the surface element (i.e., ds) and the freestream.

Letting the subscripts u and l designate upper and lower surfaces, the lift per unit span is then

$$L' = \int_{\text{LE}}^{\text{TE}} p_l \cos\theta\, ds - \int_{\text{LE}}^{\text{TE}} p_u \cos\theta\, ds^*$$

Since, from Figure 8.11, $\cos\theta\, ds = dx$, where dx is in the freestream direction,

$$L' = \int_{\text{LE}}^{\text{TE}} p_l\, dx - \int_{\text{LE}}^{\text{TE}} p_u\, dx$$

*The abscissa in Figure 8.10 must be converted to a physical dimension by multiplying by $c \cos\alpha$; or else, in equation 8.11, x can be nondimensional; i.e., x = distance from the leading edge, measured parallel to the freestream, divided by the chord. Then the limits of integration are 0 to $c \cos\alpha$ and $c = 1.0$.

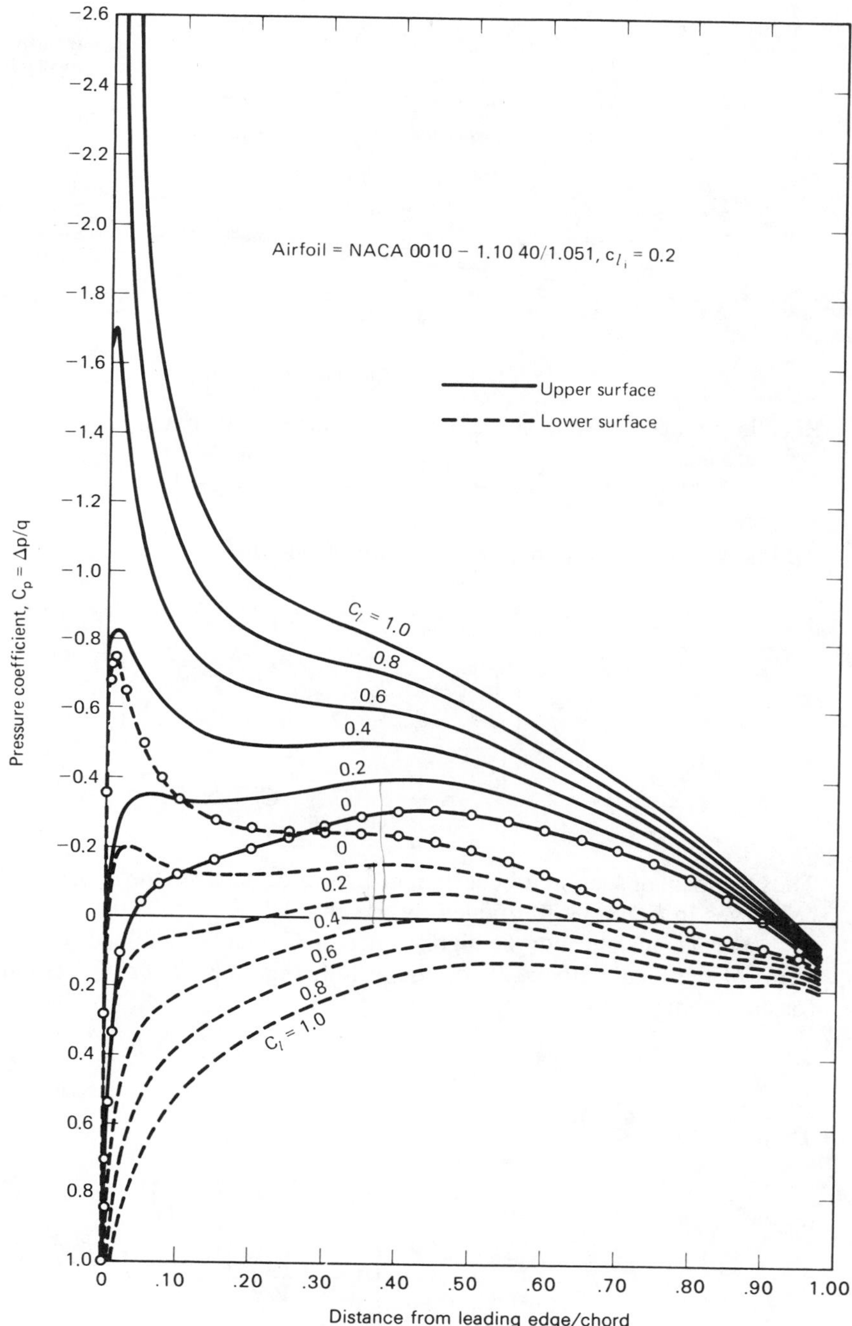

Figure 8.10 Airfoil pressure distribution.

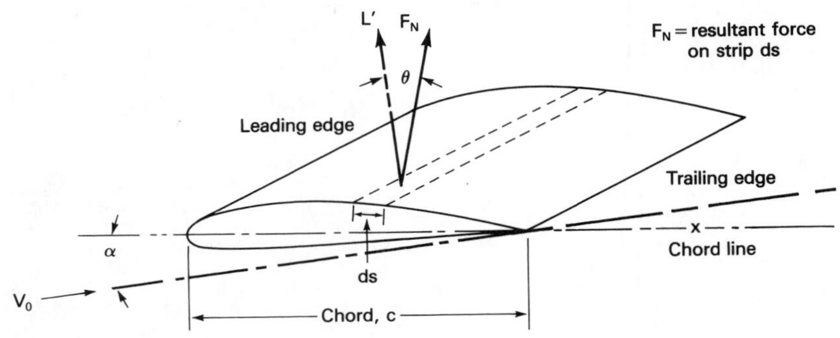

Figure 8.11 Lift forces on an airfoil.

Recalling that $L' = c_l qc$ on a unit span element, we have

$$c_l = \frac{1}{qc}\left[\int_{\text{LE}}^{\text{TE}} p_l\,dx - \int_{\text{LE}}^{\text{TE}} p_u\,dx\right]$$

Adding and subtracting ambient pressure p_0 yields

$$c_l = \frac{1}{qc}\left[\int_{\text{LE}}^{\text{TE}} (p_l - p_0)\,dx - \int_{\text{LE}}^{\text{TE}} (p_u - p_0)\,dx\right]$$

$$= \frac{1}{c}\left[\int_{\text{LE}}^{\text{TE}} \frac{p_l - p_0}{q}\,dx - \int_{\text{LE}}^{\text{TE}} \frac{p_u - p_0}{q}\,dx\right]$$

$$\boxed{c_l = \frac{1}{c}\int_{\text{LE}}^{\text{TE}} (C_{p_l} - C_{p_u})\,dx} \tag{8.11}$$

Thus the local or section lift coefficient is the area between the upper and lower surface C_p curves in Figure 8.10 divided by the chord.

Further insight into the significance of C_p can be obtained by expressing C_p in terms of the local velocity V_L on the airfoil, Figure 8.12. From Bernoulli's equation for an incompressible fluid,

$$p_L + \frac{\rho}{2}V_L^2 = p_0 + \frac{\rho}{2}V_0^2$$

Then

$$\Delta p = p_L - p_0 = \frac{\rho}{2}\left(V_0^2 - V_L^2\right)$$

$$= \frac{\rho}{2}V_0^2\left[1 - \left(\frac{V_L}{V_0}\right)^2\right]$$

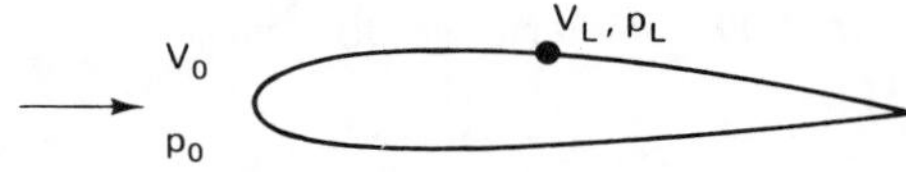

Figure 8.12 Local velocity and pressure at a point on an airfoil.

and

$$C_p = \frac{\Delta p}{q} = \left[1 - \left(\frac{V_L}{V_0}\right)^2\right] \tag{8.12}$$

C_p is a function of the velocity ratio V_L/V_0. In an incompressible perfect fluid flow, the velocity ratio at any point on the airfoil depends only on the airfoil shape and attitude. Thus, if C_p is measured on a small model wing, it can be applied to any wing regardless of size, speed, or altitude. The pressure difference from ambient is then $C_p \cdot q$.

As the Mach number increases above about $M = 0.3$, the C_p's begin to increase significantly even at the same angle of attack; that is, the local differences in pressures, $p_l - p_0$ and $p_u - p_0$, increase somewhat faster than q. Approximate relations for this increase given by Prandtl and Glauert are

$$C_p = \frac{C_{P_{\text{INC}}}}{\sqrt{1 - M_0^2}}$$

and (8.13)

$$c_l = \frac{c_{l_{\text{INC}}}}{\sqrt{1 - M_0^2}}$$

where M_0 is the freestream Mach number and $(\cdot)_{\text{INC}}$ signifies incompressible conditions at $M = 0$.

This expression is quite good up to Mach numbers of 0.7 to 0.8. Obviously, as $M_0 \rightarrow 1.0$, $C_p \rightarrow \infty$, and since nature does not tolerate infinities, the equation breaks down. A better but more complex approximation was given by von Kármán and Tsien as

$$C_p = \frac{C_{P_{\text{INC}}}}{\sqrt{1 - M_0^2} + \frac{1}{2}(1 - \sqrt{1 - M_0^2})C_{P_{\text{INC}}}} \tag{8.14}$$

The variations of local pressure coefficient with freestream Mach number according to the approximations of Prandtl–Glauert and Kármán–Tsien are shown in Figure 8.13. Also shown are the values of the local pressure coefficient C_p required to accelerate the local flow to the speed of sound. This value of C_p is called the critical pressure coefficient, $C_{P_{\text{CR}}}$. The derivation and significance of the critical pressure coefficient are discussed in Chapter 12.

A typical experimental variation of c_l with Mach number on airfoils at constant angle of attack is shown in Figure 8.14. c_l increases with Mach number until the shock strength causes flow separation and the lift starts to decrease at the *lift divergence* Mach number.

Another important result of airfoil theory is that the lift due to angle of attack acts at the airfoil quarter-chord point (i.e., the point 25% of the chord aft of the leading edge). This has important effects on the required aircraft center of gravity location, as will be discussed later.

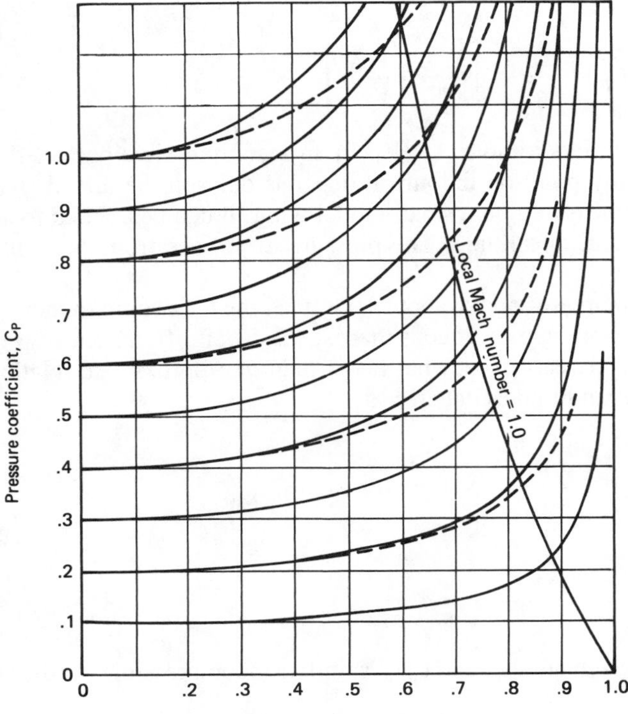

Figure 8.13 Variation of local pressure coefficient with freestream Mach number according to the approximations of Prandtl-Glauert and Kármán-Tsien.

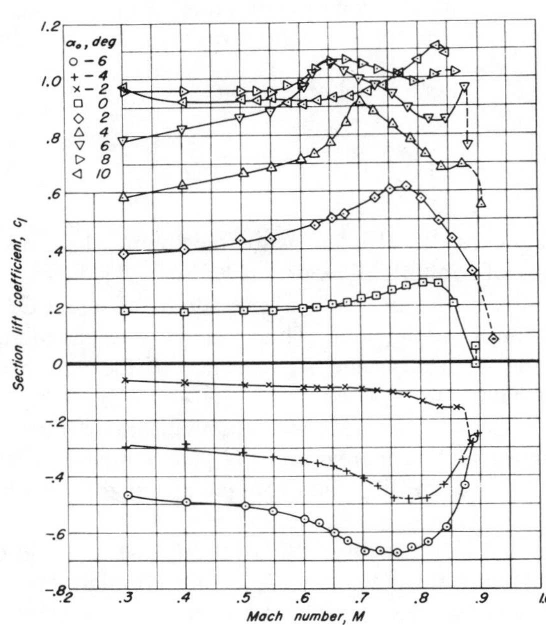

Figure 8.14 Variation of section lift coefficient with Mach number for the NACA 64-210 airfoil. (Courtesy of NASA. From NASA TN 2670.)

Example 8.1

In Example 7.3, we were given an airplane flying at a Mach number of 0.85 at a pressure altitude determined in the problem to be 10,692 m. The outside air temperature was given as 232 K. The pressure at a point on the wing upper surface was 20,100 N/m^2. What was the pressure coefficient C_p at this point? If this C_p were the average for the upper surface, and if the average C_p on the lower surface were half the magnitude and of opposite sign, determine the airfoil lift coefficient. If the wing area is 327 m^2, determine the lift.

Solution: From Example 7.3, the ambient pressure is 23,825 N/m^2 and the ambient density is 0.3578 kg/m^3. The true speed is 259.54 m/s. From equation 8.10,

$$\text{upper surface } C_p = \frac{p_u - p_0}{\rho(V_0^2/2)} = \frac{20{,}100 - 23{,}825}{\dfrac{[(0.3578)(259.54)^2]}{2}} = \frac{-3725}{12{,}050.88}$$

$$= \underline{-0.309}$$

The average C_p on the lower surface is given as $-(-0.309)/2 = +0.155$. From equation 8.11,

$$\text{lift coefficient } C_L = \frac{1}{c}\int_{\text{LE}}^{\text{TE}} [0.155 - (-0.309)]\,dx = \frac{0.464}{c}(c) = \underline{0.464}$$

Note that $\int_{\text{LE}}^{\text{TE}} dx = c \cos \alpha$. For the small angles of attack in cruising flight, $\cos \alpha$ is approximately equal to 1.0.

$$\text{lift} = C_L qS = (0.464)(12{,}050.88)(327)$$

$$= \underline{1{,}828{,}456 \text{ N}}$$

$$= \underline{186{,}577 \text{ kg}}$$

$$= \underline{411{,}074 \text{ lb}}$$

PROBLEMS

8.1. An airplane is flying at a pressure altitude of 6000 ft at a speed of 300 mph. The ambient temperature is 517°R.

(a) What is the local velocity at a point on the wing for which the pressure coefficient C_p is -0.3?

(b) What is the local Mach number at this point?

8.2. An airplane is cruising at a density altitude of 38,000 ft at a speed of 500 mph. The ambient temperature is 409°R.

(a) At a point on the wing the local velocity is 14% higher than the freestream velocity. What is the pressure coefficient at this point? If the C_p is equal to the average upper surface C_p, and if the upper surface provides three-quarters of the lift, what is the wing lift coefficient? Note that at this Mach number the compressible fluid equations must be used. Equation 8.12 is not valid.

8.3. What is the Kutta–Joukowski condition? Why is it important?

8.4. What is the cause of a curved path sometimes followed by a golf ball, baseball, or tennis ball?

8.5. A large executive jet is cruising at 0.78 Mach number at a pressure altitude of 41,000 ft. The outside air temperature is 12°C above standard. A static orifice is located on the upper surface of the wing at 35% of the chord from the leading edge. The reading from this orifice is analyzed and shown to indicate a C_p (pressure coefficient) of -0.38.

(a) What is the local Mach number and the local velocity at this point on the wing?

(b) If the average pressure coefficient equals the value at 35% chord and if two-thirds of the lift is generated by the upper surface, what is the wing lift coefficient?

(c) What is the airplane wing loading?

REFERENCES

8.1. Millikan, Clark B., *Aerodynamics of the Airplane,* Wiley, New York, 1941.

8.2. Glauert, H., *Elements of Aerofoil and Airscrew Theory,* Cambridge University Press, Cambridge, England, 1937.

8.3. Prandtl, L., and Tietjens, O. G., *Fundamentals of Hydro- and Aeromechanics,* McGraw-Hill, New York, 1934.

8.4. Kuethe, Arnold M., and Chow, Chuen-Yen, *Foundations of Aerodynamics,* Wiley, New York, 1976.

8.5. Bertin, John J., and Smith, Michael L., *Aerodynamics for Engineers,* Prentice-Hall, Englewood Cliffs, N.J., 1979.

8.6. Abbott, Ira H., and von Doenhoff, Albert E., *Theory of Wing Sections,* Dover, New York, 1959.

9

THE FINITE WING

We have learned how lift is generated on a two-dimensional infinitely long cylinder or airfoil whose axis is perpendicular to the freestream flow direction. The infinite-length assumption assures that the flow is the same in every plane perpendicular to the axis of the cylinder. On a real wing, however, the lift must decrease toward the tips. Since a finite pressure difference cannot exist at a point in the fluid, there cannot be a pressure difference between the upper and lower surfaces at the wing tips, and here the lift must become zero. Thus the lift and therefore the flow pattern over the airfoil section vary with position along the span. Over much of the wing, the rate of change of the flow field with spanwise position is sufficiently small so that each small spanwise section of the wing can be assumed to act as a two-dimensional section. The overall effect of the variation of lift across the span is very important, however, and leads to the creation of a drag force known as the *induced drag*.

In two-dimensional flow, a vortex was defined as the point at the center of a circulatory flow where the velocity becomes infinite. In three dimensions, the line connecting vortex points in the fluid is called a *vortex filament*. The strength of the vortex at any point along the filament is defined as the magnitude of the circulation around the filament. The circulation will be the same around any path enclosing the vortex filament.

HELMHOLTZ VORTEX THEOREMS

Some basic properties of vortices are fundamental to the study of three-dimensional flow. These vortex laws, first given by Helmholtz in 1858 and called the *Helmholtz vortex theorems,* are as follows:

1. A vortex filament cannot end in the fluid. It must extend to infinity or form a closed path.
2. The strength of a vortex filament is constant along its length.
3. Vortices in a fluid always remain attached to the same particles of fluid.

An additional theorem attributed to William Thomson is:

4. The circulation around any path that always encloses the same fluid particles is independent of time.

These theorems are applicable to a perfect fluid, but are good approximations to viscous fluid flows in regions where viscosity may be neglected. They are based on the concept that no tangential forces can be applied to a fluid particle and that the angular velocity of the particles must therefore remain constant.

SIMULATING THE WING WITH A VORTEX LINE

Three-dimensional wing theory was originally based on the assumption by Ludwig Prandtl, a leading German aeronautical scientist, that the physical wing could be replaced, mathematically speaking, by a vortex filament (Figure 9.1). A similar physical concept had been developed earlier by an English engineer, F. W. Lanchester. Since lift could be explained only by a circulatory flow generated by the proper vortex strength, the wing could be simulated by a properly placed vortex filament. Prandtl called this vortex a *lifting line*. It is also known as the *bound vortex,* since it is inescapably bound to the inside of the wing. The bound vortex, with its infinite velocities, does not really exist in the fluid. The fluid around the wing, however, behaves as if the vortex were really there. Furthermore, Prandtl realized that the Helmholtz vortex laws applied as if the bound vortex really existed. Then, from theorem 1, the bound vortex could not just disappear when the lift dropped to zero at the wings tips, but it would continue into the fluid and produce actual free vortices in the fluid at the tips. From theorem 2, the free vortices would have the same

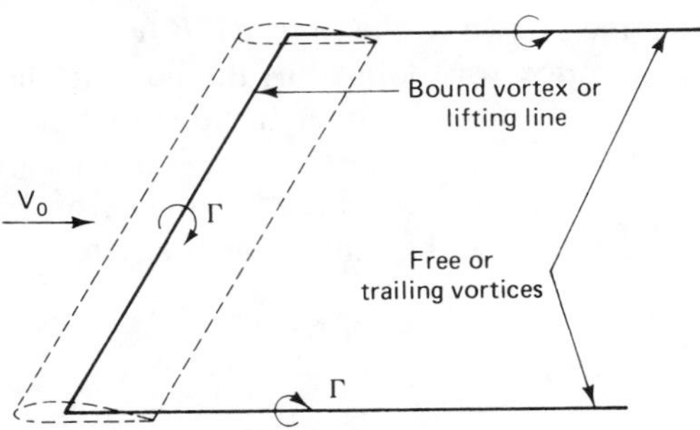

Figure 9.1 Prandtl's horseshoe vortex system.

strength as the bound vortex. From theorem 3, the vorticity would remain attached to the same air particles initially involved at the wing tips and therefore would trail behind the airplane. For this reason, these vortices are called *trailing vortices* and in a steady motion would trail downstream to infinity. This pattern of vortices, as shown in Figure 9.1, is called a *horseshoe vortex*. From the direction of the circulatory flow required for lift, it can be seen that the effect of the trailing vortices is to produce downward flow of air at and directly behind the wing. This flow is called the *downwash*.

DOWNWASH

The trailing vortices and the flow field they create, particularly the downwash, have profound effects on the performance and the stability of aircraft. The simple horseshoe vortex pattern in Figure 9.1 would produce the flow field viewed from the front of the wing represented in Figure 9.2. Each trailing vortex causes a downflow at and behind the wing and an upflow outboard of the wing. Figure 9.2 shows the streamlines in the vortex flow in the top half of the diagram and the vertical velocity or downwash w at the wing in the lower half. The downwash velocity component w at each section across the span is added vectorially to the freestream velocity V_0 and results in an effective resultant velocity V_r approaching the wing at that section, as shown in Figure 9.3. The resultant velocity V_r is tilted downward by the angle whose tangent is w/V_0. This angle is the downwash angle ϵ.

Since the lift on a two-dimensional wing section of unit span is $\rho V_0 \Gamma$, the force R' on the wing section on the finite wing will be $\rho V_r \Gamma$. The downwash velocity w is normally very small compared to V_0, so the magnitude of the resultant velocity is not significantly different from V_0. Referring to Figure 9.3, the resultant force vector R', which is perpendicular to the effective velocity direction approaching the wing, will now be perpendicular to V_r. It will, therefore, be tilted aft by the downwash angle ϵ and will produce a lift L' perpendicular to the freestream direction and equal to $R' \cos \epsilon$. There will also be a drag component D_i' on the section equal to $L' \tan \epsilon$. Remember that drag is defined as the force component parallel to the freestream direction and opposing the airplane's motion. The subscript i is used with D_i' because this drag contribution is called the *induced drag* and is the result of the flow induced by the wing lift.

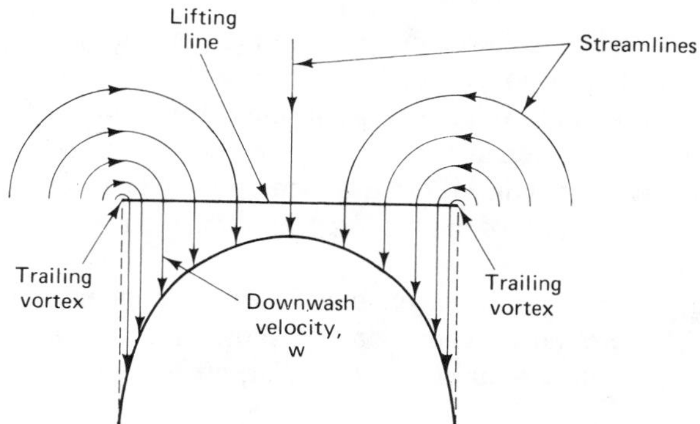

Figure 9.2 Streamlines and downwash produced by trailing vortices as viewed from front of wing.

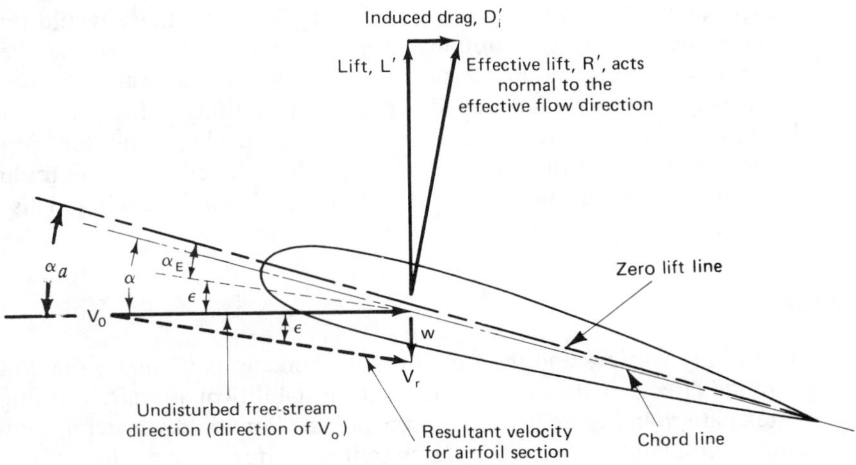

Figure 9.3 Airfoil section flow with downwash.

Since ϵ is small, $\cos \epsilon$ may be assumed to be 1.0 and $\tan \epsilon = \epsilon$, where ϵ is in radians. Then $L' = R'$ and

$$D_i' = L' \tan \epsilon = L'\epsilon = L' \frac{w}{V_0} \tag{9.1}$$

It can be seen from Figure 9.3 that in addition to tilting the force vector aft, the downwash reduces the effective angle of attack. The apparent or *geometrical angle of attack,* α, is the angle between the chord line, which may be somewhat arbitrarily defined in the airfoil, and the freestream velocity. The *absolute angle of attack* α_a is the angle of attack with respect to the angle at which the lift is zero. The *effective angle of attack* α_E is α_a reduced by the angle difference between the freestream velocity V_0 and the resultant velocity V_r. It is this angle that determines the force produced. The airfoil section acts as if it were a two-dimensional section on a wing of infinite span at an angle of attack α_E, where

$$\boxed{\alpha_E = \alpha_a - \frac{w}{V_0}} \tag{9.2}$$

A significant aspect of the flow field causing induced drag is that only a perfect fluid has been considered. This drag is not caused by friction, although we shall soon see that friction drag is very important. When energy in the form of thrust from the engines must be supplied to overcome a drag, the energy consumed must end up somewhere in the fluid. In the case of induced drag, the energy appears in the kinetic energy of the trailing vortices. The aft tilt of the lift force vector is the means by which the drag is transmitted to the wing.

A physical way of understanding the reasonableness of the concept of trailing vortices is to examine the flow around a lifting wing as shown in the front view in Figure 9.4. The pressures on the upper surface of the wing are shown to be below

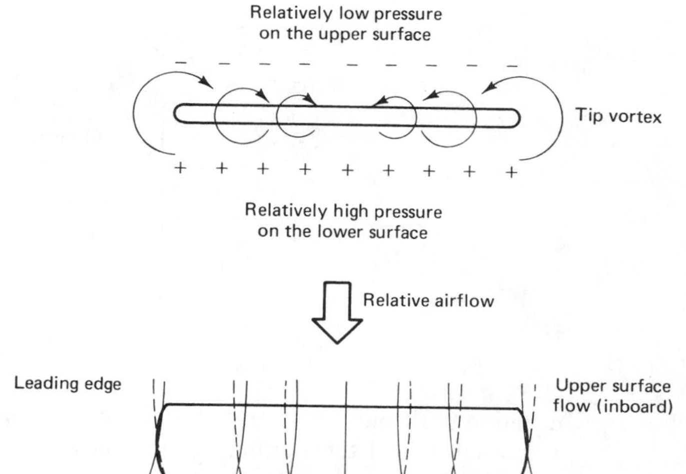

Figure 9.4 Lateral spanwise flow and tip vortices. From John J. Bertin and Michael L. Smith, *Aerodynamics for Engineers,* © 1979. Reprinted by permission of Prentice-Hall, Inc., Englewood Cliffs, N.J.

ambient pressure, while below the wing the pressures are higher than ambient. It is clear that the pressure gradient will cause air to flow laterally from the lower surface around the tips to the upper surface. The flow will encounter the small radius at the wing tips and will require a circulatory flow for a steady-state condition in which the centrifugal and pressure forces are balanced. Thus the tip vortices are formed.

We have been discussing the effect of tip vortices on a typical airfoil section on a real finite wing. We have also been assuming that all the wing circulation continued to the wing tips and then trailed downstream in the fluid. To obtain a useful result, we must first evaluate the amount of downwash produced at the wing. Equation 8.4 gave the velocity produced by a two-dimensional vortex as $V = \Gamma/2\pi r$, where r is the distance from the vortex center. This two-dimensional flow is present if a vortex filament extends to infinity in both directions perpendicular to the plane in which the flow exists. If we examine the downwash produced by the trailing vortices at a point far downstream of the wing, the vortices do effectively extend to infinity in both forward and aft directions. A point more than half the span behind the wing qualifies as being far downstream. But when we are evaluating the downwash at the wing produced by the trailing vortices, we find that the trailing vortices extend to infinity in only one direction, downstream. The flow velocity associated with a three-dimensional vortex filament element is found from the *Biot–Savart law:*

$$dv = \frac{\Gamma}{4\pi}\frac{\cos\beta\,ds}{h^2} \tag{9.3}$$

where dv is the increment of velocity induced at point A by a small section of vortex filament ds, as shown in Figure 9.5; h is the distance between A and the vortex element; and β is the angle between the length h and the line r normal to the vortex filament. The velocity produced by an infinite vortex filament is found by integrating

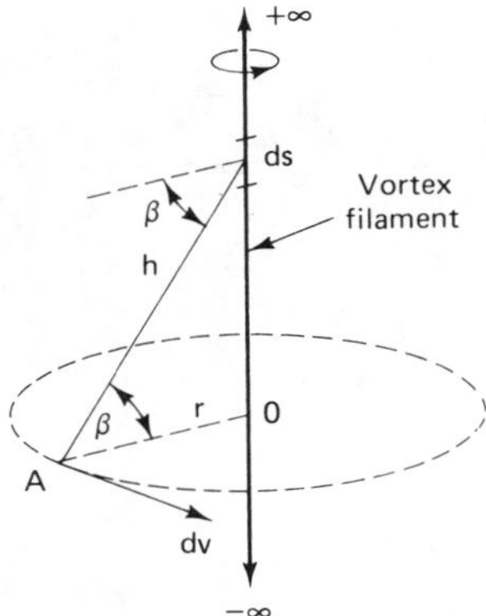

Figure 9.5 Velocity increment dv induced by element ds of a vortex filament.

equation 9.3 from $+\infty$ to $-\infty$. This is accomplished by noting that $h = r \sec \beta$ and $s = r \tan \beta$ and substituting in equation 9.3. Then

$$V = \frac{\Gamma}{4\pi} \int_{-\infty}^{\infty} \frac{\cos \beta\, ds}{h^2} = \frac{\Gamma}{4\pi} \int_{-\pi/2}^{\pi/2} \frac{(\cos \beta) r\, d(\tan \beta)}{r^2 \sec^2 \beta} = \frac{\Gamma}{4\pi r} \int_{-\pi/2}^{\pi/2} \frac{\cos \beta \sec^2 \beta\, d\beta}{\sec^2 \beta}$$

$$= \frac{\Gamma}{4\pi r} \int_{-\pi/2}^{\pi/2} \cos \beta\, d\beta \qquad (9.4)$$

$$= \frac{\Gamma}{2\pi r}$$

The integration yields the two-dimensional result of equation 8.4. Integrating between zero and $+\infty$ yields half this velocity, so the downwash at the wing is one-half of the downwash far downstream.

When Prandtl tried to solve the downwash problem over the wing span, he encountered serious difficulties. First, if the bound vortex extended to the tip, the velocities induced by the trailing vortices would become infinite at the tip. Furthermore, as the downwash became very large near the tip, the effective angle of attack α_E would become negative. If α_E were negative, the lift would be negative, and circulation direction would reverse. It was finally realized that the circulation had to vary smoothly across the span and that the single horseshoe vortex had to be replaced by a large number of horseshoe vortices whose span and strength could be varied to attain any desired spanwise lift distribution, as indicated in Figure 9.6. The bound vortices or lifting lines are superimposed on one another, and the circulation at any point is the sum of the individual circulations. The process can be carried to the limit, in which an infinite number of horseshoe vortices, each of infinitesimal strength, are added together. The flow behind the wing is then a trailing vortex sheet with a continuous distribution of circulation and therefore of lift across the span (Figure 9.7). No single vortex has finite strength, and therefore the infinite values of downwash do not appear.

It is an intricate and lengthy problem to adjust the lift distribution so that the lift at every spanwise point agrees with the section effective angle of attack, and the downwash at every point, which determines the effective angle of attack, is in accordance with the trailing vortices arising from the lift distribution. For one particular

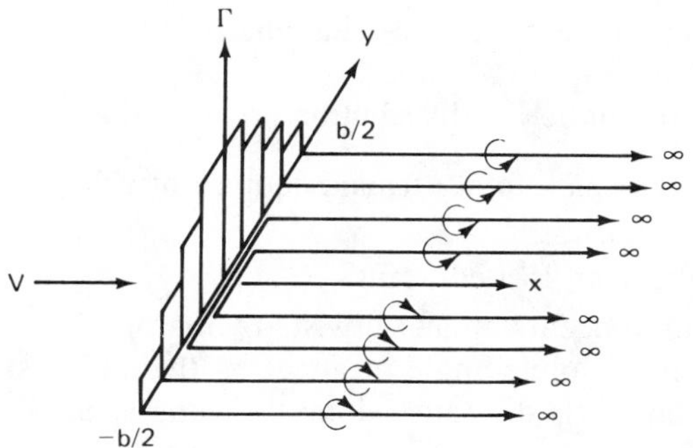

Figure 9.6 Superposition of finite-strength horseshoe vortices.

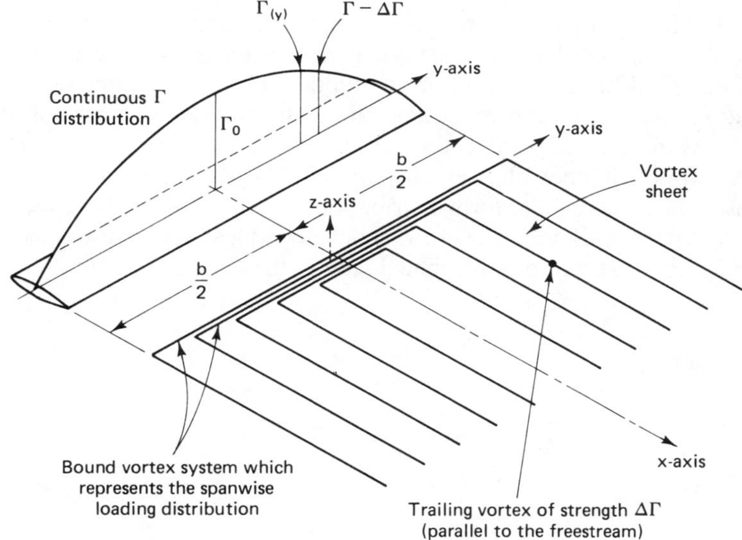

Figure 9.7 Continuous distribution of circulation and trailing vorticity. From John J. Bertin and Michael L. Smith, *Aerodynamics for Engineers,* © 1979. Reprinted by permission of Prentice-Hall, Inc., Englewood Cliffs, N.J.

spanwise lift distribution, however, Prandtl found simple and remarkably important results and published them in 1918.

ELLIPTIC LIFT DISTRIBUTION

Prandtl found that when the spanwise lift distribution was elliptic the induced drag was a minimum. An elliptic lift distribution is one in which the lift per unit span L' at any spanwise position y measured from the centerline of the wing is given by

$$L' = L'_0 \sqrt{1 - \left(\frac{y}{b/2}\right)^2} \tag{9.5}$$

where L'_0 is the value of L' at the center of the wing and b is the wing span. Equation 9.5 is obviously that of an ellipse.

The characteristics of an elliptic lift distribution are as follows:

1. For a planar wing of given span, total lift, and equivalent velocity, the induced drag is a minimum.
2. The downwash along the span is a constant.
3. The elliptic lift distribution occurs on an untwisted wing of elliptical planform. Untwisted means that the absolute angle of attack of the airfoil sections, with respect to the zero lift angle, is the same along the wing span.
4. Since the absolute angle of attack α_a and the downwash angle ϵ are the same along the span, it follows that the effective angle of attack α_E and the local lift coefficient c_l are constant along the span.

An important aspect of wing lift distribution in general is that deviations from an elliptical distribution cause downwash changes over the wing that tend to restore an elliptical lift distribution. If an elliptical planform produces an elliptical lift distribution, one might expect a conventional tapered wing to have a lift distribution that reflects the tapered planform. Figure 9.8 shows a tapered wing with a *taper ratio* (i.e., the ratio of tip chord to root chord) of 0.333. Also shown are wing lift distributions based on a tapered planform distribution, the elliptical, and one assumed to be midway between them. Schrenk (Ref. 9.4) developed a simple approximate method of calculating lift distributions or span loadings based on this "midway" assumption and

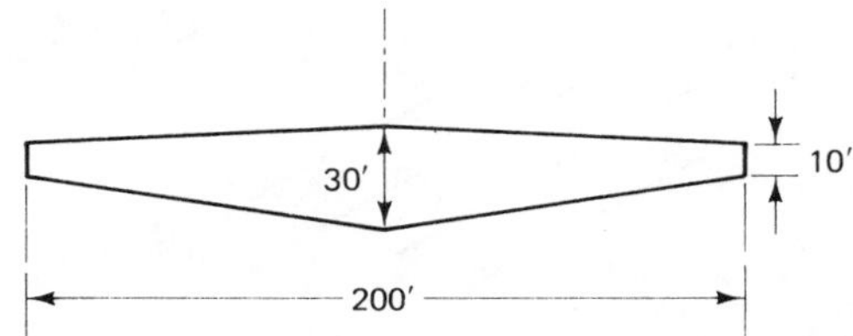

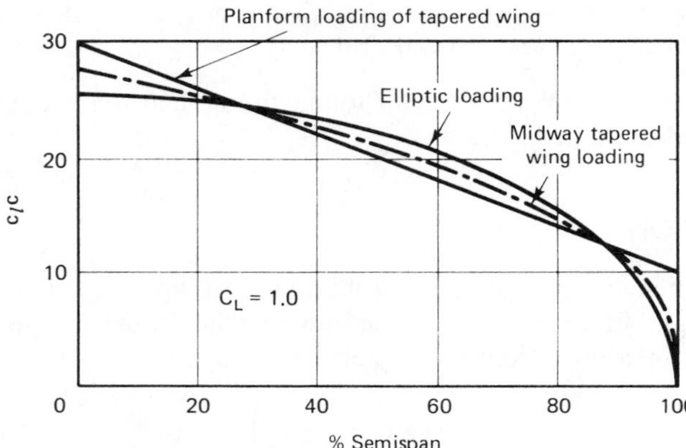

Figure 9.8 Comparison of lift distributions for a tapered wing.

showed that the resulting load distribution is a good first approximation. By twisting a tapered wing (i.e., by varying the angle of attack between the root and tip of the wing), it is usually possible to achieve lift distributions very close to the elliptical.

Since the downwash is constant along the span of a wing with elliptical distribution, the aft tilt angle of the resultant force vector at every section is the same. Thus equation 9.1 applies identically to every section. Summing all the section lifts and induced drags would give the relation between total lift and drag:

$$D_i = L\left(\frac{w}{V_0}\right) = L\epsilon \tag{9.6}$$

Dividing both sides by $(\rho/2)\,V^2S$, we obtain

$$C_{D_i} = C_L\left(\frac{w}{V_0}\right) = C_L\epsilon \tag{9.7}$$

The Prandtl wing theory shows that, for elliptic lift distribution, the value of the downwash angle is

$$\boxed{\epsilon = \frac{w}{V_0} = \frac{C_L}{\pi \cdot \mathrm{AR}}} \tag{9.8}$$

where the *aspect ratio*

$$\mathrm{AR} = \frac{b^2}{S} = \frac{(\text{span})^2}{\text{wing area}}$$

The aspect ratio can also be defined as the ratio of the span to the mean arithmetic chord.

Substituting equation 9.8 in equations 9.7 and 9.2 gives the expressions for the induced drag coefficient and the effective angle of attack, respectively:

$$\boxed{C_{D_i} = \frac{C_L^2}{\pi \cdot \mathrm{AR}}} \tag{9.9}$$

$$\boxed{\alpha_E = \alpha_a - \frac{C_L}{\pi \cdot \mathrm{AR}}} \tag{9.10}$$

Another form of the induced drag equation can be developed by multiplying both sides of equation 9.9 by qS. Then

$$\boxed{\text{induced drag } D_i = \frac{C_L^2 qS}{\pi \cdot \mathrm{AR}}\,\frac{qS}{qS} = \frac{L^2}{\pi q b^2}} \tag{9.11}$$

Equations 9.9 and 9.11 are extremely useful and, although derived from the elliptical loading assumption, need only a simple correction to make them applicable to most wings. We have mentioned the tendency of wings to produce lift distributions that do not deviate greatly from the elliptical. The small increase in induced drag due to nonelliptic distribution has been determined by detailed calculations for a wide range of aspect ratios and taper ratios (i.e., tip chord divided by root chord) and presented in the form of a correction factor* applied to the elliptic drag formula. Thus

$$C_{D_i} = \frac{C_L^2}{\pi \cdot \mathrm{AR} \cdot u} \tag{9.12}$$

For unswept wings, u is usually between 0.98 and 1.00 for most practical planforms. Only for low taper ratios below 0.2 or for high taper ratios where the wing planform approaches the rectangular are the values of u substantially lower. The airplane designer can always select a taper ratio that minimizes the loss in induced drag due to nonelliptic lift distribution. Furthermore, proper wing twist will further reduce induced drag. However, other design criteria affect taper ratio and twist. As we discuss later in considering wing high-lift characteristics, wings must be designed to prevent initial stall from occurring over the outer portion of the wing. Both taper and twist affect tip stall. Taper ratio also affects wing structural weight. Fortunately, little drag or weight penalty is normally required to solve the stall problem in high-performance airplanes. Nevertheless, many small aircraft are designed with rectangular wings that pay an induced drag penalty of about 2% to 5% because of the savings in manufacturing costs due to avoiding taper and twist.

Highly sweptback wings show larger deviations from the drag of equations 9.9 and 9.11 unless they are twisted. Since twist is always required on such wings to avoid tip stall, *washout* (i.e., twist to reduce angle of attack at the tips) is always used for both stall and induced drag purposes.

Except for the indicated compromises for manufacturing cost, u can reasonably be assumed to be 0.99. There is a further loss in induced drag, however, due to the disturbance to the wing lift distribution caused by fuselage interference. The fuselage correction factor s to be applied in addition to the u factor is about 0.98, 0.968, and 0.954 for ratios of fuselage width to wing span of 0.100, 0.125, and 0.150, respectively (see Figure 11.7).

In Chapter 11 we shall find that there is also a variation of parasite drag with C_L^2, which can conveniently be considered as a further correction to the u factor.

LIFT AND DRAG: MOMENTUM AND ENERGY

We have explained lift by considering the effect of circulatory flow around an airfoil. Induced drag has been shown to be due to the effect of downwash on the inclination of the lift vector. The downwash results from the trailing vortices associated with the circulation around the airfoil. Although this approach is the only method for obtaining quantitative values of lift and drag, some of the most important qualitative results can be derived from simple considerations of momentum and energy.

*There is also a secondary correction factor v. This factor is omitted here for simplicity.

According to *Newton's third law,* every action produces an equal and opposite reaction. Therefore, a wing generates lift by exerting an equal and opposite force on the air. It pushes the air downward. In accordance with *Newton's second law,* the force applied equals the rate of change of momentum of the air in the downward direction.

The quantity of air acquiring the downward momentum is not so clear, but it obviously is directly dependent on the span b of the wing. If we double the span, the wing will "touch" twice as much air per second as it moves through the fluid with a velocity V. We do not know the depth of the region of affected air, or even the shape of the region, but from geometric similarity considerations, the depth must be some percentage of the wing span. So let us call the depth $k \times b$. The volume of air affected per unit time is then $k \times b \times b \times V$, and its mass is $\rho k b^2 V$. Thus the mass of affected air increases with the square of the span.

The affected air is given a downward velocity. The downward velocity is not necessarily uniform over the entire region. In fact, it varies with distance below the wing. Nevertheless, let us assume that the air affected by the wing is given a uniform downward velocity w. The kinetic energy left in the stream by the passing wing must be the result of work done by the wing, specifically the work done in overcoming the induced drag. Work performed per unit time is force times velocity. The kinetic energy left in the air per unit time is

$$\frac{1}{2} \times \text{air mass affected per second} \times w^2 = \frac{\rho k b^2 V w^2}{2} \qquad (9.13)$$

and the work done per unit time overcoming the induced drag is

$$\text{thrust} \times \text{velocity} = TV = D_i V \qquad (9.14)$$

Equating equations 9.13 and 9.14 and solving for induced drag yields

$$D_i = \frac{\rho k b^2 w^2}{2} \qquad (9.15)$$

The lift L is equal to the rate of change of downward momentum of the air. This equals the product of the air mass affected per unit time and downward velocity w. Then

$$L = \rho k b^2 V w \qquad (9.16)$$

From this lift equation (9.16) we find that the assumed constant downward velocity is

$$w = \frac{L}{\rho k b^2 V} \qquad (9.17)$$

Substituting for w in equation 9.15 results in equation 9.18 as an expression for induced drag:

$$D_i = \frac{L^2}{2\rho k b^2 V^2} = \frac{1}{4kq}\left(\frac{L^2}{b^2}\right) \qquad (9.18)$$

Comparing with equation 9.11, we find that equation 9.18 is identical if we let k equal $\pi/4$. Then the effective cross section of the region of air affected by the wing is $(\pi/4)b^2$, the area of a circle whose diameter is the wing span. Be aware that this is an interesting delusion, not to be taken as a physical fact.

The following important conclusions can be drawn from this momentum–energy analysis:

1. The force created from the interaction of a body and a fluid varies linearly with the velocity imparted to the fluid (equation 9.16).
2. The energy per unit time expended in producing this force depends on the square of the velocity imparted (equation 9.15).
3. Both the force and the energy vary linearly with the mass of fluid affected per unit time (equations 9.15 and 9.16).
4. As a result, for any specified force produced on the body, the energy rate is always less when a greater mass of fluid is affected and a lower velocity increment is produced in the fluid. For any given lift, the induced drag is less when the span is greater because a greater mass of air is affected (equation 9.18).
5. Although the essential significance of the variables has been shown using momentum and energy concepts, the assumption of a constant downwash velocity throughout the affected fluid is not correct. We have seen previously that an elliptical lift distribution along the span produces a constant downwash at the wing. However, the downwash varies above and below the wing. Therefore, the result that the mass affected is bounded by a circle with its diameter equal to the span is not correct. The analysis simply shows that, if the downwash were constant, an air mass with a volume equal to this cylindrical shape would be affected.

We shall see in the discussion of propulsion that the same essential characteristics apply; that is, moving more air with a lower velocity is a more efficient way to obtain thrust.

SLOPE OF THE FINITE WING LIFT CURVE

We have seen that the downwash angle ϵ reduces the effective angle of attack α_E below the geometric absolute angle of attack α_a of the airfoil with respect to the freestream. Thus, on an airfoil section on an elliptic wing,

$$\alpha_E = \alpha_a - \epsilon = \alpha_a - \frac{w}{V_0} = \alpha_a - \frac{C_L}{\pi \cdot \text{AR}}$$

and

$$C_L = a_0 \alpha_E = a_0\left(\alpha_a - \frac{C_L}{\pi \cdot \text{AR}}\right)$$

where

$$a_0 = \text{section lift curve slope} \\ = 2\pi\eta$$

Then

$$C_L = \frac{a_0\alpha_a}{1 + a_0/(\pi \cdot \text{AR})} \tag{9.19}$$

and the wing lift curve slope is

$$\boxed{\frac{dC_L}{d\alpha} = a = \frac{a_0}{1 + a_0/(\pi \cdot \text{AR})}} \tag{9.20}$$

where α is in radians and a_0 is section c_l per radian. If α is in degrees,

$$\boxed{\frac{dC_L}{d\alpha} = a = \frac{a_0}{1 + 57.3a_0/(\pi \cdot \text{AR})}} \tag{9.21}$$

where α is in degrees and a_0 is c_l per degree.

If AR $= \infty$, $a = a_0$ and the wing lift curve slope is the airfoil section lift curve slope. The lower the aspect ratio, the higher the downwash at a given C_L and the higher the geometrical angle of attack required to achieve that C_L.

Equations 9.19, 9.20, and 9.21 require knowledge of the section lift curve slope a_0. We have said that $a_0 = 2\pi\eta$. This result is for incompressible flow. Equations 8.13 showed that the effect of Mach number is to raise the pressure coefficients at a given angle of attack, and therefore the lift coefficients, by approximately $1/\sqrt{1 - M_0^2}$. Thus the effect of Mach number may be estimated by increasing a_0 by $1/\sqrt{1 - M_0^2}$. A simpler approximation assuming that the entire wing lift coefficient varies with $1/\sqrt{1 - M_0^2}$ is often used. Thus at a given angle of attack

$$C_L = \frac{C_{L_{\text{INC}}}}{\sqrt{1 - M_0^2}} \tag{9.22}$$

To determine the required angle of attack for a given lift coefficient, the equivalent incompressible lift coefficient, $C_{L_{\text{INC}}}$, is found by multiplying by $\sqrt{1 - M_0^2}$. $C_{L_{\text{INC}}}$ is then divided by the finite wing lift curve slope from equations 9.20 or 9.21, using $a_0 = 2\pi\eta$.

We shall see in Chapter 13 that wing sweepback also affects the lift curve slope.

VERIFICATION OF THE PRANDTL WING THEORY

Equations 9.9 and 9.10, resulting from the elliptical wing theory, can be used to predict the changes in induced drag and angle of attack due to aspect ratio variations for any given lift coefficient. The best thing that can be done with any theory is to prove its validity by test. Prandtl ran tests of a series of wings with differing spans, but with identical airfoil sections and wing chords, thereby keeping the Reynolds number the same for all tests. The aspect ratios varied from 1 to 7. The drag coefficients for two aspect ratios should differ at a given C_L only by the differences in induced drag. From equation 9.9,

$$C_{D_1} - C_{D_2} = \frac{C_L^2}{\pi \cdot \mathrm{AR}_1} - \frac{C_L^2}{\pi \cdot \mathrm{AR}_2} = \frac{C_L^2}{\pi}\left(\frac{1}{\mathrm{AR}_1} - \frac{1}{\mathrm{AR}_2}\right) \tag{9.23}$$

The differences in angle of attack for a given C_L can be found from equation 9.10 to be

$$\alpha_1 - \alpha_2 = \frac{C_L}{\pi}\left(\frac{1}{\mathrm{AR}_1} - \frac{1}{\mathrm{AR}_2}\right) \tag{9.24}$$

Using these equations, Prandtl converted all the results to an aspect ratio of 5. The results are shown in Figure 9.9 and display excellent agreement. The lifting line

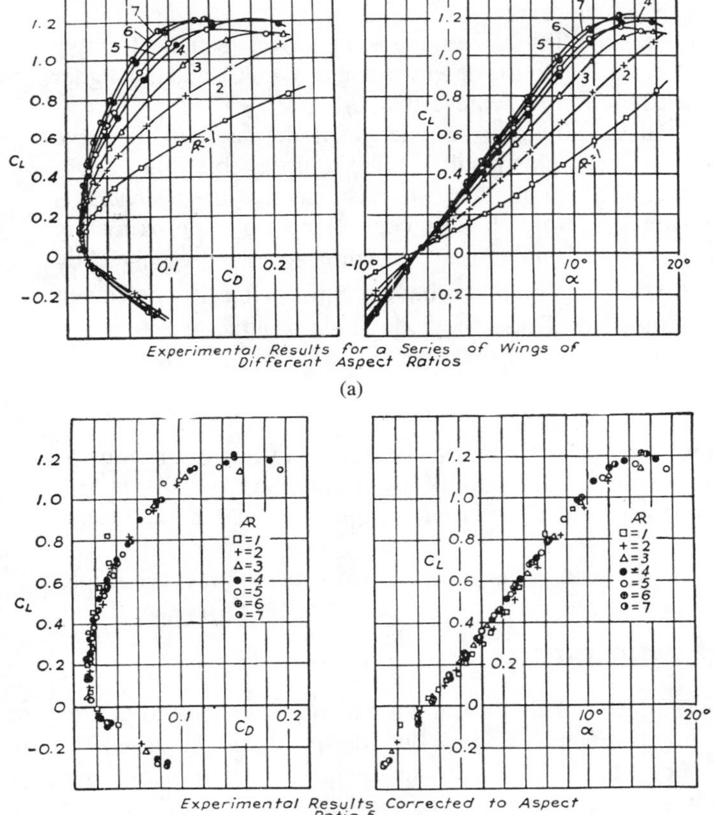

Figure 9.9 Experimental verification of Prandtl wing theory: (a) experimental results for a series of wings of different aspect ratios; (b) experimental results corrected to aspect ratio 5. From C.B. Millikan, *Aerodynamics of the Airplane*. Reprinted by permission of John Wiley & Sons, New York.

theory, which ignores chordwise effects, might be expected to work well only at higher aspect ratios. Therefore, the good agreement in Figure 9.9 at the low aspect ratios is particularly remarkable.

ADDITIONAL EFFECTS OF WING VORTICES

The reality and impact of the wing vortex pattern appear in many ways. Parallel vortex filaments with circulation in the same direction attract each other. Therefore, the vortex sheet originating at the wing trailing edge changes as it proceeds downstream behind the wing from a continuous distribution along the span to a pair of vortices generally similar to the flow pattern in Figure 9.2. The vortex sheet is said to "roll up" into a pair of discrete vortices. The distance between the rolled-up vortices is less than the wing span. For elliptically loaded wings, the distance is $\pi/4$ times the wing span.

The vorticity shed from the wing is normally greatest near the tips because the lift per unit span decreases most rapidly there. Therefore, the roll up of the sheet proceeds very rapidly near the tip. The decreased pressure near the center of the vortex leads to a decrease in temperature in accordance with equation 7.2. If the humidity is high, the condensation of water vapor caused by the temperature drop makes the helical flow of the trailing vortex clearly visible either to a passenger in the airplane or an observer on the ground. Similar visible vortices are shed at the outer end of the wing flaps, where there is a large decrease in lift. The phenomenon is seen most frequently in takeoff and landing, when the lift coefficient is high and high humidity is more likely.

If an airplane flies directly into the trailing vortex shed by a preceding airplane, the circulatory flow will cause a drop in lift on one side of the wing and an increase on the other. The result is a rolling moment that can place the aircraft in a dangerous attitude. This is particularly true if the following airplane is much smaller. For this reason, Federal Aviation Administration (FAA) air traffic rules require aircraft following larger transport aircraft in a landing pattern to keep a separation distance of 5 miles. This separation is reduced if the following airplane is the same size or larger. Although in a perfect fluid the vortices continue to infinity, they become unstable and break up due to viscous effects in a real fluid. Nevertheless, the vortices persist for a long enough time to be a serious limitation on the capacity of runways.

The downwash caused by the trailing vortices affects a large region of air extending well below and above the airplane. At the ground, a boundary condition must exist requiring the flow component perpendicular to the ground to be zero. Therefore, when an airplane flies very close to the ground, some of the downwash is suppressed. One result is that an airplane gliding to a landing tends to "float" as it sinks to a height less than one-third to one-half of the wing span. The downwash reduction leads to an increased effective angle of attack, an increase in lift, and a decreased induced drag. The rate of sink and/or the rate of deceleration is significantly decreased. These phenomena are said to be due to *ground effect*. In takeoff, the induced drag is much less when the airplane is very close to the ground. One attempt to fly the Atlantic Ocean in the late 1920s ended in disaster when the badly overloaded aircraft lifted from the runway to a height of about 5 ft but could not climb higher because of increasing induced drag. The pilot was deluded into thinking he could fly because he had lifted off. He did not realize that he had already reached his absolute ceiling (i.e., his maximum attainable altitude).

The decrease in induced drag due to ground effect can be applied to equation 9.12 as a correction factor. Then the induced drag coefficient is

$$C_{D_i} = K_L \frac{C_L^2}{\pi \cdot \text{AR} \cdot u} \tag{9.25}$$

where K_L is the ground-effect correction given in Figure 9.10 as a function of height/span.

Some transoceanic aircraft have been proposed that would take advantage of ground effect by flying very close to the surface of the water. Designed with a very large span, perhaps 200 ft, they would fly at an altitude of about 20 ft. Although the reduction in induced drag would be valuable, other practical considerations have prevented any of these ideas from being developed.

A fighter aircraft approaching a tanker aircraft from the rear for refueling flies in the downwash of the tanker. The engine thrust of the fighter must be markedly increased to overcome the additional drag due to this downwash.

Our last example deals not with downwash but with upwash. From Figure 9.2, it is seen that outboard of a lifting wing there is an upwash. Just as downwash tilts the resultant force vector aft and causes drag, so *upwash* tilts the force vector forward and

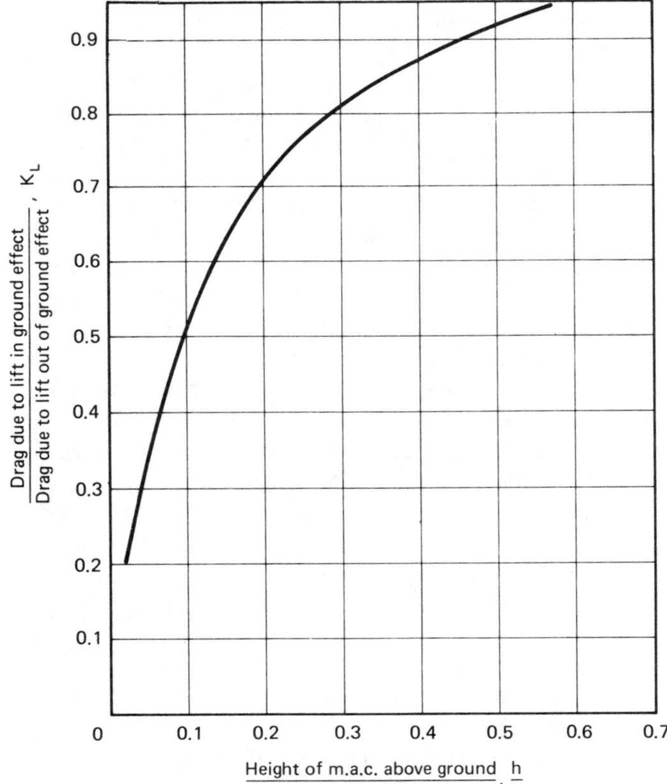

Figure 9.10 Effect of ground on drag due to lift.

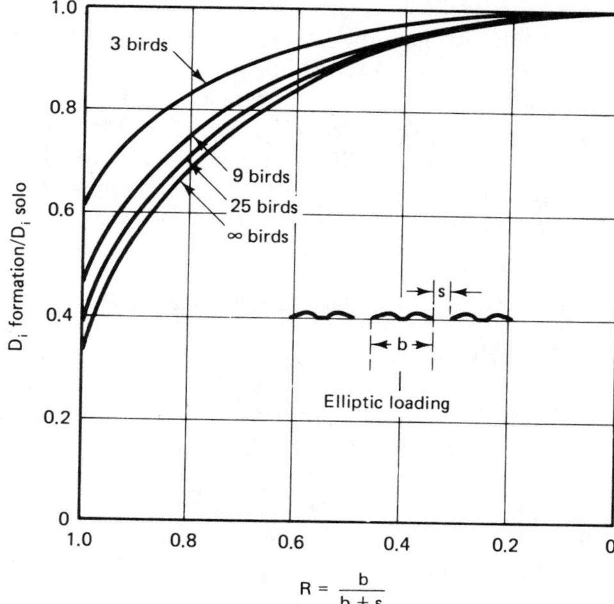

Figure 9.11 Induced drag reduction due to bird formation flight. From Lissaman and Shollenberger, "Formation Flight of Birds," *Science,* vol. 168, no. 3934, May 22, 1970, pp. 1003–5. Copyright 1970 by the American Association for the Advancement of Science.

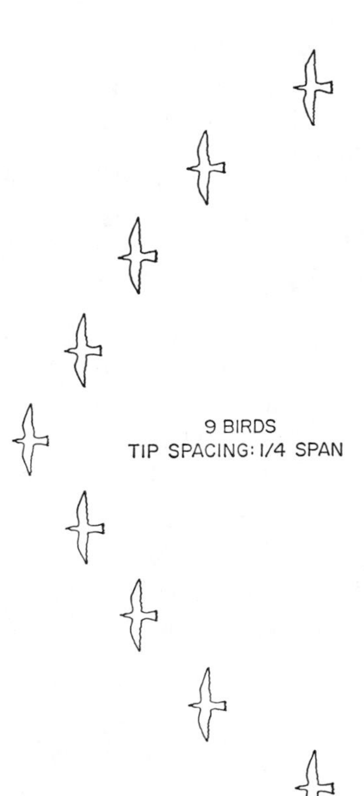

Figure 9.12 Optimal V formation. From Lissaman and Shollenberger, "Formation Flight of Birds," *Science,* vol. 168, no. 3934, May 22, 1970, pp. 1003–5. Copyright 1970 by the American Association for the Advancement of Science.

decreases drag or even creates a forward thrust. This is why birds fly in a V formation. Each bird gets the benefit of the upwash of the birds flying to its side. Figure 9.11 shows the ratio of the ideal induced drag of an entire bird formation to the sum of the individual induced drags of the birds flying solo. Figure 9.12 shows the ideal formation, close to a V, for which all birds have the same power required. According to Lissaman and Shollenberger (Ref. 9.6), proof that birds tend to fly in exactly this formation is lacking, but the general angle of observed V formations is similar to the calculated angle.

THE SEARCH FOR REDUCED INDUCED DRAG

The history of airplane design includes many attempts to significantly reduce induced drag. Unfortunately, induced drag (i.e. vortex drag) is a fact of life that will not go away. It is as omnipresent as death and taxes. For a planar wing, the Prandtl equation for an elliptical lift distribution, equation 9.9, defines the minimum induced drag. Less induced drag can be achieved for a given span by nonplanar configurations. A biplane is an example. Another example is a wing with endplates, vertical surfaces at the tip that act to increase effective aspect ratio. These reduce induced drag by permitting significant lift to extend all the way to the tips, instead of dropping to zero at the tips. One can think of endplate and other nonplanar configurations as affecting a deeper region of air and thereby obtaining the lift with less downward velocity imparted to the air. The result is less energy loss. Simple endplates have both parasite drag and weight, so the net drag gain has never justified their use on wings. Vertical tail surfaces placed at the tips of the horizontal tail do increase the effective aspect ratio of the horizontal tail. In this case, most of the weight and parasite drag would be present in a conventional vertical tail anyway. The most notable application of this use of endplates was the Lockheed Constellation (Figure 1.29). Even here the multiple vertical tails may have been chosen to reduce the overall airplane height in order to minimize the required height of the hangars, or possibly just to look unique.

Winglets, developed over the last 10 years, are a special form of endplate. Designed by modern theoretical methods, winglets are generally superior to the simple endplates studied decades ago. The detail design of winglets to optimize induced drag for a given wing weight is very complicated. The aerodynamic interference between winglet and wing, the induced drag of the winglet itself, and the bending moments introduced into the wing by both the higher lift on the outboard wing panel and the direct bending moment of the winglet itself are dependent on winglet shape, size, and angle of incidence. The details are beyond our scope, but some general conclusions can be discussed here.

First, winglets reduce induced drag. Second, the amount of induced drag reduction from well-designed winglets is about the same as extending the original wing span by an amount equal to approximately half of the winglet height. Third, the expected advantage of winglets is that the induced drag decrease is obtained with less wing bending moment and resultant wing weight increase than with a span increase. Objective general studies, (Ref. 9.7), starting with optimized wings, show that this gain is usually very small. The reduction in induced drag due to winglets, compared to a planar wing of greater span but the same wing weight, is about -1% to $+2\%$. The decision on whether to use winglets is likely to be dominated by special circumstances. Among these is a nonoptimum span loading on the original wing, minimum practical skin gauges already used on the outer panel so that the bending moment of the winglet attachment does not require more material, the desire to avoid a span increase for reasons of airport ramp or hanger space, and the possibility of flow separation at the winglet root at high lift coefficient because of flow interference. A last factor is sometimes the desire to look modern. Figure 9.13 shows an airplane equipped with winglets.

Example 9.1

An airplane with a wing area of 450 ft², a span of 60 ft, and a mean aerodynamic chord (m.a.c.) of 8 ft is flying at a pressure altitude of 6000 ft at a speed of 300 mph. The ambient temperature is 517°R. The wing has a geometric angle of attack of 4 degrees above the angle for zero lift. Assume that the two-dimensional lift curve slope is 95% of the theoretical value. Determine:

(a) Lift in pounds.
(b) Induced drag in pounds, assuming that the drag differs from that for the ideal elliptic distribution only by the correction factor u. Assume that $u = 0.98$.

Solution: Summarize the given data, always a useful procedure.

300 mph

4°

$h_p = 6000$ ft
$T_o = 517°$ R

Wing area $= 450 \text{ ft}^2$
Span $= 60$ ft
Aspect ratio, $AR = \dfrac{b^2}{S} = \dfrac{3600}{450}$
$= 8.0$
$\eta = 0.95$

At 6000 ft pressure altitude $p_0 = 1696 \text{ lb/ft}^2$.

$$\rho_0 = \frac{p_0}{RT_0} = \frac{1696}{(1718)(517)} = 0.001909 \text{ slug/ft}^3$$

$$q = \frac{\rho}{2}V^2 = \frac{0.001909}{2}\left(300 \times \frac{88}{60}\right)^2 = 184.79 \text{ lb/ft}^2$$

Figure 9.13 McDonnell-Douglas MD-11 transport airplane.

From equation 9.21,

$$C_{L_{\text{INC}}} = \frac{dC_L}{d\alpha}\alpha_a = \frac{a_0}{1 + 57.3a_0/(\pi \cdot \text{AR})}(\alpha_a)$$

where a_0 is the two-dimensional incompressible lift curve slope per degree,

$$a_0 = \frac{2\pi(0.95)}{57.3} = 0.1042$$

Then

$$C_{L_{\text{INC}}} = \frac{0.1042}{1 + (57.3)(0.1042)/\pi(8)}(\alpha_a) = (0.0842)(4) = 0.3368$$

Also

$$a = \sqrt{\gamma RT} = \sqrt{(1.4)(1718)(517)} = 1115.1 \text{ ft/s}$$

$$M = \frac{V}{a} = \frac{300 \times (88/60)}{1115.1} = 0.395$$

so that

$$C_L = \frac{C_{L_{\text{INC}}}}{\sqrt{1 - M^2}} = \frac{0.3368}{\sqrt{1 - (0.395)^2}} = 0.366$$

(a) Lift $L = C_L qS = (0.366)(184.79)(450) = \underline{30{,}435 \text{ lb}}$.
(b) From equation 9.12, with u given as 0.98,

$$C_{D_i} = \frac{C_L^2}{\pi \cdot \text{AR} \cdot u} = \frac{(0.366)^2}{\pi(8)(0.98)} = 0.00544$$

induced drag $D_i = C_{D_i} qS = (0.00544)(184.79)(450) = \underline{452.3 \text{ lb}}$

PROBLEMS

9.1. An airplane with a wing planform area of 650 ft^2 and a span of 80 ft is flying at a density altitude of 38,000 ft at a speed of 500 mph. The ambient temperature is 409°R. The gross weight is 52,000 lb. Determine:
(a) Lift coefficient.
(b) Induced drag in pounds, assuming that the drag due to lift differs from that for the ideal elliptical distribution only by having $u = 0.99$.
(c) What is the wing angle of attack (above zero lift). Assume $\eta = 0.95$.

9.2. An airplane with a wing area of 350 ft^2 and a span of 56 ft is flying at a pressure altitude of 10,000 ft at a speed of 320 mph. The ambient temperature is 495°R. The wing has a geometric angle of attack of 4 degrees, above the angle for zero lift. Assume the two-dimensional lift curve slope is 95% of the theoretical value. Determine:
(a) Lift in pounds.
(b) Induced drag in pounds.

9.3. Determine the variation of angle of attack with aspect ratio for an airplane with a cruise lift coefficient of 0.4. Study aspect ratios from 4 to 12. Graph the results. Assume $\eta = 0.97$.

9.4. Describe the cause of induced drag both in terms of the overall wing flow field and from the perspective of a single airfoil section on the wing.

9.5. On October 24, 1982, at 8:00 A.M., Marion Johnston observed a DC-8-63 airplane climbing out of runway 19L at San Francisco International Airport (SFO). The airplane had accelerated to a speed of 200 knots. Marion could see the wing tip vortices because the air held so much moisture, the morning humidity being very high, that the cooling of the air near the vortex centers condensed the water vapor. Marion realized that the helical vapor paths were portraits of circulation in action. Estimating that the takeoff weight of the DC-8-63 was 315,000 lb and knowing that a DC-8-63 has a wing area of 2883 ft^2 and a span of 148.3 ft, and that the SFO runway altitude was close to sea level, Marion calculated the value of the circulation around the wing in ft^2/s. Temperature was 59°F. Marion assumed that the wing lift was uniformly distributed along the span, an imperfect assumption but one that permitted a fairly good approximation. Also calculated was the ideal induced drag.

(a) What were Marion's answers for:

1. Circulation (ft^2/s).
2. Induced drag (lb).

(b) Derive an equation for circulation, Γ, in terms of C_L and $\bar{c}$, where $\bar{c}$ is the mean wing chord.

(c) What was the velocity in the trailing vortices at points 4 ft laterally from the vortex filament at a distance of one span behind the wing?

9.6. An airplane with a wing area of 450 ft^2 and a span of 70 ft is flying at a pressure altitude of 8000 ft at a speed of 280 mph. The ambient temperature is 500°R. The wing has a geometric angle of attack of 4 degrees above the angle for zero lift. Assume the two-dimensional lift curve slope is 95% of the theoretical value. Determine:

(a) Lift in pounds.

(b) Parasite drag (lb) if the value of C_{D_p} is 0.0190.

(c) The induced drag coefficient.

(d) The induced drag in pounds.

(e) The total drag in pounds.

(f) The ratio of lift to drag.

REFERENCES

9.1. Millikan, Clark B., *Aerodynamics of the Airplane,* Wiley, New York, 1941.

9.2. Kuethe, Arnold M., and Chow, Chuen-Yen, *Foundations of Aerodynamics,* Wiley, New York, 1976.

9.3. Bertin, John J., and Smith, Michael L., *Aerodynamics for Engineers,* Prentice-Hall, Englewood Cliffs, N.J., 1979.

9.4. Schrenk, O., *A Simple Approximation Method for Obtaining the Spanwise Lift Distribution,* NACA TM 1910, 1940.

9.5. Prandtl, L., *Applications of Modern Hydrodynamics to Aeronautics,* NACA Technical Report 116, 1921.

9.6. Lissaman, P. B. S., and Shollenberger, Carl A., "Formation Flight of Birds," *Science,* Vol. 168, No. 3934, May 22, 1970, pp. 1003–1005.

9.7. Kroo, Ilan, "A General Approach to Multiple Lifting Surface Design and Analysis," AIAA Paper 84-2507, presented to the Aircraft Design, Systems, and Operations Meeting, San Diego, Calif., October 1984.

9.8. Jones, Robert T., and Lasinski, T. A., "Effects of Winglets on the Induced Drag of Ideal Wing Shapes," NASA TM 81230, September 1980.

10

EFFECTS OF VISCOSITY

Our discussions thus far of the development of the theory and equations describing fluid flow, lift, and drag have repeatedly emphasized the assumption of a perfect or inviscid fluid. We have stressed that in the flow around wings and bodies this assumption is valid except in a narrow region immediately adjacent to the surface of a body. This region is called the *boundary layer.* Within the boundary layer, viscosity causes a significant drag in normal flight even when the basic external flow is almost unaffected. The presence of the boundary layer completely changes the total flow picture at extreme flight attitudes. The drag increase is called *parasite drag,* in coefficient form, C_{D_P}. The major flow changes are associated with separation of the flow from the body or wing. When this separation first occurs to a significant extent on a wing, the lift is decreased and the wing is said to be *stalled.*

The existence and the characteristics of a boundary layer are due to viscosity of the fluid. The nature and the effects of viscosity are the subjects of this chapter.

THE BOUNDARY LAYER

The starting point of boundary layer theory is the assumption that the layer of fluid immediately adjacent to a surface in a moving fluid sticks to the surface without any "slip" between that layer and the surface. This empirically verified assumption is called the *condition of no slip*. It is the opposite of the perfect fluid assumption that there is no retardation of the fluid next to a solid surface. The next layer of fluid has some small velocity and the next layer a still higher velocity. Figure 10.1 illustrates two schematic velocity profiles. The diagrams represent flow on a thin flat plate parallel to the flow so that the plate itself produces no disturbances to the fluid pressure or flow direction. A boundary layer velocity profile is a graph of the velocity v parallel to the plate, taken at some point along the plate, versus y, the distance perpendicular to the surface. The undisturbed velocity at this point is V. In Figure 10.1a with perfect fluid flow, the velocity V exists in the entire fluid region above the surface. In Figure 10.1b, a real fluid profile shows the velocity v to be zero at the surface and to increase slowly until, at some distance from the surface, the velocity equals the undisturbed velocity. The distance from the surface to the point where $v = V$ is the boundary layer thickness δ. If the viscosity is larger, layers of fluid farther from the surface will be slowed by the dragging action of the plate and the boundary layer will be thicker, as illustrated in Figure 10.2.

Near the leading edge of a body, the boundary layer is very thin, the fluid particles move in parallel planes, or laminae, and the interaction between planes is due only to the random molecular energy exchange. The boundary layer is then described as being *laminar*. At some point farther back on the surface, this steady shear flow becomes unstable. The flow undergoes a transition to a "churning" motion characterized by the presence of many very small vortices or eddies. These eddies greatly increase the interchange of energy between the layers, increasing the velocity near the wall and decreasing it farther from the wall. The point on the surface where the boundary layer flow becomes turbulent is called the *transition point*. The boundary layer is then said to be a *turbulent* boundary layer.

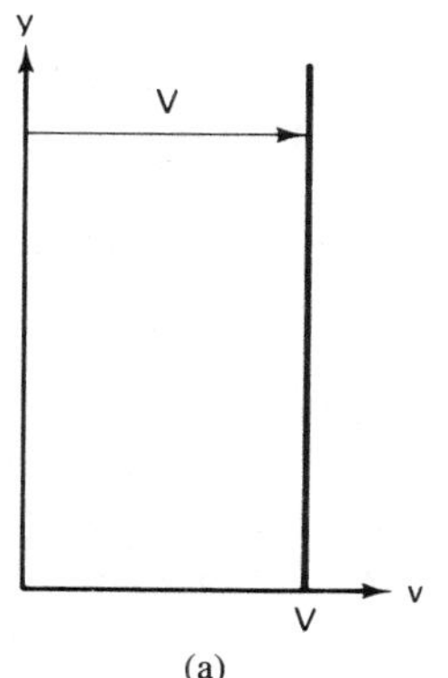

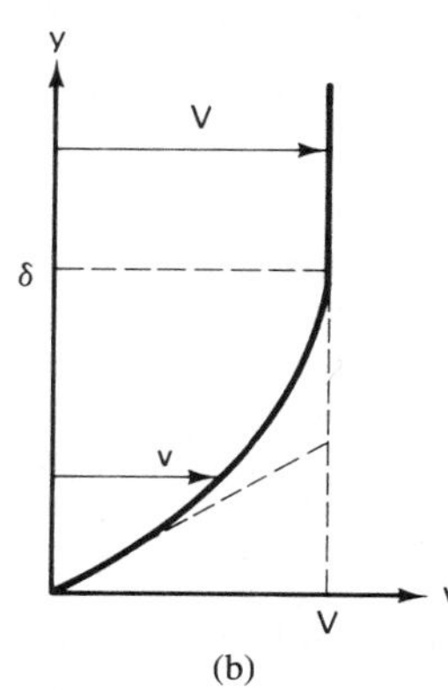

Figure 10.1 Boundary layer velocity profiles: (a) perfect fluid; (b) real fluid.

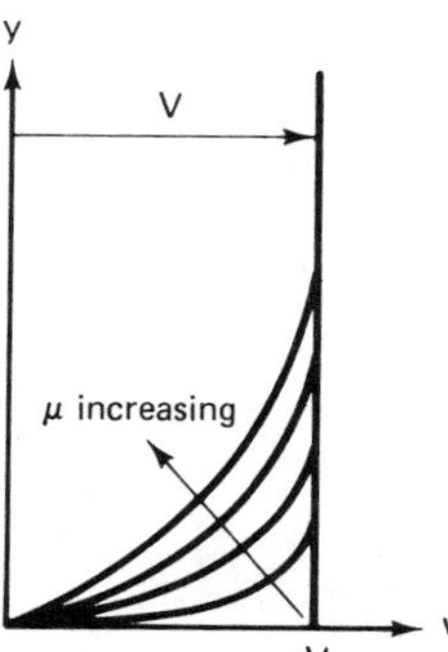

Figure 10.2 Effect of increasing viscosity on boundary layer velocity profiles.

Boundary layer theory is very complex. Solutions to the laminar boundary layer were found long ago, but the understanding of turbulent boundary layers is still incomplete. Nevertheless, the effects of boundary layers on the flow field and on the forces on wings and bodies are well known, although much of the knowledge is empirical. The shear stress τ at the wall is given by the Newtonian expression, as mentioned in Chapter 3:

$$\tau = \mu\left(\frac{dv}{dy}\right)_w$$

As the viscosity continually decreases (i.e., $\mu \rightarrow 0$), the boundary layer will get thinner and dv/dy will increase (i.e., $dv/dy \rightarrow \infty$). Thus, in the limit when $\mu = 0$,

$$\tau = 0 \cdot \infty$$

This is indeterminate in itself but has been shown for laminar boundary layers to be proportional to $1/\sqrt{\text{RN}}$. No matter how small the viscosity becomes, until μ is actually zero, a finite shear stress remains. It has also been shown that the boundary layer thickness δ varies with $1/\sqrt{\text{RN}}$. Specifically,

$$\frac{\delta}{l} \sim \frac{1}{\sqrt{\text{RN}}} \tag{10.1}$$

where l is the distance downstream from the leading edge of a body. These two concepts lead to the conclusion that at high values of Reynolds number the boundary layer may be very thin, so the flow around it is little changed by its presence, but the skin friction contribution to drag still exists.

On a flat plate or a well-streamlined body, the effect of the boundary layer is primarily the skin friction. However, there is an additional drag known as the pressure drag that results from the presence of the boundary layer. The slower-moving air in the boundary layer forces some of the streamlines outside the boundary layer to move away from the surface to make room for the boundary layer. The thickness of a boundary layer with zero velocity that would displace the flow outside the boundary layer by the same amount as the real boundary layer is called the *displacement thickness* (Figure 10.3). For a laminar boundary layer, the displacement thickness is about one-third of the thickness required to reach the velocity outside the layer. When the streamlines around the airfoil are moved away from the surface due to the boundary layer, the velocity and pressure distributions become those of a body with greater thickness at each station along the body. Since the boundary layer thickness increases along the body from the leading edge to the trailing edge, the external flow sees a body that does not completely close at the trailing edge, as shown in Figure 10.3. Therefore,

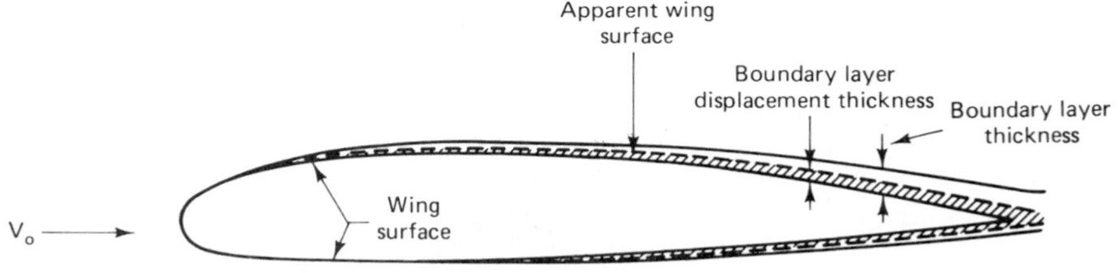

Figure 10.3 Boundary layer displacement thickness.

the flow does not produce a stagnation point at the trailing edge of the body, and all the pressures in the aft region are somewhat less positive than would be the case without a boundary layer. This lower pressure acting on an aft-facing surface causes a drag called a *pressure drag*. The magnitude of the pressure drag depends on the body thickness and the Reynolds number, both of which influence the boundary layer thickness. The reason for the influence of body thickness on boundary layer development will become clear in the next section.

BOUNDARY LAYERS ON BLUFF BODIES

The existence of a pressure gradient along a surface has a large and very significant effect on the development of a boundary layer. Consider the flow around a circular cylinder in a linear flow, as shown in Figure 10.4. Boundary layer profiles, greatly exaggerated in thickness, are shown in Figure 10.4 at typical points around the upper half of the cylinder. At each point the x-axis lies tangent to the cylinder at that point, while the y-axis is perpendicular to it. V represents the velocity in the perfect fluid flow just outside the boundary layer and V_0 is the freestream velocity far from the cylinder. As discussed in Chapter 6, the velocity varies from zero, or stagnation, at the front of the cylinder to a maximum at the top of the cylinder. The velocity then decreases along the aft face, reaching zero at the aft stagnation point. The corresponding pressures are maximum at the front and rear stagnation points and a minimum at the top of the cylinder.

At this time we must introduce a basic consideration in boundary layer theory: that the pressure is constant across a boundary layer and equal to the value just outside the boundary layer. This follows from the fact that the decreasing velocities within the boundary layer are due to viscous shearing effects and not due to the exchange of pressure and kinetic energy associated with velocity changes in the essentially perfect fluid region.

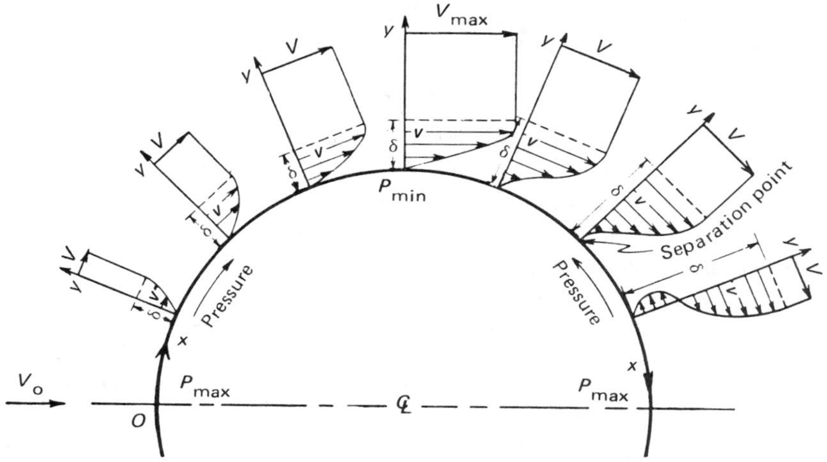

Figure 10.4 Development of boundary layer profiles around a circular cylinder. From C. B. Millikan, *Aerodynamics of the Airplane*. Reprinted by permission of John Wiley & Sons, New York.

The boundary layer profiles in Figure 10.4 start at the forward stagnation point with a shape similar to that of a flat plate. As the flow accelerates outside the boundary layer due to the decreasing pressure, the same *favorable pressure gradient* acts on the boundary layer fluid. Pressure gradients tending to accelerate the flow are called favorable gradients, while those tending to slow the flow are called *adverse gradients*. Layers of fluid closer to the surface are moving slower, and therefore the favorable gradients act on them for a longer time. Thus these slower layers acquire a larger velocity increment, since $\Delta v = a\Delta t$. The result is that the boundary layer profile becomes somewhat fuller, and the total thickness becomes less in a favorable pressure gradient. Aft of the top of the cylinder, the pressure starts to increase (i.e., the pressure gradient becomes adverse). In the stream just outside the boundary layer, the velocity and the associated kinetic energy of the fluid are just high enough to convey the fluid against the adverse gradient to the rear stagnation point. In the boundary layer the velocity and kinetic energy are less, however, since the viscous friction has been consuming kinetic energy which appears as heat. The fluid velocity reaches zero before getting to the rear stagnation point. Furthermore, the slower the fluid near the wall, the greater the deceleration it suffers because of the longer exposure to the adverse gradient. Thus the shape of the velocity profile changes, producing a reversal in curvature near the wall. When the profile becomes vertical at the wall (i.e., v becomes zero for some distance out from the wall), the flow can no longer continue around the body. This is known as the *separation point*. Beyond this, the flow along the surface will actually be reversed in direction, since no fluid flowing from the front of the body is filling the space. The main flow leaves the body and flows downstream, as indicated in Figure 10.5.

Downstream from the separation point behind the reversed-flow region, the flow is very unstable and breaks up into eddies. This eddying region is called the *wake*. The pressure on the body aft of the separation point is less than freestream instead of becoming positive as shown by perfect fluid theory. This results in "suction" on the rear of the body and causes a large drag force.

This general description of the flow around a circular cylinder applies to any body sufficiently blunt that the pressure gradient on the rear portion is high enough to cause separation. In fact, we can define a bluff body as one in which separation occurs well ahead of the trailing edge. Conversely, a streamline body is a body with a pressure increase downstream of the minimum pressure point sufficiently gradual so that separation does not occur until the trailing edge is almost reached. Airfoils at normal flight attitudes are streamline bodies. If, however, the airfoil angle of attack is increased until the pressure gradient downstream of the peak suction is so high that separation occurs well before the trailing edge, the airfoil is said to be stalled and it has really become a bluff body (Figure 10.6).

At the stall, the Kutta condition for the location of the rear stagnation point at the trailing edge is no longer met, and the circulation and therefore the lift decrease. Airfoils that stall with a separation that first occurs just forward of the trailing edge and slowly moves forward as the angle of attack is increased above the angle for initial stall have a relatively gradual loss of lift. Airfoils for which the first separation is far forward on the airfoil suffer a large immediate loss of lift, which leads to a more violent stall.

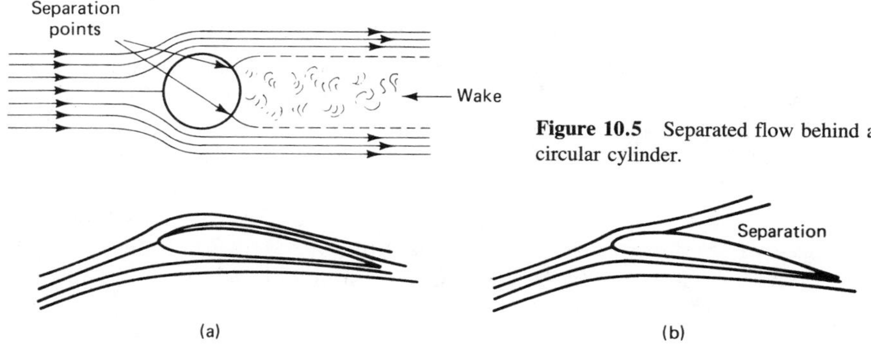

Figure 10.5 Separated flow behind a circular cylinder.

Figure 10.6 Flow around an airfoil: (a) unseparated flow; (b) separated, stalled flow.

The drag of a streamline body is roughly 80% skin friction, whereas the drag of a bluff body is mostly wake drag resulting from the deviation in aft surface pressures from the perfect fluid ideal.

TWO-DIMENSIONAL FLOW SUMMARY

A schematic summary of the various types of flow showing the effects of viscosity and circulation on two-dimensional flow is given in Figure 10.7. The picture would differ in three dimensions only by the additional of the induced drag due to downwash.

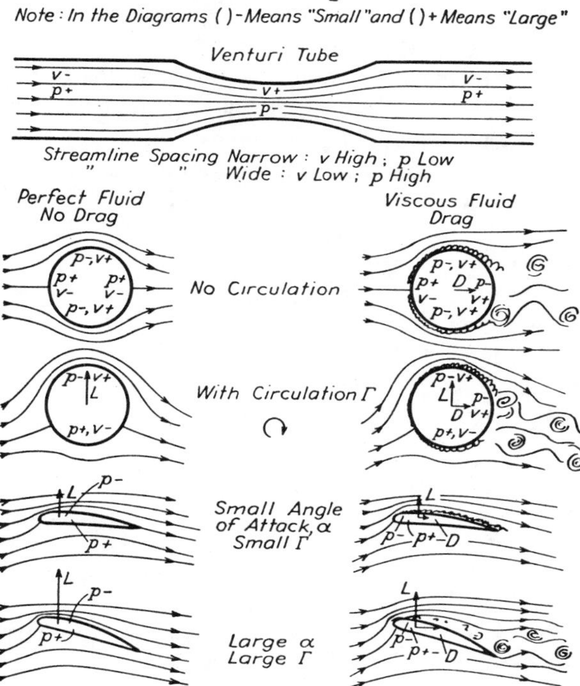

Figure 10.7 Summary of perfect and real fluid flow characteristics; two-dimensional motion. From C. B. Millikan, *Aerodynamics of the Airplane*. Reprinted by permission of John Wiley & Sons, New York.

THE CREATION OF CIRCULATION

There is one more important effect of viscosity that needs discussion. It is the role of viscosity in creating circulation. We have mentioned that Kutta assumed that flow over an airfoil must proceed smoothly to the trailing edge since the infinite velocities involved in flow from the lower to the upper surface around the sharp trailing edge were impossible. Actually, when an airfoil starts from rest, the flow around the trailing edge does occur for the first instant, as would be expected from potential theory results for flow without circulation. This type of flow was shown in Figure 8.9a. The enormous velocities create very high shear stresses in the boundary layer and cause the flow to separate and start a circulatory flow. A vortex is thus set up that trails aft in the flow. This vortex is called the *starting vortex*. Let us now consider a contour surrounding our airfoil and located quite far from the airfoil (Figure 10.8). The Thomson vortex theorem discussed in Chapter 9 stated that the circulation around any path enclosing the same fluid particles is independent of time. Before the airfoil in Figure 10.8 started moving, the circulation around the contour was zero. The contour moves with the same set of fluid particles but, being far from the airfoil, still encloses it at some time after the airfoil has been set in motion. A starting vortex now exists, but the circulation around the contour must still be zero. To satisfy this, an equal and opposite circulation must exist, and it is this opposite circulation that is generated around the airfoil. As long as any flow exists around the trailing edge, additional starting vortices are produced. When smooth flow is attained at the trailing edge, no further vortex strength is generated. The proper circulation around the airfoil has been established.

This description of the starting vortex and the associated formation of circulation around the airfoil has been shown to exist by Prandtl, who photographed the flow around an airfoil starting from rest. Figure 10.9 shows successive photographs of an airfoil moving through a fluid from the instant motion is started until the full development of smooth flow from the trailing edge. The initial rear stagnation point can be seen on the upper surface. The starting vortex trails behind the airfoil in successive photos as the flow changes to a smooth departure from the trailing edge.

It should be noted that, whenever the wing lift is changed either by angle of attack or speed change, starting vortices are created to adjust the wing circulation. When the lift decreases, the starting vortex will have the opposite direction (i.e., some of the bound vorticity will leave the wing and trail downstream). The reader does not need a high-speed moving picture camera and a wind tunnel to prove the validity of this discussion of the starting vortex and airfoil circulation. A reasonably curved spoon moving through a bowl of soup will be quite adequate. The soup should have chopped

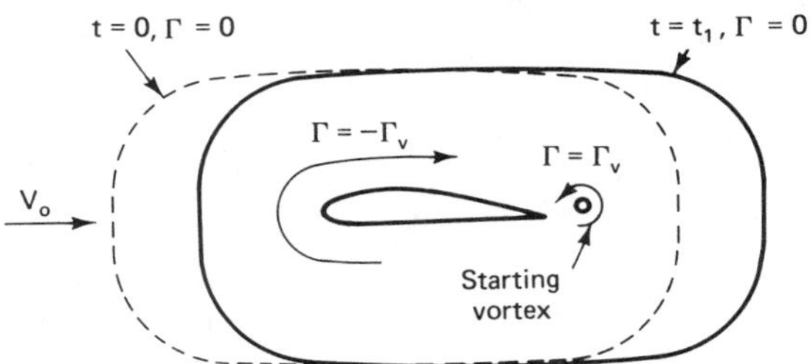

Figure 10.8 Fluid contour enclosing a starting airfoil.

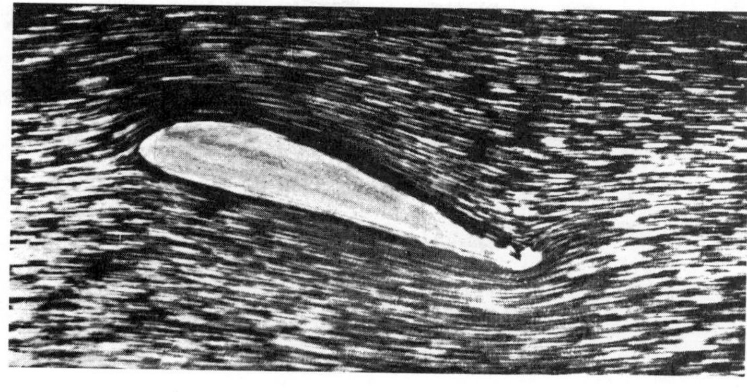

Figure 10.9 Consecutive stages of flow around an airfoil starting from rest. From Prandtl and Tietjens, *Applied Hydro- and Aeromechanics*. Reprinted by permission of McGraw-Hill Book Co., New York.

parsley, particles of fat, or the equivalent to aid the flow visualization. Moving the airfoil (i.e., the spoon) from one side of the bowl to another with a sudden start, a smooth motion, and then a sudden stop will show the starting vortex and, after the stop, the lifting vortex peeling off the spoon.

The starting vortex fills in the last gap in our finite wing discussion. We have said that a vortex must extend to infinity or form a closed path. From Figure 10.10,

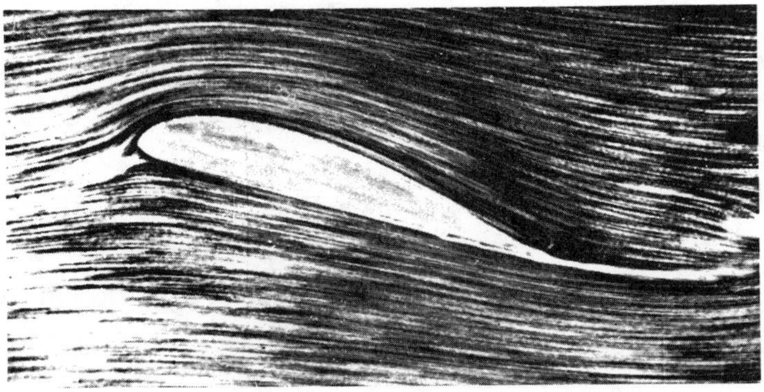

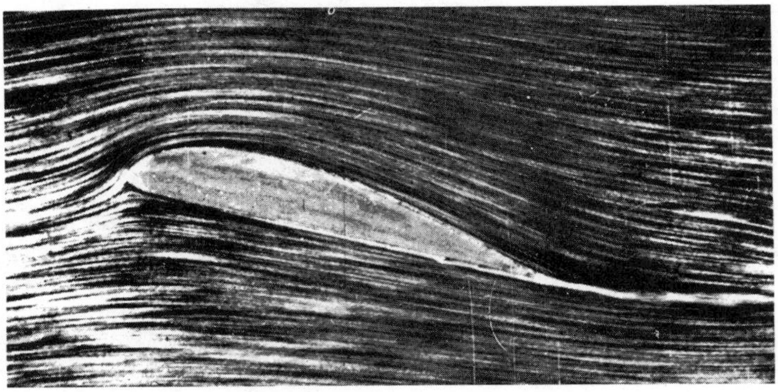

Figure 10.9 (Continued).

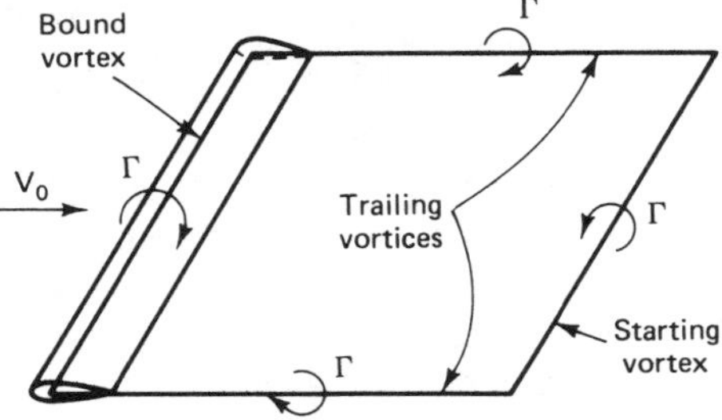

Figure 10.10 Idealized closed-path vortex pattern.

we can see that the starting vortex ends in the trailing vortices that stream forward to the bound vortex in the wing, thus forming the completely closed vortex picture.

LAMINAR AND TURBULENT BOUNDARY LAYERS: SKIN FRICTION

We have mentioned two types of boundary layers. At the leading edge of a surface, the adjacent layers, or *laminae,* move smoothly past one another, with each layer having increasing velocity as the distance from the boundary surface increases. These are laminar boundary layers. When the local Reynolds number, based on the distance from the leading edge, reaches some critical value (normally between 200,000 and 1,500,000), the *laminar flow* becomes unstable and is supplanted by the turbulent flow mentioned previously. Turbulent boundary layer flow has an irregular, eddying, or fluctuating nature. The eddies move irregularly, producing rapid fluctuations in local velocity that can only be measured by hot-wire anemometers or laser-based instrumentation. The average velocity is, however, of primary interest to the aerodynamicist. The eddies transfer momentum from layer to layer, with the result that the characteristics of the turbulent boundary layer are quite different from those of the laminar layer.

These differences are as follows:

1. The turbulent boundary layer is much thicker than the laminar one, and the thickness increases more rapidly with distance from the start of the layer.
2. The turbulent velocity profile is much fuller near the wall and flatter some distance from the wall, as shown in Figure 10.11.
3. The skin friction is much larger in turbulent than in laminar flow.

The process of changing from laminar to turbulent flow is called *transition,* and the Reynolds number at which laminar flow breaks down is known as the *transition critical Reynolds number,* $RN_{l_{crit}}$. The value of $RN_{l_{crit}}$ varies with turbulence in the freestream, surface roughness, vibration, and pressure distribution. If dp/dx is negative (i.e., a favorable gradient with decreasing pressure in the flow direction), the boundary layer is stabilized as discussed earlier, and $RN_{l_{crit}}$ increases. If the air faces

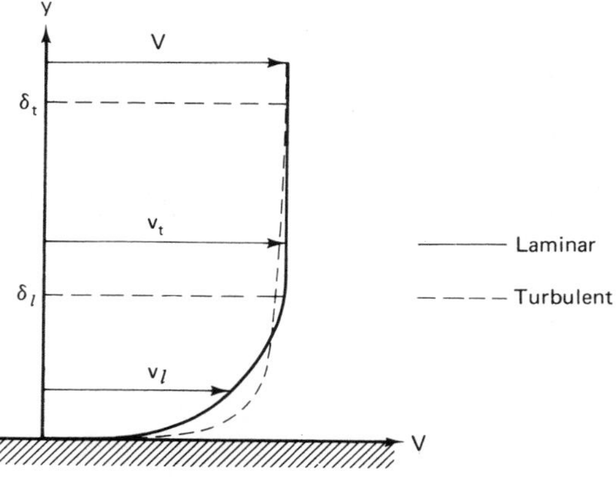

Figure 10.11 Laminar and turbulent velocity profiles.

an adverse gradient where dp/dx is positive, slowing the flow in the boundary layer, transition occurs earlier. For a given $RN_{l_{crit}}$, transition occurs farther along on the surface for a low RN per foot of length, and closer to the leading edge for a high RN per foot of length, as indicated in Figure 10.12.

Boundary layer theory gives the following important results.

Laminar Boundary Layers

$$\delta = \frac{5.2l}{\sqrt{RN_l}} \tag{10.2}$$

$$C_{f_l} = \frac{0.664}{\sqrt{RN_l}} \tag{10.3}$$

$$C_f = \frac{1.328}{\sqrt{RN}} \quad \text{Blasius friction law} \tag{10.4}$$

where

δ = boundary layer thickness

l = distance from the leading edge of the surface (i.e., from the start of the boundary layer)

RN_l = Reynolds number based on $l = \dfrac{\rho V_0 l}{\mu} = \dfrac{V_0 l}{\nu}$

C_{f_l} = local smooth flat-plate skin friction drag coefficient at a point l units from the leading edge $= \tau/q = \dfrac{\tau}{(\rho V_0^2)/2}$

C_f = total smooth flat-plate skin friction drag coefficient, the average drag coefficient over the surface from the leading to the trailing edge

RN = Reynolds number based on total surface length $= \dfrac{\rho V_0 L}{\mu}$

L = distance from the leading to the trailing edge

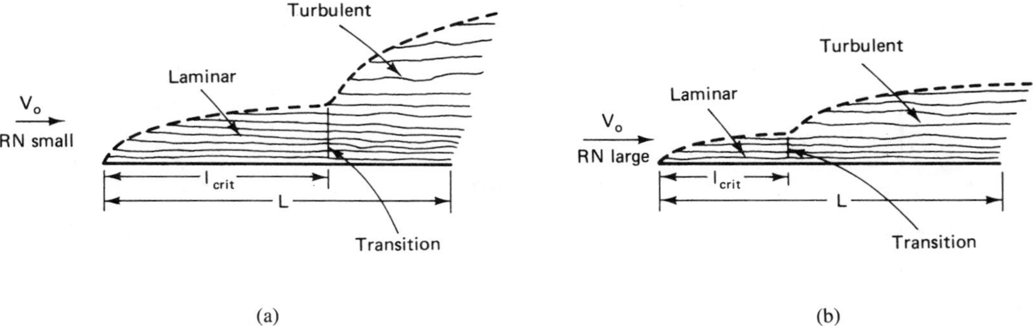

Figure 10.12 Effect of freestream velocity (or Reynolds number) on transition.

It is very important to understand that C_f is based on wetted area; that is,

$$\text{drag (lb)} = C_f \frac{\rho V_0^2}{2} S_{\text{wetted}}$$

where S_{wetted} is the surface area in contact with the air. For a surface such as a wing or tail, the wetted area equals twice the exposed planform area (i.e., the top and the bottom) increased by a small factor, say 1.02, to correct for curvature.

Turbulent Boundary Layers

$$\delta = \frac{0.37l}{(\mathrm{RN}_l)^{0.2}} \tag{10.5}$$

$$\frac{0.242}{\sqrt{C_f}} = \log_{10}(\mathrm{RN} \cdot C_f)$$ von Kármán analytical logarithmic equation with a constant developed by E. Schoenherr (Ref. 10.1) (10.6a)

or

$$C_f = \frac{0.455}{(\log_{10} \mathrm{RN})^{2.58}}$$ Schlichting empirical formula (Ref. 10.1) (10.6b)

The laminar equations are derived analytically and show that the boundary layer thickness δ grows with the square root of the distance from the leading edge, $\sqrt{l}$. The local skin friction coefficient C_{f_l} decreases with increase in $\sqrt{l}$. This follows from the fact that, as δ grows in going downstream, (dv/dy) in the boundary layer and, more specifically, $(dv/dy)_w$ at the wall decrease. Since $\tau = \mu(dv/dy)_w$, the skin friction decreases.

The turbulent boundary layer problem is much more difficult to solve theoretically. The equations derived from theory must be calibrated from experimental evidence. These somewhat approximate but very useful equations show that δ grows as $l^{4/5}$ rather than as $l^{1/2}$ for the laminar case. Similarly, turbulent skin friction is much higher, as shown in Figure 10.13.

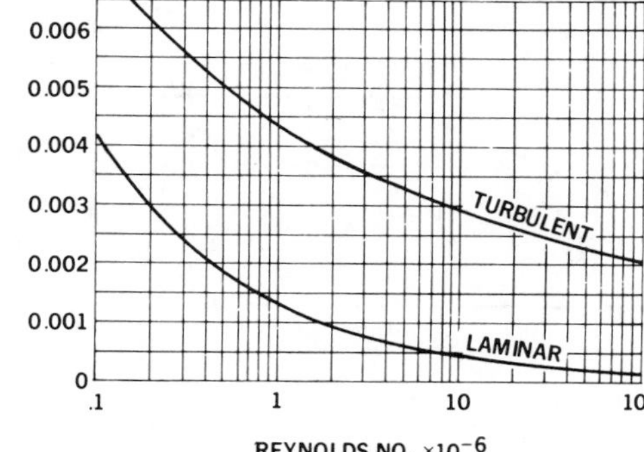

Figure 10.13 Laminar and turbulent smooth flat-plate skin friction curves.

The effective viscosity for the turbulent boundary layer case is strongly affected by the eddying activity, which transfers momentum between layers. However, RN_l or RN is still based on the fluid viscosity μ. Note that, although μ decreases with higher temperature in liquids, it increases with higher temperatures in gases. For air, as T increases, μ increases. For air at standard sea-level temperature,

$$\mu = 1.7894 \times 10^{-5} \text{ kg/m s} = 3.7373 \times 10^{-7} \text{ lb-s/ft}^2$$

Note that lb-s/ft^2 = slug/ft-s.

The variation of μ with temperature is given in Figure 10.14. The equation for μ versus °R is

$$10^{10}\mu = 0.3170(°\text{R})^{3/2}\left(\frac{734.7}{°\text{R} + 216}\right) \tag{10.7}$$

where μ is in lb-s/ft^2. Viscosity is given in a normalized form in Figure 10.15.

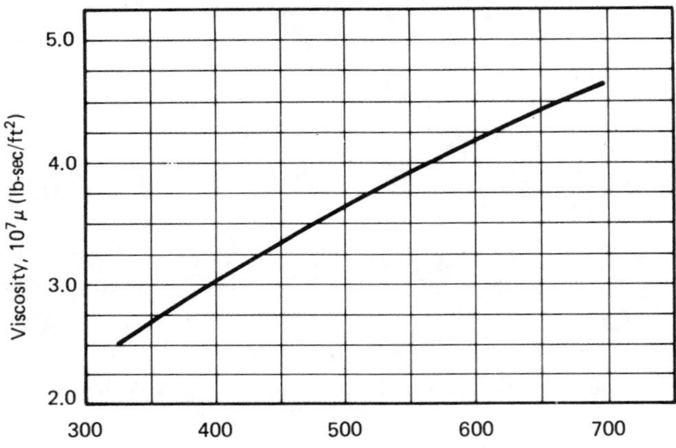

Figure 10.14 Absolute viscosity of air.

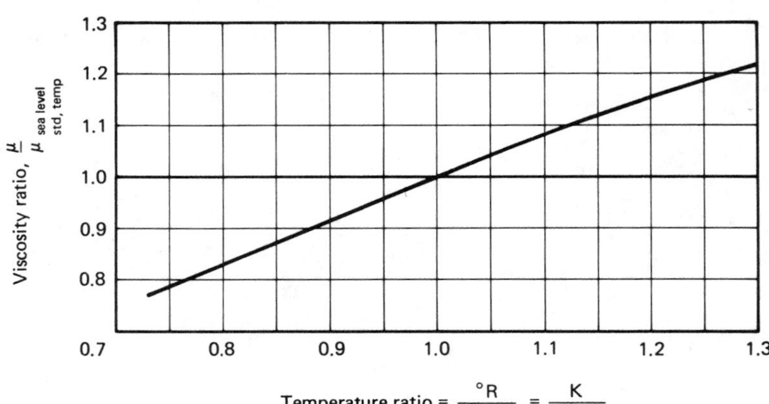

Figure 10.15 Relative viscosity versus temperature ratio.

THE NATURE OF THE REYNOLDS NUMBER

It can be shown that the Reynolds number has the nature of the ratio of inertia forces to viscous forces. Thus

$$\mathrm{RN} = \frac{\rho VL}{\mu} = \frac{\rho VL}{\mu}\left(\frac{V}{V}\right) = \frac{\rho V^2}{\mu(V/L)}$$

ρV^2 is proportional to the kinetic energy of the fluid. (V/L) is related to V/δ, which is similar to dV/dy. Then $\mu(V/L)$ is associated with $\mu(dV/dy) = \tau$. Thus the higher the Reynolds number, the greater the ratio of dynamic forces to viscous forces. As $\mathrm{RN} \to \infty$, the importance of $\mu \to 0$.

An essential part of the process of drag determination is calculating the Reynolds number. Since the given conditions often involve nonstandard temperatures, it is generally necessary to use equation 3.7 or 5.13 to compute the Reynolds number. For standard atmospheric conditions, however, Figure 10.16 can be used to determine graphically the Reynolds number per foot of characteristic length. It is particularly helpful in quickly estimating Reynolds number for various-size aircraft at different speed and altitude conditions.

Example 10.1

A Piper Cherokee is flying at 6000 ft altitude at 120 mph on a standard day. The gross weight is 1700 lb. The wing has an area of 160 ft^2, with 85% of the wing area exposed, is of rectangular planform, and has a span of 30 ft. Assuming completely turbulent boundary layer flow, determine:

(a) Flat-plate skin friction drag of the wing in both coefficient and force terms.
(b) Boundary layer thickness at the trailing edge of the wing.
(c) Lift coefficient.
(d) Induced drag (assuming elliptical lift distribution) in both coefficient and force terms.

Solution: At 6000 ft altitude on a standard day, $T = 497.3°\mathrm{R}$, $\rho = 0.001987$, and $\nu = 0.0001820$ from Table A.2. The wing chord = area/span, since the platform is rectangular.

$$\text{Reynolds number} = \frac{\rho VL}{\mu} = \frac{VL}{\nu} = \frac{(120 \times 88/60)(160/30)}{0.0001820} = 5{,}157{,}510$$

or $\mu = 3.62 \times 10^{-7}$ from Figure 10.14, and

$$\mathrm{RN} = \frac{(0.001987)(120 \times 88/60)(160/30)}{0.000000362} = 5{,}152{,}295$$

Then $C_f = 0.0033$ from Figure 10.13, or

$$C_f = \frac{0.455}{(\log_{10} \mathrm{RN})^{2.58}} = \frac{0.455}{(\log_{10} 5{,}157{,}510)^{2.58}} = 0.00335$$

$$\text{(a) } C_{D_p} = \frac{C_f S_{\text{wetted}}}{S_{\text{wing}}} = \frac{(0.00335)(160 \times 0.85 \times 2 \times 1.02)}{160} = \underline{0.00581}$$

Note that the wetted area is reduced by the factor 0.85, since 85% of the wing is exposed. The other 15% of the nominal or gross wing area is inside the fuselage and is not scrubbed by the air. The 1.02 factor accounts for the airfoil curvature.

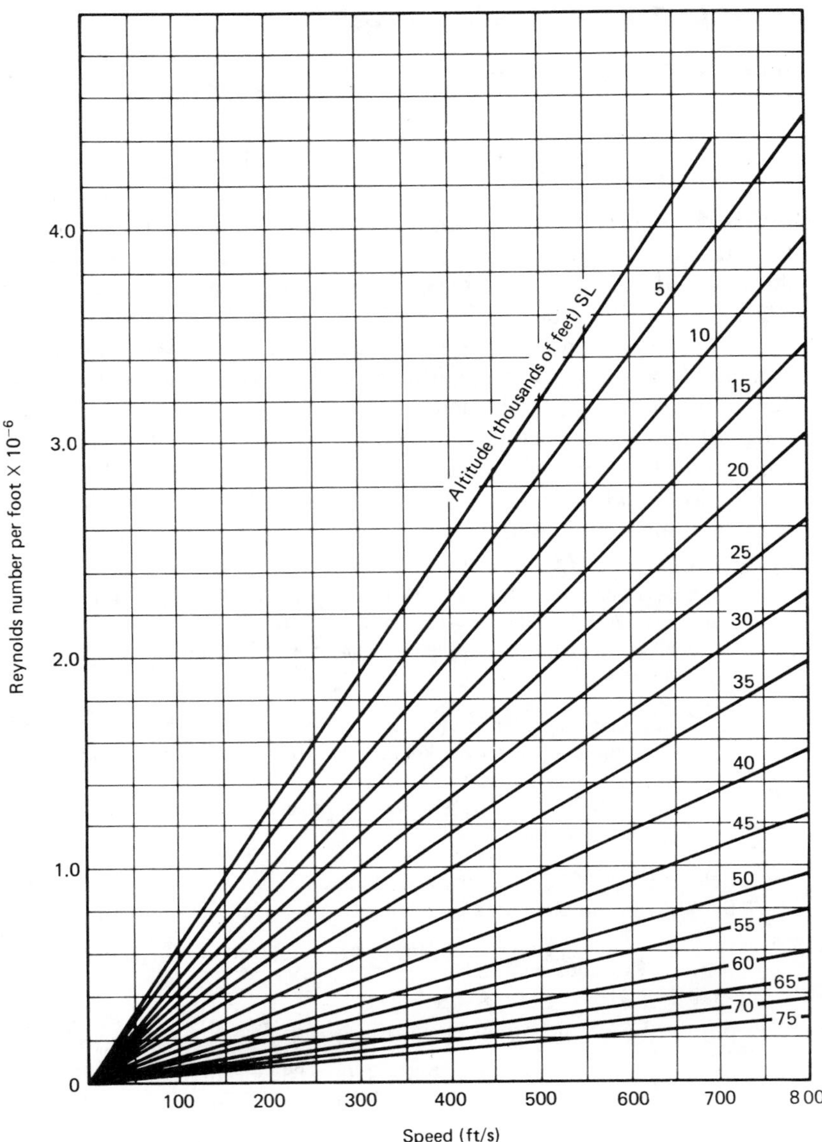

Figure 10.16 Reynolds number per foot as a function of altitude and speed for the standard atmosphere.

$$D_p = C_{D_p}\frac{\rho}{2}V^2S = 0.00581\left(\frac{0.001987}{2}\right)\left(120 \times \frac{88}{60}\right)^2(160)$$

$$= \underline{28.60 \text{ lb}} \qquad \text{smooth flat-plate friction drag of wing}$$

(b) $\delta = 0.37L/(\text{RN})^{0.2}$ for the turbulent boundary layer. At the trailing edge of the wing, L = chord = $160/30 = 5.33$ ft.

$$\delta = \frac{(0.37)(5.33)}{(5{,}157{,}510)^{0.2}} = \underline{0.0896 \text{ ft}} = \underline{1.076 \text{ in.}}$$

$$\text{(c) } C_L = \frac{W}{(\rho/2)V^2S} = \frac{1700}{(0.001987/2)(120 \times 88/60)^2(160)} = \underline{0.345}$$

$$\text{(d) } C_{D_i} = \frac{C_L^2}{\pi \cdot \text{AR}} = \frac{(0.345)^2}{\pi(5.625)} = \underline{0.0067}$$

Note that $\text{AR} = b^2/S = 30^2/160 = 5.625$.

$$D_i = 0.0067\left(\frac{0.001987}{2}\right)\left(120 \times \frac{88}{60}\right)^2(160)$$

$$= \underline{32.99 \text{ lb}}$$

EFFECT OF TURBULENT BOUNDARY LAYERS ON SEPARATION

Although turbulent boundary layers have much higher skin friction values than those of laminar boundary layers, they have some redeeming features. In Figure 10.4, the process of boundary layer separation on a bluff body was shown schematically. If we assume that the velocity profiles were specifically drawn for laminar layers, the turbulent case would have higher velocities near the wall, as indicated in Figure 10.11. The higher kinetic energy of the fluid near the wall would permit the flow to proceed farther along the body before the velocity near the wall slows to zero and separation occurs. The separated region and the associated drag will therefore be smaller, as shown in Figure 10.17. The drag coefficients of cylinders and spheres are shown versus Reynolds number in Figure 10.18. The drag coefficients are based on the frontal areas of the bodies and show a rapid drop above RN = 200,000, reaching a lower plateau above RN = 500,000. Transition from laminar to turbulent flow at the separation point is starting at the 200,000 region, and by RN = 500,000 to 600,000 the flow is fully turbulent at separation. There are practical cases, such as a small circular strut on a landing gear, in which it is beneficial to add roughness to ensure the presence of a turbulent boundary layer before the separation point. Dimples on golf balls serve the same function, ensuring that a turbulent boundary layer exists to reduce the separated area and the drag.

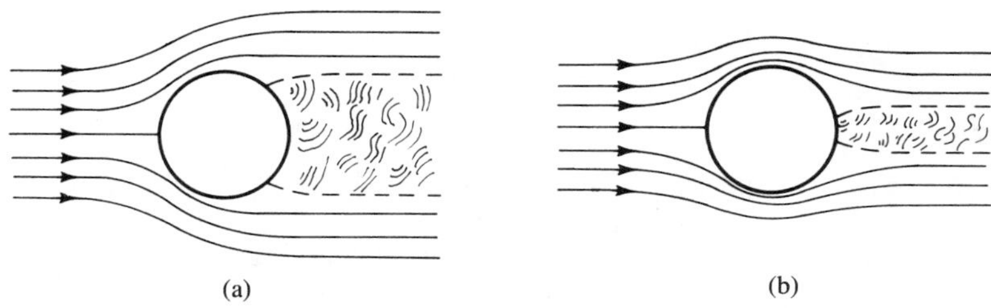

Figure 10.17 Flows around a sphere having laminar and turbulent boundary layers: (a) laminar boundary layer, early separation; (b) turbulent boundary layer, delayed separation.

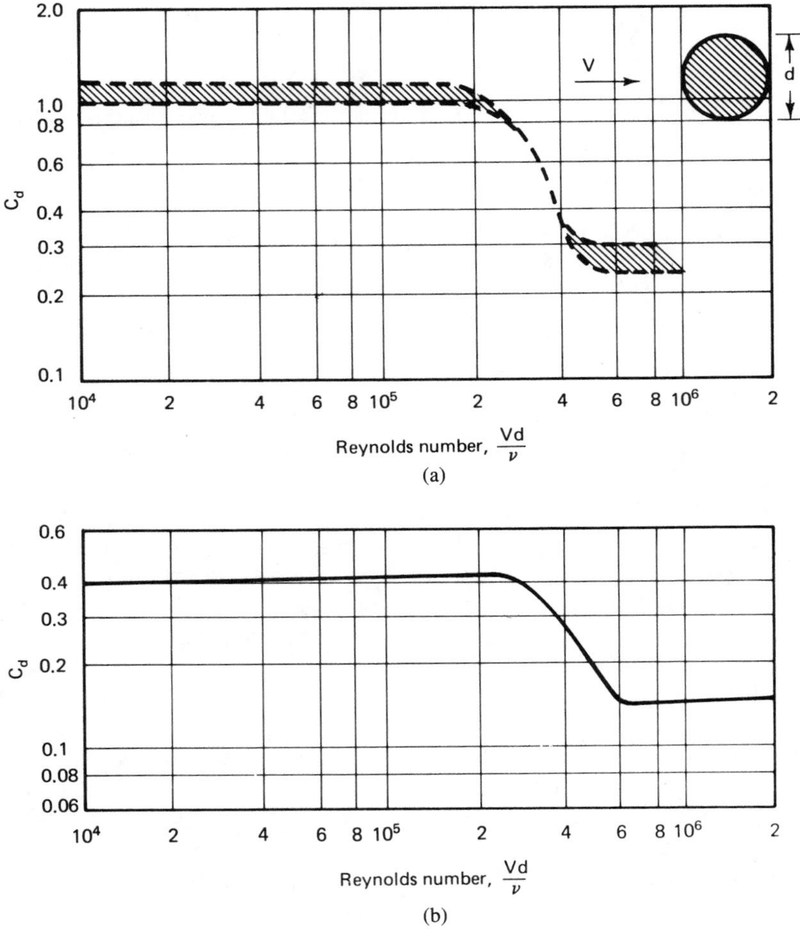

Figure 10.18 Effect of Reynolds number on the drag coefficients of cylinders and spheres: (a) two-dimensional circular cylinders; (b) spheres. From Barnes McCormick, *Aerodynamics, Aeronautics, and Flight Mechanics*. Reprinted by permission of John Wiley & Sons, New York.

PROBLEMS

10.1. Consider a high-speed subsonic wind tunnel. The conditions in the large-diameter section upstream of the test section are $V = 228$ mph and $T = 540°R$. At the test section, $T = 473°R$ and the pressure is two standard sea-level atmospheres.

(a) What is the Mach number in the test section?

(b) What is the pressure in the large upstream section?

(c) If a wind tunnel model, placed in the test section, has a wing chord of 12 in., what is the test Reynolds number based on that chord?

(d) What is the overall smooth flat-plate skin friction coefficient of the model wing if the boundary layer is turbulent? What is the C_{D_p}?
(e) Determine the boundary layer thickness at the trailing edge of the model wing.
(f) Calculate the skin friction drag in pounds. Model span is 3 ft.

10.2. Consider a high-speed subsonic wind tunnel. The conditions in the large-diameter section upstream of the test section are $V = 102$ m/s and $T = 320$ K. At the test section, $T = 280$ K and the pressure is two standard sea-level atmospheres.
(a) What is the Mach number in the test section?
(b) What is the pressure in the large upstream section?
(c) If a wind tunnel model, placed in the test section, has a wing chord of 30 cm, what is the test Reynolds number based on that chord?
(d) What is the boundary layer thickness at the trailing edge of the model wing if the boundary layer is turbulent?
(e) What is the value of the overall smooth flat-plate skin friction coefficient, C_f?
(f) What is the skin friction drag in newtons? (The model span is 1 m.)

10.3. A Piper Cherokee is flying at 4000 ft altitude at 120 mph on a standard day. The gross weight is 1850 lb. The wing has an area of 160 ft^2, 85% of which is exposed, is of rectangular planform, and has a span of 30 ft. Assume completely turbulent boundary layer flow. Pressure drag and surface roughness add 25% to wing parasite drag (above pure skin friction). Wing parasite drag is 38% of the total airplane parasite drag.
(a) Determine the smooth flat-plate skin friction drag of the wing in both coefficient and force terms.
(b) Determine the total parasite drag of the wing and the airplane in both coefficient and force terms.
(c) Calculate the induced drag, assuming $u = 0.985$.
(d) What is the boundary layer thickness of the wing trailing edge?

10.4. A Cessna 310 twin engine airplane is flying at 7000 ft altitude at 205 mph. Outside air temperature is 50°F. The gross wing area is 179 ft^2 and the span is 37 ft. The mean aerodynamic chord (m.a.c.), upon which the wing Reynolds number is based, is 5% greater than the average chord and 82% of the wing is exposed. The boundary layer is completely turbulent. Pressure drag and surface roughness add 33% to the wing drag (above pure skin friction). Assume wing parasite drag is 40% of the total. The airplane gross weight is 4800 lb.
(a) Determine the flat-plate skin friction drag of the wing in both coefficient and force terms.
(b) Determine the total parasite drag of the wing and the airplane in both coefficient and force terms.
(c) Determine the induced drag in both coefficient and force terms. Assume $u = 0.975$.
(d) What is the boundary layer thickness at the trailing edge of that section of the wing on which the m.a.c. lies?
(e) What is the absolute angle of attack in degrees from the angle for zero lift? Assume the airfoil lift curve η to be 0.96.

10.5. A DC-8-63 is flying at 35,000 ft on a standard day at $M = 0.82$. The fuselage is 183 ft long. Assuming that the turbulent boundary layer grows as on a smooth flat plate, find the boundary layer thickness at the rear of the fuselage.

10.6. Repeat Problem 10.3, parts (a), (b), and (d), with a temperature 25°F above standard at the cruise altitude.

10.7. Repeat Problem 10.3, parts (a), (b), and (d), assuming that a laminar boundary layer exists. What are the ratios of the laminar to the turbulent skin friction drag and boundary layer thickness?

REFERENCES

10.1. von Kármán, Theodore, "Turbulence and Skin Friction," *Journal of the Aeronautical Sciences,* Vol. 1, No. 1, 1934.

10.2. Millikan, Clark B., *Aerodynamics of the Airplane,* Wiley, New York, 1941.

10.3. Prandtl, L., and Tietjens, O. G., *Applied Hydro- and Aeromechanics,* McGraw-Hill, New York, 1934.

10.4. McCormick, Barnes W., *Aerodynamics, Aeronautics, and Flight Mechanics,* Wiley, New York, 1979.

11

DETERMINATION OF TOTAL INCOMPRESSIBLE DRAG

Figure 11.1 shows typical three-component wind tunnel results for an airplane. The three components almost always measured are lift, drag, and pitching moment. The other three components, not shown here, are the side force, rolling moment, and yawing moment. Figure 11.1 summarizes and verifies much of the theory we have developed so far. First, the lift is linear in angle of attack except at very high angles of attack, where the boundary layer is beginning to thicken markedly at the trailing edge, and then finally at the stall, when the lift actually decreases. Second, there is a drag even at zero lift. This is mostly the skin friction plus a little pressure drag due to boundary layer thickness. Third, the drag variation with lift is parabolic as C_D varies with C_L^2. The linear curve of pitching moment coefficient versus lift coefficient verifies that the lift due to angle of attack acts through one point on the chord. We have previously noted that wing lift due to angle of attack theoretically acts at the quarter-chord point.

One of the most important tasks of the aerodynamicist is the prediction of drag. In Chapter 9, we learned how to estimate the induced drag (i.e., the drag due to lift) for an elliptical wing or for a wing designed to achieve close to elliptical loading no matter what its planform shape may be. The planform must be reasonable, of course. In Chapter 10, we gave the equations for the skin friction drag of a smooth flat plate for either laminar or turbulent boundary layer flow. In this chapter, we add to our growing collection of tools the additional data necessary to account for the variation of boundary layer drag with body thickness and angle of attack. The use of these methods for a complete airplane drag prediction will be described. Since our goal is

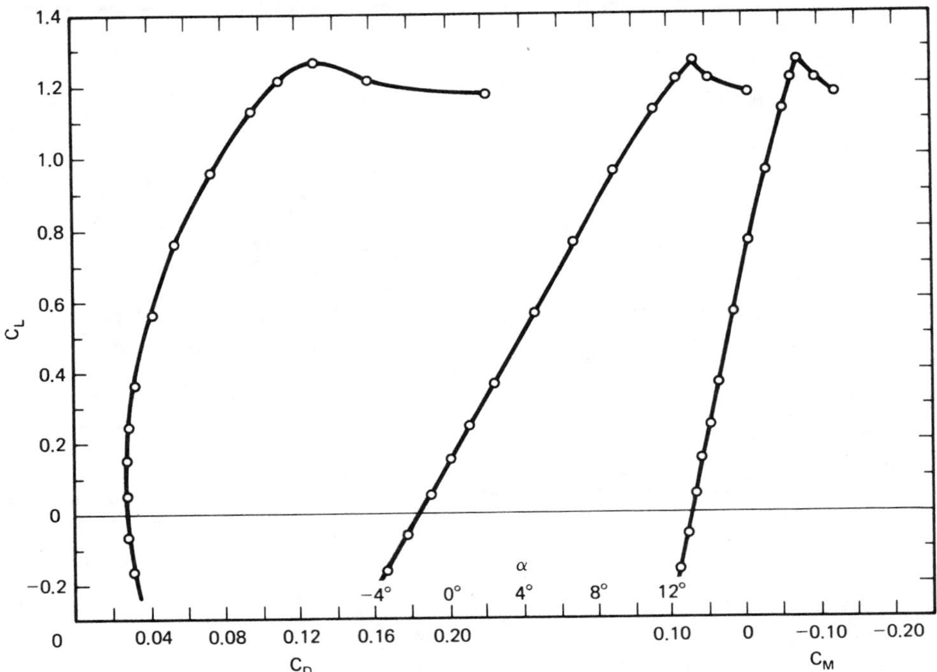

Figure 11.1 Typical three-component wind tunnel test results.

to supply as clear an understanding of these methods as possible, we will make a few simplifications. Thus the method will not yield the most accurate drag prediction possible, but it will come quite close, with a great reduction in complexity.

PARASITE DRAG

Although laminar boundary layer skin friction is much lower, in practical cases the boundary layer is almost entirely turbulent. Exceptions are very slow, small, very well maintained aircraft such as sailplanes. Maintenance of laminar flow requires a very smooth surface, a favorable pressure gradient, and a low Reynolds number. Sucking off the boundary layer through porous or slotted wing surfaces can also maintain laminar flow. The problems of building such porous laminar flow control (LFC) wings at an acceptable cost and keeping them free of dirt and insects have not yet been solved. Recent research has shown that new composite materials such as graphite-epoxy or fiberglass may provide the very smooth surfaces and freedom from waviness required for laminar flow. The dirt and insect problems still must be solved. For the present, it is realistic to assume a turbulent boundary layer.

The turbulent skin friction is usually calculated from equations such as 10.6 or read from a graph such as Figure 11.2. The curve in Figure 11.2 labeled "smooth surfaces" gives the turbulent skin friction coefficient based on wetted area for a smooth

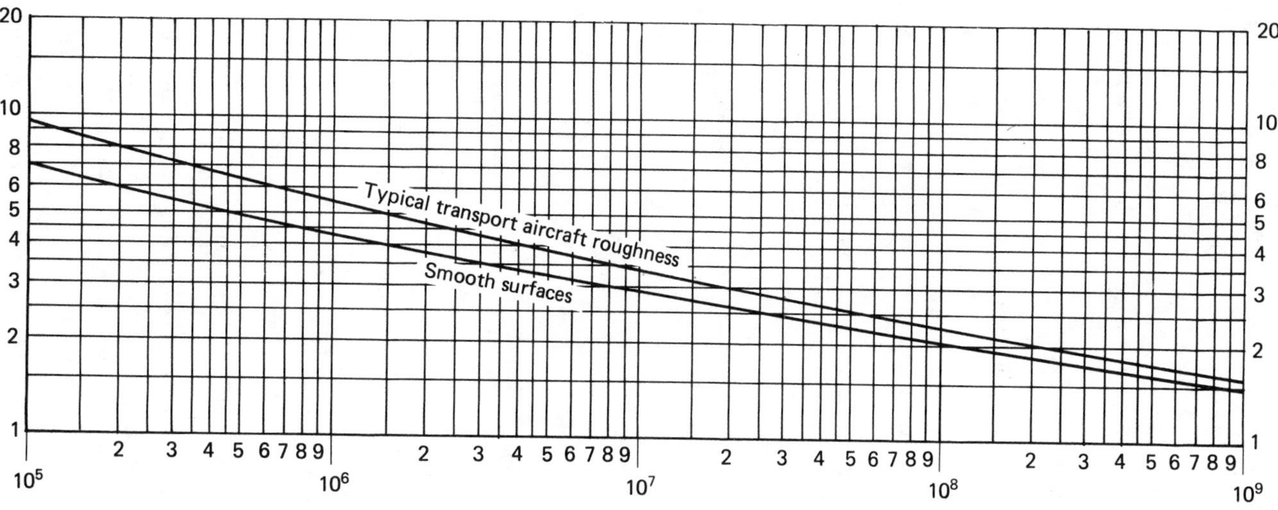

Figure 11.2 Flat-plate skin friction coefficient; turbulent boundary layer; $M = 0.50$.

flat plate according to the von Kármán approximation (equation 10.6 a) adjusted to a Mach number of 0.5 (see Figure 12.5). Airplane surfaces do not have ideal smoothness, however. Rivets, skin joints, windows, and doors are all less than perfectly smooth. Since explicitly accounting for the drag of every little disturbance is not possible, an *equivalent sand grain roughness* is often used. This relates the roughness associated with a certain type of construction with the drag increment produced by a uniform coverage of the surface with sand grains of a certain size. Some small roughness is permissible without any drag penalty because the grains lie within what is known as the *laminar sublayer* of the turbulent boundary layer. The determination of the equivalent roughness and the appropriate skin friction penalty is more complex than is suitable for our purposes. Therefore, we have included on Figure 11.2 a single typical curve with roughness appropriate to transport aircraft to indicate the nature of the effect. Transport aircraft have an equivalent sand grain roughness of about 0.0016 in. *In using the graph for practical purposes, the curve with roughness should generally be used.* Less sophisticated aircraft could have skin friction drag higher by 10% to 20%.

Determining the skin friction drag from Figure 11.2 involves computing the Reynolds number, $\rho VL/\mu$. The characteristic length L is the length of the body for fuselages and nacelles and the mean aerodynamic chord, m.a.c. or $\bar{c}$, for wings and other surfaces. The m.a.c. is the chord of an imaginary wing, with constant chord, having the same aerodynamic characteristics as the actual wing. The m.a.c. used for skin friction is the mean chord of the *exposed* wing weighted for the area affected by the chord of each spanwise increment. For a tapered wing with straight leading and trailing edges, the m.a.c. is given by

$$\text{m.a.c.} = \frac{2}{3}\left(C_R + C_T - \frac{C_R C_T}{C_R + C_T}\right) \tag{11.1}$$

$$= \frac{2}{3} C_R\left(1 + \sigma - \frac{\sigma}{1 + \sigma}\right) \tag{11.2}$$

where

C_R = exposed root chord at centerline of the wing or the side of the fuselage as appropriate
C_T = tip chord
σ = taper ratio = tip chord/root chord

Note that for stability and control purposes (see Chapter 16) m.a.c. is defined by the gross wing area, including the area buried in the fuselage.

In addition to skin friction, pressure drag exists. Because the boundary layer effectively changes the shape of the airfoil or body, particularly over the aft portion where the boundary layer is thicker, the theoretical, perfect fluid pressures are not quite obtained. This largely appears as a failure to obtain as high (positive) pressures over the rear portion of the body as theory would predict—and as would be obtained if the boundary layer were absent. The result is a drag, a loss of air pressure pushing on an aft-facing surface. This pressure drag is small at low normal flight angles of attack but is still significant.

We may consider the pressure drag as having two portions. The first portion occurs even at zero lift and depends on the airfoil thickness ratio or the body fineness ratio. A thicker airfoil or body will have a greater adverse pressure gradient on the rear portion. As discussed in Chapter 10, this will thicken the boundary layer and increase the pressure drag. Also, the thicker airfoil or body will provide increased vertical projected area on which the pressure increment can act to produce drag. The second part of the pressure drag arises from the existence of lift on the wing. The higher the lift, the greater will be the adverse pressure gradient on the upper surface, the thicker the boundary layer, and the greater the pressure drag on the upper surface.

There is a third drag source that we include with pressure drag for convenience. The thickness of a body causes local velocities over the body that are higher than free-stream. This higher velocity increases the skin friction drag. The drag effect of the higher local velocities is a linear function of the thickness ratio.

Therefore, parasite drag D_p, the portion of drag independent of lift, consists of friction drag plus pressure drag and is calculated as

$$D_p = C_{D_p} q S_{\text{REF}} \tag{11.3}$$

where

$$C_{D_p} = \sum_i \frac{K_i C_{f_i} S_{\text{WET}_i}}{S_{\text{REF}}} \tag{11.4}$$

D_p = parasite drag, lb

C_f = skin friction coefficient from Figure 11.2

S_{WET} = wetted area (ft^2), the actual area in contact with the air; for surfaces, the wetted area is twice the exposed planform area plus a small correction (about 2%) for airfoil curvature; then $S_{\text{WET}} = S_{\text{exposed}} \times 2 \times 1.02$

K = correction factor for pressure drag and increased local velocities, a function of airfoil or body thickness expressed as the ratio of maximum thickness to chord t/c for airfoils and length to maximum diameter L/D for bodies; K is primarily empirical and is found from Figure 11.3 for airfoils and Figure 11.4 for bodies; for surfaces, K is also a function of the wing sweepback angle, $\Lambda_{c/4}$ (see Chapter 12)

$(\cdot)_i$ = signifies the ith component (i.e., wing, nacelle, horizontal tail, etc.)

S_{REF} = reference area, usually the wing planform area (ft^2)

An additional possible source of drag is interference drag, the drag increase that may occur due to higher induced velocities or adverse gradients caused by the effect of the flow field of one airplane component on another. Interference drag might occur, for example, at the intersections of the wing and fuselage, the horizontal tail and the fuselage, the horizontal tail and the vertical tail, or a pylon and the wing. Careful design and relative placement of components and the use of fillets can keep interference drag to a minimum. A classic example of a wing-fuselage fillet can be seen in the photograph of the DC-3 in Figure 1.20. Any interference drag that remains in good designs is implicitly accounted for in the determination of the equivalent sand grain roughness.

$$M_0 = 0.5$$

$$K = [1 + Z(t/c) + 100(t/c)^4]$$

where

$$Z = \frac{(2 - M_0^2)\cos \Lambda_{C/4}}{\sqrt{1 - M_0^2 \cos^2 \Lambda_{C/4}}}$$

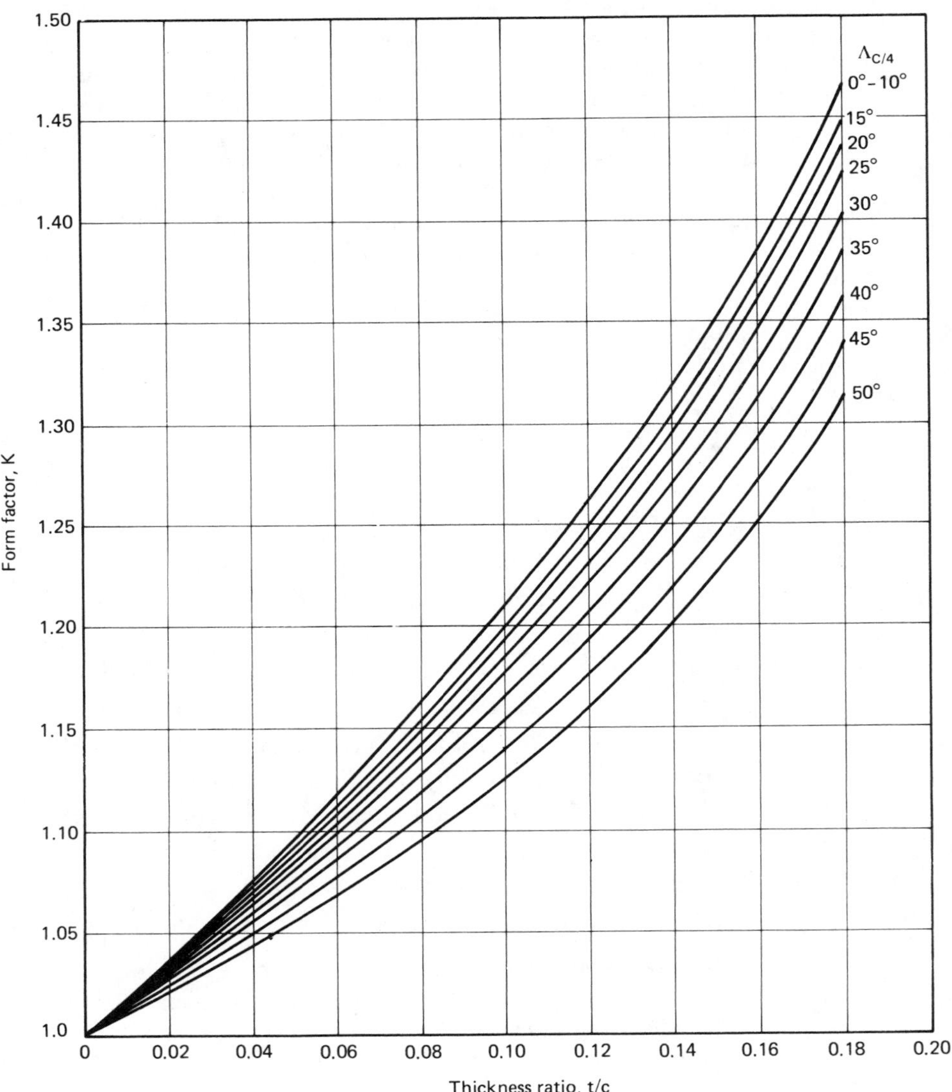

Figure 11.3 Aerodynamic surface form factor.

For most airplanes for which the bodies and surfaces are clearly defined, as in the design shown in Figure 11.5, equation 11.4 contributes about 90% to 94% of a satisfactory parasite drag prediction. The remaining drag is the result of wing twist,

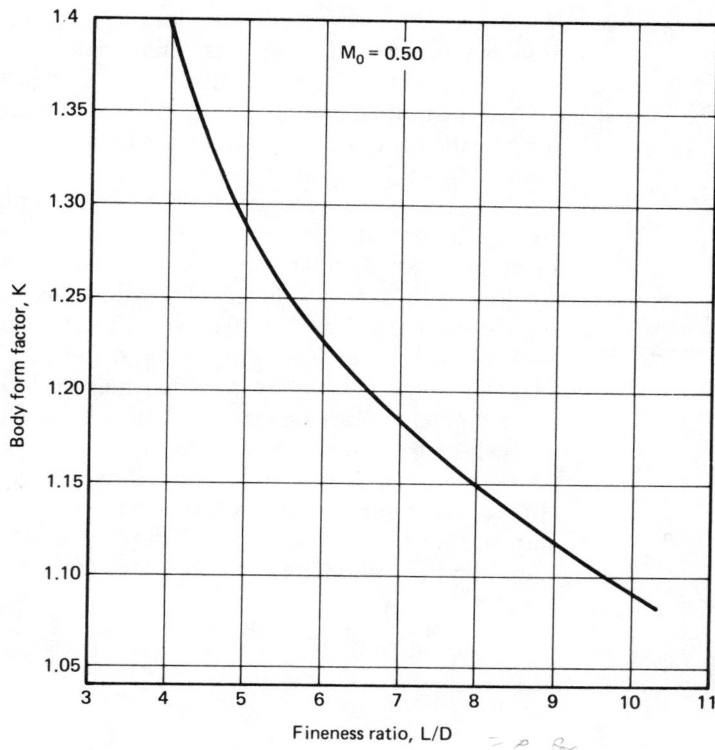

Figure 11.4 Effect of fineness ratio on body form factor.

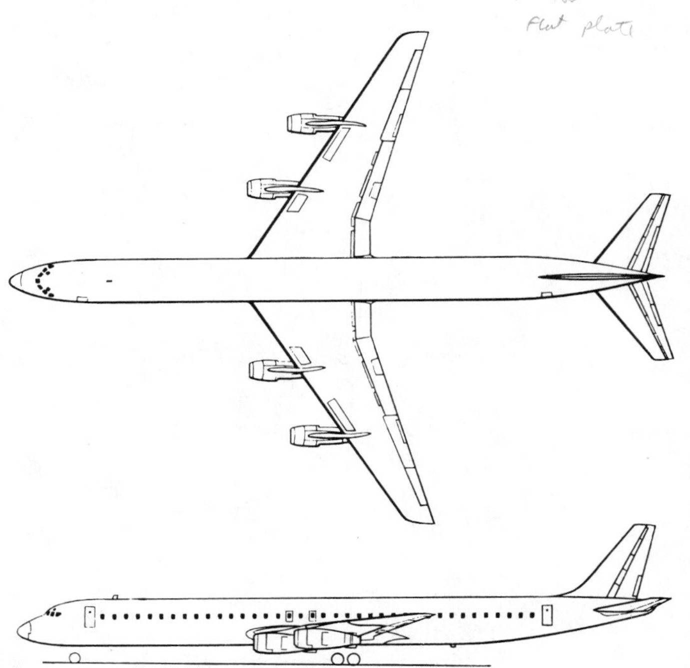

Figure 11.5 DC-8-61 transport aircraft. (Courtesy of Douglas Aircraft Company.)

which produces some induced drag even at zero lift; upsweep of the rear of the fuselage; protuberances such as drains, radio antennae, and air scoops for cooling air; drag due to control surface gaps; and base drag due to blunt rims around the exhaust nozzles of turbine engines. Fully powered controls or sealed aerodynamic nose balance on the control surfaces (Chapter 16) results in minimum control surface gap drag, about 1% of the total parasite drag. Unsealed aerodynamically balanced control surfaces contribute about 5% to the parasite drag. An approximation to the total of these additional drag items for turbine-powered aircraft is to increase the drag coefficient from equation 11.4 by 6% with sealed or fully powered control surfaces or by 10% with aerodynamically balanced, unsealed control surfaces. Reciprocating-engine-powered aircraft suffer a significant engine cooling airflow drag that may be 10% of the total parasite drag. Drag of wing struts, if any, must be included using methods such as those given by Hoerner (Ref. 11.1).

Fighter aircraft are often such blended configurations that it is difficult to decide just where the wing ends and the fuselage or nacelles begin (Figure 11.6). In such cases, the fineness ratio is hard to define. The integration of the propulsion system is difficult, so there is often excessive base and boat-tail pressure drag at the rear of the airplane. For this reason, fighter airplane designers often rely on original estimates based on the wetted area ratio between the new design and a previous one, plus the

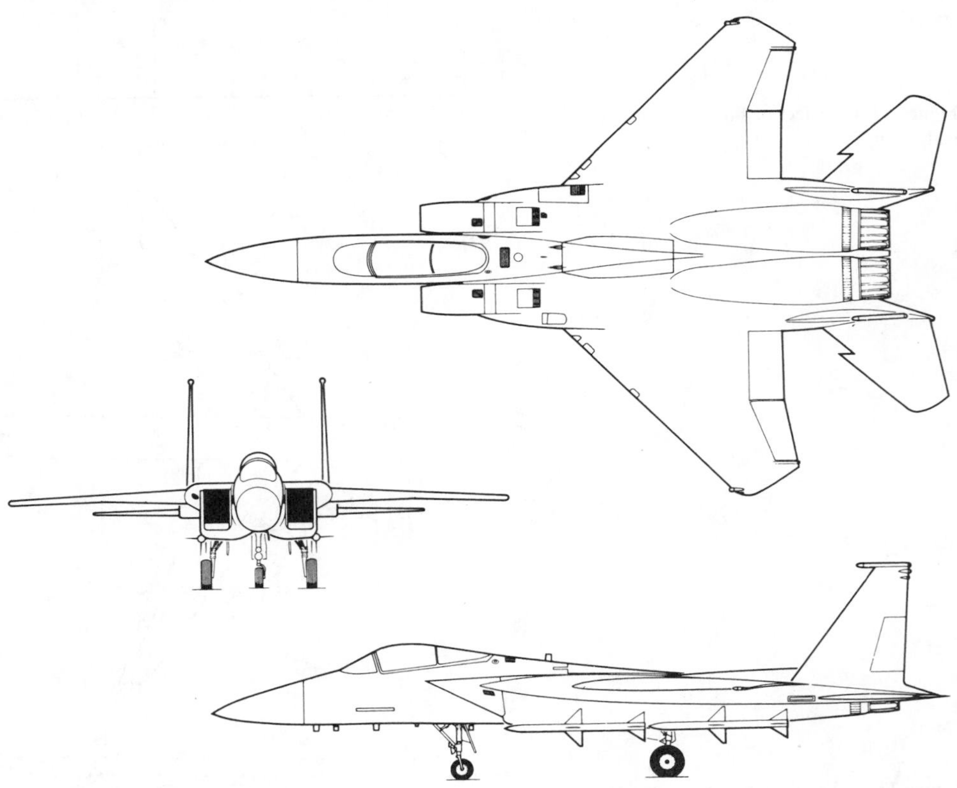

Figure 11.6 F-15A fighter aircraft. (Courtesy of McDonnell Aircraft Company.)

flight results of the previous airplane. Wind tunnel tests are then used extensively to study the drag.

Parasite drag is often expressed in terms of *equivalent parasite drag area f*. The equivalent parasite drag area is the area of a mythical flat plate perpendicular to the freestream, having a drag coefficient of 1.0 and the same total drag as the airplane. The equivalent parasite drag area is

$$f(\text{ft}^2) = C_{D_P} S_{\text{REF}} \tag{11.5}$$

where S_{REF} is in square feet.

The advantage of the parameter f is that it is a quantitative dimensional value expressing the total parasite drag. The relative parasite drag of various aircraft can be understood just by knowing f. For example, approximate values of f for transport aircraft are 20.6 ft^2 for the DC-9-30, 42 ft^2 for the stretched DC-8-63, 57 ft^2 for the DC-10, and 77 ft^2 for the 747. One example of aircraft design progress is the fact that the f of the pioneering DC-3 was 23.7 ft^2. The DC-9-30 carries 105 passengers compared to the 21 for which the DC-3 was designed, five times the passenger load for 13% less f.

The parasite drag D_p in units of force is simply

$$D_p = fq = f\left(\frac{\rho}{2}\right)V^2 \tag{11.6}$$

If f is expressed in square feet, the usual practice, all other units must be in the English engineering system. In the SI system, f would be stated in square meters.

DRAG DUE TO LIFT

Many empirical studies have shown that airfoil pressure drag not only increases with C_L, but that it approximately increases parabolically with C_L; that is,

$$\Delta C_{D_p} = kC_L^2$$

where k is empirical. We can now write for the total drag coefficient C_D

$$C_D = \underbrace{C_{D_P}}_{\substack{\text{parasite,}\\ \text{independent}\\ \text{of } C_L}} + \underbrace{kC_L^2}_{\substack{\text{parasite,}\\ \text{varying}\\ \text{with } C_L}} + \underbrace{\frac{C_L^2}{\pi \cdot \text{AR} \cdot us}}_{\text{induced}} \tag{11.7}$$

Here s is the correction for fuselage interference discussed in Chapter 9 (an approximation for s is given in Figure 11.7). Combining terms, we obtain

$$C_D = C_{D_P} + \left(k + \frac{1}{\pi \cdot \text{AR} \cdot us}\right)C_L^2$$

It is convenient to define the total variation of drag with lift as

$$C_{D_i} = \frac{C_L^2}{\pi \cdot \text{AR} \cdot e} = C_L^2\left(k + \frac{1}{\pi \cdot \text{AR} \cdot us}\right)$$

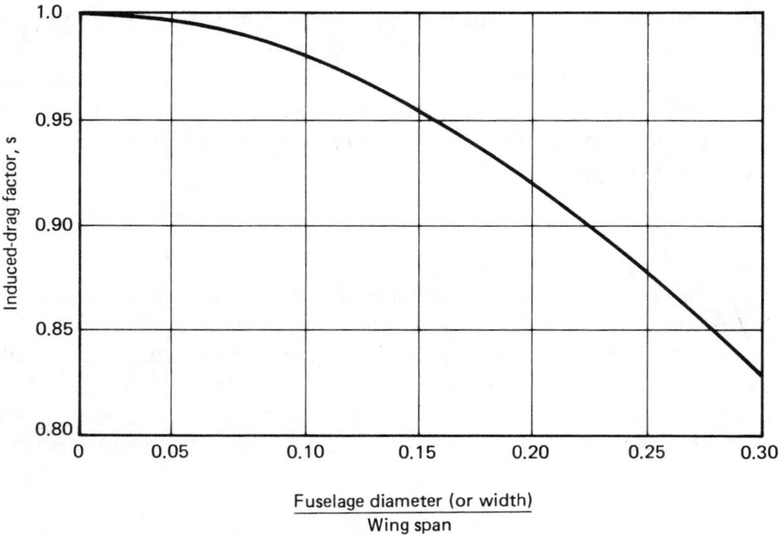

Figure 11.7 Lift-dependent drag factor to account for fuselage interference.

Then

$$\frac{1}{\pi \cdot \text{AR} \cdot e} = k + \frac{1}{\pi \cdot \text{AR} \cdot us}$$

Solving for e yields

$$e = \frac{1}{(\pi \cdot \text{AR} \cdot k) + 1/us} \tag{11.8}$$

We have now included the variation of parasite drag with C_L^2 in the induced drag. It is accounted for in the constant e. Note that C_{D_i} has become the total drag coefficient due to lift. We use $(\cdot)_i$, which stands for "induced," although C_{D_i} now represents both the induced drag due to downwash and the variation of C_D with C_L due to viscous effects. e is a total correction factor to the ideal elliptic drag due to lift. It accounts for any deviations from elliptic lift distribution and for parasite drag variations with C_L; e is known as the *airplane efficiency factor*. It is also called the *Oswald efficiency factor* after W. B. Oswald, who first used it. A Douglas Aircraft unpublished study based on analyzing the flight tests of many aircraft showed average values of k to be 0.38 C_{D_P} for unswept wings, 0.40 C_{D_P} for 20 degrees sweptback wings, and 0.45 C_{D_P} for 35 degrees sweptback wings. A rather good approximation to e can be found from Figure 11.8. The basic chart is based on equation 11.8, assuming $u = 0.99$, $s = 0.975$, and zero sweep. A correction for wing sweep angle is given in the insert. In practical aircraft, e is usually between 0.75 and 0.90.

Aircraft with propellers mounted ahead of the wing have a further reduction in e due to the downwash behind inclined propellers. The exact effect is difficult to calculate, but a reduction of about 4% is usually reasonable.

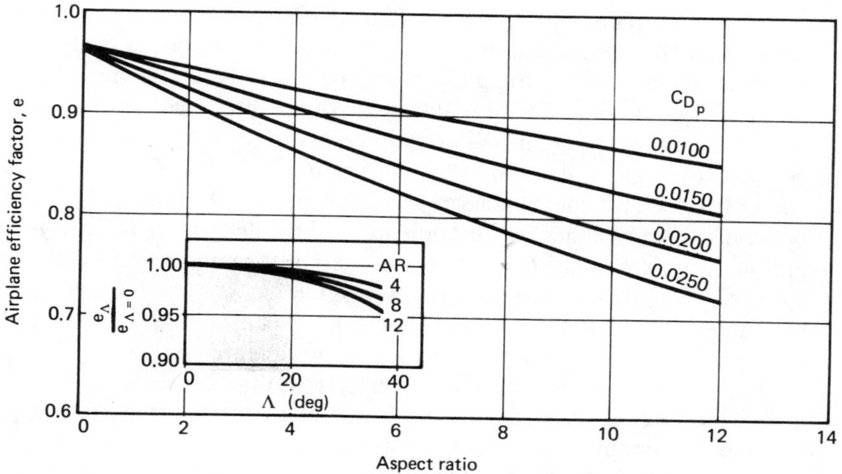

Figure 11.8 Airplane efficiency factor, e.

Thus the total incompressible drag coefficient is

$$C_D = C_{D_P} + \frac{C_L^2}{\pi \cdot \text{AR} \cdot e} \tag{11.9}$$

Since incompressible drag $D = C_D qS$,

$$D = C_{D_P} qS + \frac{C_L^2}{\pi \cdot \text{AR} \cdot e} qS \tag{11.10}$$

Remember that C_{D_P}, the portion of the drag coefficient independent of lift, is called the parasite drag coefficient even though the parasite drag varying with lift is accounted for by the airplane efficiency factor e. From equations 9.11 and 11.6, we can also write

$$D = fq + \frac{L^2}{\pi q b^2 e} \tag{11.11}$$

The method of determining incompressible drag described here is called a two-parameter system because only two parameters, C_{D_P} and e, or f and e, need be known. Equation 11.9 shows that the minimum drag coefficient is at zero lift. Airplanes often have their minimum drag coefficient at some positive lift coefficient, such as a C_L of 0.1 or 0.2. At lower C_L's the drag coefficient may increase. This results from wing section or body shapes that have their minimum parasite drag at some positive angle of attack. Highly cambered airfoils or fuselages with high aft upsweep angles are examples. With high flap deflection, minimum drag occurs at lift coefficients much above zero. Fitting an analytical curve to the entire range of such data requires a third term, usually a linear function of C_L.

The two-parameter system suffices for almost all practical cases, however. In cruise with flaps retracted, airplanes almost never fly continuously at lift coefficients below the value for minimum C_D nor above $C_L = 0.7$. The upper limit is established by the lift coefficient for most efficient flight discussed in Chapter 15. Therefore, accurate drag results are usually required only in the portion of the drag curve that can be fitted by a parabola. Flight test curves of C_D versus C_L are fitted with parabolic curves between lift coefficients of about 0.2 and 0.7. With flaps deflected, the airplane is flown at much higher lift coefficients, and the drag curve is fitted in the appropriate region. The extrapolation of the fitted analytical curves to zero lift gives an apparent value of C_{D_P}. This is a useful effective value but may be lower than the actual zero lift value. Since we do not fly at that low lift coefficient, it is of no importance. Of course, a high-speed, low-altitude attack airplane may utilize the low C_L region, and for such an airplane a three-parameter system may be necessary.

The methods described in this chapter for determining C_{D_P} and e have been validated by comparison with flight test drag curves derived as described above.

For incompressible flow this is the whole drag story. As we approach $M = 1.0$, however, additional drag due to compressibility arises, and to this we now turn our attention. First, however, a brief discussion of the significance of the aspect ratio is in order.

IMPORTANCE OF THE ASPECT RATIO

Equations 11.9 and 11.11 show that the drag coefficient and the drag itself are reduced by using large aspect ratio. Remember that aspect ratio is b^2/S, so for a given wing area S a large aspect ratio means a large span. From a pure drag standpoint, the larger the span can be, the better the airplane design will be. However, a large span means larger bending moments in the wing structure because the lift loads are acting farther from the root of the wing. Furthermore, a large span with a fixed area means shorter wing chords all along the span and, therefore, thinner wings. The wing acts as a beam, and a shallow beam requires heavier material on the top and bottom of the structure to withstand a given bending moment. Thus a high-aspect-ratio wing has a heavier structure. The higher wing weight raises the average flying weight and, therefore, increases the drag, counteracting some of the aerodynamic drag gain. Also, a thinner wing with a longer span has less internal volume for fuel. The selection of the optimum aspect ratio is based on a balance of these opposing factors. The most efficient wing depends on the range, design cruise speed, and cost of fuel. Choosing the best aspect ratio is a major factor in airplane design. Sometimes, especially for twin-engine aircraft, which must be able to climb with only one operative engine after one engine fails, a high aspect ratio is chosen to improve low-speed climb performance even though it is greater than optimum for cruising flight. In low-speed climbing flight, the induced drag may be 75% of the total drag, and the aspect ratio has an enormous effect on performance.

Example 11.1

A large transport aircraft has a wing planform area of 3699 ft^2, a root chord at the airplane centerline of 34.13 ft, a root chord at the side of the fuselage of 31.22 ft, an overall (centerline to wing tip) taper ratio of 0.275, and a span of 170 ft. The average weighted exposed airfoil thickness ratio is 10.2% and the wing has 36° of sweepback at the 25% chordline. Eighty-two percent of the wing planform area is exposed. Airfoils are the conventional peaky type.

The fuselage is 175 ft long and has a diameter of 20 ft and a wetted area of 9800 ft^2. The wing and body produce 72% of the total parasite drag. The airplane is cruising at a pressure altitude of 37,000 ft on a standard day with a wing loading of 105 lb/ft^2. The cruise Mach number is 0.83. Determine:

(a) Parasite drag coefficient.
(b) Induced drag coefficient.
(c) Total incompressible drag coefficient.
(d) What is the lift/drag ratio, still neglecting any additional drag due to compressibility?

Solution:

(a) *Parasite drag:* Since h_p = 37,000 ft on a standard day, we find from Table A.2 that $p_0 = 453.86$ lb/ft^2, $\rho_0 = 0.000678$, $T_0 = 389.99°R$, $\nu_0 = 0.0004379$, $a_0 = 968.08$ ft/s, and $V_0 = 0.83 \times 968.08 = 803.9$ ft/s.

For the wing, the overall taper ratio is $\sigma = 0.275$. Then the tip chord $C_T = 0.275 \times 34.13 = 9.39$ ft. Exposed $\sigma = 9.39/31.22 = 0.301$.

$$\text{exposed m.a.c.} = \frac{2}{3}C_R\left(1 + \sigma - \frac{\sigma}{1+\sigma}\right) = \frac{2}{3}(31.22)\left(1 + 0.301 - \frac{0.301}{1.301}\right)$$
$$= 22.26 \text{ ft}$$

Then

$$\text{RN} = \frac{V_0 L}{\nu} = \frac{(803.9)(22.26)}{0.0004379} = 40{,}865{,}000$$

From Figure 11.2 with typical roughness,

$$C_f = 0.00265$$

From Figure 11.3 at $\Lambda = 36$ degrees and $t/c = 0.102$, $K = 1.168$.

$$\text{wing wetted area } S_{\text{WET}} = 3699 \times 0.82 \times 2 \times 1.02$$
$$= 6188 \text{ ft}^2$$

Then

$$\Delta C_{D_{P\,\text{wing}}} = \frac{C_f K S_{\text{WET}}}{S_{\text{REF}}} = \frac{(0.00265)(1.168)(6188)}{3699}$$
$$= 0.00518$$

For the fuselage,

$$\text{RN} = \frac{V_0 L}{\nu} = \frac{(803.9)(175)}{0.0004379} = 3.2 \times 10^8$$

From Figure 11.2, $C_f = 0.00188$.

$$\text{body}\frac{L}{D} = \frac{175}{20} = 8.75$$

From Figure 11.4, $K = 1.125$. Then

$$\Delta C_{DP\,\text{fuselage}} = \frac{(0.00188)(1.125)(9800)}{3699}$$

$$= 0.00560$$

$$C_{D_P}(\text{wing} + \text{fuselage}) = 0.00518 + 0.00560 = 0.01078$$

Since wing and fuselage produce 72% of total parasite drag,

$$\text{total } C_{D_P} = \frac{C_{DP\,\text{wing}} + C_{DP\,\text{fuselage}}}{0.72} = \frac{0.01078}{0.72}$$

$$= 0.0150$$

(b) *Induced drag:*

$$\text{aspect ratio} = \frac{b^2}{S} = \frac{(170)^2}{3699} = 7.813$$

$$\text{weight} = \text{wing area} \times \text{wing loading} = 3699 \times 105$$

$$= 388{,}400 \text{ lb}$$

$$C_L = \frac{W}{qS} = \frac{388{,}400}{(\gamma/2)(453.86)(0.83)^2(3699)} = 0.480$$

$$C_{D_i} = \frac{C_L^2}{\pi \cdot \text{AR} \cdot e}$$

From Figure 11.8, with AR = 7.813, $C_{D_P} = 0.0150$, and $\Lambda = 36$ degrees,

$$e = e_{\Lambda=0}\frac{e_\Lambda}{e_{\Lambda=0}} = (0.851)(0.97) = 0.825$$

or, from equation 11.8 with $k = 0.45\ C_{D_P}$, $u = 0.99$ and s (from Figure 11.7) = 0.972.

$$e = \frac{1}{\pi(7.813)(0.45)(0.0150) + [1/(0.99)(0.972)]} = 0.830$$

$$C_{D_i} = \frac{(0.480)^2}{\pi(7.813)(0.83)} = 0.0113$$

(c) *Total incompressible drag coefficient*

$$C_D = 0.0150 + 0.0113$$

$$= 0.0263$$

$$D = C_D qS = 0.0263(\gamma/2)(453.86)(.83)^2(3699)$$

$$= 21{,}292 \text{ lb}$$

(d)

$$\frac{\text{lift}}{\text{drag}} = \frac{388{,}400}{21{,}292} = 18.24$$

or

$$\frac{\text{lift}}{\text{drag}} = \frac{C_L}{C_D} = \frac{0.480}{0.0263} = 18.25$$

The small round-off differences are trivial.

PROBLEMS

11.1. A twin turbofan transport airplane is cruising at 31,000 ft pressure altitude at a Mach number of 0.78. Outside air temperature is −60°F. The airplane gross weight is 98,000 lb. The airplane has unsealed aerodynamically balanced control surfaces. Following are the airplane dimensional data:

Wing	
Span	= 93.2 ft
Planform area	= 1000 ft^2
Average t/c	= 0.106
Sweepback angle	= 24.5 deg
Taper ratio	= 0.2
Root chord	= 17.8 ft
Wing area covered by fuselage	= 17%

Fuselage	
Length	= 107 ft
Diameter	= 11.5 ft
Wetted area	= 3280 ft^2

Horizontal Tail	
Exposed planform area	= 261 ft^2
t/c	= 0.09
Sweepback	= 31.6 deg
Taper ratio	= 0.35
Root chord	= 11.1 ft

Vertical Tail	
Exposed planform area	= 161 ft^2
t/c	= 0.09
Sweepback	= 43.5 deg
Taper ratio	= 0.80
Root chord	= 15.5 ft

Pylons	
Total wetted area	= 117 ft^2
t/c	= 0.06
Sweepback	= 0 deg
Taper ratio	= 1.0
Chord	= 16.2 ft

Nacelles	
Total wetted area	= 455 ft^2
Effective fineness ratio	= 5.0
Length	= 16.8 ft

Flap Hinge Fairings	
Δf	= 0.15 ft^2

Determine

(a) Incompressible parasite drag coefficient and equivalent flat-plate area.
(b) Induced drag coefficient.
(c) Total incompressible drag coefficient.
(d) Total incompressible drag in pounds.
(e) Ratio of lift to drag, neglecting compressibility.

11.2. At speeds where compressibility may be ignored,

$$C_D = C_{D_P} + \frac{C_L^2}{\pi \text{AR} e}$$

(a) Determine the C_L for best C_L/C_D. (Try for minimum C_D/C_L).
(b) What is the maximum ratio of lift to drag (C_L to C_D) in terms of C_{D_P}, AR, and e?

11.3. A transport aircraft has a wing planform area of 2500 ft^2, a root chord at the airplane centerline of 28.06 ft, a root chord at the side of the fuselage of 25.67 ft, an overall (centerline to tip) taper ratio of 0.275, and a span of 140 ft. The average weighted exposed airfoil thickness ratio is 11.5%, the wing has 25 degrees of sweepback at the 25% chord line, and 82% of the wing planform area is exposed. The wing creates 38% of the total parasite drag. The airplane is cruising at a pressure altitude of 39,000 ft on a standard day with a wing loading of 90 lb/ft^2. Cruise Mach number is 0.80.

(a) Determine the total incompressible drag coefficient and the drag in pounds.

(b) What is the ratio of lift to drag?

REFERENCE

11.1. Hoerner, S. F., *Fluid-Dynamic Drag,* Hoerner Fluid Dynamics, Brick Town, N. J., 1965.

12

COMPRESSIBILITY DRAG

In exploring compressibility drag, we first limit the discussion to unswept wings. The effect of sweepback will then be introduced. For aspect ratios above 3.5 to 4.0, the flow over much of the wing span can be considered to be very similar to two-dimensional flow. Therefore, we will be thinking at first in terms of flow over two-dimensional airfoils.

From our previous discussion, it is clear that lift is created by the occurrence of velocities higher than freestream on the upper surface of the wing and lower than freestream on much of the lower surface. As the flight speed of an airplane approaches the speed of sound (i.e., $M > 0.65$), the higher local velocities on the upper surface of the wing may reach and even substantially exceed $M = 1.0$. The existence of supersonic local velocities on the wing is associated with an increase of drag due to a reduction in total pressure through shock waves and due to thickening and even separation of the boundary layer due to the local but severe adverse pressure gradients caused by the shock waves. The drag increase is generally not large, however, until the local speed of sound occurs at or behind the *crest* of the airfoil, or the *crestline,* which is the locus of airfoil crests along the wing span. The crest is the point on the airfoil upper surface to which the freestream is tangent (Figure 12.1). The occurrence of substantial supersonic local velocities well ahead of the crest does not lead to

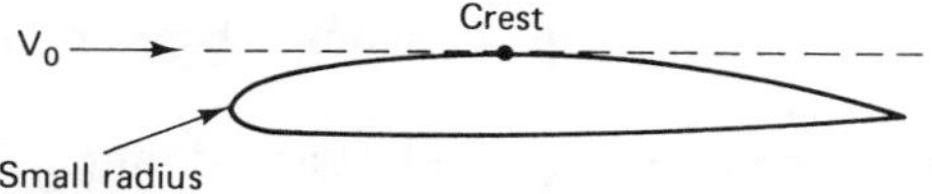

Figure 12.1 Definition of the airfoil crest.

significant drag increase provided that the velocities decrease below sonic forward of the crest. The incremental drag coefficient due to compressibility is designated ΔC_{DC}.

It was mentioned in Chapter 7 that a shock wave is a thin sheet of fluid across which abrupt changes occur in p, ρ, V, and M. In general, air flowing through a shock wave experiences a jump toward higher density, higher pressure, and lower Mach number. The effective Mach number approaching the shock wave is the Mach number of the component of velocity normal to the shock wave. This component Mach number must be greater than 1.0 for a shock to exist. On the downstream side, this normal component must be less than 1.0. In a two-dimensional flow, a shock is usually required to bring a flow with $M > 1.0$ to $M < 1.0$. Remember that the velocity of a supersonic flow can be decreased by reducing the area of the channel or streamtube through which it flows (equation 7.30). When the velocity is decreased to $M = 1.0$ at a minimum section and the channel then expands, the flow will generally accelerate and become supersonic again. A shock just beyond the minimum section will reduce the Mach number to less than 1.0, and the flow will be subsonic from that point onward.

Whenever the local Mach number becomes greater than 1.0 on the surface of a wing or body in a subsonic freestream, the flow must be decelerated to a subsonic speed before reaching the trailing edge. If the surface could be shaped so that the surface Mach number is reduced to 1.0 and then decelerated subsonically to reach the trailing edge at the surrounding freestream pressure, there would be no shock wave and no shock drag. This ideal is theoretically attainable only at one unique Mach number and angle of attack. In general, a shock wave is always required to bring supersonic flow back to $M < 1.0$. A major goal of transonic airfoil design is to reduce the local supersonic Mach number to as close to $M = 1.0$ as possible before the shock wave. Then the fluid property changes through the shock will be small and the effects of the shock may be negligible. When the Mach number just ahead of the shock becomes increasingly larger than 1.0, the total pressure losses across the shock become greater, the adverse pressure change through the shock becomes larger, and the thickening of the boundary layer increases.

Near the nose of a lifting airfoil, the streamtubes close to the surface are sharply contracted, signifying high velocities. This is a region of small radius of curvature of the surface (Figure 12.1), and the flow, to be in equilibrium, responds like a vortex flow (i.e., the velocity drops off rapidly as the distance from the center of curvature is increased). Thus the depth, measured perpendicular to the airfoil surface, of the flow with $M > 1.0$ is small. Only a small amount of fluid is affected by a shock wave in this region, and the effects of the total pressure losses caused by the shock are, therefore, small. Farther back on the airfoil, the curvature is much less, the radius is larger, and a high Mach number at the surface persists much farther out in the stream. Thus a shock affects much more fluid. Furthermore, near the leading edge the boundary layer is thin and has a full, healthy, velocity profile. Toward the rear of the wing, the boundary layer is thicker, its lower layers have a lower velocity, and it is less able to keep going against the adverse pressure jump of a shock. Therefore, it is more likely to separate.

For the reasons cited, supersonic regions can be carried on the forward part of an airfoil almost without drag. Letting higher supersonic velocities create lift forward allows the airfoil designer to reduce the velocity at and behind the crest for any required total lift, and this is the crucial factor in postponing compressibility drag on the wing.

The unique significance of the crest in determining compressibility drag is largely an empirical matter, although many explanations have been advanced. One is that the crest divides the forward-facing portion of the airfoil from the aft-facing portion. Supersonic flow and the resulting low pressures (suction) on the aft-facing surface would contribute strongly to drag. Another explanation is that the crest represents a minimum section when the flow between the airfoil upper surface and the undisturbed streamlines some distance away is considered (Figure 12.2). Thus, if $M = >1.0$ at the crest, the flow will accelerate in the diverging channel behind the crest; this leads to a high supersonic velocity, a strong suction, and a strong shock.

The freestream Mach number at which the local Mach number on the airfoil first reaches 1.0 is known as the *critical Mach number,* M_c. The freestream Mach number at which $M = 1.0$ at or behind the airfoil crest is called the *crest critical Mach number,* M_{cc}. The locus of the airfoil crests from the root to the tip of the wing is the crestline.

Empirically, it is found for all airfoils except the supercritical airfoil, to be discussed briefly later, that at 2% to 4% higher Mach number than that at which $M = 1.0$ at the crest the drag rises abruptly. The Mach number at which this abrupt drag rise starts is called the *drag divergence Mach number,* M_{DIV}. This is a major design parameter for all high-speed aircraft. The lowest-cost cruising speed is either at or slightly below M_{DIV}, depending on the cost of fuel.

Since C_P at the crest increases with C_L, M_{DIV} generally decreases at higher C_L. At very low C_L, the lower surface usually becomes critical and M_{DIV} decreases, as shown in Figure 12.3.

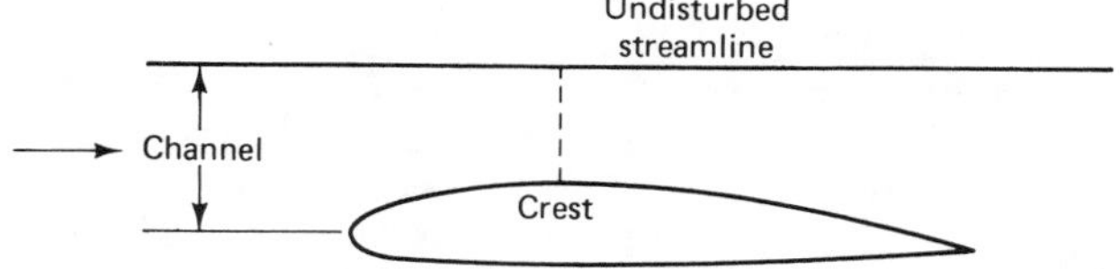

Figure 12.2 One view of the airfoil crest.

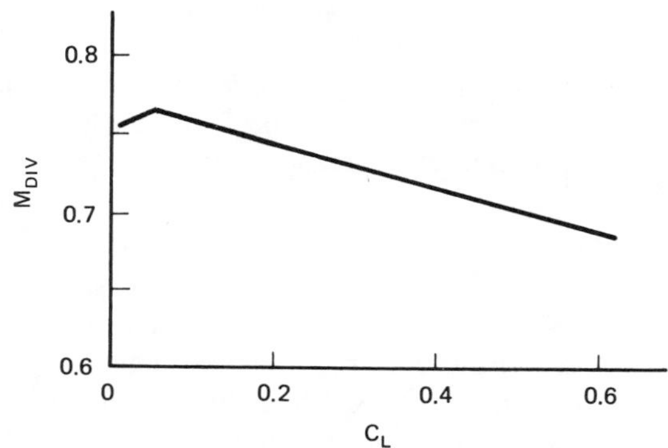

Figure 12.3 Typical variation of airfoil M_{DIV} with C_L.

The drag usually rises slowly somewhat below M_{DIV} due to the increasing strength of the forward, relatively benign shocks and to the gradual thickening of the boundary layer. The latter is due to the shocks and the higher adverse pressure gradients resulting from the increase in airfoil pressures because C_P at each point rises with $1/\sqrt{1 - M_0^2}$, as noted in Chapter 8. The nature of the early drag rise is shown in Figure 12.4.

There is also one favorable drag factor to be considered as the Mach number is increased. The skin friction coefficient decreases with increasing Mach number, as shown in Figure 12.5. Below Mach numbers at which shock waves first appear and above about $M = 0.5$, this reduction just about cancels the increased drag from the higher adverse pressure gradient due to the Mach number. Therefore, the net effect on

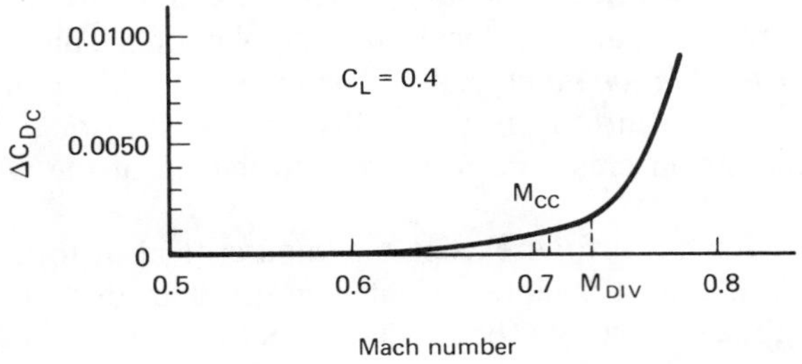

Figure 12.4 Typical variation of ΔC_{Dc} with Mach number.

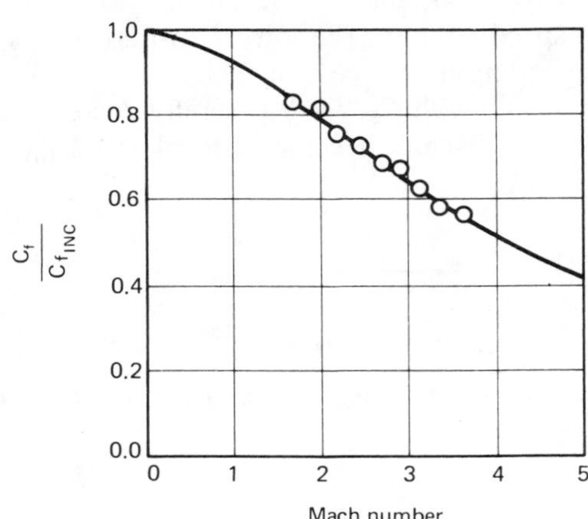

Figure 12.5 Ratio of the skin friction coefficient in compressible turbulent flow to the incompressible value at the same Reynolds number. From J. Bertin and M. Smith, *Aerodynamics for Engineers*, © 1979, p. 272. Reprinted by permission of Prentice-Hall, Inc., Englewood Cliffs, N.J.

drag coefficient due to increasing the Mach number above $M = 0.5$ is usually negligible until some shocks occur on the wing or body. The favorable effect of Mach number on skin friction is very significant at supersonic Mach numbers, however.

PREDICTION OF M_{DIV}

Since M_{DIV} is 2% to 4% above M_{cc} (we shall see that the "2% to 4%" is dependent on wing sweepback angle), we can predict the drag rise Mach number M_{DIV} if we can predict M_{cc}. If we can identify the pressure decrease, or, more conveniently, the local pressure coefficient C_p required on an airfoil to accelerate the flow locally to exactly the speed of sound, measured or calculated crest pressures can be used to determine the freestream Mach numbers at which $M = 1.0$ at the crest. If p is the pressure at a point on an airfoil of an unswept wing, the pressure coefficient is

$$C_p = \frac{p - p_0}{q_0} = \frac{p_0}{p_0}\left(\frac{p}{q_0}\right) - \frac{p_0}{q_0} = \frac{p_0}{q_0}\left(\frac{p}{p_0} - 1\right) \tag{12.1}$$

Note that $q_0 = (\gamma/2)p_0 M_0^2$; $(\cdot)_0$ corresponds to freestream conditions. Also,

$$\frac{p_T}{p} = \left(1 + \frac{\gamma - 1}{2}M^2\right)^{\gamma/(\gamma-1)}$$

and

$$\frac{p_T}{p_0} = \left(1 + \frac{\gamma - 1}{2}M_0^2\right)^{\gamma/(\gamma-1)}$$

Dividing the left and right sides of these two equations yields

$$\frac{p}{p_0} = \left(\frac{1 + \frac{\gamma - 1}{2}M_0^2}{1 + \frac{\gamma - 1}{2}M^2}\right)^{\gamma/(\gamma-1)} \tag{12.2}$$

so that

$$C_p = \frac{p_0}{q_0}\left(\frac{p}{p_0} - 1\right) = \frac{p_0}{(\gamma/2)p_0 M_0^2}\left[\left(\frac{1 + \frac{\gamma - 1}{2}M_0^2}{1 + \frac{\gamma - 1}{2}M^2}\right)^{\gamma/(\gamma-1)} - 1\right]$$

By definition, when the local Mach number $M = 1.0$, $C_p = C_{p_{\mathrm{CR}}}$, the critical pressure coefficient. Thus

$$C_{p_{\mathrm{CR}}} = \frac{2}{\gamma M_0^2}\left[\left(\frac{2 + (\gamma - 1)M_0^2}{\gamma + 1}\right)^{\gamma/(\gamma-1)} - 1\right] \tag{12.3}$$

A graph of this equation is shown in Figures 8.13 and 12.6. If the C_p at the crest is known as a function of M_0, the value of M_0 for which the speed of sound occurs at the crest can be determined immediately. Equation 12.3 applies to unswept wings and must be modified for wings with sweepback.

It will be noted from Figure 12.6 that the airfoil information required is $C_{p_{\mathrm{crest}}}$ versus M_0. In Figure 12.6, typical wind tunnel airfoil crest C_p variations with M_0 are shown for several angles of attack. M_{cc} occurs when the $C_{p_{\mathrm{crest}}}$ versus M_0 curve for a given angle of attack intersects the curve of $C_{p_{\mathrm{CR}}}$ versus M_0. A few percent above this speed, the abrupt drag rise will start at M_{DIV}. The approximate relationship between M_{DIV} and M_{cc} is given in the next section.

If the airfoil pressure distribution is calculated by one of various complex theoretical methods at $M_0 = 0$, the value of the crest C_p can be plotted versus M_0 using the Prandtl–Glauert approximation, $C_p = C_{p_{\mathrm{INC}}}/\sqrt{1 - M_0^2}$, or the more accurate von

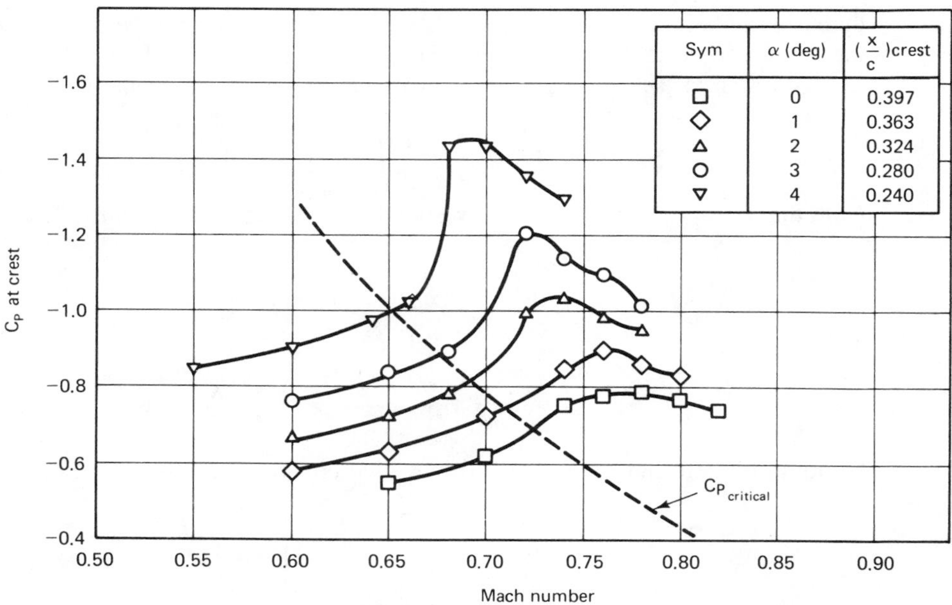

Figure 12.6 Variation of pressure coefficient at the crest with Mach number on a modern peaky airfoil, $t/c = 0.104$, RN = 14.5 million. (Courtesy of Douglas Aircraft Company.)

Kármán–Tsien relationship, equation 8.14, to estimate M_{cc} and M_{DIV}. This is illustrated in Figure 8.13.

The value of C_p at the crest is an important design characteristic of high-speed airfoils. In general, $C_{p_{\text{crest}}}$ at a given C_L depends on the thickness ratio (ratio of the maximum airfoil thickness to the chord) and the shape of the airfoil contour.

We have been describing a method of predicting M_{cc} that is useful in evaluating a particular airfoil design and in understanding the nature of the process leading to the occurrence of significant additional drag on the wing. Often in an advanced design process the detailed airfoil pressure distribution is not available. The airfoil is probably not even selected. It is still possible to estimate the M_{cc} closely from Figure 12.7. This graph displays M_{cc} as a function of airfoil mean thickness ratio t/c and C_L for unswept wings. It is based on studies of the M_{cc} of various airfoils representing the best state of the art for conventional "peaky"-type airfoils typical of all existing late-model transport aircraft. The significance of the term "peaky" is discussed in Chapter 13. Use of the chart assumes that the new aircraft will have a well-developed "peaky" airfoil and that the upper surface of the wing is critical for compressibility drag rise. Implied in the latter assumption is a design that assures that elements other than the wing (i.e., fuselage, nacelles, etc.) have a higher M_{DIV} than the wing. Up to design Mach numbers greater than 0.92 to 0.94, this is attainable. Furthermore, it is assumed that the lower surface of the wing is not critical. This assumption is always valid at the normal cruise lift coefficients, but may not be true at substantially lower lift coefficients. Here the wing twist or washout, designed to approach elliptical loading at cruise and to avoid first stalling at the wing tips, may lead to very low angles of attack on the outer wing

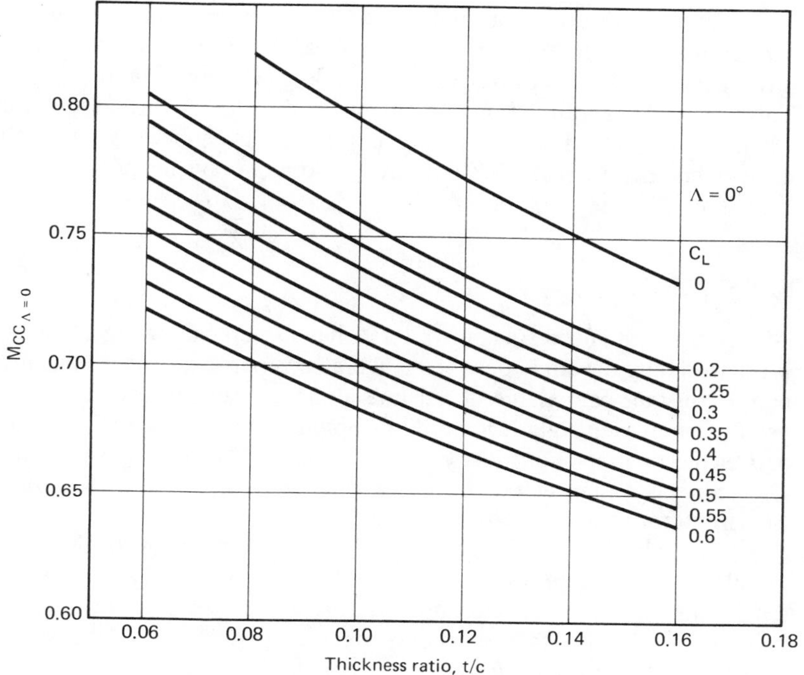

Figure 12.7 Crest critical Mach number, zero sweep.

panel. The highest $C_{p_{\text{crest}}}$ may then occur on the lower surface, a condition not considered in developing Figure 12.7. Thus the chart may give optimistic values of M_{cc} at lift coefficients more than 0.1 to 0.15 below the design cruise lift coefficients.

In Figure 12.7, the mean thickness ratio t/c is the average t/c of the exposed wing weighted for wing area affected, just as the mean aerodynamic chord, m.a.c., is the average chord of the wing weighted for wing area affected. The mean thickness ratio of a wing with linear thickness distribution is given by

$$\left(\frac{\bar{t}}{c}\right)_{\substack{\text{weighted}\\ \text{for chord}}} = \frac{\int_0^{b/2} (t/c)c\,dy}{\int_0^{b/2} c\,dy} = \frac{\int_0^{b/2} t\,dy}{(b/2)[(C_R + C_T)/2]} = \frac{t_R + t_T}{C_R + C_T}$$

This equation for $\overline{t/c}$ is based on a linear thickness (not linear t/c) distribution. This results from straight-line fairing on constant percent chord lines between airfoils defined at root and tip. The same equation is valid on a portion of wing correspondingly defined when the wing has more than two defining airfoils. The entire wing $\overline{t/c}$ can then be determined by averaging the $\overline{t/c}$ of these portions, weighting each $\overline{t/c}$ by the area affected. Note that C_R and C_T are the root and tip chords, while t_R and t_T are the root and tip thicknesses; b is the wing span and y is the distance from the centerline along the span.

Figure 12.7 does not apply directly to the new class of airfoils called *supercritical airfoils*. These are discussed in Chapter 13. The effective M_{cc} of supercritical airfoils can be estimated, however, by finding M_{cc} from Figure 12.7 and adding 0.06. This increment applies to a full supercritical airfoil with maximum aft camber (see Chapter 13). Note that Figure 12.7 is for zero sweepback. Most high-speed aircraft have sweptback wings. We shall now explore the physics of the sweptback wing and the effect of sweepback on M_{cc}.

SWEPTBACK WINGS

Almost all high-speed subsonic aircraft have sweptback wings. The amount of sweep is measured by the angle between a lateral axis perpendicular to the airplane centerline and a constant percentage chord line along the semispan of the wing. The latter is usually taken as the quarter-chord line both because subsonic lift due to angle of attack acts at the quarter-chord and because the crest is usually close to the quarter-chord.

Span is defined aerodynamically perpendicular to the centerline or x-axis, although the structural span is $b/\cos\Lambda$.

Sweepback increases M_{cc} and M_{DIV}. The component of the freestream velocity parallel to the structural span direction, $V_{\parallel}$, as shown in Figure 12.8, does not encounter the airfoil curvatures that produce increased local velocities, reduced pressures, and therefore lift. Only the component perpendicular to the swept span, $V_{\perp}$, is effective. Thus on a wing with sweep angle Λ,

$$\begin{aligned} V_{0\,\text{effective}} &= V_0 \cos\Lambda \\ M_{0\,\text{effective}} &= M_0 \cos\Lambda \\ q_{0\,\text{effective}} &= q_0 \cos^2\Lambda \end{aligned} \tag{12.4}$$

When the shock waves associated with compressibility drag occur, they form in planes parallel to the *isobars*, or lines of constant pressure. Shock waves cannot exist across isobars since this would require varying conditions along the shock, an unstable state. Thus the meaningful crest critical Mach number M_{cc} is the freestream Mach number at which the component of the local Mach number at the crest perpendicular to the isobars first reaches 1.0. These isobars coincide with constant percent chord lines on a well-designed wing.

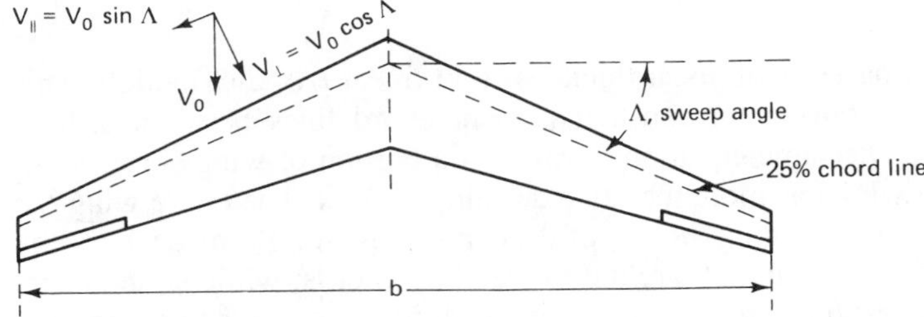

Figure 12.8 Velocity components affecting a sweptback wing.

Since $q_{0\,\text{effective}}$ is reduced, the angle of attack must be increased for a given lift. Thus

$$\alpha_{\perp c/4} = \frac{\alpha_{\Lambda=0}}{\cos^2\Lambda} \tag{12.5}$$

where

$\alpha_{\perp c/4}$ = α measured in a plane perpendicular to the quarter-chord line
$\alpha_{\Lambda=0}$ = α required for same lift and zero sweep

Because of this, the C_L based on $q_{0\,\text{effective}}$ and the C_p at the crest, based on $q_{0\,\text{effective}}$, will increase, and M_{cc} and M_{DIV} will be reduced. Furthermore, the sweep effect discussion so far has assumed the thickness ratio to be defined perpendicular to the quarter-chord line. Usual industry practice defines the thickness ratio parallel to the freestream. This corresponds to sweeping the wing by shearing in planes parallel to the freestream, rather than by rotating the wing about a point on the wing centerline. When the wing is swept with a constant freestream thickness ratio, the thickness ratio perpendicular to the quarter-chord line increases. The physical thickness is constant, but the chord decreases. The result is a further decrease in sweep effectiveness below the pure cosine variation. Thus there are several opposing effects, but the favorable one is dominant.

From the pure sweep effect at zero lift with constant effective t/c, M_{cc} with sweep is given by

$$M_{cc_\Lambda} = \frac{M_{cc_{\Lambda=0}}}{\cos\Lambda}$$

Because of the increased C_p based on $q_{0\,\text{effective}}$ at a given airplane lift and the increased t/c perpendicular to the quarter-chord line, the effectiveness of sweep is reduced:

$$M_{cc_\Lambda} = \frac{M_{cc_{\Lambda=0}}}{\cos^m\Lambda} \tag{12.6}$$

where m is an approximate function of C_L, as shown in Figure 12.9. The derivation of m is based on defining the thickness ratio of the swept wing in the freestream direction.

In addition to increasing M_{cc}, sweepback slightly increases the speed increment between the occurrence of the speed of sound at the crest and the start of the abrupt increase of drag at M_{DIV}. Using a definition for M_{DIV} as the Mach number at which the slope of the curve of C_D versus M_0 is 0.05 (i.e., $dC_D/dM_0 = 0.05$), the following empirical expression closely approximates M_{DIV}:

$$M_{\text{DIV}} = M_{cc}[1.02 + (1 - \cos\Lambda)(0.08)] \tag{12.7}$$

This analysis is based on two-dimensional sweep theory and applies exactly only to a wing of infinite span. It also applies well to most wings of aspect ratio greater than 4, except near the root and tip of the wing where significant interference effects occur.

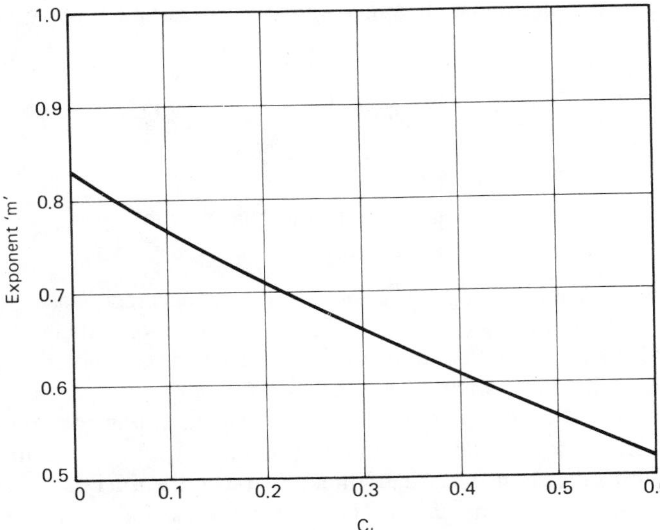

Figure 12.9 Exponent m versus C_L.

The effect of the swept wing is to curve the streamline flow over the wing as shown in Figure 12.10. The curvature is due to the deceleration and acceleration of the flow in the plane perpendicular to the quarter-chord line. Near the wing tip the flow around the tip from the lower surface to the upper surface obviously alters the effect of sweep. The effect is to unsweep the spanwise constant-pressure lines, known as *isobars*. To compensate, the wing tip may be given additional structural sweep (Figure 12.11).

It is at the wing root that the straight fuselage sides more seriously degrade the sweep effect by interfering with the curved flow of Figure 12.10. Airfoils are often modified near the root to change the basic pressure distribution to compensate for the distortions to the swept wing flow. Since the fuselage effect is to increase the effective airfoil camber, the modification is to reduce the root airfoil camber and in some cases to use negative camber. The influence of the fuselage then changes the altered root airfoil pressures back to the desired positive camber pressure distribution existing farther out along the wing span.

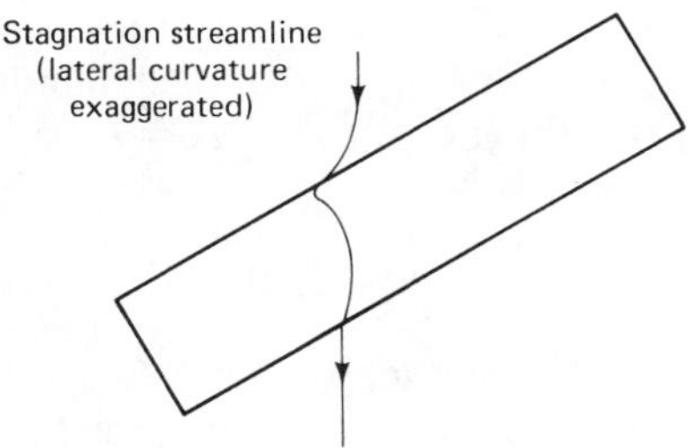

Figure 12.10 Stagnation streamline with sweep.

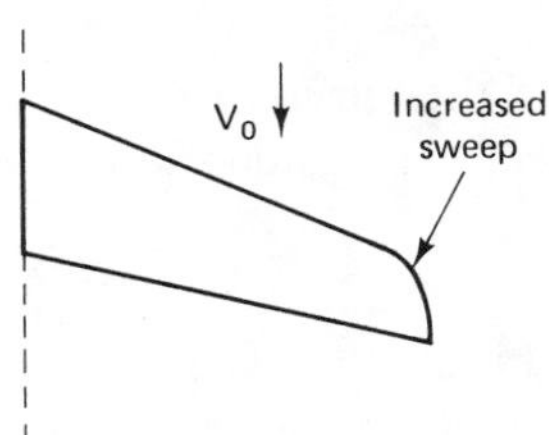

Figure 12.11 Highly swept wing tip.

This same swept wing root compensation can be achieved by adjusting the fuselage shape to match the natural swept wing streamlines. This introduces serious manufacturing and passenger cabin arrangement problems, so the airfoil approach is used for transports. Use of large fillets or even fuselage shape variations is appropriate for fighters.

The designing of a fuselage with variable diameter for transonic drag reasons is sometimes called "coke-bottling" (Figure 12.12). At $M = 1.0$ and above, there is a definite procedure for this minimization of shock wave drag. It is called the *area rule* and aims at arranging the airplane components and the fuselage cross-sectional variation so that the total aircraft cross-sectional area, in planes perpendicular to the line of flight, has a smooth and prescribed variation in the longitudinal (flight) direction. This is discussed further later in the chapter.

The estimates provided by Figure 12.7 and equation 12.6 for M_{cc} and by equation 12.7 for M_{DIV} assume that the wing root intersection has been designed to compensate for the "unsweeping" effect of the fuselage either with airfoil or fuselage fairing treatment. If this is not done, M_{DIV} will be reduced and/or there will be a substantial drag rise at Mach numbers lower than M_{DIV}. For all aircraft, there is some small increase in drag coefficient due to compressibility at Mach numbers below M_{DIV}, as illustrated in Figure 12.4. The increment in drag coefficient due to compressibility, ΔC_{D_C}, from its first appearance to well beyond M_{DIV} can be estimated from Figure 12.13, where ΔC_{D_C} is normalized by dividing by $\cos^3\Lambda$ and plotted against the ratio of freestream Mach number M_0 to M_{cc}. Actual aircraft may have slightly less drag rise than indicated by this method if very well designed. A poor design could easily have a higher drag rise. The differences arise from early shocks on some portion of the wing or other parts of the airplane. Figure 12.13 is an empirical average of existing transport aircraft data.

Wing sweepback affects lift as well as drag. The wing aerodynamic center is moved aft and the lift curve slope is reduced by sweepback, as discussed in Chapter 13.

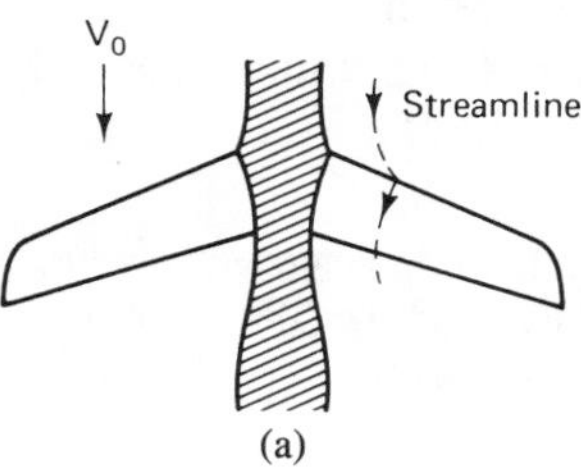

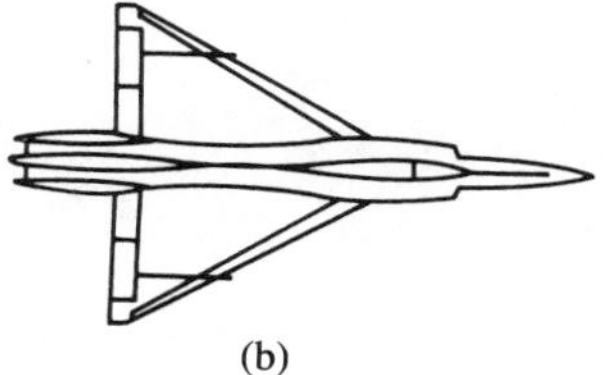

Figure 12.12 "Coke-bottled" fuselage: (a) the concept; (b) practical application, the Convair F 102A. From J. Bertin and M. Smith, *Aerodynamics for Engineers,* © 1979, p. 291. Reprinted by permission of Prentice-Hall, Inc., Englewood Cliffs, N.J.

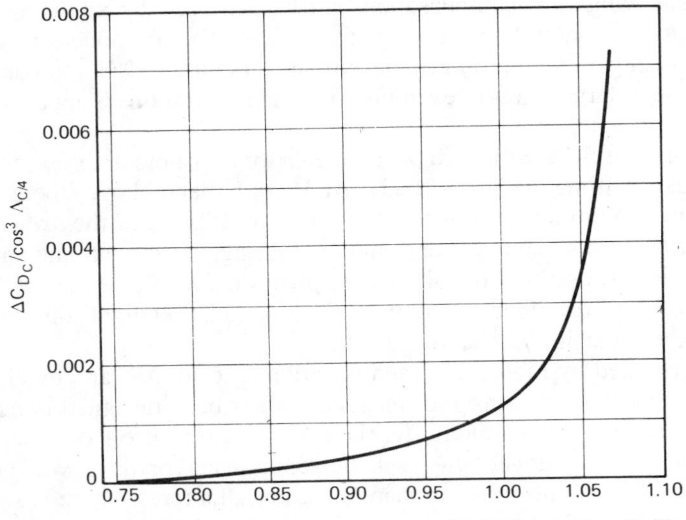

Figure 12.13 Incremental drag coefficient due to compressibility.

TOTAL DRAG

The total drag coefficient can now be written as

$$C_D = C_{D_p} + \frac{C_L^2}{\pi \cdot \text{AR} \cdot e} + \Delta C_{D_C} \tag{12.8}$$

parasite induced compressibility

A complete set of drag curves for a large wide-bodied trijet is given in Figure 12.14 in a format using M_0 as the abscissa. The same data can be presented in the form of drag *polars,* a term designating graphs of C_L versus C_D at constant M, as shown schematically in Figure 12.15.

Example 12.1

Using the airplane of Example 11.1, determine the incremental drag coefficient due to compressibility. Find the total drag and the lift/drag ratio, including compressibility.

Solution: From Example 11.1, $t/c = 0.102$, $\Lambda = 36$ degrees, and $C_L = 0.48$. From Figure 12.7 for zero sweepback, $M_{cc_{\Lambda=0}} = 0.703$. From Figure 12.9 at $C_L = 0.48$, $m = 0.575$. Then, from equation 12.6,

$$M_{cc_\Lambda} = \frac{M_{cc_{\Lambda=0}}}{\cos^m \Lambda} = \frac{0.703}{(0.809)^{0.575}} = 0.794$$

$$\frac{M_0}{M_{cc_\Lambda}} = \frac{0.83}{0.794} = 1.045$$

From Figure 12.13,

$$\frac{\Delta C_{D_C}}{\cos^3 \Lambda} = 0.0030$$

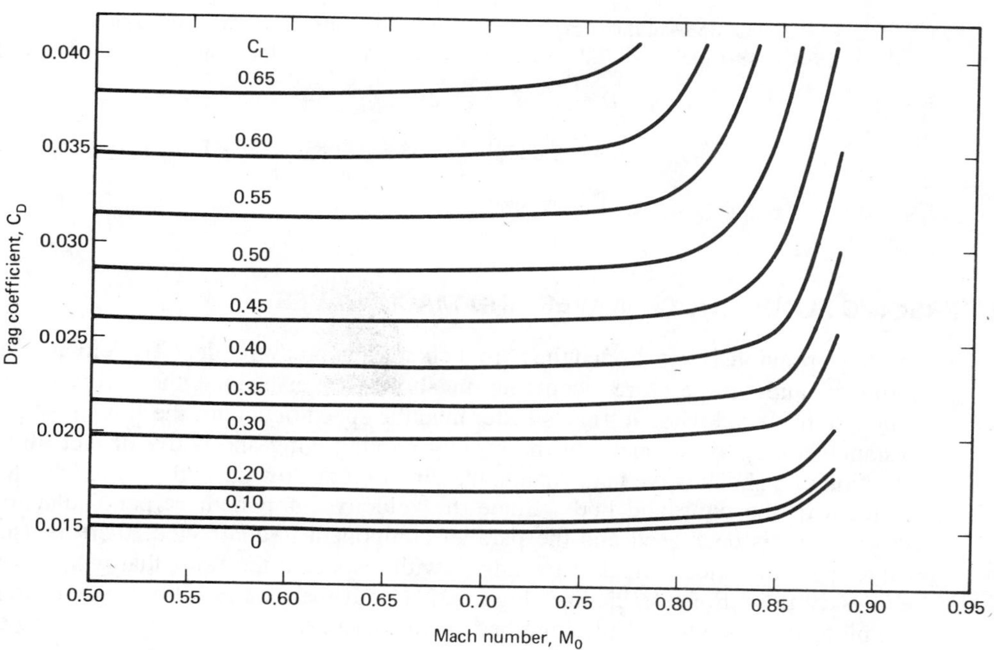

Figure 12.14 Drag coefficient versus Mach number.

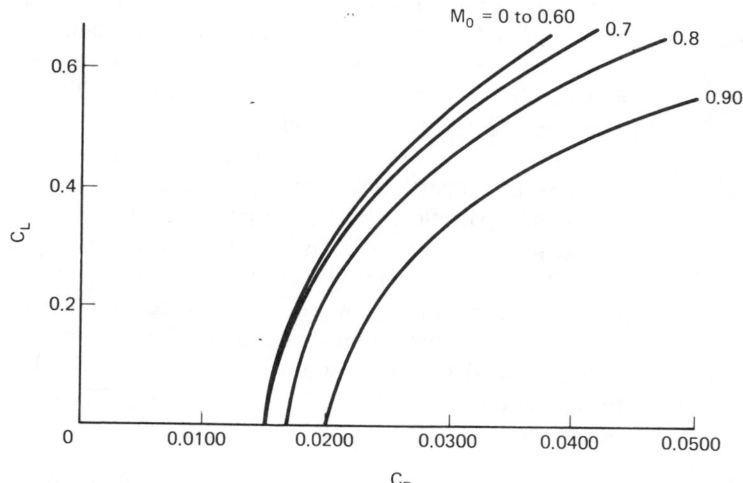

Figure 12.15 Drag characteristics (polars) for various Mach numbers.

Then

$$\Delta C_{Dc} = 0.0030 \cos^3 36° = \underline{0.0016}$$

From Example 11.1, the parasite drag coefficient is 0.0150 and the induced drag coefficient is 0.0113. Thus the total drag coefficient is 0.0150 + 0.0113 + 0.0016 = $\underline{0.0279}$.

The compressibility drag is $0.0016/0.0279 \times 100 = 5.7\%$ of the total. The lift/drag ratio is $0.48/0.0279 = \underline{17.2}$. Ambient pressure is 453.86 lb/ft^2 at 37,000 ft. The drag is

$$\begin{aligned} D &= C_D q S \\ &= 0.0279\left[\frac{1.4}{2}(453.86)(0.83)^2\right](3699) \\ &= \underline{22{,}587 \text{ lb}} \end{aligned}$$

SUPERSONIC FLOW: SHOCK WAVES AND MACH WAVES

We have previously mentioned the abrupt changes in pressure, density, velocity, Mach number, and temperature occurring in shock waves. Since the aircraft cannot "telegraph" ahead when it travels faster than the speed of sound, the flow must adjust instantaneously to the shape of the rapidly moving wing and body. In fact, the real function of a shock wave is to change the direction of flow abruptly to conform to the surfaces of the wing and body. Since the velocity component perpendicular to the shock wave is decreased and the parallel component is unaffected, there is usually some specific oblique shock wave angle, with respect to the flow, that will cause the necessary flow direction change (Figure 12.16). If the required flow angle change is too high, the wave will detach and become a curved shock ahead of the leading edge (Figure 12.17). At the plane of symmetry, the curved shock is perpendicular to the freestream flow. This portion of the curved shock is a straight shock behind which the flow is subsonic. Then the flow around the wing or body has the quality of subsonic flow, but with the initial conditions existing behind the straight shock. Freestream streamlines farther from the axis of symmetry face a lesser required deflection angle and are able to conform to the necessary flow with increasingly oblique shock waves. This process forms the curved shock wave.

It is of interest first to determine the wave angle of a very weak disturbance. Suppose that a particle is moving to the left at velocity V, as in Figure 12.18. At time $t = t_1$, the particle is located at point p. At $t = 0$, it lay a distance Vt_1 to the right; at that instant it caused a disturbance that (assuming it to be small) spread out at the speed of sound a, forming a circle whose radius at $t = t_1$ is at_1. At successively later instants it created other waves, as indicated by the various circles. The wave production is actually continuous, of course. The region thus disturbed by the particle, and thus informed that a body is moving through the fluid, will lie inside a wedge that forms the envelope of this family of circles. The semivertex angle β of this wedge, or cone in three dimensions, is given by

$$\sin\beta = \frac{at_1}{Vt_1} = \frac{a}{V} = \frac{1}{M} \quad \text{or} \quad \tan\beta = \frac{at_1}{\sqrt{(Vt_1)^2 - (at_1)^2}} = \frac{1}{\sqrt{M^2 - 1}} \tag{12.9}$$

$$\beta = \sin^{-1}\frac{1}{M} \qquad \text{or} \qquad \beta = \tan^{-1}\frac{1}{\sqrt{M^2 - 1}}$$

This wave angle due to a small disturbance is called the *Mach angle*. For any finite angle through which the flow must be turned, the wave angle will be greater than

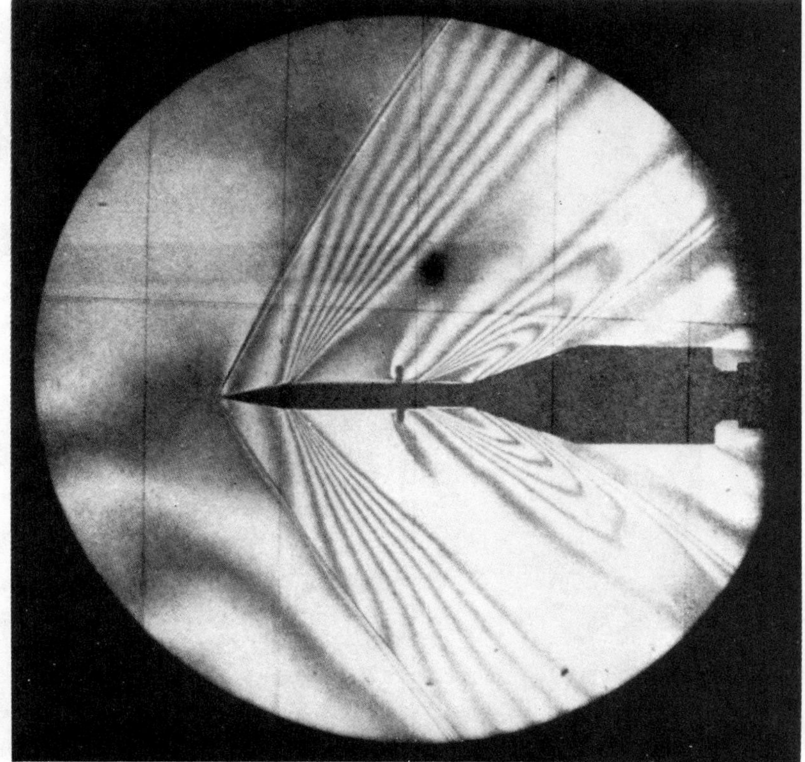

Figure 12.16 Attached shock wave of a wedge. Mach number 1.45. The details of the flow field are made visible by the use of an interferometer. Light and dark fringes indicate surfaces of equal air density. (Courtesy of Guggenheim Aeronautics Laboratory, California Institute of Technology.)

the Mach angle. When the wave angle equals the Mach angle, the Mach number perpendicular to the wave is 1.0, and the shock strength is zero. Referring to Figure 12.18, we get

$$\tan(90 - \beta) = \frac{\sqrt{(Vt_1)^2 - (at_1)^2}}{at_1} = \sqrt{\left(\frac{V}{a}\right)^2 - 1} \qquad (12.10)$$

$$= \sqrt{M^2 - 1}$$

The angle $(90 - \beta)$, defined between the Mach wave and a plane perpendicular to the flight direction, is similar to our means of defining wing sweepback angle. If a wing is swept behind the Mach angle, it operates such that the flow ahead of each airfoil leading edge is aware of the pressure changes caused by the airfoils farther inboard. This approaches subsonic flow, where the flow can start adjusting to the body before it actually reaches the leading edge. Such a wing is said to have a *subsonic*

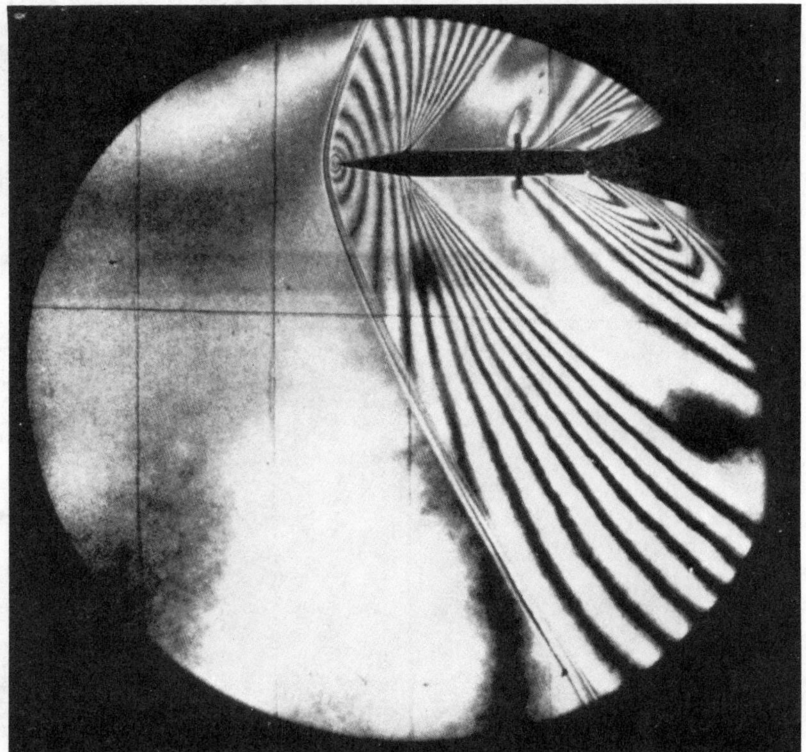

Figure 12.17 Detached shock wave of a wedge. Mach number 1.32. The optical technique is the same as in Figure 12.16. (Courtesy of Guggenheim Aeronautics Laboratory, California Institute of Technology.)

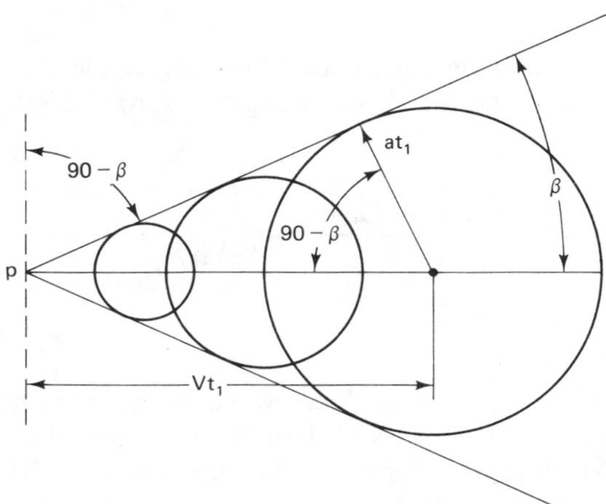

Figure 12.18 Development of a Mach wave.

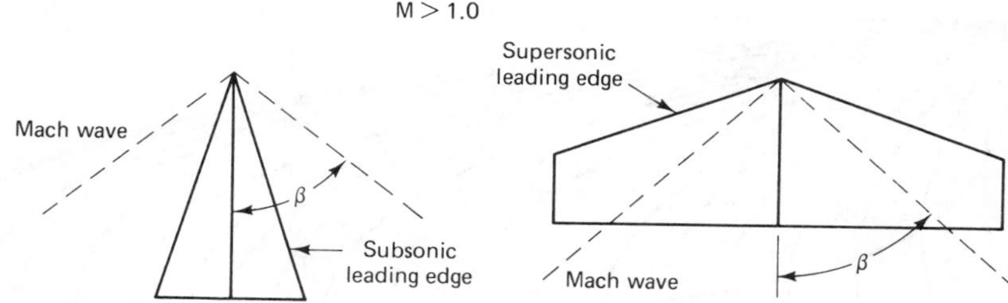

Figure 12.19 Subsonic and supersonic leading edges.

leading edge. A wing with leading edge sweep less than the sweep of the Mach wave has a *supersonic leading edge* (Figure 12.19).

Defining the sweep of the Mach waves as $(90° - \beta)$, when

$M = 1.0$	sweep of Mach wave $= 0°$
$M = 1.15$	sweep of Mach wave $= 30°$
$M = 2.0$	sweep of Mach wave $= 60°$
$M = 3.0$	sweep of Mach wave $= 70.5°$

It is worth restating the fundamental difference between a Mach and a shock wave. A Mach wave is a boundary between the fluid that is affected by pressure disturbances from a moving body of very small thickness and fluid that is unaware of the approach of the body. The Mach number perpendicular to the Mach wave is 1.0, and there are infinitesimally small, essentially zero, changes to the fluid properties through the Mach wave. Therefore, there is no change in flow direction. A shock wave has a greater angle with respect to the body axis (i.e., less sweepback) than a Mach angle. The Mach number perpendicular to the shock wave is greater than 1.0, and significant changes in fluid properties, including change in direction, occur through the shock wave.

The flow through a shock wave is analyzed using the continuity, momentum, and energy equations. The rather involved mathematics can be found in Refs. 12.4, 12.5, and 12.6 and leads to a relationship between the upstream Mach number, the deflection angle, and the shock wave angle. These results are summarized in Figure 12.20. For any value of the upstream Mach number, there are two shock wave angles that produce the same flow deflection. The wave defined by the larger angle is called the *strong shock*. The wave given by the smaller angle is the *weak shock*. Experience shows that the weak shock (i.e., the smaller shock angle) usually occurs in external aerodynamic flows. For any value of Mach number, there is a maximum flow deflection angle. If the deflection angle required by the body exceeds this maximum, there is no shock angle that will produce the required deflection. A detached shock then occurs (Figure 12.17).

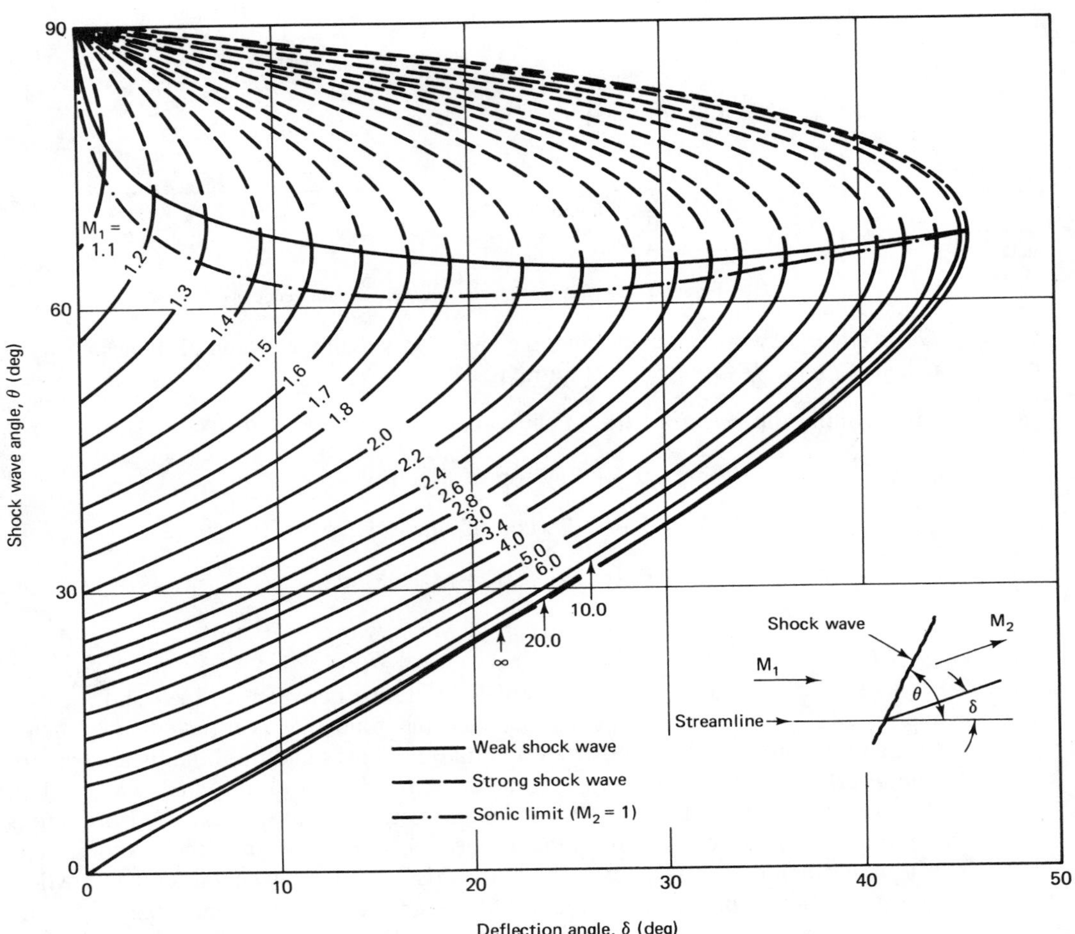

Figure 12.20 Variation of shock wave angle with flow deflection angle for various upstream Mach numbers, $\gamma = 1.4$. From J. Bertin and M. Smith, *Aerodynamics for Engineers,* © 1979, p. 260. Reprinted by permission of Prentice-Hall, Inc., Englewood Cliffs, N.J.

SUPERSONIC WING LIFT AND DRAG

Since an oblique shock wave always causes a pressure increase, the pressure is increased on the front surfaces of a body such as the double wedge airfoil at zero angle of attack in Figure 12.21. On a forward-facing surface, this increases drag. This type of drag is called *wave drag*. On the rear surfaces the flow must again abruptly change direction, but here the flow is turned away from the freestream through an expansion wave, a fan-shaped region through which the pressure decreases. This process is known as a *Prandtl–Meyer expansion*. The result is a decreased pressure on aft-facing surfaces, again a source of drag and, indeed, another component of wave drag. At the

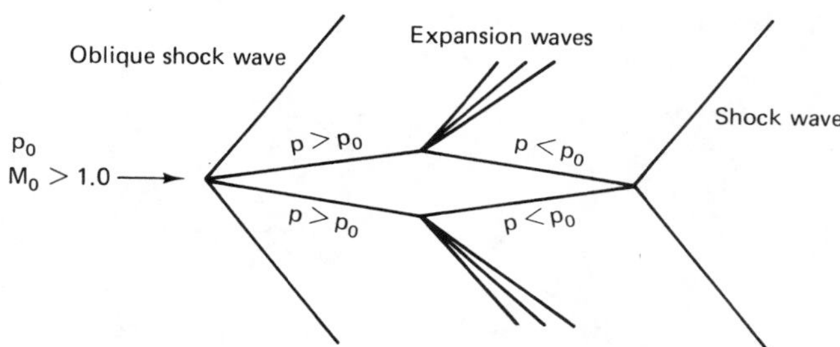

Figure 12.21 Thickness wave drag on a double wedge airfoil.

trailing edge the flow is again turned abruptly, this time through a compression or shock wave, so that the flow direction is restored to the freestream direction.

The supersonic lifting process can best be visualized by a flat plate at an angle of attack as in Figure 12.22. As the flow turns away from the freestream at the leading edge of the upper surface, an expansion wave reduces the pressure. On the lower surface, a shock wave raises the pressure and turns the flow into the freestream. At the trailing edge, the process reverses to turn the flows to the freestream direction. The net result is a lift resulting half from the upper surface reduced pressure and half from the lower surface increased pressure. The resultant force is perpendicular to the plate, so there is a drag due to lift equal to lift times the sine of the angle of attack.

An important supersonic result is the existence of a drag due to lift even for the two-dimensional case and a drag due to thickness without consideration of friction. Neither of these wave drag effects, or any equivalent, exists subsonically. Although vortex drag exists on a supersonic wing only at the tips or if the wing leading edge is swept behind the Mach wave from the most forward part of the wing, the two-dimensional drag coefficient due to lift is greater than a vortex drag coefficient would be. Thus, for a given lift, supersonic drag is much greater than the subsonic drag of a reasonable wing.

The results of two-dimensional supersonic theory for thin airfoils at small angles of attack are surprisingly simple. Only a brief summary will be given here.

1. The pressure coefficient at any point on the surface with an angle θ to the flow is

$$C_p = \frac{2}{\sqrt{M_0^2 - 1}}\theta \tag{12.11}$$

where θ is in radians, negative for an expansion, positive for a compression.

2.

$$C_L = \frac{4}{\sqrt{M_0^2 - 1}}\alpha \tag{12.12}$$

where α is the angle of attack in radians.

3.

$$C_{D\,\text{wave, due to lift}} = \frac{4\alpha^2}{\sqrt{M_0^2 - 1}} \tag{12.13}$$

4. $C_{D\,\text{wave},\,t/c} = 0$ flat plate

$$= \frac{4}{\sqrt{M_0^2 - 1}}\left(\frac{t}{c}\right)^2 \quad \text{double-wedge airfoil} \tag{12.14}$$

$$= \frac{16}{3\sqrt{M_0^2 - 1}}\left(\frac{t}{c}\right)^2 \quad \text{biconvex or circular arc airfoil (Figure 12.23)} \tag{12.15}$$

where t is the airfoil maximum thickness, c the chord, t/c the thickness ratio, and $C_{D\,\text{wave},\,t/c}$ is the wave drag coefficient due to thickness.

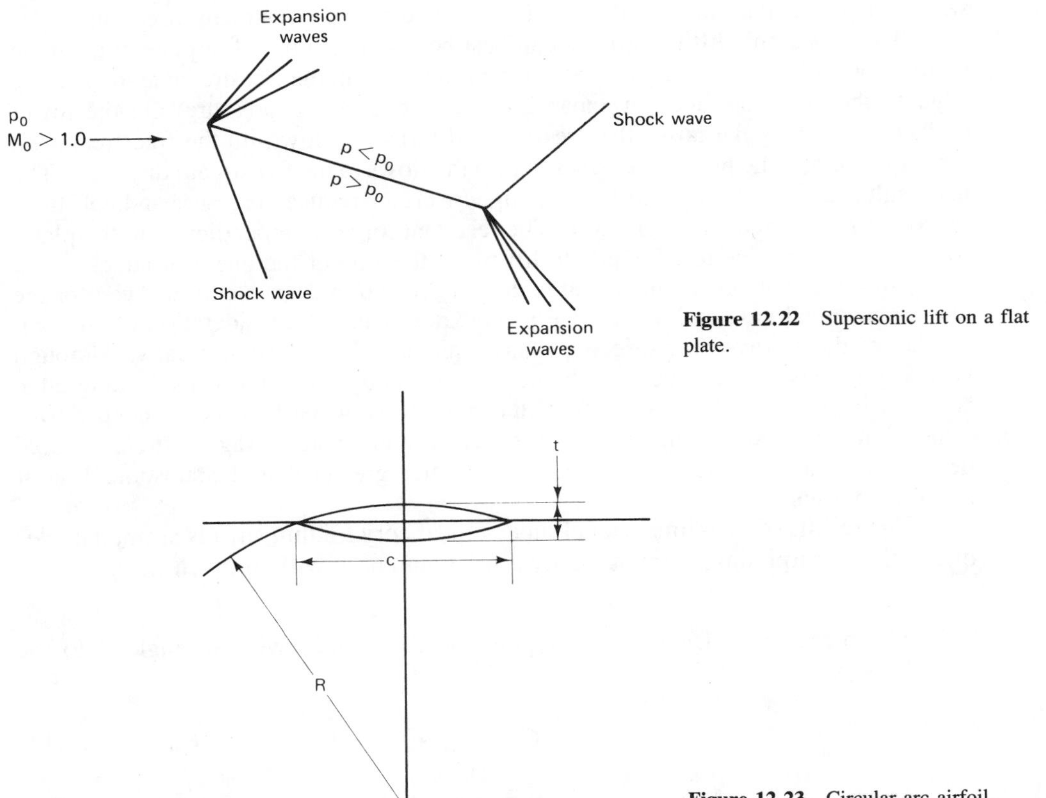

Figure 12.22 Supersonic lift on a flat plate.

Figure 12.23 Circular arc airfoil.

SUPERSONIC BODY DRAG

The pressures on fuselages and other bodies at supersonic speeds also depend on the angle through which the flow must be deflected. Since the shock pattern is a three-dimensional cone rather than the two-dimensional wedge-shaped shock pattern of the two-dimensional airfoil, the equations are somewhat different. The usual long, slender body is swept well behind the Mach waves so that for small disturbances, the wave drag coefficient is not dependent on the Mach number. The drag coefficient is based on the body maximum cross-sectional area measured in a plane perpendicular to the body axis. As with a wing, positive pressures on the forward-facing surfaces and negative pressures (with respect to the freestream) acting on the aft-facing surfaces contribute to the drag. For a slender paraboloid, pointed at both the front and rear ends (Figure 12.24), the wave drag coefficient at zero lift is approximately

$$C_{D_w} = \frac{10.67}{(\mathrm{FR})^2} \tag{12.16}$$

where FR is the fineness ratio, the ratio of the body length to the maximum diameter. This is analogous to the inverse of the thickness ratio for airfoils.

The minimum wave drag for a given diameter and length is obtained with a special shape known as the Sears–Haack body (Figure 12.24); see Ref. 12.7. For this shape,

$$C_{D_w} = \frac{9.87}{(\mathrm{FR})^2} \tag{12.17}$$

The fuselages of supersonic aircraft will usually be designed with high fineness ratios because their wave drag varies inversely as the square of the fineness ratio, as shown in equations 12.16 and 12.17.

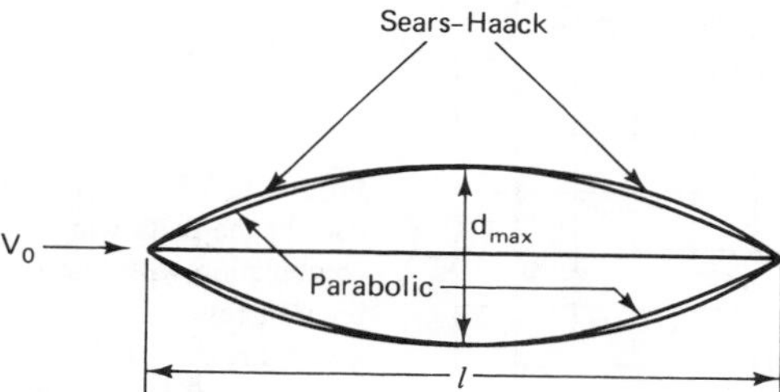

Figure 12.24 Sears-Haack optimum and parabolic bodies of revolution.

THE AREA RULE

The theories of transonic and supersonic flow with small disturbances (i.e., small thickness ratios, large fineness ratios, and low angles of attack) were extended in a very practical way by Whitcomb (Ref. 12.3) in his development of the area rule. We have mentioned that the presence of a disturbance due to a body was felt in the region behind the Mach wave. At $M = 1.0$, the Mach wave lies in a plane normal to the line of flight, so all disturbances caused in that plane are transmitted to other parts of the body in that plane. The area rule states that all aircraft that have the same longitudinal distribution of cross-sectional area in the planes normal to the flight direction, including the wing, fuselage, nacelles, tail surfaces, or any other components, will have the same wave drag at zero lift. Thus a complex airplane will have the same zero-lift wave drag as a body of revolution whose cross-sectional area is the same at every longitudinal station as the real aircraft. This is illustrated in Figure 12.25. The concept allows relatively simple methods for estimating the wave drag of a body of revolution to be used for a real configuration by applying the method to the "equivalent" body.

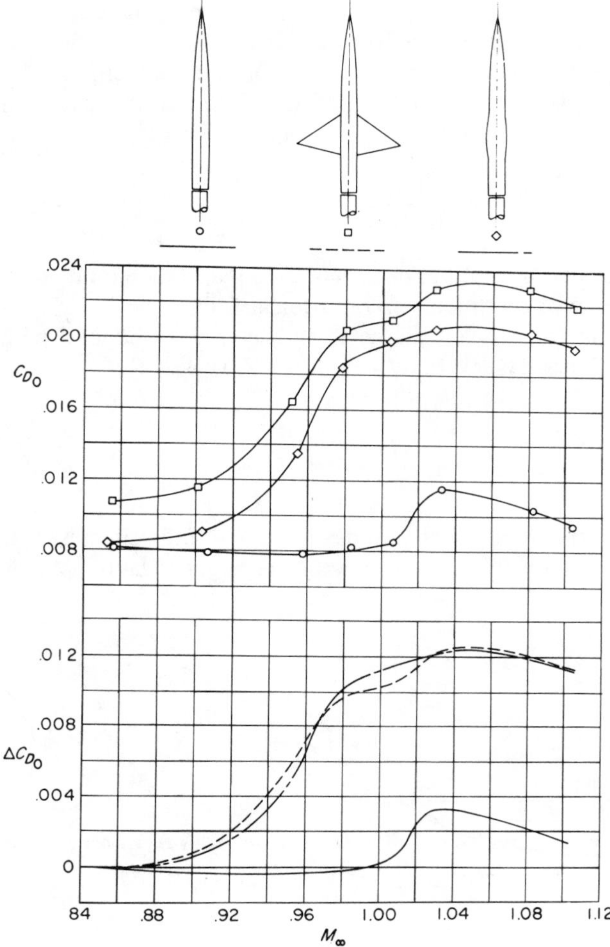

Figure 12.25 Comparisons of the drag rise for the delta wing-cylindrical body combination with that for the comparable body of revolution and cylindrical body alone. From R. T. Whitcomb, "A Study of the Zero-Lift Drag-Rise Characteristics of Wing-Body Combinations near the Speed of Sound," *NACA* TR 1273, 1956.

Furthermore, the drag can be minimized by shaping and locating the airplane components so that the shape of the equivalent body will be as close as possible to the minimum-drag Sears–Haack body. The concept can be seen to be related to the "coke-bottling" mentioned for high subsonic speeds.

Designing a transonic airplane configuration so that the equivalent body of revolution has a cross-sectional variation identical with the minimum-drag Sears–Haack shape is usually impossible. The goal then is to obtain a shape as close as possible to the ideal. One rule of thumb sometimes used in preliminary evaluation, before a detailed analysis is achieved, is to assume that the zero-lift wave drag will be about 50% greater than the theoretical minimum. Thus the first estimate is

$$C_{D_w} = \frac{9.87}{(\text{FR})^2} \cdot 1.5 \tag{12.18}$$

where C_{D_w} is the drag coefficient based on the maximum cross-sectional area of the equivalent body and FR is the fineness ratio of the equivalent body.

The area rule description above is strictly applicable for $M = 1.0$. It is very helpful, however, in the entire transonic region from about $M = 0.95$ to $M = 1.3$. When the Mach number exceeds 1.0, there is a supersonic procedure corresponding to the one described here. The method is too complex for our purposes, but involves measuring the cross-sectional areas in planes that are parallel to the Mach waves at each Mach number, rather than normal to the flight direction.

SUPERSONIC AIRCRAFT

Because of the high wave drag, supersonic aircraft have higher drag than subsonic aircraft. We shall see in the discussion of range performance that the ratio of lift to drag, L/D, is a crucial parameter. Figure 12.26 shows the variation of L/D with Mach number for transport aircraft using optimized designs at each Mach number. The Anglo-French Concorde, a design finalized by the mid-1960s, has a significantly lower L/D. After M_{DIV} is exceeded, the L/D drops dramatically and continues to decrease at a lesser rate as the Mach number increases above 1.2. Furthermore, the drag would be even higher if the supersonic designs did not employ very thin wings to minimize wave drag, as required by equations 12.14 and 12.15. Thin wings have reduced structural depth. Depth is an important design parameter that provides leverage critical to the bending strength of the wing. With reduced depth, the wing skins must carry increased tension and compression loads. To permit this, the skin thickness must be increased, causing substantial increases in wing weight. The high drag requires large engines, which further increase drag and weight. The combined weight and drag difficulties are major reasons why an economically feasible supersonic transport has not yet been attainable.

Another difficulty facing designers is that a supersonic transport must also fly at subsonic speeds during climb, descent, and cruising flight over land when required by sonic boom environmental concerns. Since subsonic induced drag is minimized by high span while the weight of the thin wings is reduced by low span, difficult compromises must be made.

Example 12.2

A straight-wing supersonic fighter with a 3% thick double-wedge wing is flying at a Mach number of 1.5 at 40,000 ft on a standard day. It has a wing area of 200 ft^2 and a span of 28 ft. The total weight is 12,000 lb; 80% of the wing area is exposed. Assume that only the exposed wing carries lift supersonically. Neglect tip effects. The fuselage has a maximum

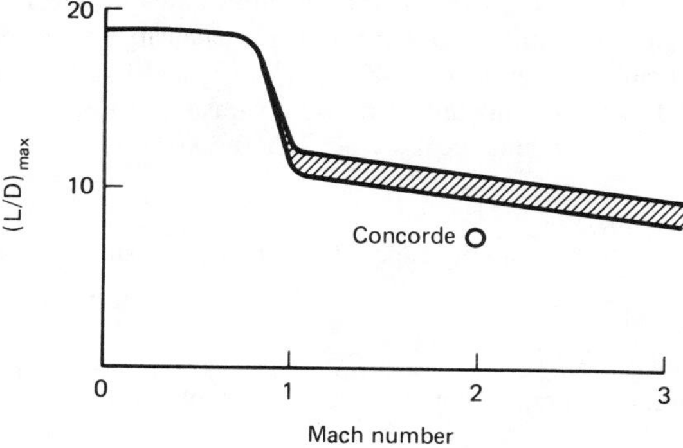

Figure 12.26 L/D_{max} versus Mach number for transport aircraft.

cross-sectional area of 9 ft^2, a fineness ratio of 8.5, and 20% more wave drag than a Sears–Haack body. What is the total wing wave drag due to both thickness and lift? What is the fuselage wave drag?

Solution: At 40,000 ft, $p = 393.12$ lb/ft^2.

$$\text{dynamic pressure } q = \frac{\gamma}{2} pM^2 = (0.7)(393.12)(1.5)^2 = 619.16 \text{ lb/ft}^2$$

$$C_L = \frac{W}{qS} = \frac{12{,}000}{(619.16)(200 \times 0.8)} = 0.1211$$

Also,

$$C_L = \frac{4}{\sqrt{M^2 - 1}} \alpha$$

Then

$$\alpha = \frac{C_L\sqrt{M^2 - 1}}{4} = \frac{0.1211\sqrt{1.5^2 - 1}}{4} = 0.03386 \text{ rad}$$

and

$$C_{D\,\text{wave, lift}} = \frac{4\alpha^2}{\sqrt{M^2 - 1}} = \frac{4(0.03386)^2}{\sqrt{1.5^2 - 1}} = 0.00410$$

$$\text{wing } C_{D\,\text{wave}, t/c} = \frac{4}{\sqrt{M^2 - 1}}\left(\frac{t}{c}\right)^2 = \frac{4}{\sqrt{1.5^2 - 1}}(0.03)^2 = 0.00322$$

$$\text{wing total } C_{D\,\text{wave}} = 0.00410 + 0.00322 = 0.00732$$

(based on exposed planform area)

$$\text{wing } D_{\text{wave}} = C_{D\,\text{wave}}\, qS = (0.00732)(619.16)160 = \underline{725.16 \text{ lb}}$$

↑ exposed area in this case

For the fuselage, from equation 12.17,

$$C_{DW} = 1.2\frac{9.87}{(\text{FR})^2} = 1.2\left(\frac{9.87}{72.25}\right) = 0.164$$

$$\text{fuselage } D_{\text{wave}} = (0.164)(619.16)(9) = \underline{913.5 \text{ lb}}$$

PROBLEMS

12.1. An airplane with an unswept, 12% thick wing, a wing planform area of 450 ft^2, a span of 60 ft, and a mean aerodynamic chord (m.a.c.) of 8 ft is flying at a density altitude of 28,000 ft at a speed of 400 mph. The ambient temperature is 430°R. The gross weight is 30,000 lb. The exposed wing area is 80% of the total wing area. The wing parasite drag is 35% of the total parasite drag. The airfoil is a conventional peaky type. Determine

(a) Lift coefficient.
(b) Total parasite drag in pounds.
(c) Boundary layer thickness at the trailing edge of the m.a.c., assuming the flat-plate turbulent boundary layer equation.
(d) Induced drag in pounds.
(e) Crest critical Mach number, M_{cc}.
(f) Compressibility drag and total drag in pounds and the ratio of lift to drag.

12.2. A turbofan transport aircraft has an equivalent parasite drag area of 39 ft^2. Wing area is 2900 ft^2, span is 148.5 ft, wing thickness ratio is 0.108, and the sweep of the quarter-chord line is 30.5 degrees. A peaky airfoil is used. The airplane is cruising at a Mach number of 0.82 at 35,000 ft pressure altitude. Ambient temperature is −51°C. Gross weight is 300,000 lb. Determine

(a) Crest critical Mach number, M_{cc}.
(b) ΔC_{D_c}, increment in drag coefficient due to compressibility.
(c) M_{DIV}.
(d) Induced drag coefficient (including effect of e).
(e) Total airplane drag in coefficient and force terms.
(f) Total lift-to-drag ratio.

12.3. A straight wing supersonic fighter with a circular arc airfoil is flying at $M = 2.0$ at 45,000 ft on a standard day. It has a wing area of 320 ft^2 and a span of 36 ft. The total weight is 21,000 lb. Eighty percent of the wing area is exposed. Assume only the exposed wing carries lift supersonically. Neglect tip effects. The total wing wave drag due to thickness and lift is 1630 lb. What is the wing thickness ratio?

12.4. A straight wing supersonic fighter with a 3.5% thick biconvex airfoil wing is flying at $M = 2.0$ at 47,000 ft on a standard day. It has a wing area of 170 ft^2 and a span of 25.2 ft. The total weight is 17,000 lb. Eighty percent of the wing area is exposed. Neglect tip effects. Assume only the exposed wing carries lift supersonically. What is the total wing wave drag due to both thickness and lift?

12.5. A straight wing supersonic fighter with a 3% thick diamond airfoil wing is flying at $M = 1.5$ at 40,000 ft pressure altitude. Temperature is 405°R. Wing area is 200 ft^2, of which 80% is exposed. Wing span is 28 ft. The total weight is 18,000 lb. Assume only the exposed wing carries lift supersonically. Neglect tip effects. Determine the total wave drag due to both wing thickness and lift.

12.6. A 410,000 lb DC-10 is flying at a pressure altitude of 35,000 ft on a standard day at a Mach number of 0.83. Wing area is 3520 ft^2, wing span = 155 ft, equivalent parasite area, f, = 57 ft^2, wing sweep is 35°, and t/c is 0.102. The airfoils are the peaky type. What is the total drag including parasite, induced, and compressibility drags?

REFERENCES

12.1. Nitzberg, G. E., and Crandall, Stewart, *A Study of Flow Changes Associated with Airfoil Section Drag Rise at Supercritical Speed,* NACA TN 1813, February 1949.

12.2. Shevell, R. S., and Bayan, F. P., *Development of a Method for Predicting the Drag Divergence Mach Number and the Drag Due to Compressibility for Conventional and Supercritical Wings,* SUDAAR 522, Stanford University, Department of Aeronautics and Astronautics, July 1980.

12.3. Whitcomb, R. T., *A Study of the Zero-Lift Drag-Rise Characteristics of Wing-Body Combinations near the Speed of Sound,* NACA TR 1273, 1956.

12.4. McCormick, Barnes W., *Aerodynamics, Aeronautics, and Flight Mechanics,* Wiley, New York, 1979.

12.5. Bertin, John J., and Smith, Michael L., *Aerodynamics for Engineers,* Prentice-Hall, Englewood Cliffs, N. J., 1979.

12.6. Anderson, John D., Jr., *Fundamentals of Aerodynamics,* McGraw-Hill, New York, 1984.

12.7. Sears, William R., "On Projectiles of Minimum Wave Drag," *Quarterly of Applied Mathematics,* Vol. 4, No. 4, Jan. 1947.

13

AIRFOILS AND WINGS

Airfoils are the cross-sectional shapes of wings as defined by the intersections with planes parallel to the freestream and normal to the plane of the wing. On sweptback wings, airfoils are sometimes defined in a plane perpendicular to the quarter-chord line. The definitions of the significant airfoil geometric parameters are shown in Figure 13.1.

The essential characteristics of airfoils can be derived from the fluid mechanics fundamentals we have discussed. First, the leading edge should be rounded, with the radius of curvature sufficiently large to avoid excessive suction. Such high suctions would have to be followed by a long and/or strong adverse pressure gradient and could lead to flow separation. Second, the trailing edge must be sharp in order to establish the Kutta–Joukowski condition. A substantial radius at the trailing edge of an airfoil at an angle of attack could allow the fluid to flow part of the way from the lower surface to the upper surface without excessive velocities. This would reduce the circulation and lift. In addition, there are structural considerations in airfoil design. A thicker wing makes a better structural beam, permitting the same load to be carried with less structural weight. Furthermore, the wing usually serves as the fuel tank, so maximum volume is desired. On the other hand, a large wing thickness leads to higher induced velocities due to thickness. This in turn causes higher adverse pressure gradients on the aft portion, thicker boundary layers, and greater pressure drag. For high-speed aircraft, the higher local velocities, due to greater thickness, reach the speed of sound at a lower freestream Mach number and reduce the M_{DIV} of the airfoil. The detailed shaping of the airfoil is also affected by the desire to get the greatest possible lift on the airfoil while still having the smallest possible negative pressure coefficient at the upper surface crest to maximize M_{DIV}.

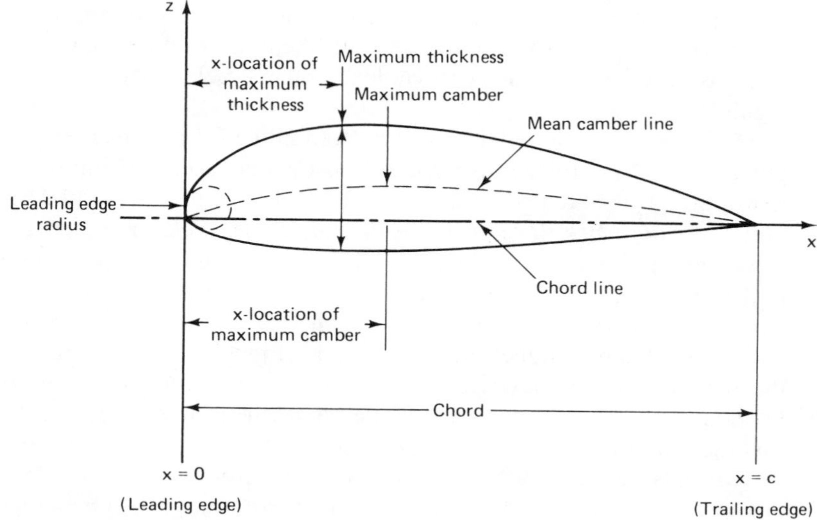

Figure 13.1 Airfoil geometric parameters. From J. Bertin and M. Smith, *Aerodynamics for Engineers,* 1979, p. 77. Reprinted by permission of Prentice-Hall, Inc., Englewood Cliffs, N.J.

Another important airfoil design goal is to achieve a nose shape that permits the highest possible lift coefficient before large-scale separation occurs. This maximum lift coefficient $C_{L_{\text{MAX}}}$ determines the minimum flight speed and is a primary factor in establishing the takeoff and landing runway-length requirements.

EARLY AIRFOIL DEVELOPMENT

Early airfoil researchers developed systematic methods in which mathematical curves described the thickness distribution for a symmetrical airfoil. Different mathematical equations described the curvature of the mean line between the upper and lower surfaces. The amount of this curvature is called the *camber* and is usually expressed in terms of the maximum mean line ordinate as a percent of chord. Given the thickness distribution equation and the equation of the mean line, the entire airfoil could be specified. The early development of airfoils is shown in Figure 13.2.

Designation	Date	Diagram	Designation	Date	Diagram
Wright	1908		Gottingen 387	1919	
Bleriot	1909		Clark Y	1922	
R.A.F. 6	1912		M-6	1926	
R.A.F. 15	1915		R.A.F. 34	1926	
U.S.A. 27	1919		N.A.C.A 2412	1933	
Joukowsky (Göttingen 430)	1912		N.A.C.A. 23012	1935	
Göttingen 398	1919		N.A.C.A. 23021	1935	

Figure 13.2 Early airfoil development. From C. B. Millikan, *Aerodynamics of the Airplane,* 1941. Reprinted by permission of John Wiley & Sons, New York.

The purpose of camber is to increase the airfoil maximum lift coefficient, $c_{l_{max}}$, (see Chapter 14) and to raise the lift coefficient for minimum parasite drag. The latter effect is only found at lower Reynolds numbers, below about 5 million, as shown in Figure 13.12.

The NACA (National Advisory Committee for Aeronautics, predecessor of the present NASA) airfoil series were the most widely used. Definitions of their four- and five-digit airfoils are shown in Figure 13.3. The term c_{l_i} used in defining the camber of the five-digit airfoils refers to the design, or ideal, lift coefficient for which the flow direction just ahead of the leading edge of the mean line is tangent to the mean line at the leading edge.

In the early 1940s, a new series of airfoils was developed known as the *six series* and designated by numbers such as 65–212. Figure 13.4 illustrates the significance of the numbers in this designation. The six series airfoils were designed to maintain laminar flow over a large part of the chord. This was to be achieved by having a favorable pressure distribution over much of the airfoil; that is, the pressure became increasingly negative, with respect to ambient, from the leading edge to the minimum pressure point well back on the chord. Favorable pressure gradients increase the transition Reynolds numbers and permit laminar flow to exist. The airfoils were very successful in wind tunnels, but with the amount of real-life surface roughness, dust, and insects, little significant laminar flow was obtained in operational flights. The one exception is a very low Reynolds number airplane with a very well maintained wing, a situation to be found primarily in sailplanes. As noted in Chapter 11, future use of graphite-epoxy or equivalent very smooth skins may improve the possibility of achieving practical laminar flow.

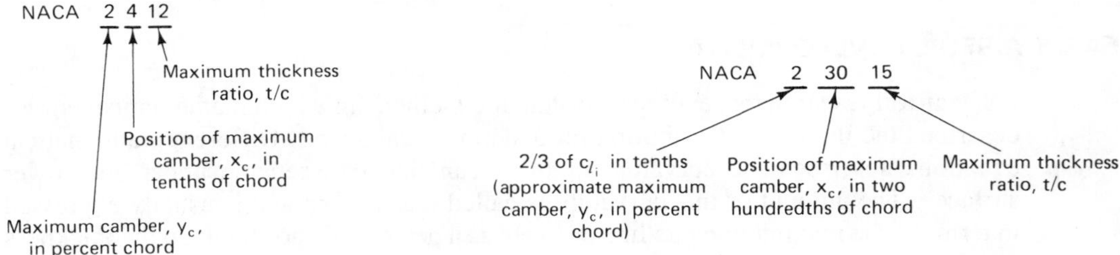

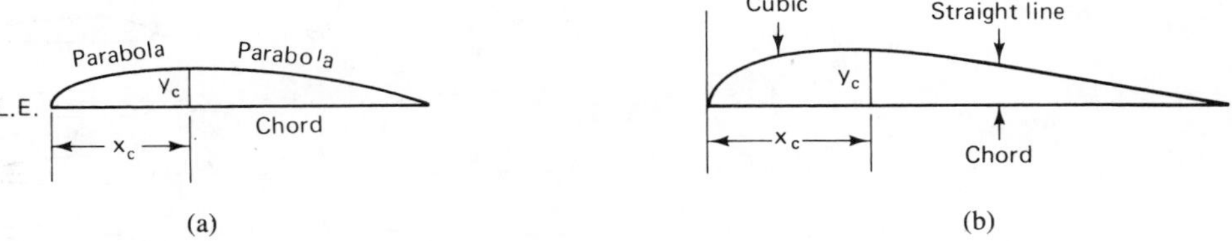

Figure 13.3 Definitions of NACA four- and five-digit series airfoils: (a) NACA four-digit series notation; (b) NACA five-digit series notation. From C. Perkins and R. Hage, *Airplane Performance, Stability and Control,* 1949. Reprinted by permission of John Wiley & Sons, New York.

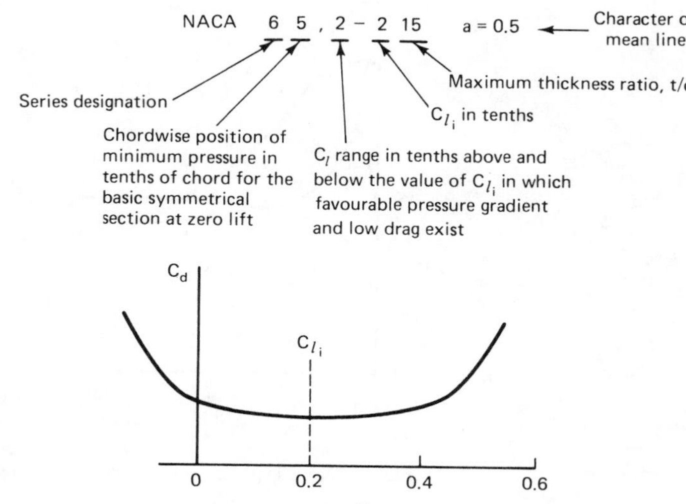

Figure 13.4 NACA six-series airfoils. From C. Perkins and R. Hage, *Airplane Performance, Stability and Control,* 1949. Reprinted by permission of John Wiley & Sons, New York.

MODERN AIRFOILS

When jet engines were invented and aircraft began to enter the high-subsonic-speed region where local sonic velocities became a problem, the six series came into their own. It was thought at the time that it was necessary to avoid sonic velocity anywhere on the airfoil. Since the six series tried to maintain a relatively flat top pressure distribution, it was exactly the type of pressure distribution needed to get the greatest possible lift with a given maximum negative pressure coefficient anywhere on the airfoil. Therefore, all the early jet aircraft used six-series airfoils. In the early 1950s, it became apparent that the six series was not the best high-speed airfoil and that, as has been discussed under compressibility drag, it is perfectly permissible and even desirable to have considerable supersonic velocity on the forward part of an airfoil. Therefore, the design criterion changed from avoiding supersonic velocity anywhere on the airfoil to avoiding supersonic velocity at or behind the crest. It was also necessary to have the peak negative pressure coefficient very far forward so that the slope of the pressure coefficient versus chordwise position was gentle for some distance ahead of the crest. This led to airfoils that were really closer to the old-fashioned NACA four-digit airfoil in terms of having nose peaks, but different from them in having rather sharp nose peaks. These airfoils are called *peaky airfoils*. They are found on most modern transport aircraft. Figure 13.5 shows the difference between the old-fashioned pressure distribution, the "laminar flow" six series, and contemporary peaky airfoils. Only the upper surface pressures are shown.

Many general aviation aircraft were designed with six-series airfoils starting in the late 1940s, either to take advantage of the alleged laminar flow or just to look modern. The laminar flow was never attained, and the six-series airfoils had poor maximum lift. Even if laminar boundary layers had been attainable in general, the propeller slipstream would probably have caused transition on the wing immersed in the slipstream anyway.

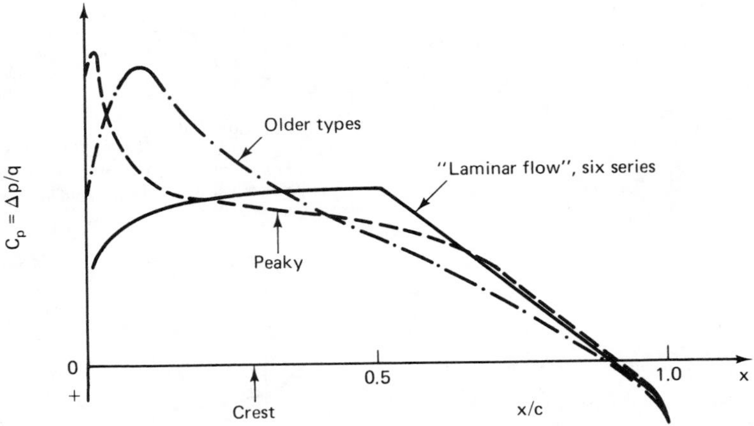

Figure 13.5 Types of airfoil upper surface pressure distribution, $M = 0$.

Today, practically all airfoils are tailormade for their particular design requirement. The old NACA four-digit, five-digit, and six-series airfoils are seldom used except by general aviation aircraft manufacturers, who have often been quite unsophisticated in the airfoil business. In recent years, NASA has undertaken a large educational effort to introduce the most modern airfoil concepts to the general aviation field, and future designs will probably show considerable improvement for this reason.

The most promising high-speed airfoil development for the future is called the *supercritical airfoil*. For any given thickness ratio, this airfoil provides a higher M_{DIV} than conventional airfoils. The advantage in M_{DIV} is about 0.06 above the best peaky airfoils and 0.09 above the six series.

The supercritical airfoil was pioneered by Richard Whitcomb at NASA's Langley Research Center. The "supercritical" title is a misnomer since present peaky airfoils all operate at cruise with large regions of supercritical ($M_{local} > 1.0$) flow with little drag penalty. The supercritical airfoils exploit this effect to a greater extent by having very small curvature over much of the upper surface so that the aft-facing surface has very little vertical projected area for a considerable distance behind the crest. Furthermore, by cambering the aft portion of the airfoil to carry more load aft, the C_p at the crest can be lowered for any given C_L, thereby raising M_{cc}. Another feature is the tangency of the upper and lower surfaces at the trailing edge. This reduces the high adverse pressure gradient at the rear and permits the aft cambering without excessive pressure drag. Figures 13.6 and 13.7 illustrate the character and M_{DIV} advantage of the supercritical airfoil.

As in so many technical advances, there are some negative aspects to the supercritical wing. The very thin trailing edge is a structural problem, although honeycomb construction can probably provide the required strength and stiffness without excess weight penalty. The aft camber leads to a large negative pitching moment, which must be balanced by the tail and causes trim drag, as discussed later in this chapter.

At the wing root, the large adverse pressure gradient affects not only the airfoil boundary layer, for which the airfoil has been designed, but also the fuselage boundary layer. Thus wing root flow separation is a problem at higher angles of attack. On swept

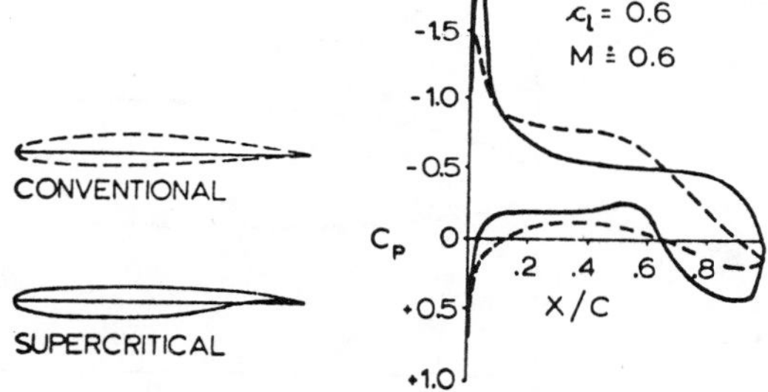

Figure 13.6 Supercritical airfoil. From R. S. Shevell, "Technological Development of Transport Aircraft, Past and Future." Reprinted from AIAA Paper No. 78-1530.

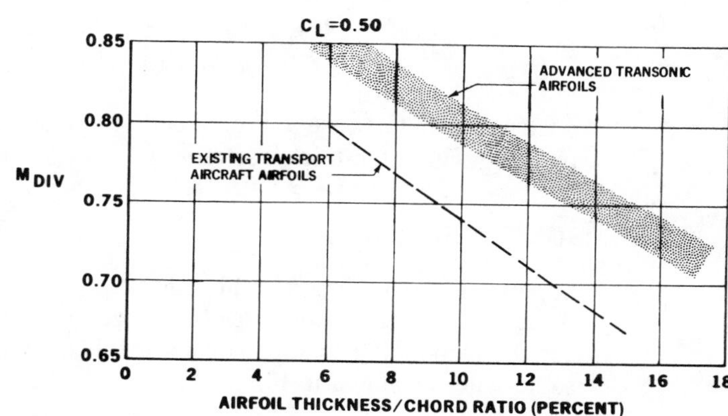

Figure 13.7 Advanced transonic (supercritical) airfoil performance. From R. S. Shevell, "Technological Development of Transport Aircraft, Past and Future." Reprinted from AIAA Paper No. 78-1530.

wings, this is less serious because the fuselage–wing interference due to sweep requires less camber on the root airfoils anyway.

In spite of these problems, the supercritical wing can either permit greater speed with a given wing sweep and thickness or less sweep and/or greater thickness for a given M_{DIV}. Less sweep and greater thickness reduce wing weight. Less structural weight leads to a smaller wing area, reduced fuel required, and lower operating costs.

As an extension of the work on the supercritical wing, NASA has developed special airfoils for low-speed general aviation aircraft. Here the objective was reduced drag at the higher lift coefficients at which climbing flight is performed and an increased $c_{l_{\max}}$. The result of this effort was the GAW-2 airfoil (Figure 13.8). Although the drag curve was much improved at high C_L compared to the six-series airfoils, which never should have been used on general aviation aircraft, it is little better than the older NACA airfoils such as the 2415 and the 23015. However, the $c_{l_{\max}}$ is superior for the GAW-2. The GAW-2 attains this higher $c_{l_{\max}}$ by using a large nose radius and by carrying more lift on the aft portion of the airfoil. The latter leads to a steeper adverse pressure gradient, which the airfoil can sustain but which may cause serious flow separation at the fuselage juncture. At the root, the fuselage boundary layer mixes with the wing boundary layer and reduces the ability of the flow to travel to the trailing edge without separation. This is an example of the need to consider the whole airplane and not just one component in making design decisions. However, a modified GAW-2 may yet turn out to be a successful airfoil design. Further work on general aviation airfoils is being carried out with a new airfoil family designation in which the GAW-2 is now called the LS-0413. The "04" signifies the camber by specifying the lift

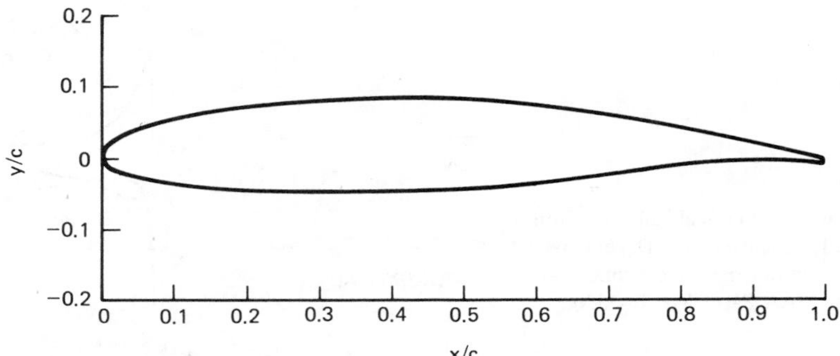

Figure 13.8 NASA GAW-2 airfoil.

coefficient, in this case 0.4, obtained at zero angle of attack of the chord line. The last two digits give the thickness ratio.

SUPERSONIC AIRFOILS

Supersonic airfoils are very thin since $C_{D_{\text{wave}}}$ varies with $(t/c)^2$. A highly swept supersonic wing with subsonic leading edges is a very complex cambered and twisted surface responding to sophisticated mathematical analyses. An example of a supersonic wing is shown in Figure 13.9.

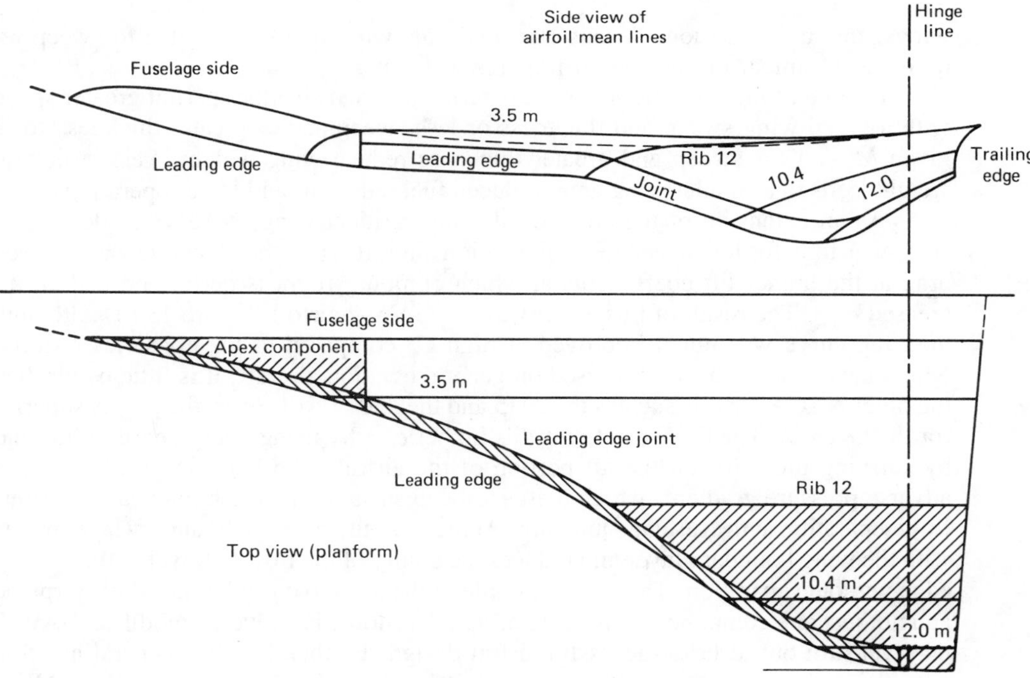

Figure 13.9 Concorde wing camber and twist distribution. From J. Rech and C. S. Leyman, "Concorde Aerodynamics and Associated Systems Development." Reprinted from the AIAA Professional Study Series by permission of the American Institute of Aeronautics and Astronautics.

AIRFOIL PITCHING MOMENTS

We have mentioned that subsonic airfoil theory shows that lift due to angle of attack acts at a point on the airfoil 25% of the chord aft of the leading edge. This location is called the *quarter-chord point*. The point through which this lift acts is the aerodynamic center (a.c.). In tests, the a.c. is usually within 1% or 2% chord of the quarter-chord point until the Mach number increases to within a few percent of the drag divergence Mach number. The aerodynamic center then slowly moves aft as the Mach number is increased further.

Lift due to the usual airfoil mean line curvature, camber, acts aft of the quarter-chord. For example, a circular arc mean line produces lift at the 50% chord point. Figure 13.10 shows such an airfoil at the angle of attack for zero lift. The lift due to angle of attack L_α is negative, canceling the lift due to camber L_c.

It is clear that even at zero lift a pitching moment exists. This moment is a couple, $M_{\text{a.c.}} = L_c(0.25c - 0.50c)$. In coefficient form, the moment at zero lift is $C_{M_{\text{a.c.}}} = M_{\text{a.c.}}/qSc$ and with positive camber is always negative. A pitching moment is negative when it tends to pitch the nose down.

It will be seen in Chapter 16 that the airplane center of gravity (c.g.) must be close to the aerodynamic center or excessive moments will be created. The c.g. is usually within 10% $\bar{c}$ of the a.c., although it may be varied $\pm 15\%\ \bar{c}$ to 20% $\bar{c}$ from the a.c. in certain designs.

Even if the center of gravity is located at the a.c. so that no moments about the c.g. occur due to lift due to angle of attack, $C_{M_{\text{a.c.}}}$ will still exist unless the airfoil is symmetrical (i.e., no camber). Since $C_{M_{\text{a.c.}}}$ is almost always negative (i.e., nose down pitching moment), almost all airplanes fly with some download on the tail to balance the $C_{M_{\text{a.c.}}}$. The tail download, together with the additional wing lift required, causes a small but measurable drag penalty called *trim drag*. The tail download also affects the tail structural weight. Thus a large $C_{M_{\text{a.c.}}}$ is to be avoided, although the camber that creates it is necessary to achieve high $C_{L_{\text{MAX}}}$ and good high-speed characteristics at cruise. Another design compromise!

It is interesting that the trim drag penalty, usually on the order of 1% to 2%, would be much larger if the wing downwash did not tilt the negative tail lift vector forward, providing significant favorable wing–tail interference.

The equation for trim drag may be written as

$$D_{\text{trim}} = \frac{L_W^2}{\pi q b^2 e} + L_H \tan \epsilon_H + \frac{L_H^2}{\pi q b_H^2 e_H} - \frac{L^2}{\pi q b^2 e} \tag{13.1}$$

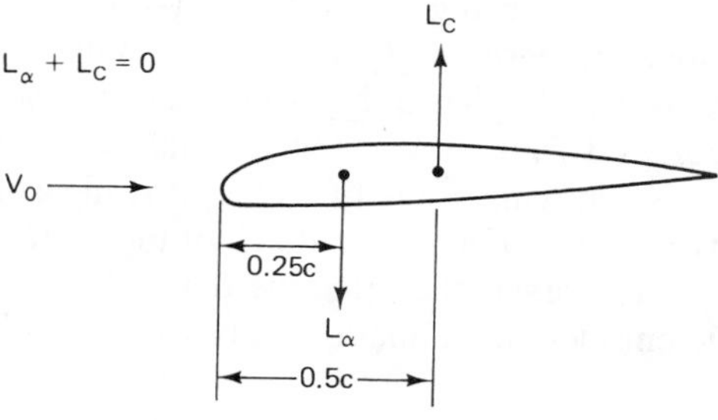

Figure 13.10 Cambered airfoil at zero lift.

The first term is the induced drag of the wing with lift, L_W. Airplane weight, or total lift, is the sum of the wing lift and tail lift, L_H. Since the tail lift is generally negative (downward), L_w is larger than L. The second term is the result of wing downwash tilting the tail lift vector. ϵ_H is the angle of downwash at the tail due to wing lift. With negative tail lift, the downwash produces a forward tilt and the term is a negative drag. The third term is the induced drag of the horizontal tail due to its own lift. This is usually much smaller than the second term. The sum of the first three terms is the induced drag of the trimmed airplane. The last term is the induced drag if there is no tail load.

If the airplane is flying with some compressibility drag, another term must be added. The higher wing lift will lower M_{DIV} and cause a small increase in compressibility drag.

The drag methods given in Chapters 11 and 12 include an average trim drag since the empirical factors, such as (1) the equivalent sand grain roughness, (2) the k in equation 11.8 for airplane efficiency factor, e, and (3) the incremental compressibility drag coefficient in Figure 12.13, are based on the flight drags of various aircraft without explicitly accounting for the trim drag. This is reasonable unless a design has an unusual tail trim load, in which case an estimate of the difference in trim drag from a normal amount must be included.

The aerodynamic center of supersonic airfoils is at the midchord point. This can be easily seen from equation 12.11 and Figure 12.22. The pressures due to angle of attack on both the upper and lower surfaces of the flat plate airfoil are constant along the chord. Therefore, the center of lift is always at the center of area. This illustrates one of the problems of supersonic airplane design. As the airplane goes from subsonic to supersonic speeds, the aerodynamic center moves from the 25% chord point to the 50% chord point. A large change in moment occurs and the control surfaces must be designed to accommodate it. The resulting large load on the control surfaces adds significant trim drag. A solution to this problem is to shift the fuel between tanks to move the center of gravity of the airplane close to the center of lift for whatever speed regime is appropriate. One advantage of highly swept delta wings with subsonic leading edges is that the shift in the position of the center of lift between the subsonic and supersonic speed regimes is much reduced.

EFFECT OF SWEEPBACK ON LIFT

Wing sweepback affects lift as well as drag. Sweepback reduces the slope of the lift curve, $dC_L/d\alpha$, and moves the aerodynamic center aft.

Equations 8.13 showed that, according to the Prandtl–Glauert approximation, the airfoil section lift curve slope increased as $1/\sqrt{1 - M_0^2}$. With sweepback angle, Λ, the effective Mach number perpendicular to the swept isobars is $M_0 \cos \Lambda$, so the Mach number correction becomes $1/\sqrt{1 - M_0^2 \cos^2\Lambda}$. Analysis of the geometry of a swept wing shows that a change in wing angle of attack, $\delta\alpha$, measured at the plane of symmetry yields an effective change of angle of attack of the section perpendicular to the sweep line of $\delta\alpha/\cos \Lambda$. However, the effective dynamic pressure is reduced by $\cos^2\Lambda$. Thus the lift coefficient increment due to $\delta\alpha$, based on freestream dynamic pressure, is changed by $1/\cos \Lambda \cdot \cos^2\Lambda$ or simply reduced by $\cos \Lambda$. The section lift curve slope is then

$$\frac{dc_l}{d\alpha} = \frac{a_0 \cos \Lambda}{\sqrt{1 - M_0^2 \cos^2\Lambda}} \tag{13.2}$$

Letting $a_0 = 2\pi\eta$ and substituting equation 13.2 in equation 9.20, the Prandtl equation for the lift curve slope of a finite wing, yields

$$\frac{dC_L}{d\alpha} = \frac{2\pi \text{AR}}{2 + \sqrt{(\text{AR}^2/\eta^2)(1 + \tan^2\Lambda - M_0^2)}} \tag{13.3}$$

For aspect ratios usually used for transports and general aviation aircraft, this equation works quite well. A modified equation, the development of which is given in Refs. 13.5 and 13.6, has been shown to give better agreement with experiment at lower aspect ratios, especially at aspect ratios below about 6. This equation is

$$\frac{dC_L}{d\alpha} = \frac{2\pi \text{AR}}{2 + \sqrt{(\text{AR}^2/\eta^2)(1 + \tan^2\Lambda - M_0^2) + 4}} \tag{13.4}$$

Equation 13.4 is probably the most widely used method for estimating lift curve slope when a complete computational aerodynamic analysis is not available. Note that equation 13.4 gives the rate of change of C_L per radian.

In discussing the effect of wing sweepback angle on drag, the sweep angle has been defined as the sweep of the quarter-chord line of the wing. On sweptback tapered wings, typical of almost all high-speed aircraft, the sweep angle depends on which constant percent chord line is used to establish the sweep. The leading edge has more sweep than the trailing edge. For the lift equations (13.2, 13.3, and 13.4), the sweep angle defined by the 50% chord line has been established as the appropriate sweep angle.

The aft movement of the aerodynamic center with increase in sweepback angle occurs because the effect of the downwash pattern associated with a swept wing is to raise the lift coefficient on the outer wing panel relative to the inboard lift coefficient. Since sweep moves the outer panel aft relative to the inner portion of the wing, the effect on the center of lift is an aft movement. The effect of wing sweepback on aerodynamic center position is shown in Figure 13.11 for aspect ratios of 7 and 10 and for taper ratios of 0.25 and 0.50. For a taper ratio of 0.25 and with 35 degrees of sweep, typical of many jet transports, the aerodynamic center is moved aft to 32% to 35% m.a.c., depending on the aspect ratio. With 25 degrees of sweep, the aerodynamic center is moved aft to about 29% to 31% m.a.c. Note that sweep angle in Figure 13.11 is defined in the usual manner as the sweep of the $c/4$ line.

In Chapter 14, it is shown that the higher outer panel lift coefficients must be carefully considered in designing for safe wing stall characteristics. In Chapter 16, we shall see that the change in aerodynamic center position is important in the analysis of airplane stability.

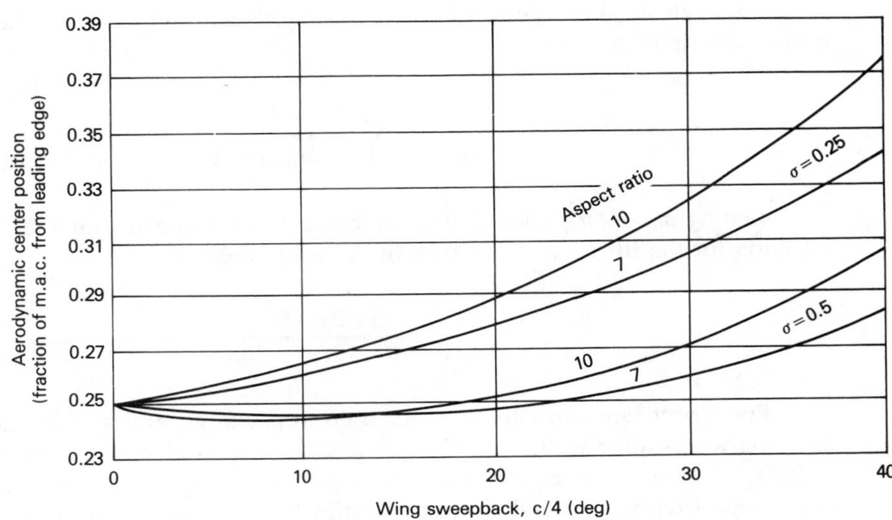

Figure 13.11 Effect of wing sweepback on aerodynamic center position for several aspect ratio and taper ratio combinations (Courtesy of Ilan Kroo.)

AIRFOIL CHARACTERISTICS

Airfoil characteristics are determined both by test and calculation. Airfoil tests are conducted in wind tunnels, usually with a constant chord model that spans the tunnel from wall to wall to provide two-dimensional flow. The results are presented in the form of the force coefficients c_l and c_d, the moment coefficient c_m, and the angle of attack α, as shown in Figure 13.12. The lowercase letters used for the coefficients signify that the characteristics are section or two-dimensional data, rather than the results for a finite wing.

Because of the effects of Reynolds number, particularly on c_d and $c_{l_{\max}}$, the data are often determined for several Reynolds numbers. In Figure 13.12, the drag curves display a reduced rate of increase of c_d with c_l as Reynolds number increases from 3 to 9×10^6. This drag increase is due to increased local velocities and pressure drag. It is not related to downwash from trailing vortices because a two-dimensional airfoil has none. The higher Reynolds number thins the boundary layer and reduces the pressure drag. The result is a lower value of k in equation 11.7. In addition, the minimum drag level is reduced with higher Reynolds number. This is due to the effect of Reynolds number on the skin friction coefficient shown in Figure 11.2. There is a counteracting effect, however, that tends to increase the drag. As the airfoil Reynolds number increases, the transition critical Reynolds number occurs farther forward on the wing. The laminar flow region is reduced and the average drag increased. Since the wind tunnel model is very smooth and always cleaned before a test, considerable laminar flow may exist on the airfoil. In real aircraft at higher Reynolds number and with normal construction and maintenance, the transition will always occur close to the leading edge. Care must be taken in interpreting wind tunnel tests to sort out these differences. This is usually done by applying a roughness strip to the leading edge of

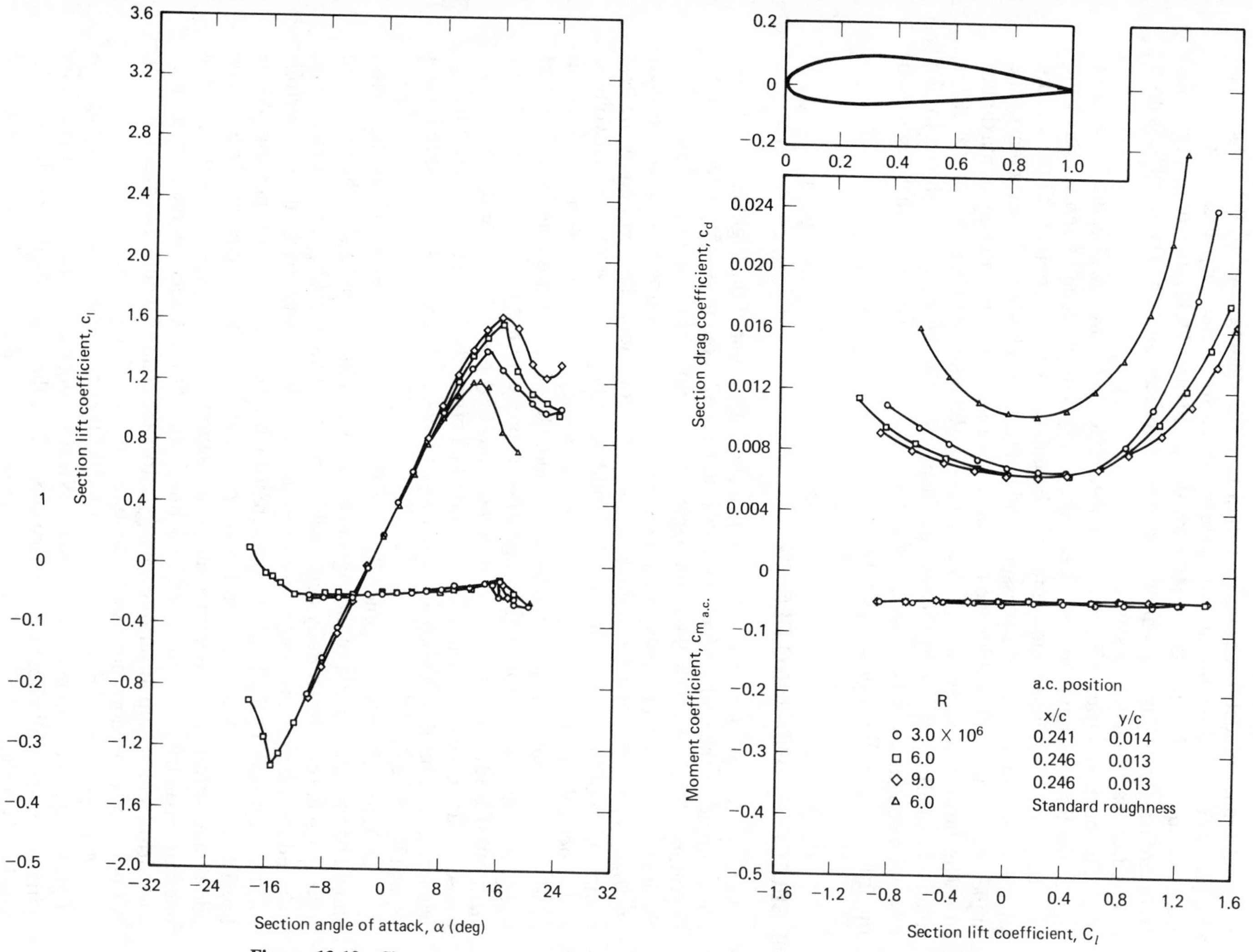

Figure 13.12 Characteristics of the NACA 2415 airfoil section. From I. Abbott and A. von Doenhoff, *Theory of Wing Sections*, 1959. Reprinted by permission of Dover Publications, Inc., New York.

the wind tunnel model to ensure transition near the leading edge of the wing. This roughness must be the minimum to cause transition so that the strip does not raise the drag measurably due to its own thickness. In Figure 13.12, the drag curve for "standard roughness" is much higher than the smooth airfoil data. The roughness used in that test was clearly excessive.

The other significant Reynolds number effect in Figure 13.12 is the increase in $c_{l_{max}}$ as the Reynolds number increases. Since $c_{l_{max}}$ is determined by separation of the boundary layer, a thinner, more energetic boundary layer can continue to the trailing edge against a higher adverse pressure gradient (i.e., at a higher lift coefficient). The higher Reynolds number thins the boundary layer and raises the attainable maximum lift. The further increase in $c_{l_{max}}$ due to raising Reynolds number above 9×10^6 is relatively small, as shown by the data in Chapter 14. The lowest $c_{l_{max}}$ is shown for the case with roughness, which, as noted previously, was excessive. Further discussion on airfoil $c_{l_{max}}$ is given in Chapter 14.

AIRFOIL SELECTION AND WING DESIGN

Selection of airfoils for a wing starts with a clear statement of the flight requirements. For example, if the cruise Mach number is to be 0.82, compressibility effects will be important. Since, as we shall see in Chapter 15, the cruise Mach number for jet aircraft should be close to but probably slightly less than M_{DIV}, the airfoil should have as high an M_{DIV} as possible for a given thickness and lift coefficient. This will allow the best combination of thickness ratio, which should be as large as possible for fuel volume and structural weight reasons, and sweepback angle, which must be high enough to meet the M_{DIV} requirements but otherwise should be as low as possible for weight and $C_{L_{MAX}}$ reasons. The airfoil with a high M_{DIV} is probably not the airfoil with the best $c_{l_{max}}$, but if leading edge devices are to be used, they can often make up for too low a nose radius on the basic airfoil. $c_{l_{max}}$ is important primarily during takeoff and landing, when the high lift devices would be deployed anyway. Flaps-up stalling speed is not especially critical.

Different airfoils will generally be selected at the wing root, where the airfoil should be thicker because of the higher structural bending loads, and at the tip, where a thinner section is structurally permissible, may raise the local $c_{l_{max}}$, and, in addition, may reduce the pressure drag caused by thickness. With sweepback, the root airfoils must be adjusted for the fuselage interference with the flow of the swept wing. At the tip airfoil, the tip flow effect on the swept isobars must be considered. At least one additional airfoil at an intermediate span station of a swept wing, usually about one-third span from the root, must be defined to limit the influence of the compensated root airfoil and to help in varying the $c_{l_{max}}$ across the span as required to avoid tip stall. An unswept wing does not usually require a third defining airfoil.

A proper variation of $c_{l_{max}}$ and the local lift coefficient across the wing span is of great importance in avoiding tip stall. When a wing reaches the angle of attack for initial stall, the stall should occur first inboard on the wing. At that angle the outer panel airfoils should be 2 to 3 degrees below their local stall angle. With smooth air flow over the ailerons, the pilot retains good roll control. If the outer panel stalls first, a large rolling moment is created with little roll control available to counteract it. The roll occurs because airplanes are not usually perfectly symmetrical or because of a small yaw angle. In addition, the change in lift distribution resulting from an inboard stall decreases the downwash over the tail. The download on the tail is decreased, and an airplane nose down or negative pitching moment is produced. The airplane tends to pitch down out of the high angle of attack region causing the stall. With a tip stall the opposite is true, and the airplane is pitched deeper into the stall. Complete loss of control and an entry into a tail spin could result.

Tip stall on a swept wing is even more serious. If the outboard section of a swept wing stalls, the lift loss is behind the aerodynamic center of the wing. The inboard portion of the wing ahead of the aerodynamic center maintains its lift and produces a strong pitch-up moment, tending to throw the airplane deeper into the stall. Combined with the effect of tip stall on the pitching moment produced by the tail, this effect is very dangerous and must be avoided by airfoil selection and wing twist.

Taper ratio, the ratio of tip chord to root chord, is an important parameter. A wing with a low taper ratio (i.e. a highly tapered wing) tends to have higher lift coefficients on the outer portion as the downwash pattern changes to move the lift distribution toward the elliptical. This tends to encourage tip stall. On the other hand, a low taper ratio leads to larger chords and physical wing thickness inboard where the bending moments are the highest. It also moves the lift inboard, reducing the aerodynamic bending moments. Both of these effects are favorable for wing structural weight. The most efficient wings usually have low taper ratios, particularly when the wings are swept. The tendency toward tip stall is handled, as noted previously, by wing twist and airfoil variation along the span. Taper ratios below 0.2 are not generally used because the tip stall tendency becomes excessive. The exceptions are delta, or triangular, wings. These wings are of such low aspect ratio that the flow is very different from the flows discussed here.

A low-speed aircraft design is not concerned with M_{DIV} so that low-speed drag, wing weight, and $c_{l_{\max}}$ may dominate the study. Some airfoils provide better fuel volume than others, even at the same thickness ratio. In all airfoils, the pitching moment coefficient about the aerodynamic center is important both because of wing structural torsional loads and because of the load required on the tail to balance the moment.

Another airfoil characteristic is the shape of the lift curve at and beyond the stall angle of attack. An airfoil with a gentle drop in lift after the stall, rather than an abrupt, rapid lift loss, leads to a safer stall from which the pilot can more easily recover. In general, airfoils with high thickness and/or camber, in which the separation is associated with the adverse gradient on the aft portion rather than the nose pressure peak, have a more gradual loss of lift. Unfortunately, the best airfoils in this regard tend to have lower maximum lift coefficients and low M_{DIV}. An initial stall at the wing root progressing slowly toward the tip will yield the same flight characteristics and permit the use of the higher-lift airfoils.

Selecting airfoils is part of the overall wing design. In addition to airfoil selection, the wing will often require some twist or washout to avoid tip stalling and to produce a lift distribution close to the ideal elliptical loading. The problem may be further complicated by the presence of engine nacelles and pylons whose flow field may disturb a carefully designed wing. Obviously, the whole procedure is intricate and requires a good understanding of the physical processes plus the availability of high-quality wind tunnel results to confirm the design. In recent years the enormous increase in the capability of digital computers has resulted in analytical tools being developed to calculate the entire aircraft flow field. The methods of computational aerodynamics can estimate lift and drag, including, with considerable but not perfect success, the effects of viscosity and shock waves. Wings can be designed much faster and with an improved chance that the first wind tunnel model built will be close to the final configuration. The tunnel tests are still required for final validation of a design.

Selection of the wing incidence angle is an important aspect of wing–body integration. Wing incidence angle is typically the angle between the wing root chord line and the fuselage reference line. The fuselage reference line lies in the plane of symmetry and is usually defined parallel to the cabin floor. The prime criterion for wing incidence selection is that a level cabin floor is desirable for comfort reasons in cruising flight. The passengers prefer a level floor, and cabin crews do not enjoy pushing heavily laden food carts uphill through the cabin. In addition, fuselages usually have the least drag at a fuselage angle of attack close to zero. The design procedure is to estimate the average lift coefficient to be expected in cruising flight. Knowing the angle of the wing root chord line for zero wing lift (generally, a function of airfoil camber and wing twist), and calculating the lift curve slope from equation 9.21 or the equivalent more complex expression for swept wings, equation 13.4, the wing root chord angle of attack required to obtain the desired lift coefficient can be determined. The wing is then set on the fuselage so that it has this angle when the fuselage angle of attack is zero.

Fuselages with large upsweep over the rear portion to accept aft cargo doors may have their minimum drag at a small positive angle of attack. In such cases, the wing incidence will be reduced accordingly. Another, less fundamental, consideration is that stopping performance during landing or an aborted takeoff is improved with less wing lift. For this case, it is important to get as much weight on the braked wheels as possible. Thus, there is a benefit to reducing the wing incidence slightly to the extent that the change is not felt significantly in the cabin. Reducing the nose gear length also helps. This technique is limited because a level cabin floor is also desirable on the ground. For combat aircraft, the level floor is not a consideration.

There is another airfoil and wing characteristic that must be considered by the designer of high-speed aircraft. It is the effect of Mach number on the maximum lift coefficient. The maximum lift coefficient reduces sharply as the Mach number increases above about 0.25. Because of the large increase in local velocity over the nose of an airfoil at high angle of attack, sonic velocity may be exceeded locally even though the freestream velocity is low. Shock waves then result that cause early separation and stall. Since stalling speeds are primarily critical for takeoff and landing operations, the determination of airplane stalling speed in flight should be done at altitudes representative of the highest airport elevations anticipated. This ensures that realistic Mach numbers were experienced in the test. In addition to this effect on low-speed operations, the reduction of maximum lift coefficient at high Mach number cruise is very important. At a lift coefficient somewhat below the stall, some separation and buffet occur. *Buffet* is a shaking of the airplane due to the unsteady separated flow on the wing. The maximum allowable cruise lift coefficient must be selected to allow reasonable maneuvering, with a lift up to about 30% above the level flight value, without encountering initial buffet. The curve of lift coefficient for buffet onset versus Mach number is called the *buffet boundary* (Figure 13.13). Pilot operating handbooks include a graph showing the combination of airplane weight, altitude, and Mach number corresponding to the buffet boundary for level flight and for various amounts of excess maneuvering lift, i.e., various load factors; see Chapter 18. Wings must be designed so that the same requirements for avoiding tip stall and pitch up are met at the high Mach number stall as for the low-speed case.

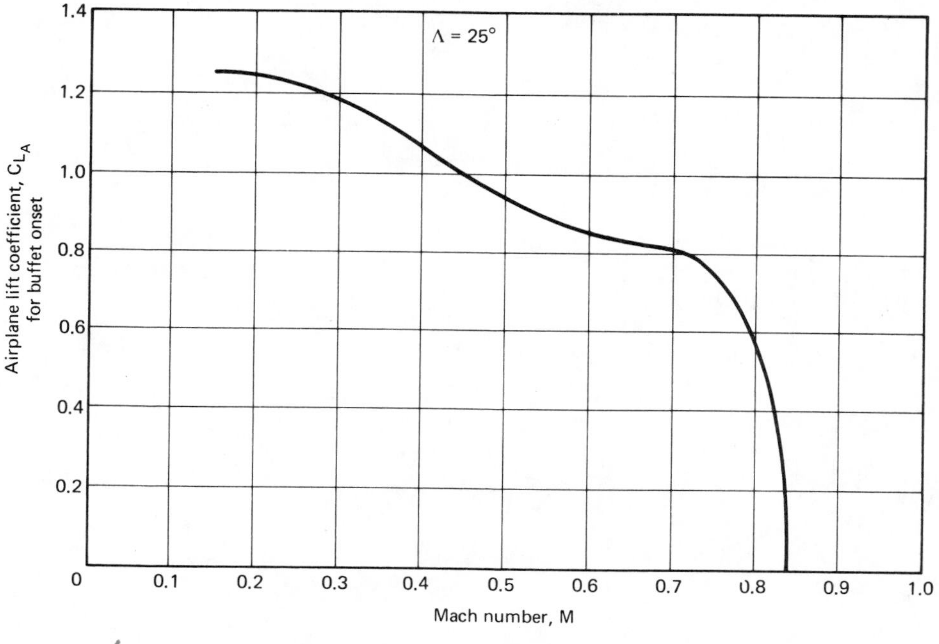

Figure 13.13 Typical buffet boundary.

PROBLEMS

13.1. Discuss the criteria for the selection of an airfoil. What are the desirable aerodynamic characteristics? Under what circumstances might the desirable structural properties conflict with aerodynamic goals?

13.2. What are the defining geometric characteristics of airfoils?

13.3. What are the advantages and disadvantages of the supercritical airfoil?

13.4. Is there a single optimum airfoil for each airplane wing? Justify your answer by discussing the factors involved in wing design.

REFERENCES

13.1. Millikan, Clark B., *Aerodynamics of the Airplane,* Wiley, New York, 1941.

13.2. Perkins, Courtland D., and Hage, Robert E., *Airplane Performance, Stability and Control,* Wiley, New York, 1949.

13.3. Rech, J., and Leyman, C. S., "Concorde Aerodynamics and Associated Systems Development," presented at AIAA Professional Study Workshop, Washington, D.C., June 1980.

13.4. Abbott, Ira H., and von Doenhoff, Albert E., *Theory of Wing Sections,* Dover, New York, 1959.

13.5. McCormick, Barnes W., *Aerodynamics, Aeronautics, and Flight Mechanics,* Wiley, New York, 1979.

13.6. *USAF Stability and Control Datcom,* Flight Control Division, Air Force Flight Dynamics Laboratory, Wright-Patterson Air Force Base, Ohio, Oct. 1960 (revised Jan. 1975).

14

HIGH-LIFT SYSTEMS

The runway lengths required for the takeoff and landing of aircraft depend directly on the square of the speed that must be attained at liftoff for takeoff or maintained down to a specified height, usually 50 ft, in landing. The lower the speed, the shorter the runway can be. These minimum flight speeds are almost always selected as a percentage above the stall speed in order to provide a margin of safety. The stall speed is given by

$$V_S = \sqrt{\frac{2W}{\rho S C_{L_{\mathrm{MAX}}}}}$$

where

W = airplane gross weight
S = wing area
$C_{L_{\mathrm{MAX}}}$ = maximum attainable lift coefficient

The stall speed is the lowest speed at which steady controllable flight can be maintained. Any further increase in angle of attack will cause flow separation on the wing upper surface, a drop in lift, a large increase in drag, and, in a well-designed airplane, a strong pitch-down moment. The pitch-down automatically reduces the angle of attack to assist a quick recovery from the stalled altitude. The C_L at the stall speed is $C_{L_{\mathrm{MAX}}}$.

High-lift systems consist of leading and trailing edge devices whose primary purpose is to obtain the highest $C_{L_{\text{MAX}}}$. The secondary goal of high-lift systems is to obtain the highest possible ratio of lift to drag, L/D, at the high C_L used at takeoff. The objective is to obtain high $C_{L_{\text{MAX}}}$ at the takeoff flap deflection, but suffer the least possible drag at the moderately lower C_L at which the airplane normally flies during climb after takeoff. The lift coefficient used in the climb immediately after takeoff is chosen to provide an adequate speed margin above the stall speed. A frequently used value is

$$C_{L_{\text{TO. CLIMB}}} = \frac{C_{L_{\text{MAX}}}}{(1.2)^2} = 0.7C_{L_{\text{MAX}}}$$

This corresponds to 1.2 V_S, the lowest allowable climb speed for jet aircraft and twin-engine propeller-driven aircraft. Four-engine propeller airplanes may climb at 1.15 V_S. Then

$$C_{L_{\text{TO. CLIMB}}} = \frac{C_{L_{\text{MAX}}}}{(1.15)^2} = 0.76\,C_{L_{\text{MAX}}}$$

A third requirement is to develop reasonably high drag at the lift system settings for landing in order to produce an adequately steep angle of descent in final approach.

The high-lift system must meet these aerodynamic requirements with the least possible structural weight and a high degree of mechanical reliability. The best starting point for a good high-lift system is an airfoil with the highest $c_{l_{\max}}$ consistent with other requirements, such as high M_{cc}.

AIRFOIL $c_{l_{\max}}$

One of the most satisfactory airfoil $c_{l_{\max}}$ prediction methods is a blend of theory and empiricism and is based on correlating the theoretical pressure difference between the peak nose pressure and the pressure at 90% chord with $c_{l_{\max}}$ (Ref. 14.1). The method is limited to thin or moderately thick airfoils. Another approximate but educational method of estimating airfoil $c_{l_{\max}}$ is shown in Figure 14.1, where the airfoil $c_{l_{\max}}$ is plotted versus a nose radius parameter, Δy. Δy is the change in height of the upper surface coordinate between 0.15% chord, a point very close to the leading edge, and 6% chord. Δy is expressed in percent of the chord and is a simple measure of the effective radius of the top surface of the airfoil nose. Since a large radius is associated with lower velocities, a higher Δy means that a higher angle of attack can be tolerated before the boundary layer finds the upper surface adverse gradient too great to overcome all the way to the trailing edge. Data are shown for many kinds of airfoils, and although the scatter band is $\pm 10\%$, most of the data are within 5% of the average fairing. Although leading edge shape is not completely responsible for airfoil $c_{l_{\max}}$, Figure 14.1 shows that it is the dominant parameter. Because a thicker wing has a larger leading radius, assuming that only the thickness distribution about the mean line

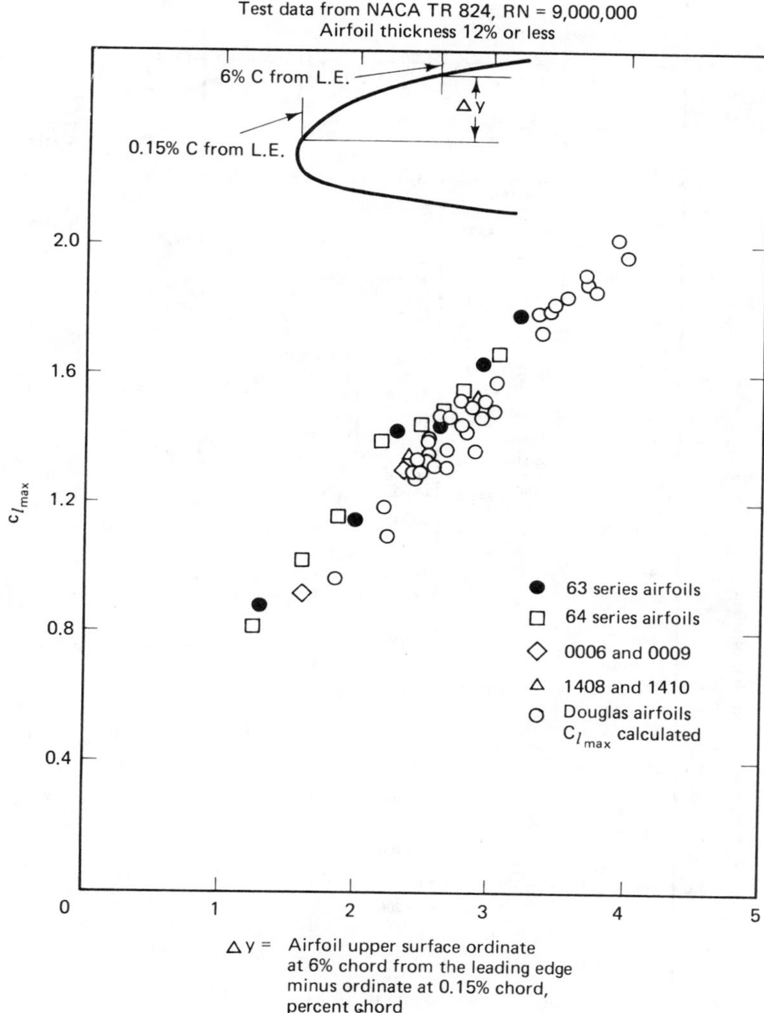

Figure 14.1 Relationship between airfoil nose shape and section maximum lift coefficient. (Courtesy of Douglas Aircraft Company.)

is changed, a higher thickness ratio is generally associated with higher $c_{l_{max}}$. This is true until the larger adverse gradient due to the thickness becomes dominant and starts decreasing $c_{l_{max}}$. That is why Figure 14.1 is labeled "airfoil thickness 12% or less."

Figure 14.2 is a summary of the $c_{l_{max}}$ for many airfoils. $c_{l_{max}}$ is plotted against the ideal lift coefficient. Since mean-line camber raises the ideal lift coefficient, the data show that in general maximum lift increases with increase in camber. $c_{l_{max}}$ also increases with thickness ratio up to about 12%. The scatter is enormous, however, if only those two parameters are considered. The character of the airfoil shape is the other general factor, particularly the shape of the nose, as indicated by the reasonably successful correlation in Figure 14.1.

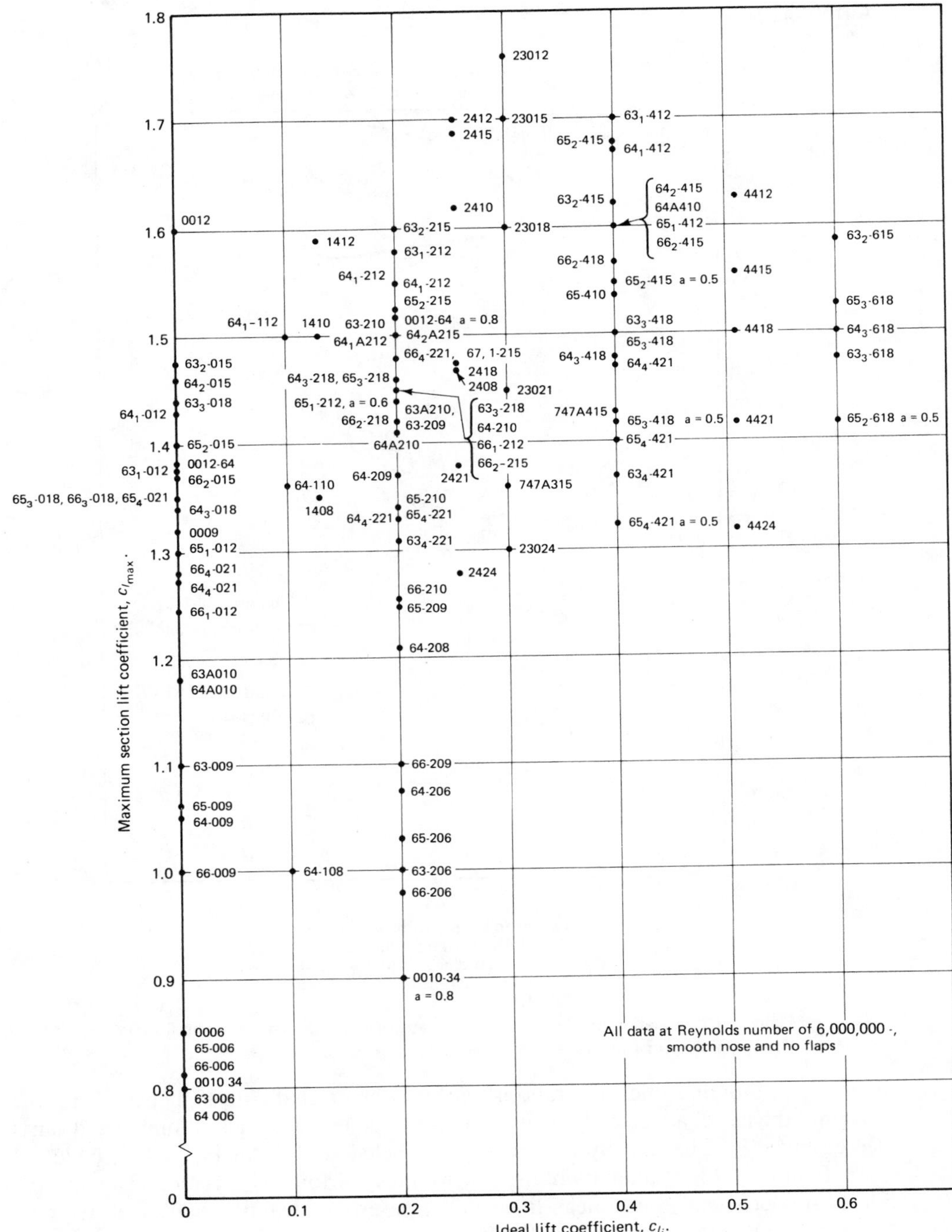

Figure 14.2 Variation in maximum lift coefficient for various airfoil section ideal lift coefficients. Data from I. Abbott and A. von Doenhoff, *Theory of Wing Sections,* 1959. Used by permission of Dover Publications, Inc., New York.

LEADING AND TRAILING EDGE DEVICES

The highest lift on a wing system must come from distributing lift as evenly as possible over the chord, so that, with Δp being the pressure increment between the lower and upper surfaces,

$$\int_{\mathrm{LE}}^{\mathrm{TE}} \Delta p \, dx$$

will be as large as possible for a given peak pressure. The wing design must avoid an excessive adverse pressure gradient over the aft portion, or energy must be added to the boundary layer to permit attached flow against large adverse gradients.

Methods of increasing wing $C_{L_{\mathrm{MAX}}}$ are then as follows:

1. Keep the nose peak pressure relatively low.
2. Increase the lift over the mid and aft sections of airfoil.
3. Add fresh air to the boundary layer behind peak pressures or, as is usually the case, replace the original boundary layer by displacing it with a stream of air from the lower surface.

Nose peak pressures are controlled by the following factors:

1. Airfoil nose radius and/or camber that lower nose peak pressures.
2. Leading edge flaps, which change nose camber only.
   (a) Simple nose flap.
   (b) Kruger flap, which rotates a portion of the lower surface forward about a simple hinge.
3. Leading edge slots, which improve boundary layer energy by ducting air from the bottom to the top of the wing. Higher peak pressures can be tolerated.
4. Leading edge slats, which provide both camber and boundary layer energy improvement.

Various leading edge devices are shown in Figure 14.3. The most effective method used on all large transports is the leading edge slat. Careful studies in wind tunnels

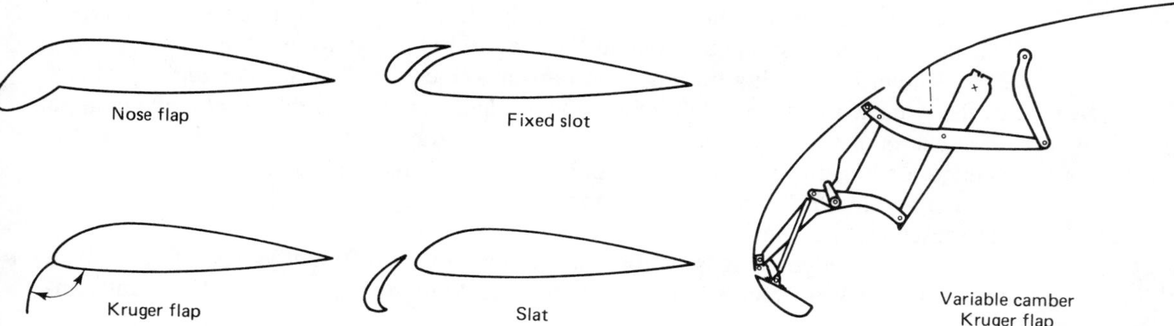

Figure 14.3 Types of leading edge devices. (Kruger flap courtesy of Boeing Commercial Airplane Co.)

are conducted to determine the optimum angle of deflection and position of the slat. An example of such a study is shown in Figure 14.4.

A variant on the leading edge slat is a variable camber slotted Kruger flap (VCK) used on the Boeing 747. Aerodynamically, this is a slat, but mechanically it is a Kruger flap. It uses a linkage that produces the slot and bends the fiber-glass skin to an ideal slat contour, permitting optimum performance. The mechanical complexity is considerable, however.

Lift over mid and aft sections of the airfoil is increased by airfoil camber and trailing edge flaps. There are many types of wing trailing edge flaps:

1. Split flap

2. Plain flap
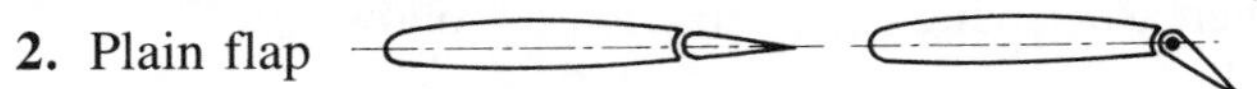
3. Single-slotted flap
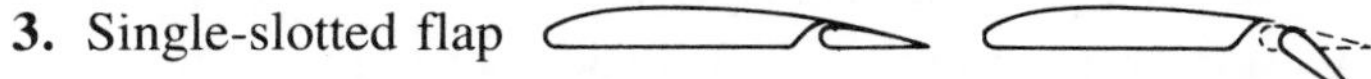
4. Fowler flap: a single slot + chord extension
5. Double-slotted flap: the usual vane + flap type is illustrated in Figure 14.5a on a wing also equipped with a slat. Figure 14.5b shows a recent form of the double-slotted flap using a two-segment flap. Both of these designs obtain the lift benefits of chord extension either by placing the hinge point below the wing, as in Figure 14.5a, or by use of a track system.
6. Triple-slotted flap (i.e., two vanes + flap or one vane + flap + auxiliary flap), as shown in Figure 14.6 on a wing with a slat. The version with two vanes was specifically designed to fold into a small space behind a landing gear strut.

Slotted flaps bring high-energy air from the lower surface to improve the energy of the upper surface boundary layer. All large aircraft use some form of slotted flap. The drag and lift of slotted flaps depend on the shape and dimensions of the vanes and flaps, their relative position, and the slot geometry. Very careful design and testing is required to obtain the best flap performance.

Mounting hinges and structure may seriously degrade flap or slat performance if not carefully designed to minimize flow separation. Typical examples are DC-8 original flap hinges and the DC-9 original slat design, both of which were redesigned during the flight test stage to obtain the required $C_{L_{MAX}}$ and low drag.

The effect of wing trailing edge flaps is to increase the effective angle of attack of the wing without actually pitching the airplane. Therefore, the lift at a given angle of attack is increased as shown in Figure 14.7, a graph of the DC-9-30 lift curves for flap deflections from 0° to 50°. Not only is the lift increased at a given angle of attack, but the maximum lift coefficient is increased. The angle of attack for stall, however, decreases as flap angle is increased.

Leading edge devices such as slats function very differently. The lift at a given angle of attack is changed very little, but the angle of attack for stall is greatly in-

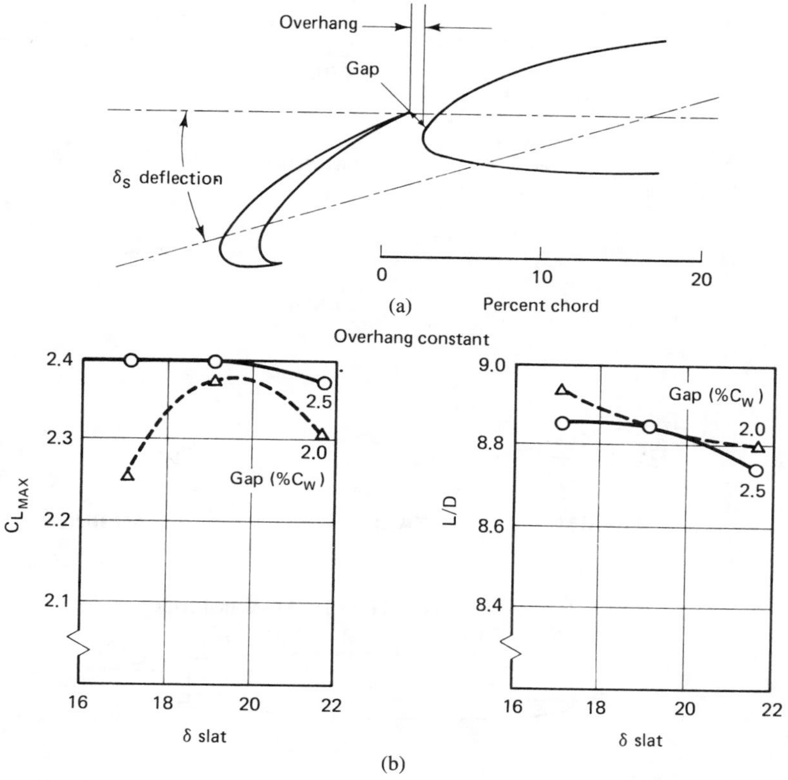

Figure 14.4 Slat optimization study: (a) slat geometry definition; (b) typical wind tunnel slat survey data. From Schaufele and Ebeling, "Aerodynamic Design of the DC-9 Wing and High-Lift System," SAE Paper No. 670846, Oct. 1967. Reprinted with permission. © 1967 Society of Automotive Engineers, Inc.

creased. Figure 14.8 shows DC-9-30 lift curves with and without slats with 0° and 50° flap angle. Obviously, the improvement in $C_{L_{MAX}}$ is very large. One disadvantage of slats is that the airplane must be designed to fly at a high angle of attack for takeoff and landing to utilize the high available lift coefficient. This affects the design of the windshield, because of visibility requirements, and the rear of the fuselage, which must be curved upward to avoid contacting the ground at liftoff and touchdown. Another approach to the latter problem is a longer landing gear, a heavy and sometimes difficult solution due to gear stowage problems. In spite of these problems, slats are so powerful in creating high lift and permitting the weight and drag savings associated with smaller wing areas that all high-speed transports designed since about 1964 use some form of slat in addition to trailing edge wing flaps. The only exceptions are aircraft that have a wing area dictated by some other requirement, such as fuel volume. These are usually small executive jets. If leading edge devices serve simply to shorten takeoff and/or landing runway lengths below the required values and wing area cannot be reduced, their weight and complexity are not justified.

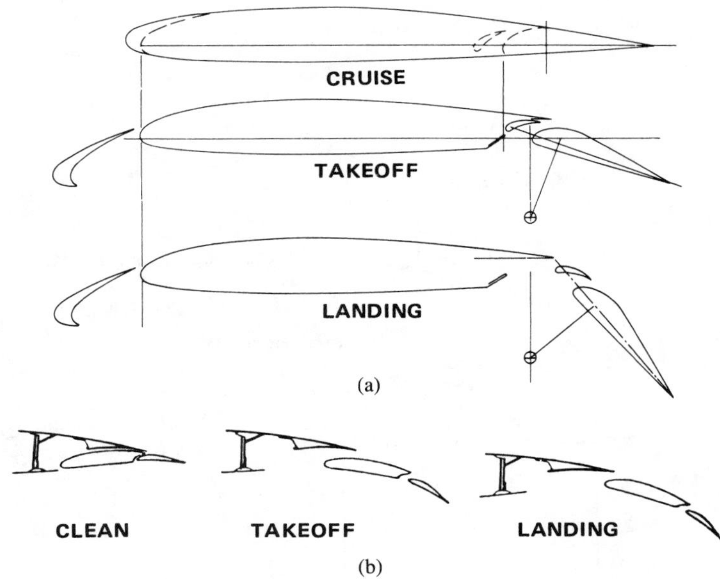

Figure 14.5 Double-slotted flaps.

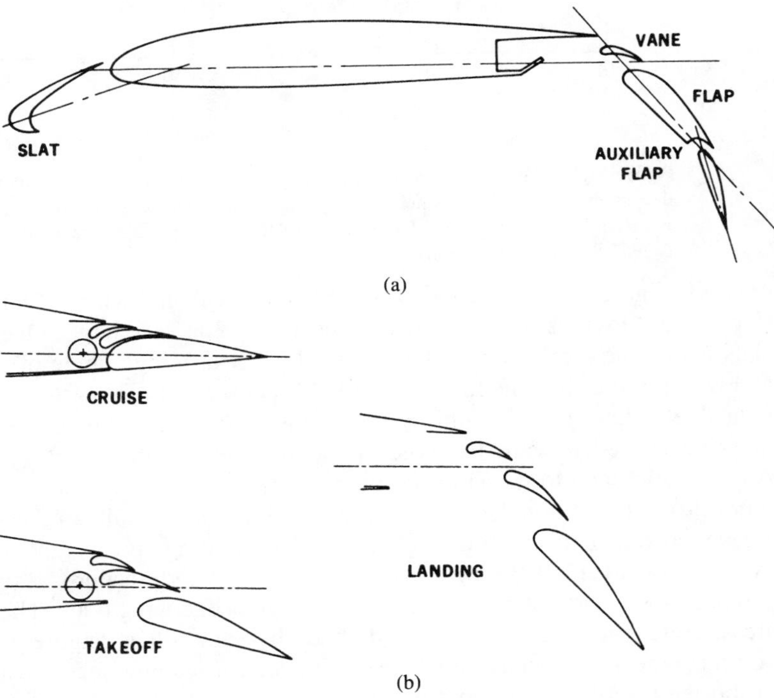

Figure 14.6 Triple-slotted flaps: (a) vane + auxiliary flap; (b) double vane. From Schaufele and Ebeling, "Aerodynamic Design of the DC-9 Wing and High-Lift System," SAE Paper No. 670846, Oct. 1967. Reprinted with permission. © 1967 Society of Automotive Engineers, Inc.

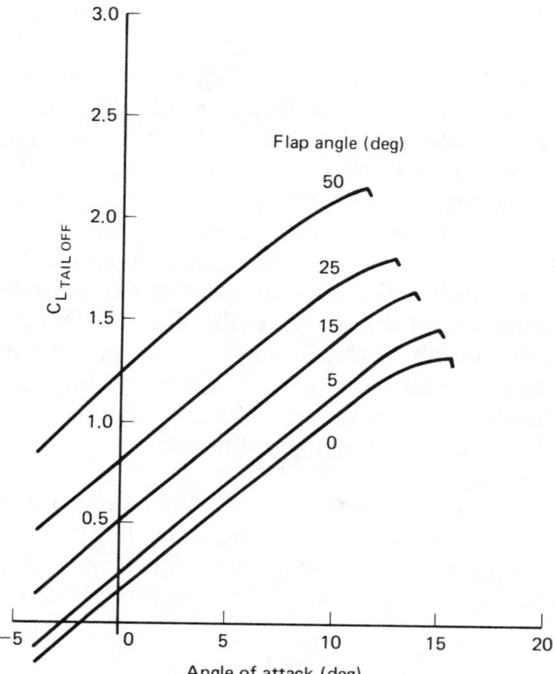

Figure 14.7 DC-9-30 lift curves: effect of flaps. (Tail-off, $M = 0.2$, slats retracted.) (Courtesy of Douglas Aircraft Company.)

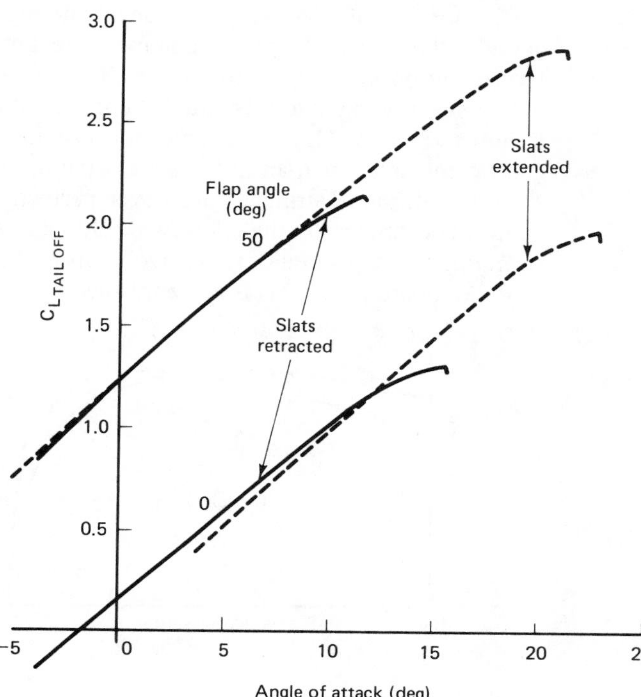

Figure 14.8 DC-9-30 lift curves: effect of slats. (Tail-off, $M = 0.2$.) (Courtesy of Douglas Aircraft Company.)

The spanwise extent of trailing edge devices depends on the amount of span required for ailerons. In general, the outer limit of the flap is at the spanwise station where the aileron begins. Although the exact span needed for ailerons depends on an airplane's moment of inertia in roll, dihedral stability, and maneuverability requirements, low-speed aircraft usually utilize about 45% of the total semispan for ailerons. This means that flaps can start at the side of the fuselage and extend to the 55% semispan station. High-speed aircraft are usually equipped with spoilers to provide much of the roll control. Ailerons on the outer wing panel then play a diminished role when the airplane is traveling at high speed. This is because the pitching moment of the airfoil with deflected ailerons may twist the thin outer panel so much that the intended lift change due to the aileron is reduced or even reversed. In such cases, the ailerons are disabled except in low-speed flight when the flaps are deflected. With spoilers available, the ailerons are generally reduced in size, and the flaps can extend to about 75% of the semispan. However, a small inboard aileron is often provided for lateral trimming and gentle maneuvers, and this serves to reduce the effective span of the flaps.

Leading edge devices intended to substantially raise the maximum lift coefficient must extend along the entire leading edge except for a small cutout near the fuselage to trigger the inboard stall. Some designs utilize a less powerful device, such as a Kruger flap, on the inboard part of the wing to ensure inboard initial stall.

EFFECT OF SWEEPBACK

Sweepback affects the maximum lift coefficient for several reasons. First, as discussed in Chapter 12, the effective q is reduced, although not by as much as in cruise, because at high lift coefficients the straightening effect of the fuselage intersection is felt much farther out on the wing. Furthermore, the effect of the downwash pattern associated with a swept wing is to raise the lift coefficient on the outer wing panel and to lower it inboard (Figure 14.9). This tends to cause the outer panel to reach its maximum lift coefficient first. Such an initial stall at the tip would cause an unstable pitch up and a rapid roll, since airplanes are never perfectly symmetrical. Furthermore, the roll control contributed by the ailerons would be severely restricted, since they would be operating in a separated flow. To avoid this problem, wings should always be designed to stall inboard first. This is accomplished by intentionally reducing the maximum lift

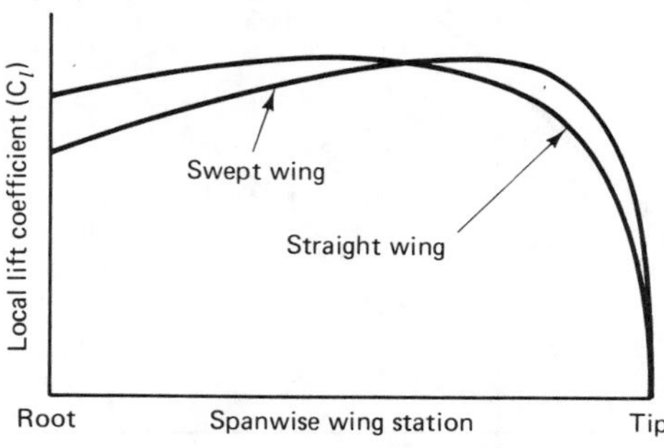

Figure 14.9 Tapered wing spanwise lift coefficient distribution with and without sweep.

coefficient of the inboard airfoils and by twisting the wing to reduce the angle of attack of the outer panel. This twist is called *wing washout*. The result in practice is always a loss in wing $C_{L_{MAX}}$ with respect to the value theoretically attainable if all airfoils could reach $C_{L_{MAX}}$ at the same time. To give the outer panel some margin when the wing stalls, all wings, including those without sweep, suffer some loss, but swept wings have a larger penalty. The airplane maximum lift coefficient with an unflapped wing can be estimated from the $c_{l_{max}}$ of the outer panel airfoils by applying the correction in Figure 14.10. The correction has been derived from the study of a family of wings of different sweep angles. The curve includes the loss in airplane lift coefficient due to the trim download on the tail for typical aircraft. Note that the thickness ratio of the outer panel airfoils is typically about 10% less than the average thickness ratio of the wing. In addition to the significant loss due to sweepback shown for the basic wing in Figure 14.10, there is a further loss in the incremental lift due to flap deflection as a result of sweepback.

It is important to understand the difference between Figure 9.8, which shows the lift distribution (i.e. the spanwise variation of the lift per unit span) and Figure 14.9, a graph of the spanwise variation of the local lift coefficient. The lift distribution is expressed as $c_l c$, the product of the local lift coefficient and the local chord, versus span. It is vital in determining the induced drag and the aerodynamic bending moments on the wing. The local lift coefficient in Figure 14.9 is of primary importance in determining at which spanwise station the wing will first stall.

In addition to the high-lift problems associated with wing sweep already discussed, there is another powerful disturbing influence due to spanwise flow in the boundary layer. Figure 14.11 shows wing chordwise pressure distributions at several spanwise locations of a swept wing. It can be seen that the peak pressures occur farther aft, with respect to a reference point on the root chord, at the more outboard wing stations. As a result, the pressure becomes increasingly negative in the spanwise direction perpendicular to the centerline of the airplane. The direction of the spanwise pressure gradient acting on the slow-moving air in the boundary layer is outboard and produces a flow in the boundary layer toward the tip. The effect is shown in Figure 14.12, a photograph of the boundary layer flow on the upper surface of a

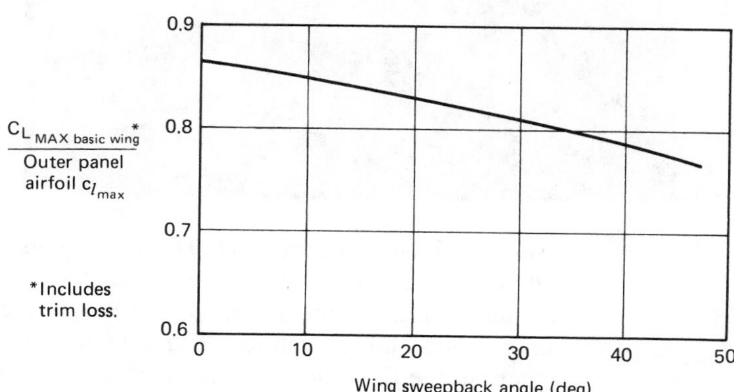

Figure 14.10 Ratio of basic wing $C_{L_{MAX}}$ to outer panel airfoil $c_{l_{max}}$.

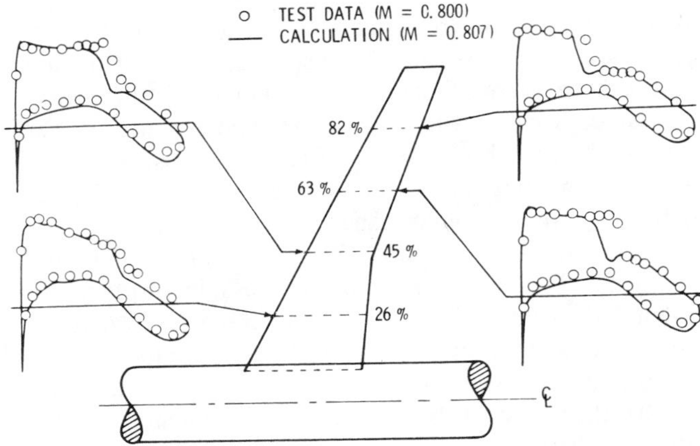

Figure 14.11 Pressure distributions on a swept wing. (Courtesy of NASA.)

Figure 14.12 Boundary layer crossflow on a swept wing wind tunnel model. (Courtesy of Douglas Aircraft Company.)

wind tunnel model swept wing. The visible contours are obtained by wetting the wing with a mixture of rapidly evaporating kerosene and lampblack or with talcum powder in water and detergent. Remember that only the "tired" boundary layer streamlines acquire such high curvatures. The velocity just outside the boundary layer is much higher, so the flow curvature is much less.

The effect of the boundary layer crossflow is to draw the boundary layer outboard, away from the root, and collect it on the outer panel. This reduces the boundary layer thickness near the root, thereby raising its $c_{l_{max}}$, and thickens the boundary layer on the outer panel, lowering the local $c_{l_{max}}$. An early tip stall, a dangerous condition, would result. It is for this reason that there is no such thing as a "clean" swept wing. Swept wings always have a leading edge fence of some sort on the leading edge of the wing, usually at about 35% of the span from the centerline. These fences interrupt the crossflow at the leading edge caused by sweep, as shown in Figure 12.10. The crossflow creates a side lift on the fence that produces a strong trailing vortex. The vortex is carried over the top surface of the wing, mixing fresh air into the boundary layer and sweeping the boundary layer off the wing and into the outside flow. The amount of boundary layer air flowing outboard at the rear of the wing is greatly reduced, and the outer panel $c_{l_{max}}$ is much improved. Pylons supporting the engines on the wing serve the purpose of the leading edge fences, but all aft-mounted engine configurations with swept wings require some leading edge device. The ideal device is the underwing fence, called a *vortilon,* on the DC-9 aircraft (Figure 14.13). This device acts as a leading edge fence at high angles of attack close to the stall, but is well behind the stagnation point and avoids interfering with the natural swept wing flow in cruise and climb attitudes. Modern pylon design emphasizes having the leading edge of the pylon intersect the wing lower surface well behind the wing leading edge for the same reason.

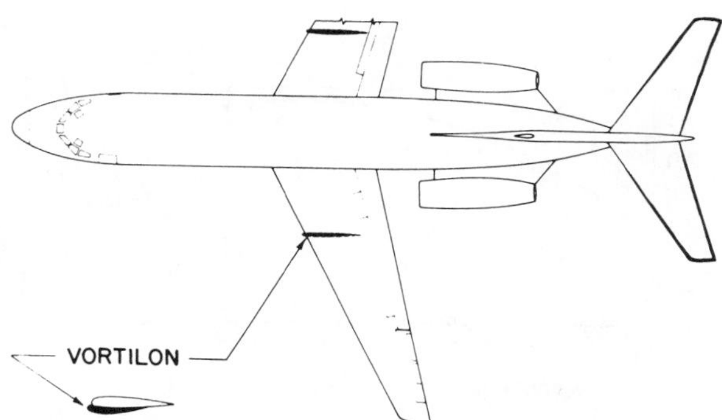

Figure 14.13 Underwing fence, or vortilon, used on the DC-9 aircraft. (Courtesy of American Institute of Aeronautics and Astronautics.)

THE DEEP STALL

Airplanes with T-tails are subject to a dangerous condition known as the deep stall, a stalled condition at an angle of attack far above the original stall angle. T-tail aircraft often suffer a severe pitching moment instability at angles well above the initial stall angle of about 12 degrees, without wing leading edge devices, or about 20 degrees with leading edge devices. If the airplane is allowed to pitch into this unstable region, it might rapidly pitch up to an angle of 35 to 45 degrees. The causes of the instability are fuselage vortices, shed from the forward portion of the fuselage at high angles of attack, and the wing and nacelle wakes. The vortices produce a high downwash on the inner part of the horizontal tail, which has a dynamic pressure close to freestream, and a high favorable upwash over the part of the tail extending outboard of the fuselage, where the wing wake greatly reduces the velocity over the tail. Thus the tail effect on stability is greatly reduced. At the higher angles the wakes from aft-mounted nacelles further reduce the velocity at the tail. Eventually, at perhaps 45 degrees, the tail emerges below the wing and nacelle wakes and the airplane becomes stable. It may then be in a stable trimmed condition with an enormous drag and a resulting high rate of descent. The elevators and ailerons have severely reduced effectiveness because they are stalled at the high angles of attack. A highly swept rudder may have no effect at all because its effective sweep is approaching 90 degrees. This is known as a locked-in deep stall, a potentially fatal state.

The design cures for a locked-in deep stall are to (1) ensure a stable pitchdown at the initial stall, (2) extend the horizontal tail span substantially beyond the nacelles, and (3) use a power system to enable full down elevator angles if a deep stall does occur. The adequacy of the design must be confirmed in a wind tunnel because analytical methods are poor or nonexistent in the stalled regions. In addition, the airplane should be well protected from the initial stall by some combination of stick shakers, lights, and horns. See Ref. 14.3 for more discussion of this phenomenon.

EFFECT OF REYNOLDS NUMBER

Just as Reynolds number affects basic airfoil maximum lift, it is also an important factor in total $C_{L_{MAX}}$ with or without deflecting flaps. Figure 14.14 illustrates the

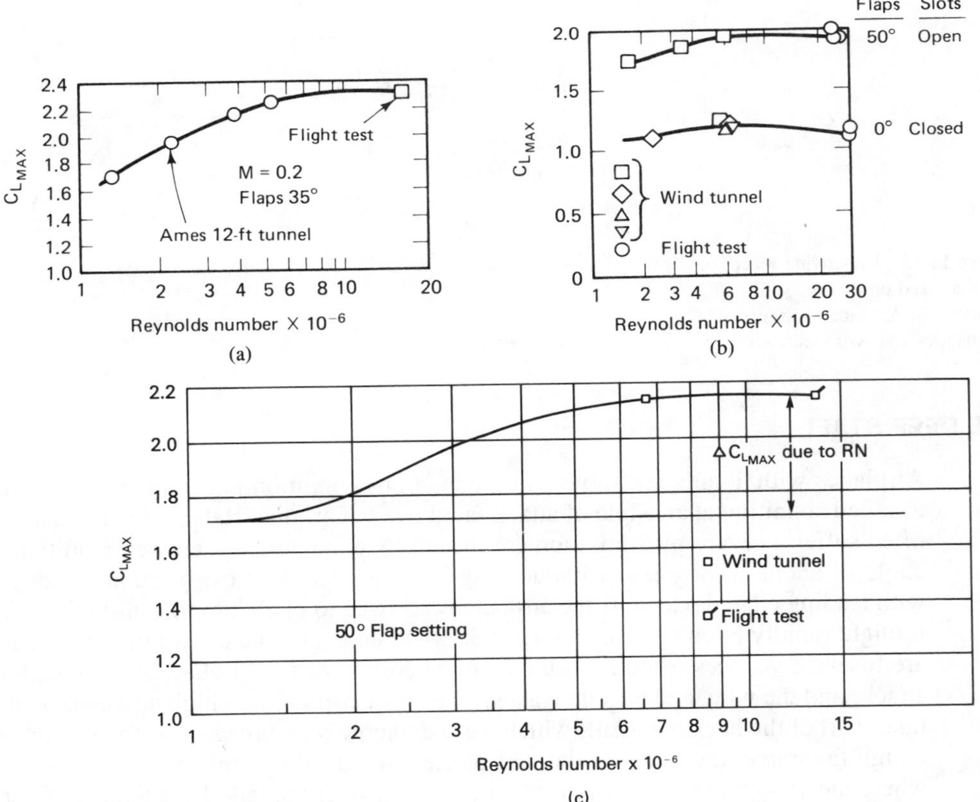

Figure 14.14 Effect of Reynolds number on airplane $C_{L_{MAX}}$: (a) Lockheed C-141; (b) DC-8; (c) DC-9-10. (Courtesy of Douglas Aircraft Company.)

impact of Reynolds number for three airplanes, the Douglas DC-8, the Lockheed C-141, a large military transport, and the DC-9-10. The higher Reynolds number data points are flight tests, and the others are wind tunnel tests run at varying Reynolds number. In general, $C_{L_{MAX}}$ increases significantly as Reynolds number increases up to

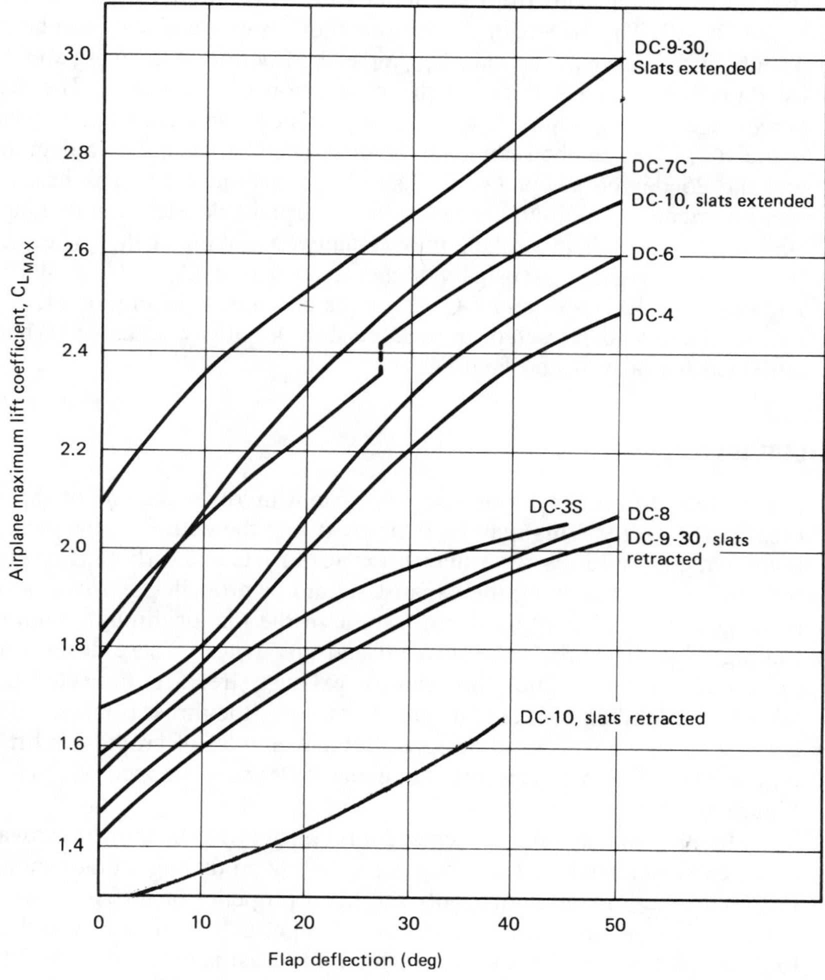

	$\frac{S_{W_F}}{S_W}$	Type of flap	Flap chord (% chord)	$\Lambda_{C/4}$
DC-3S	0.575	Split	0.174	~ 10°
DC-4	0.560	Single slotted	0.257	0°
DC-6	0.589	Double slotted	0.266	0°
DC-7C	0.630	Double slotted	0.266	0°
DC-8	0.587	Double slotted	0.288	30.5°
DC-9-30	0.590	Double slotted	0.360	25° } Slats
DC-10-10	0.542	Double slotted	0.320	35° } Slats

Figure 14.15 Airplane maximum lift coefficient based on $V_{S_{min}}$ ($dV/dt = -1$ knot/s).

6 to 9 million, with little change above 9 million. However, full-scale results may differ from high Reynolds number wind tunnel data due to differences in the details of the flap and slat supporting hardware.

Figure 14.15 shows values of $C_{L_{MAX}}$ as a function of deflection angle of the flap for various aircraft from the Super DC-3 (a 1950 modified version of the DC-3) to the DC-10. The effects of the various flap systems and slats can be glimpsed from a study of the curves. The ratio S_{W_F}/S_W is the fraction of the wing area that is affected by flaps (i.e., carries flaps on the aft portion of the wing). The flap chord as a percentage of wing chord is another important parameter. Higher sweepback angles reduce $C_{L_{MAX}}$, as can be seen from the difference between the 35 degree swept DC-10 and the 25 degree swept DC-9. The $C_{L_{MAX}}$ in Figure 14.15 is based on flight test measurements of minimum speeds with an airplane deceleration rate, approaching the stall, of 1 knot/s. The airplane may actually be sinking at $V_{S_{MIN}}$ (i.e., $L < W$). Thus this flight test $C_{L_{MAX}}$ is usually higher than a true $C_{L_{MAX}}$ ($L = W$) by about 0.2. Although the DC-9-30 and DC-10 curves are shown in Figure 14.15 for both slats extended and slats retracted, in practice slats are always extended when the flaps are deflected for takeoff and landing.

PROPULSIVE LIFT

An entirely different type of high-lift system involves the use of thrust or power to create lift directly. This may be done by tilting the entire engine or engine/propeller combination, by tilting the exhaust nozzles of jets and turbofans or the propellers of turboprops, by deflecting the exhaust stream or propeller slipstream through the use of wing flaps, and even by ducting some of the fan air through a nozzle at the wing trailing edge that directs the flow aft and downward. These designs are called *propulsive lift systems*. The slipstream or exhaust stream is deflected only during the takeoff and landing phases of flight. For cruise, the thrust is usually directed straight forward, as in conventional designs. Some examples of propulsive lift systems based on the interaction between turbofan exhaust flow and flapped wings are illustrated in Figure 14.16.

In all propulsive lift concepts, lift is created by producing downward momentum of the exhaust stream just as thrust is created by producing aft momentum of the fluid. For configurations involving only the tilted propeller or nozzle, a simple downward deflection of the propelled stream through an effective angle δ would create a lift due to thrust of $T \sin \delta$. The associated forward thrust is reduced from T to $T \cos \delta$. For configurations in which the exhaust stream or slipstream is directed over the wing and flaps, the lift due to power can be much larger than the vertical component of the deflected thrust. As shown in Figure 14.17, the total lift consists of the lift from the wing without engine operation, the lift due to deflecting the exhaust stream downward, and the additional circulation lift created by the action of the slipstream on the wing and flap system. The additional lift is called *powered circulation lift*. The physical basis of the latter can be the increased velocity over the wing and/or a larger effective flap chord created by the high-speed exhaust flow roughly parallel to and in the same plane as the flap chord. While very high lift coefficients can be produced, based on the freestream dynamic pressure, a large induced drag also results.

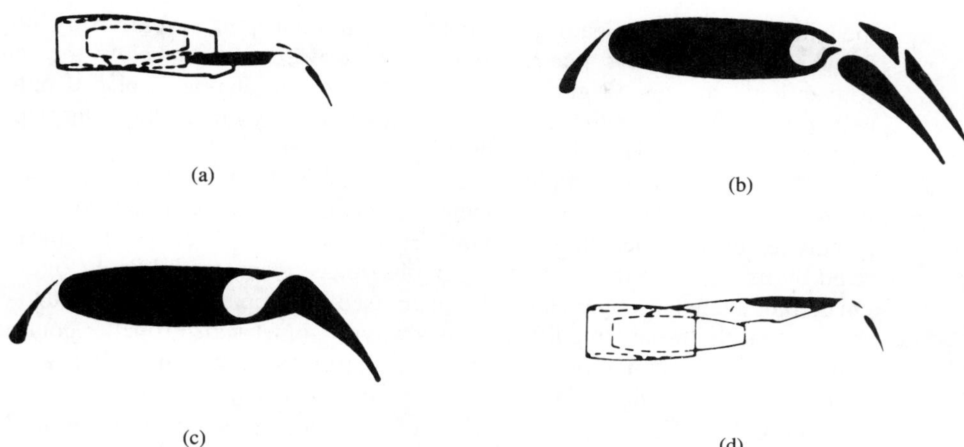

Figure 14.16 Some types of propulsive lift systems utilizing interaction between the turbofan exhaust stream and the flapped wings: (a) upper surface blown flap; (b) augmentor wing; (c) internally blown flap; (d) externally blown flap.

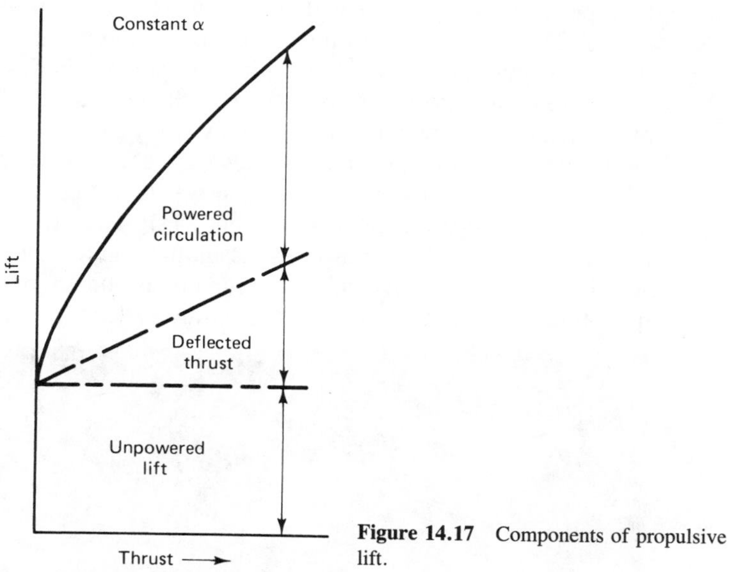

Figure 14.17 Components of propulsive lift.

One particularly interesting use of wing flaps for propulsive lift is the upper surface blown flap concept illustrated in Figure 14.16a. The upper surface design places the turbofan engine ahead of and slightly above the wing. The exhaust actually scrubs the upper surface and follows the curved contour of a specially designed flap to bend the flow direction downward. The flow "sticks" to the contour because of a phenomenon known as the *Coanda effect*. The high energy in the boundary layer prevents separation. One advantage is the use of the wing as an acoustic shield to reduce noise levels on the ground. A disadvantage is the jet exhaust friction drag on the wing in cruise. The leading competitor to this concept is the externally blown flap, in which the engine is placed on a pylon forward of and below the wing in a conventional manner (Figure 14.16d). However, the engine is placed only a little below the wing to have the exhaust stream intentionally impact the wing flaps, which deflects the stream partially downward.

Control after engine failure is an additional problem for propulsive lift aircraft. Lift is lost asymmetrically and compensating rolling moments must be available. In the prototype upper surface blown flap military transport, the Boeing C-14, this is ameliorated by using two engines with the engines placed very far inboard (Figure 14.18). On certain ducted arrangements, all fan air used for propulsive lift is fed to a common duct so that the propulsive lift force is always symmetrical. Propeller configurations may use propellers interconnected by cross-shafting so that all propellers keep turning in the event of an engine failure, although the total power is reduced.

Propulsive lift is used for a class of airplanes called *short takeoff and landing* (STOL) aircraft. STOL airplanes are generally defined as having field-length requirements of 2500 ft or less and using propulsive lift. Once heralded as the future mode for short-range operations, STOL technology has never been used in commercial practice because short field lengths are inevitably associated with higher costs and higher fuel consumption (Figure 14.19). This result follows from the fact that short field performance, with or without propulsive lift, requires more wing area and more power. This raises initial airplane cost, drag, and weight, with the resulting increases in fuel and operating costs. In addition, propulsive lift systems are themselves heavier and more complicated. On the other hand, field lengths below about 3000 ft can be more efficiently obtained with propulsive lift than by the use of more power and wing area in the conventional manner. Study of total transport systems has shown, however, that for routes of moderate or large travel densities it is cheaper to extend runway lengths than to operate the less efficient STOL aircraft. Therefore, STOL aircraft will be used only for military purposes, assuming that this eventually proves economically sensible, or for very specialized commercial operations, such as service to a small mountain community where a longer runway length is impossible.

Figure 14.18 Boeing C-14 upper surface blown wing STOL military cargo airplane. (Courtesy of Boeing Commercial Airplane Co.)

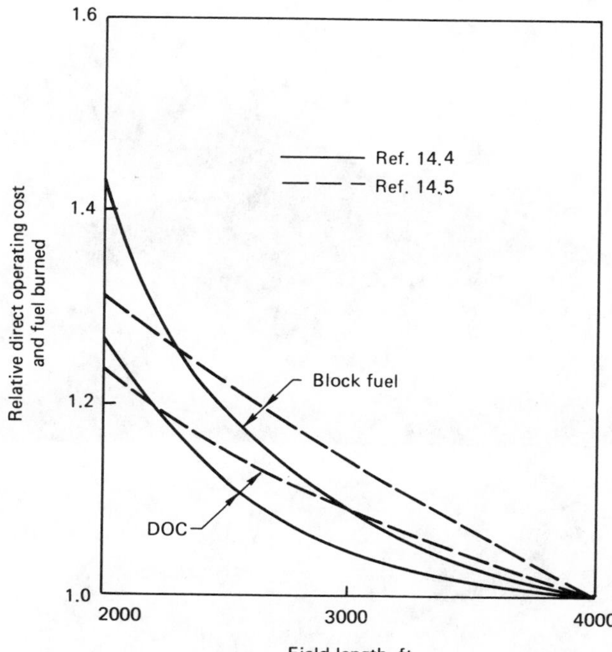

Figure 14.19 Effect of runway length on direct operating costs and fuel consumption. Reprinted from *Journal of Aircraft,* Vol. 17, No. 2, Feb. 1980, p. 67. (Courtesy of American Institute of Aeronautics and Astronautics.)

The extreme of STOL is *vertical takeoff and landing* (VTOL). Many types of VTOL aircraft have been designed and built, but the only successful approaches so far have been the helicopter, which uses a very large diameter propeller or rotor to provide vertical lift, and the AV-8B Harrier fighter (Figure 14.20), which uses variable-deflection jet engine nozzles. VTOL aircraft are very inefficient as flying machines, but they provide a service that can be achieved in no other way. When the runway length is restricted to essentially zero, the helicopter is a unique machine that competes not with conventional airplanes but with boats, mule trains, derricks, and bulldozers. The unique capability of the helicopter has made it one of the fastest-growing segments of aviation. It is, however, used primarily for industrial purposes. Use of the helicopter for the transportation of people in the conventional sense has not grown very much because of high operating costs. The technology of the helicopter is related to the discussions of the propeller but is much more complicated. The reader is referred to Ref. 14.6 for a discussion of helicopter fundamentals.

A strong candidate for the third successful VTOL configuration is the tilt rotor aircraft, the Bell/Boeing Vertol V-22 (Figure 14.21). This aircraft has a normal wing plus two large rotors driven by swiveling turboprop engines mounted at the wing tips. The engine–rotor combination can be rotated from the takeoff position, for which the engine axis and the thrust direction are vertical, to the cruise position with a conventional horizontal axis so that forward thrust is provided. This airplane is intended to combine the vertical takeoff and landing capability of the helicopter with the cruise speed capability of a conventional airplane. The cruise speed of a helicopter is normally limited by the inability to develop sufficient lift on the retreating blades at high forward speeds.

Figure 14.20 McDonnell Douglas/British Aerospace AV-8B VTOL attack airplane (Courtesy of McDonnell Douglas Corporation.)

Figure 14.21 Bell/Boeing Vertol V-22 tilt rotor airplane. (Courtesy of Boeing Vertol Co.)

Example 14.1

A commuter transport, designed to carry 30 passengers, is expected to have a maximum landing weight of 24,000 lb. The maximum landing weight is the sum of the empty weight, the maximum payload that can be carried, and the largest amount of reserve fuel anticipated. A proposed design requirement is that the stalling speed at the maximum landing weight must not exceed 90 knots at sea level. The high-lift system will utilize efficient wing flaps and is expected to produce a maximum lift coefficient of 2.9. What is the required wing area?

Solution: Since lift equals weight in steady flight,

$$L = C_L\left(\frac{\rho}{2}\right)V^2S = W$$

At the stall, $C_L = C_{L_{MAX}} = 2.9$, and the stalling speed V_S must be 90 knots.

$$V_S \text{ (ft/s)} = V_S \text{ (knots)} \times 1.69 = 90 \times 1.69 = 152.10 \text{ ft/s}$$

Then

$$S = \frac{W}{C_{L_{MAX}}(\rho/2)V_S^2} = \frac{24{,}000}{(2.9)(0.002377/2)(152.10)^2} = 301 \text{ ft}^2$$

PROBLEMS

14.1. Discuss the desirable characteristics of high-lift systems.

14.2. How do high-lift devices function to increase the maximum lift coefficient?

14.3. Describe the differences in the effects on wing lift from trailing edge flap deflection and from leading edge devices.

14.4. A turbofan airplane with 1200 ft^2 of wing area and a weight of 115,000 lb is taking off from an airport with a runway altitude of 1000 ft. Temperature is 60°F. Takeoff flap angle is 20 degrees. The flap system and wing sweep are similar to that of the DC-9-30. Determine the true takeoff climb speed.

REFERENCES

14.1. Loftin, L. K., Jr., and von Doenhoff, A. E., *Exploratory Investigation at High and Low Subsonic Mach Numbers of Two Experimental 6% Thick Airfoil Sections Designed to Have High Maximum Lift Coefficients,* NACA RM L51F06, 1951.

14.2. Abbott, Ira H., and von Doenhoff, Albert E., *Theory of Wing Sections,* Dover, New York, 1959.

14.3. Shevell, R. S., and Schaufele, R. D., "Aerodynamic Design Features of the DC-9," *Journal of Aircraft,* Vol. 3, Nov.–Dec. 1966, pp. 515–523.

14.4. Shevell, R. S., *Further Studies in Short Haul Air Transportation in the California Corridor,* NASA CR 137485 and SUDAAR No. 478, Stanford University, Department of Aeronautics and Astronautics, July 1974.

14.5. Conlon, J. A., and Bowles, J. V., "Powered Lift and Mechanical Flap Concepts for Civil Short-Haul Aircraft," *Journal of Aircraft,* Vol. 15, March 1978, pp. 168–178.

14.6. Gessow, Alfred, and Myers, Garry C., Jr., *Aerodynamics of the Helicopter,* Macmillan, New York, 1952.

15

AERODYNAMIC PERFORMANCE

Fluid mechanics is the study of the motion of fluids and the resulting forces imposed on bodies in contact with the fluids. Flight mechanics is the study of the motions of the bodies, in our case aircraft and rockets, through the fluid. Flight mechanics may be divided into stability and contròl, the science of designing for steady and controllable flight characteristics, and aerodynamic performance of the vehicle such as speed, rate of climb, range, fuel consumption, maneuverability, and runway-length requirements. In this chapter we discuss aircraft performance, leaving an explanation of rocket performance to Chapter 20. The principles of stability and control of aircraft are presented in Chapter 16.

LEVEL FLIGHT PERFORMANCE

The simplest performance condition is steady level flight cruise. Then all forces are in equilibrium as the aircraft moves at a constant speed and altitude. From Figure 15.1, equilibrium requires that

$$\text{lift } L = \text{weight } W = C_L \frac{\rho}{2} V^2 S$$

$$C_L = \frac{W}{(\rho V^2/2)S} \tag{15.1}$$

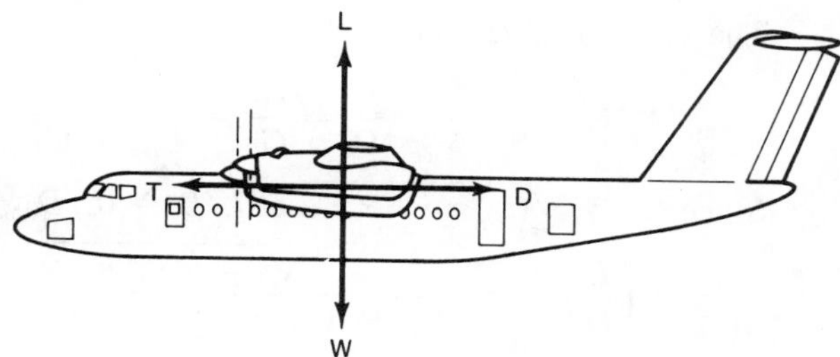

Figure 15.1 Forces on an airplane in level flight. (Courtesy of de Havilland Aircraft of Canada.)

and

$$\text{thrust } T = \text{drag } D = C_D \frac{\rho}{2} V^2 S$$

$$= \left[C_{D_P} + \frac{C_L^2}{\pi \cdot \text{AR} \cdot e} + \Delta C_{D_C} \right] \frac{\rho V^2}{2} S \qquad \text{subsonic flight} \qquad (15.2)$$

For any given weight, the C_L may be found and substituted into the induced drag term. ΔC_{D_C}, the empirical compressibility drag coefficient, is dependent on C_L and Mach number.

Several fundamental airplane characteristics can be derived from the drag equation. One major objective of airplane design is to minimize the drag for any required lift. At any altitude and speed, the ratio of drag to lift depends only on the ratio of C_D to C_L. At low Mach numbers, $\Delta C_{D_C} = 0$ and

$$C_D = C_{D_P} + \frac{C_L^2}{\pi \cdot \text{AR} \cdot e}$$

For minimum drag, C_D/C_L is a minimum. Now

$$\frac{C_D}{C_L} = \frac{C_{D_P}}{C_L} + \frac{C_L}{\pi \cdot \text{AR} \cdot e}$$

At the value of C_L for which C_D/C_L is a minimum, $d(C_D/C_L)/dC_L = 0$. Then

$$\frac{d(C_D/C_L)}{dC_L} = -\frac{C_{D_P}}{C_L^2} + \frac{1}{\pi \cdot \text{AR} \cdot e} = 0$$

and for L/D = maximum,

$$C_{D_P} = \frac{C_L^2}{\pi \cdot \text{AR} \cdot e} \qquad (15.3)$$

Thus, for minimum drag, the lift coefficient is the value for which drag due to lift is equal to parasite drag. For this condition, from equation 15.3,

$$C_{L_{(L/D)_{\max}}} = \sqrt{C_{D_P} \pi \cdot \text{AR} \cdot e} \qquad (15.4)$$

The value of (L/D) is

$$\frac{L}{D} = \frac{C_L}{C_D} = \frac{C_L}{C_{D_P} + C_L^2/(\pi \cdot \text{AR} \cdot e)}$$

Since $C_L = \sqrt{C_{D_P}\pi \cdot \text{AR} \cdot e}$ at $L/D = (L/D)_{\text{max}}$,

$$\left(\frac{L}{D}\right)_{\text{max}} = \frac{\sqrt{C_{D_P}\pi \cdot \text{AR} \cdot e}}{C_{D_P} + (C_{D_P}\pi \cdot \text{AR} \cdot e)/(\pi \cdot \text{AR} \cdot e)} = \frac{\sqrt{C_{D_P}\pi \cdot \text{AR} \cdot e}}{2C_{D_P}}$$

$$= \frac{\sqrt{\pi}}{2}\frac{b\sqrt{e}}{\sqrt{C_{D_P}S}} \qquad (15.5)$$

Since the product of the parasite drag coefficient and S, the wing area, is the equivalent parasite drag area f, $(L/D)_{\text{max}}$ becomes

$$\left(\frac{L}{D}\right)_{\text{max}} = \frac{\sqrt{\pi}}{2}\frac{b\sqrt{e}}{f^{1/2}} = \frac{0.886\, b\sqrt{e}}{f^{1/2}} \qquad (15.6)$$

$(L/D)_{\text{max}}$ depends only on span, f, and e. We can conclude that if we know f and e and the wing span, we have completely defined incompressible minimum drag. To obtain this minimum drag in flight, we must fly at the speed corresponding to the C_L given by equation 15.4. This speed is designated as $V_{(L/D)_{\text{max}}}$. Then

$$V_{(L/D)_{\text{max}}} = \sqrt{\frac{2W}{\sqrt{C_{D_P}\pi \cdot \text{AR} \cdot e}\,\rho S}} = \frac{21.79}{(fe)^{1/4}}\left(\frac{W}{\sigma b}\right)^{1/2} \qquad (15.7)$$

where $V_{(L/D)_{\text{max}}}$ is in feet per second in the English system.

We can also write, from equation 11.11,

$$D = fq + \frac{1}{\pi q}\left(\frac{L}{b}\right)^2\frac{1}{e} \qquad (15.8)$$

$$= \underbrace{f\frac{\rho}{2}V^2}_{\text{parasite}} + \underbrace{\frac{1}{\pi(\rho/2)V^2 e}\left(\frac{L}{b}\right)^2}_{\text{induced}}$$

Note that, at a given flight condition, induced drag depends primarily on the lift carried per unit span, L/b.

If we plot parasite and induced drag separately against V, the parasite drag is seen to increase directly with the square of the speed, while the induced drag decreases with the inverse of the square of the speed. At the speed for minimum drag, corresponding to $C_L = \sqrt{C_{D_P}\pi \cdot \text{AR} \cdot e}$, parasite and induced drags are equal, as shown in Figure 15.2.

The equilibrium level flight speed is the speed at which thrust equals drag, as shown in Figure 15.3.

The speed for minimum drag, $V_{(L/D)_{\text{max}}}$, is a very important flight condition, although it is important for different reasons for jet-powered aircraft than for propeller-driven planes. Jet engines deliver thrust that is almost constant with speed. Their fuel flow depends on the thrust and is expressed as specific fuel consumption c, in pounds of fuel flow per pound of thrust per hour. Minimum thrust, which in steady-state level

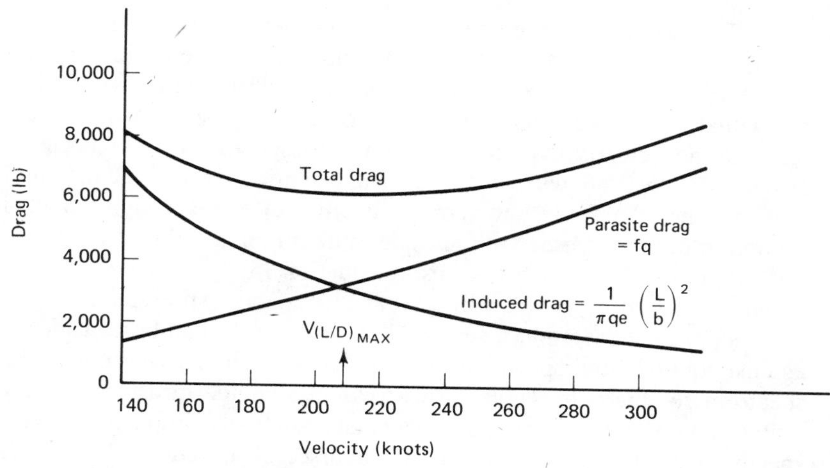

Figure 15.2 Variation of incompressible drag with speed.

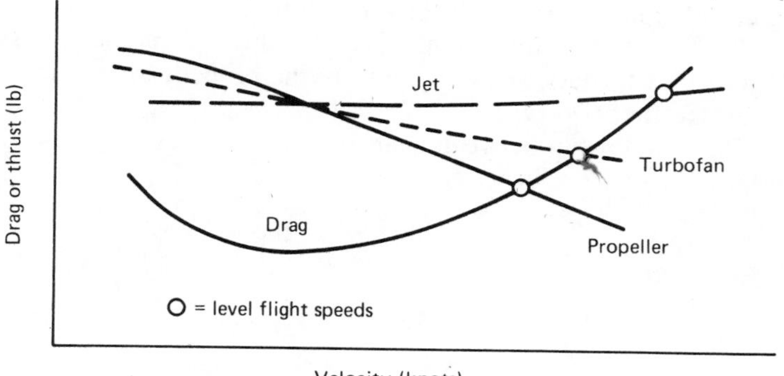

Figure 15.3 Level-flight speeds.

flight is minimum drag, will give the least fuel flow and the longest endurance. Therefore, neglecting the small variation in c with speed, $V_{(L/D)_{max}}$ is the speed for best endurance for jets. The speed for best range, at a given altitude, will be somewhat greater. This follows because, if we fly a little faster than $V_{(L/D)_{max}}$, the drag will increase trivially while the speed gain will be significant, thereby increasing the miles flown per pound of fuel, a quantity called the *specific range*. On the other hand, the minimum drag is associated with a particular value of C_L. If we increase the altitude at this constant C_L, the speed at which the weight is supported at the C_L for best L/D will increase. This altitude and speed increase can be continued until the drag is increasing, due to compressibility, by a larger percentage than the speed. We will have increased the drag a little, but will have attained a much higher speed. Such a flight condition will yield very nearly the best possible specific range and the highest speed at the same time. This is indeed where jets should be flown. We will explore the best range speed for jets further in the range discussion.

The variation of drag with speed for a series of altitudes is shown in Figures 15.4 and 15.5 for a small turbofan transport similar to the Boeing 737 or Douglas DC-9-30. Figure 15.4 shows the drag neglecting compressibility drag, while Figure 15.5 shows the actual total drag, including the compressibility contribution. The difference is dramatic and clearly explains why transport speeds have not been increased above Mach 0.80 to 0.86 without large cost penalties. Note that the minimum drag is independent of altitude until compressibility effects are significant. The speed for minimum drag increases with altitude until compressibility drag becomes dominant.

Figure 15.4 is constructed using equation 15.8 in which drag is a function of the dynamic pressure, q. Recalling that q can be expressed as $q = \rho V^2/2$ or as $q = \rho_S V_E^2/2$, we see that a sea-level curve for drag can be considered to be plotted against equivalent airspeed. In this form, that single curve represents the entire drag spectrum for incompressible flow. Available engine thrust can be plotted for each altitude against equivalent airspeed so that a relatively simple graph provides the entire speed capability of the airplane. Of course, at altitudes other than sea level, the equilibrium speeds must be converted from equivalent to true speeds in using these results. At speeds at which compressibility drag enters, the drag at any given equivalent airspeed is a function of Mach number, which varies with altitude, so separate drag calculations are required for each altitude.

The variation of thrust with speed for different propulsion systems is indicated in Figure 15.3 for jets, shaft engines driving propellers, and turbofans, which fall in between the first two types.

The relationship between shaft power and thrust is given by

$$\text{bhp} \cdot \eta = \text{thp} = \frac{TV}{550} \tag{15.9}$$

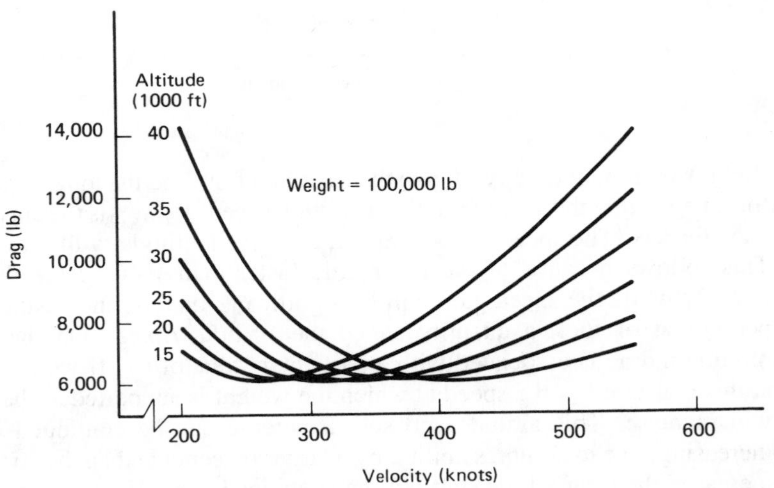

Figure 15.4 Small turbofan transport drag variation with velocity; compressibility neglected.

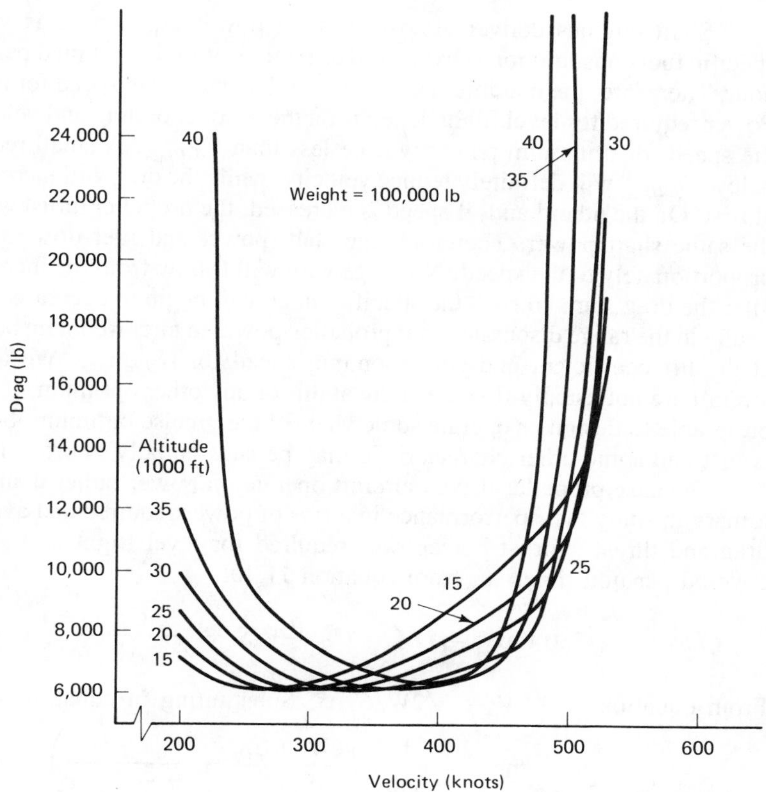

Figure 15.5 Small turbofan transport drag variation with velocity; compressibility drag included.

Then thrust becomes

$$T = \frac{\text{thp}}{V} 550 = \frac{\text{bhp} \cdot \eta}{V} 550 \tag{15.10}$$

where

bhp = engine shaft horsepower
thp = thrust horsepower
V = speed (ft/s)
T = thrust (lb)
η = propeller efficiency
550 = conversion constant to horsepower (i.e., 550 ft-lb/s = 1 hp)

η accounts for the energy losses in the propeller. These losses incurred by the rotating blades are due to skin friction, pressure drag, and "inflow" losses, exactly analogous to the corresponding downwash losses on a lifting wing.

Shaft engines deliver almost constant power as airspeed is varied, and their specific fuel consumption is expressed as pounds of fuel consumed per horsepower per hour. Therefore, the maximum endurance will occur at the speed for minimum power. Power required for level flight depends on the product of drag and velocity. Therefore, the speed for minimum power will be less than $V_{(L/D)_{max}}$. A small reduction in speed below $V_{(L/D)_{max}}$ will definitely reduce velocity, while the drag will increase only trivially at first. On the other hand, if speed is increased, the propeller thrust will decrease with the same shaft power. Therefore, the shaft power and fuel flow must be increased proportionately to the speed. No range gain will follow from the increased speed, and after the drag starts to rise, the specific range will begin to decrease. We show analytically in the range discussion that propeller-powered aircraft obtain best specific range at the lift coefficient and corresponding speed for $(L/D)_{max}$. We also discuss why aircraft are not usually flown exactly at this or any other optimum. It is almost always preferable to design or operate somewhat off the precise optimum, because very little is lost and some other characteristic may be substantially improved by doing so.

Because propeller-driven aircraft operate on power rather than thrust, it is customary to study their performance in terms of power required and available instead of drag and thrust. Thrust horsepower required for level flight is drag times distance covered per unit time, so, from equation 11.10,

$$550\ \text{thp}_{req} = DV = C_{D_P}\frac{\rho}{2}V^3S + \frac{C_L^2}{\pi \cdot \text{AR} \cdot e}\frac{\rho}{2}V^3S \tag{15.11}$$

From equation 15.1, $V = \sqrt{2W/C_L\rho S}$. Substituting in equation 15.11, we obtain

$$\text{thp}_{req} = \frac{1}{550}\sqrt{\frac{2W^3}{\rho S}}\left(\frac{C_{D_P}}{C_L^{3/2}} + \frac{C_L^{1/2}}{\pi \cdot \text{AR} \cdot e}\right) \tag{15.12}$$

The constant 550 is carried to keep thp units in horsepower and the other units in the corresponding English system units. The minimum power will be obtained when the term in parentheses is a minimum. Taking the derivative of that term with respect to C_L, equating it to zero, and defining $C_{L_{mp}}$ as the lift coefficient for minimum power required leads to

$$-\frac{3}{2}\frac{C_{D_P}}{C_{L_{mp}}^{5/2}} + \frac{1}{2}\frac{1}{\pi \cdot \text{AR} \cdot e\, C_{L_{mp}}^{1/2}} = 0$$

Therefore,

$$C_{L_{mp}}^2 = 3\pi C_{D_P} \cdot \text{AR} \cdot e$$

and

$$C_{L_{mp}} = \sqrt{3\pi C_{D_P} \cdot \text{AR} \cdot e} = \sqrt{3}\, C_{L_{(L/D)_{max}}} \tag{15.13}$$

Substituting equation 15.13 in the induced drag coefficient portion of equation 11.9 gives

$$C_{D_{i_{mp}}} = \frac{3\pi C_{D_P} \cdot \text{AR} \cdot e}{\pi \cdot \text{AR} \cdot e} = 3C_{D_P} \tag{15.14}$$

At the minimum power condition, the induced drag coefficient is three times as large as the parasite drag coefficient. This contrasts with the minimum drag condition, for which they are equal. Since, for a given total lift, the speed varies inversely as the square root of the lift coefficient, the speed for minimum power is lower than the speed for minimum drag by the ratio $1/(3)^{1/4} = 0.76$. Taking the inverse, the minimum drag speed is 1.32 times the minimum power speed.

Substituting the value of $C_{L_{mp}}$ from equation 15.13 into equation 15.12 produces the expression for minimum power required for level flight. Thus

$$\begin{aligned} \text{thp}_{\text{MIN}} &= \frac{1}{550}\sqrt{\frac{2W^3}{\rho S}}\left[\frac{C_{D_P}}{(3\pi C_{D_P}\cdot \text{AR}\cdot e)^{3/4}} + \frac{(3\pi C_{D_P}\cdot \text{AR}\cdot e)^{1/4}}{\pi\cdot \text{AR}\cdot e}\right] \\ &= \frac{1.052}{550}\sqrt{\frac{W^3}{\rho S}}\left[\frac{C_{D_P}}{(\text{AR}\cdot e)^3}\right]^{1/4} \end{aligned} \tag{15.15}$$

thp_{MIN} is in units of horsepower, W in pounds, and ρ in slugs per cubic foot.

Equation 15.5 shows that minimum drag varies directly with the square root of C_{D_P} and inversely with the span. Equation 15.15 shows that minimum power required varies directly with the one-quarter power of C_{D_P} and inversely with span to the 1.5 power. These differences have important effects on airplane design, minimum drag being the usual criteria for long-range propeller-driven airplanes and normal jet or turbofan aircraft designed to fly at optimum altitude. Long-endurance aircraft would be designed for minimum power. Aircraft designed for altitudes and speeds at which the lift coefficient is much below the lift coefficient for minimum drag would obtain less drag benefit from high span. In each case, the aerodynamic benefit of high span must be balanced against the associated structural weight penalties, as discussed in Chapter 11.

Figure 15.6 shows a graph of power required versus speed. Also plotted are representative power available curves, one for a propeller, for which this type of graph really applies, and the other for a jet, for which the power available increases with speed at constant thrust. This is a fundamental reason why jet-powered aircraft must fly fast to be efficient. As their speed increases, the thrust horsepower delivered by the engine increases with relatively little increase in the fuel flow.

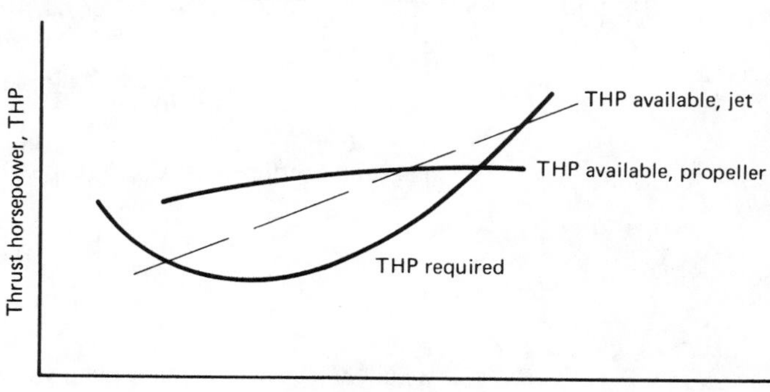

Figure 15.6 Power required and available.

Example 15.1

A two-place airplane is flying at a pressure altitude of 4000 ft at a speed of 120 mph. Outside air temperature is 58°F. The gross weight is 2000 lb. The rectangular wing has an area of 170 ft^2 with a span of 33.25 ft. The wing parasite drag is 25% greater than pure turbulent smooth flat-plate skin friction and is 43% of the total parasite drag; 90% of the wing is exposed. Assuming a propeller efficiency of 0.82, determine the required cruising brake horsepower.

Solution: At a pressure altitude h_p of 4000 ft, $p = 1827.7$ lb/ft^2.

$$T = 58°\text{F} = 518°\text{R}, \qquad \rho = \frac{p}{RT} = \frac{1827.7}{(1718)(518)} = 0.002054$$

$$\mu = 0.000000375 \qquad \text{from Figure 10.14}$$

$$\text{wing chord} = \frac{170}{33.25} = 5.11 \text{ ft}, \qquad \text{aspect ratio} = \frac{b^2}{S} = \frac{33.25^2}{170} = 6.5$$

$$\text{exposed wing area} = 0.9 \times 170 = 153 \text{ ft}^2$$

$$\text{wing wetted area} = 153 \times 2 \times 1.02 = 312 \text{ ft}^2$$

Now

$$\text{Reynolds number RN} = \frac{\rho VL}{\mu} = \frac{(0.002054)(120 \times 88/60)(5.11)}{0.000000375} = 4.93 \times 10^6$$

From equation 10.6b,

$$C_f = \frac{0.455}{(\log_{10}\text{RN})^{2.58}} = \frac{0.455}{[\log_{10}(4.93 \times 10^6)]^{2.58}} = 0.00337$$

or, from the "smooth surfaces" curve of Figure 11.2, $C_f = 0.00325$.

$$C_{D_{P\,\text{wing}}} = \frac{C_f S_{\text{wet}} K}{S_{\text{REF}}} = \frac{0.00325 \times 312 \times 1.25}{170} = 0.00746$$

$$\text{total } C_{D_P} = \frac{C_{D_{P\,\text{wing}}}}{0.43} = \frac{0.00746}{0.43} = 0.01734$$

$$C_L = \frac{W}{qS} = \frac{2000}{(0.002054/2)(120 \times 88/60)^2 170} = 0.370$$

$$C_{D_i} = \frac{C_L^2}{\pi \cdot \text{AR} \cdot e} = \frac{(0.370)^2}{\pi(6.5)(0.855)} = 0.00784$$

($e = 0.855$ from Figure 11.8 at AR $= 6.5$ and $C_{D_P} = 0.01734$)

$$\text{total drag coefficient} = C_{D_P} + C_{D_i} = 0.01734 + 0.00784 = 0.02518$$

$$\text{drag} = C_D \frac{\rho}{2} V^2 S = 0.02518\left(\frac{0.002054}{2}\right)\left(120 \times \frac{88}{60}\right)^2 (170)$$

$$= 136.2 \text{ lb}$$

$$\text{brake horsepower required, bhp} = \frac{TV}{\eta 550} = \frac{136.2 \times 120 \times 88/60}{(0.82)(550)}$$

$$= \underline{53.2 \text{ bhp}}$$

CLIMB PERFORMANCE

Figure 15.7 illustrates the forces on an airplane in steady-state constant-speed climb. The thrust is shown acting parallel to the flight path direction. In general, this is not quite true, but in conventional aircraft the effects of an inclination of the thrust vector are small enough to be neglected.

Equating forces perpendicular and parallel to the flight path,

$$\begin{aligned} L &= W \cos \gamma \\ T &= D + W \sin \gamma \end{aligned} \tag{15.16}$$

Then

$$\sin \gamma = \frac{T - D}{W} = \frac{T}{W} - \frac{D}{W} = \frac{T}{W} - \frac{D}{L} \tag{15.17}$$

γ is the flight path angle or angle of climb. We may assume that γ is sufficiently small so that $\cos \gamma$ is approximately equal to 1.0. Then $L = W$.

$$\text{rate of climb } RC = V \sin \gamma = \frac{V(T - D)}{W} \tag{15.18}$$

For propeller-driven aircraft, it is convenient to use power rather than thrust and drag. If RC is to be determined in feet per minute, the usual units, then with V in feet per second,

$$\begin{aligned} RC\,(\text{ft/min}) = 60\left(\frac{TV - DV}{W}\right) &= \frac{\text{thp}_{\text{avail}} - \text{thp}_{\text{req}}}{W}(33{,}000) \\ &= \frac{\text{thp}_{\text{excess}}}{W}(33{,}000) \end{aligned} \tag{15.19}$$

$\text{thp}_{\text{excess}}$ is the thrust horsepower available for climbing; W is in pounds.

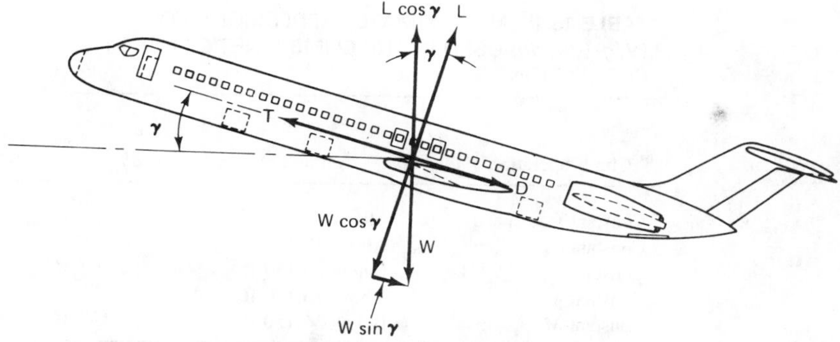

Figure 15.7 Steady-state climb. (Courtesy of McDonnell Douglas Corp.)

The preceding equations are based on an airplane climbing at constant true airspeed. In practical operations, climbing flight is done at a constant indicated airspeed or a constant Mach number. This provides the pilot with a simple guide to the proper climb speed, whereas a constant true speed would mean an ever-changing indicator reading as the altitude increases. A constant indicated speed essentially corresponds to a constant calibrated airspeed. At low Mach numbers, this is the same as a constant equivalent airspeed, since the compressibility correction to airspeed is very small.

With a constant equivalent airspeed, the airplane continually accelerates as the altitude increases. The equilibrium equation along the flight path must then include an inertial term. Thus equation 15.16 becomes

$$T = D + W \sin \gamma + \frac{W}{g}\frac{dV}{dt} \tag{15.20}$$

Since

$$\frac{dV}{dt} = \frac{dV}{dh}\frac{dh}{dt} \quad \text{and} \quad \frac{dh}{dt} = V \sin \gamma$$

we can write

$$\sin \gamma = \frac{(T - D)/W}{1 + (V/g)(dV/dh)} = \frac{T - D}{W}(\text{K.E. factor}) \tag{15.21}$$

Equation 15.21 differs from equation 15.17 in the kinetic energy correction factor $[1 + (V/g)(dV/dh)]^{-1}$. Approximate values of the term $(V/g)(dV/dh)$ are given as functions of Mach number in Table 15.1 for various types of climb paths. Note that, for constant Mach number climb below the isothermal atmosphere, the correction increases the rate of climb. In this region, constant Mach number means a decreasing velocity as altitude increases because the speed of sound is decreasing. The airplane is losing kinetic energy and trading it for increased rate of climb. When

TABLE 15.1 APPROXIMATE EXPRESSIONS FOR $(V/g)(dV/dh)$ USED IN THE CLIMB KINETIC ENERGY CORRECTION

Climb operation	Altitude	$\frac{V}{g}\frac{dV}{dh}$ (approx.)
Constant true speed	All	0
Constant V_E	Above 36,150 ft	$0.7\ M^2$
Constant V_E	Below 36,150 ft	$0.567\ M^2$
Constant M	Above 36,150 ft	0
Constant M	Below 36,150 ft	$-0.133\ M^2$

applicable, the kinetic energy correction factor is applied to gradient of climb and rate of climb calculations.

Equations 15.17 through 15.21 include a surprisingly large amount of useful information. First, the Federal Air Regulations (FAR) specify minimum permissible performance for commercial aircraft in terms of minimum climb gradients, primarily after failure of one engine. A gradient is the tangent of an angle. For small to moderate flight path angles, the sine of the angle is essentially equal to the tangent, so

$$\text{gradient } \gamma = \tan\gamma \simeq \sin\gamma = \frac{T - D}{W}\,(\text{K.E. factor})$$

$$= \left(\frac{T}{W} - \frac{D}{W}\right)(\text{K.E. factor}) \qquad (15.22)$$

Thus the gradient depends on the thrust/weight ratio minus the inverse of the lift/drag ratio. Obviously, to meet the FAR requirements and to maximize the obstacle-clearing ability of a multiengine airplane in the critical takeoff phase after an engine failure, the L/D must be as high as possible. That is why high-lift systems such as flaps and slats must emphasize minimum drag at the flap settings used for takeoff. At the slow speeds and high C_L's just after liftoff, the induced drag will be much larger than the parasite drag. Note that at the speed for minimum power, usually close to the takeoff climb speed, the induced drag is three times the parasite drag. Therefore, aircraft most sensitive to climb problems after engine failure, such as two-engine planes that lose 50% of their power when one engine fails, have higher aspect ratios to reduce induced drag. Of course, high thrust or power ratings increase gradient, too, but the larger the engine, the greater the weight and cost of the engine.

Rate of climb is important in itself because the sooner an airplane can attain its normal cruise altitude, the more efficient it will be. Rate of climb for a jet depends not only on $(T - D)$ but also on the flight speed. Therefore, the speed for best rate of climb for a jet will be faster than the speed for minimum drag. On the other hand, for a propeller that delivers almost constant power, equation 15.19 in terms of power applies. If the power available is constant, clearly the speed for best rate of climb will be the minimum power required speed, as defined by equation 15.13. In the takeoff climb speed regime, the propeller efficiency is normally improving with speed, and therefore the speed for best rate of climb will be somewhat faster than the speed for minimum power required for level flight.

Once an airplane has cleared the ground obstacles, the best choice for climb speed is faster than the speeds for which gradient of climb or the rate of climb are the highest. Going a little faster than the optimum rate of climb speed will decrease the rate of climb only a little, while the distance covered during the climb will be substantially greater. The effect is to improve the average distance flown for a given amount of fuel.

Thus there are three climb speeds, one for best gradient for obstacle clearance, the best rate of climb speed purely for gaining altitude as quickly as possible, such as for a fighter/interceptor mission, and the best efficiency climb speed, which will be the fastest. The relationship of these speeds is shown in Figure 15.8.

The climb performance elements in which pilots and engineers are most interested are the time, fuel, and distance to climb from takeoff to the cruise altitude or, in the case of fighters, to the interception altitude. The first step is to determine the rate of climb for various airplane weights as a function of altitude, as shown in Figure 15.9. The discontinuities in the curves are due to changing the climb speed from a constant indicated airspeed, so that the airplane is accelerating during the climb, to a constant Mach number, so that the airplane is decelerating below 36,200 ft or holding constant speed above that altitude. From these data and the engine thrust and fuel flow information, the time, fuel, and distance to climb each step, say 2000 ft or 5000 ft, can be determined. The sum of the steps will give the total quantities to the desired altitude. If the rate of climb, the climb speed, and the fuel flow can be expressed analytically, the time, fuel, and distance to climb can be expressed by integrals. Both the summation and integral forms are given:

$$\text{time to climb} = \int_{h_1}^{h_2} \frac{dh}{RC} = \sum_{i}^{n} \frac{\Delta h}{(RC)_i} \tag{15.23}$$

$$\text{fuel to climb} = \int_{h_1}^{h_2} w_F \frac{dh}{RC} = \sum_{i}^{n} \frac{(w_F)_i}{(RC)_i} \Delta h \tag{15.24}$$

$$\text{distance to climb} = \int_{h_1}^{h_2} V \frac{dh}{RC} = \sum_{i}^{n} \frac{(V)_i}{(RC)_i} \Delta h \tag{15.25}$$

where RC is the rate of climb, w_F the fuel flow per unit time, V the true climbing speed (a function of altitude), h_1 and h_2 the initial and final altitudes, and n the number of steps. Quantities in parentheses are average values for the step. Units must be consistent in using these equations.

Examples of the total climb performance are shown in Figures 15.10, 15.11, and 15.12 for the DC-10. The almost vertical lines, sloping slightly to the left, are lines

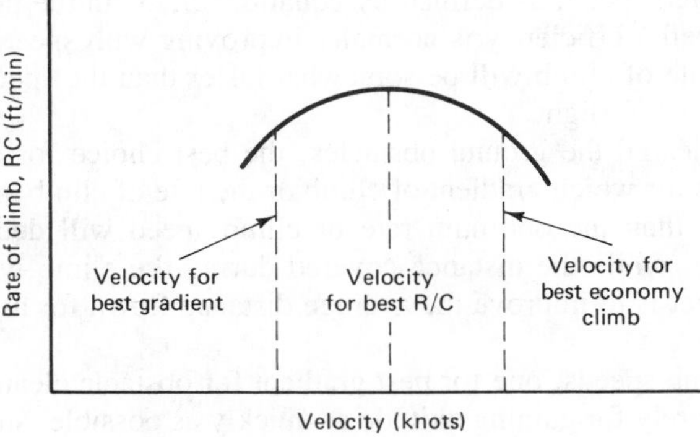

Figure 15.8 Rate of climb versus velocity: various significant climb speeds.

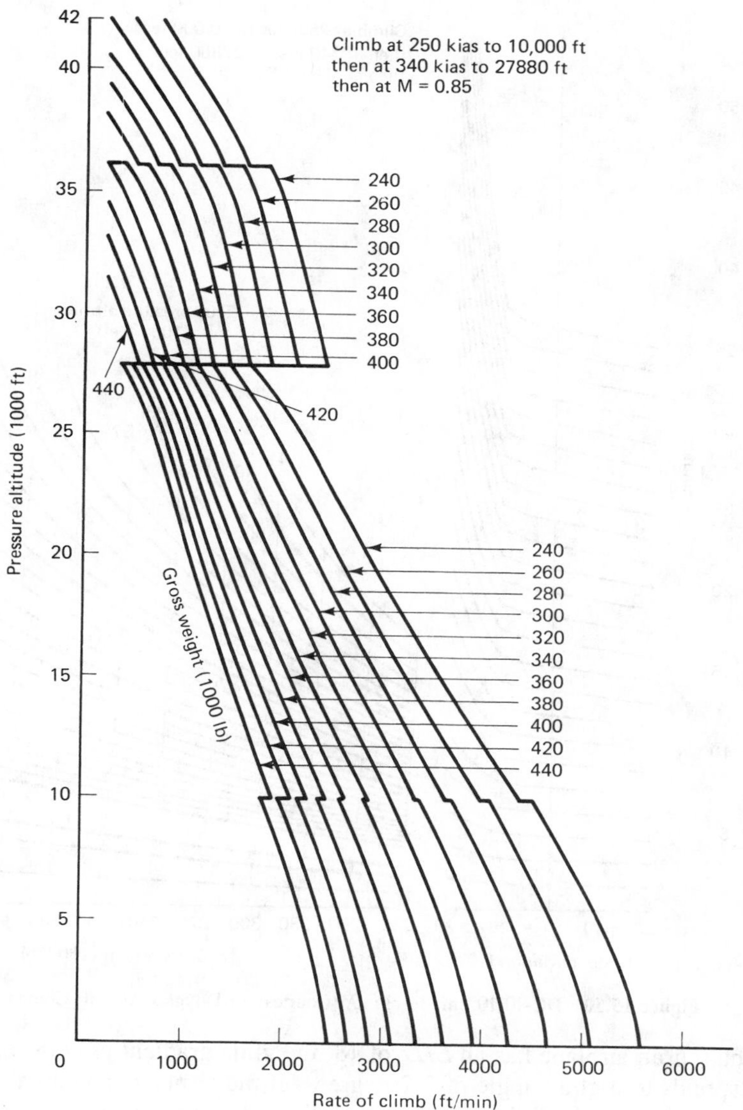

Figure 15.9 DC-10-10 rate of climb. (Courtesy of Douglas Aircraft Company.)

of constant takeoff weight and show the decrease in airplane weight as fuel is consumed during the climb.

One more important performance characteristic can be found in equation 15.17. If the thrust is zero, the flight path glide gradient is the inverse of the L/D ratio.

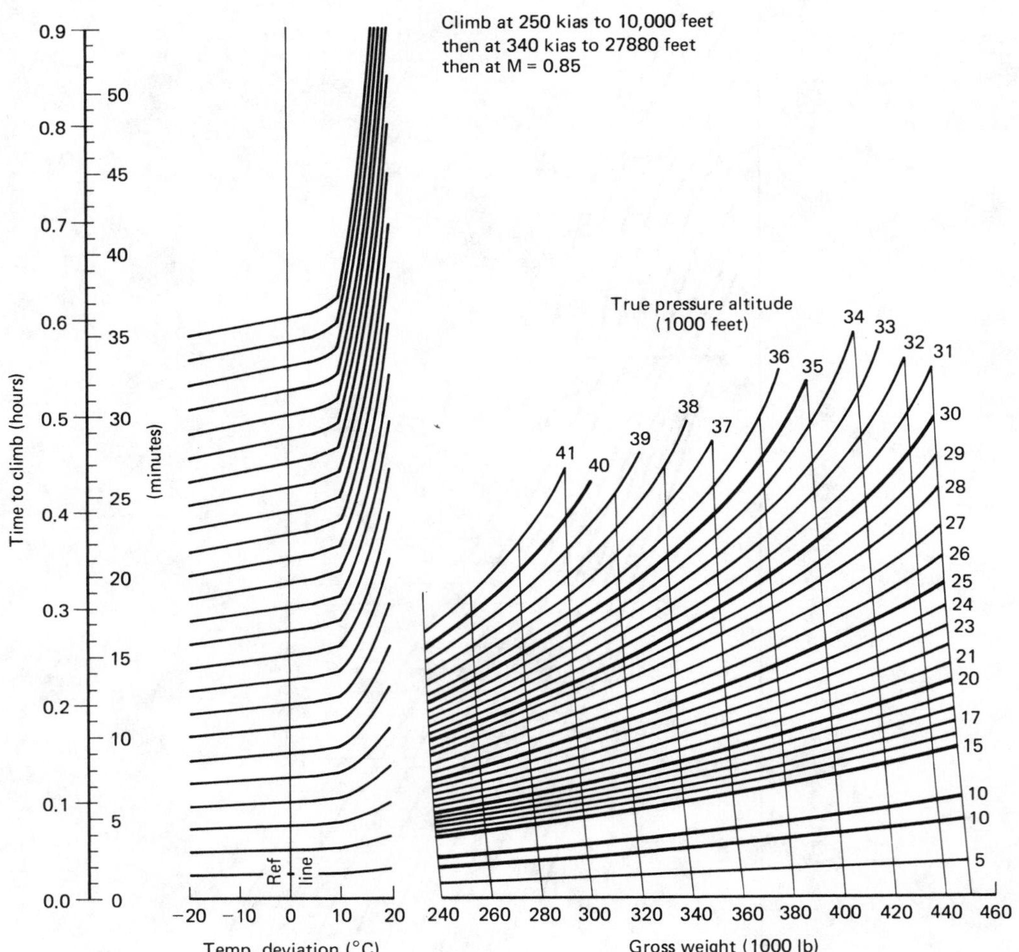

Figure 15.10 DC-10-10 time to climb. (Courtesy of Douglas Aircraft Company.)

Thus, if an airplane has an L/D of 18, the glide gradient is $1/18$, or 0.056. This corresponds to a glide angle of 3.2 degrees. If the airplane is at an altitude of 20,000 ft at the beginning of the glide, the gliding distance covered before landing will be $20{,}000 \times 18 = 360{,}000$ ft, 68 statute miles, or 110 km.

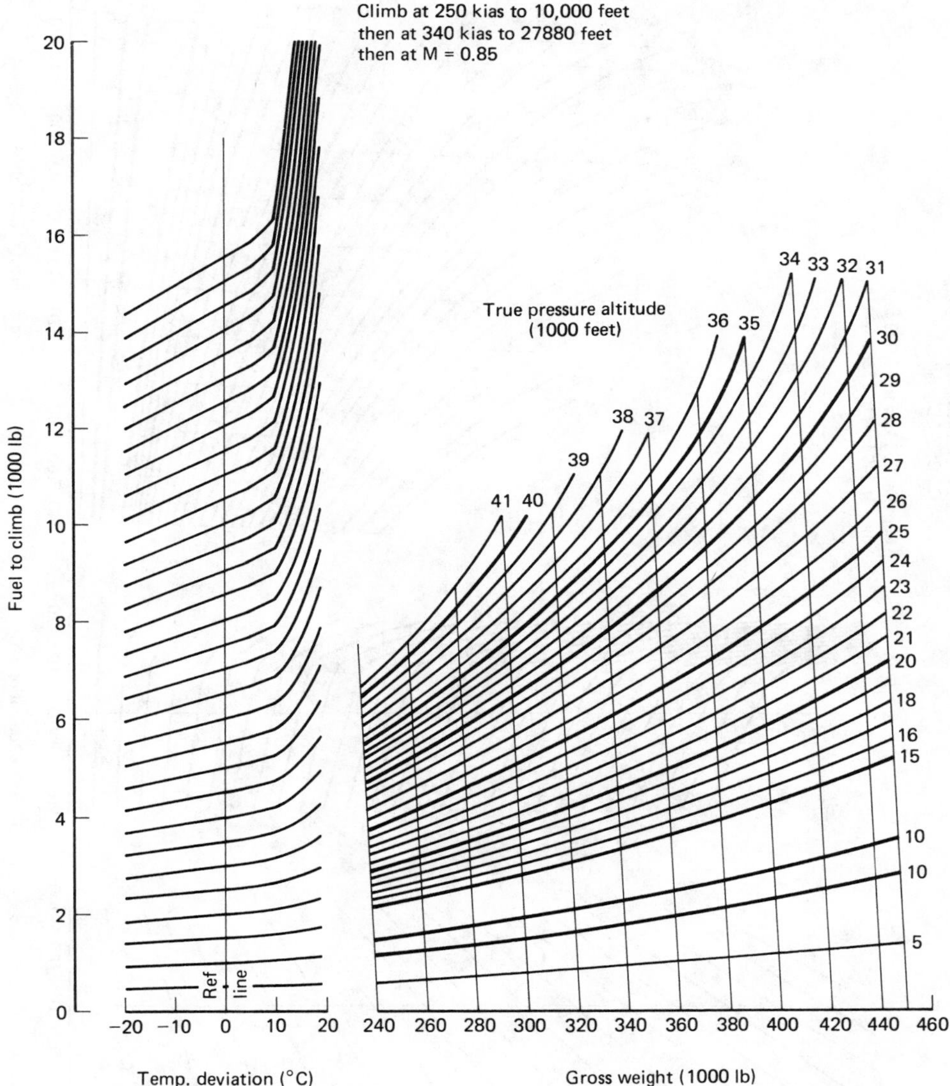

Figure 15.11 DC-10-10 fuel to climb. (Courtesy of Douglas Aircraft Company.)

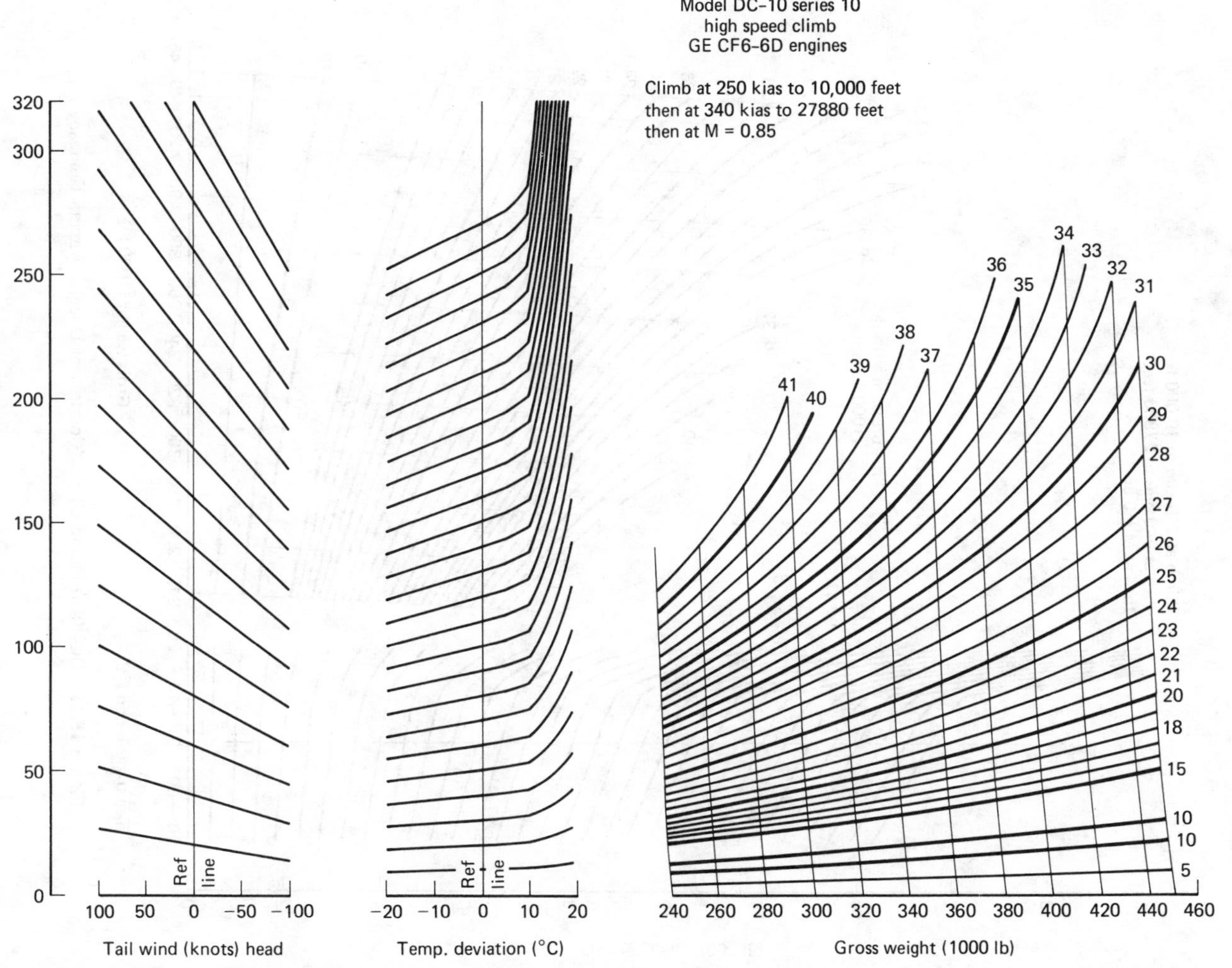

Figure 15.12 DC-10-10 distance to climb. (Courtesy of Douglas Aircraft Company.)

RANGE

The total range of an airplane consists of climb, cruise, and descent segments. Usually, the cruise segment is the longest. The cruise range of an aircraft is equal to the summation of the range increments obtained by multiplying miles flown per pound of fuel(mi/lb) at selected average aircraft weights times the appropriate incremental fuel quantity as illustrated in Figure 15.18. The value of the miles per pound is called the *specific range* and is calculated as follows:

1. For jet or turbofan aircraft,

$$\frac{\text{mi}}{\text{lb}} = \frac{\text{miles flown per hour}}{\text{fuel flow (lb/h)}} = \frac{V}{cT} = \frac{V}{cD} \tag{15.26}$$

where c is the specific fuel consumption (sfc) in pounds of fuel per pound of thrust per hour (lb/lb-h).

The specific fuel consumption used in the performance equations is the *installed* specific fuel consumption. "Installed" means that all adverse effects on sfc associated with the engine installation are included. Turbine engine installation losses include energy losses due to bleeding compressed air from the engine for cabin pressurization and air conditioning, power expended in driving generators and hydraulic pumps, and inlet and nozzle total pressure losses greater than those assumed in the bare engine performance data provided by the engine manufacturers. A typical markup on engine specification sfc for jet and turbofan transports is 3% (i.e., installed sfc equals 1.03 times bare engine sfc). In some cases, for which the engine data are based on zero inlet and nozzle losses, the markup may be twice that amount.

Also,

$$D = \frac{W}{L/D} = \frac{D}{L} \cdot W$$

Substituting in equation 15.26 yields

$$\text{mi/lb} = \frac{V}{c(D/L)W} = \frac{V}{c}\frac{L}{D}\frac{1}{W} \qquad \text{jet} \tag{15.27}$$

The term $(V/c)(L/D)$ is called the *range factor* and is a measure of the aerodynamic and propulsive system range efficiency. If V is in knots, miles per pound will be in nautical miles per pound of fuel.

2. For a propeller-driven airplane,

$$\frac{\text{mi}}{\text{lb}} = \frac{V}{c \cdot bhp}$$

where c is the specific fuel consumption in pounds of fuel per horsepower per hour (lb/bhp-h). Then the nautical miles per pound of fuel is

$$\frac{\text{mi}}{\text{lb}} = \frac{V(\text{knots})}{c(\text{thp}/\eta)} = \frac{\eta}{c}\left(\frac{V(550)}{DV \times 1.69}\right) = 325\frac{\eta}{c}\frac{1}{D}$$

$$= 325\frac{\eta}{c}\frac{L}{D}\frac{1}{W} \qquad \text{propeller} \tag{15.28}$$

The constant 325 arises from the use of 1.69 to convert V from knots to feet per second; 550 is the constant converting ft-lb/s to horsepower.

Equations 15.27 and 15.28 lead to some very interesting conclusions. In equation 15.27, we see that for a jet-powered aircraft, the specific range (i.e., miles flown per pound of fuel) depends on $(V/c)(L/D)$. Since c is approximately constant for jet engines, miles per pound is a function of speed times the ratio of lift to drag. Thus one seeks not only a high L/D in a design, but also a high speed. A slow jet is an inefficient jet with poor range. To optimize specific range, one determines the incompressible C_L for best L/D (in practice, a slightly lower C_L is used) and then increases the cruise altitude so that the increasingly high speed associated with supporting the weight at that C_L can be obtained. When the Mach number gets so high that any further altitude increase would cause a larger percentage increase in drag, due to compressibility, than the percentage increase in speed, the specific range would start to decrease, as shown by equation 15.27. The optimum long-range cruise altitude and speed have been reached.

Figures 15.13 and 15.14 show the specific range without and with compressibility drag for the turbofan-powered transport of Figures 15.4 and 15.5. The specific range is plotted versus speed for several altitudes. Without compressibility drag, both the maximum specific range and the associated speed increase continually with altitude. With real-world compressibility drag, the effect of increasing altitude is the same until the combination of speed and lift coefficient makes the compressibility drag large enough to prevent further gains at higher altitude.

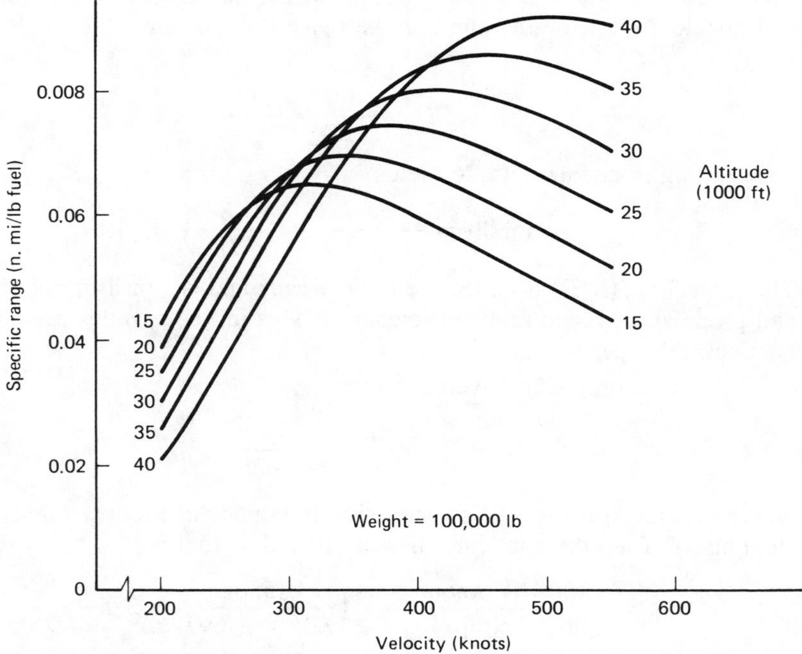

Figure 15.13 Specific range versus velocity for a small turbofan transport (without compressibility drag).

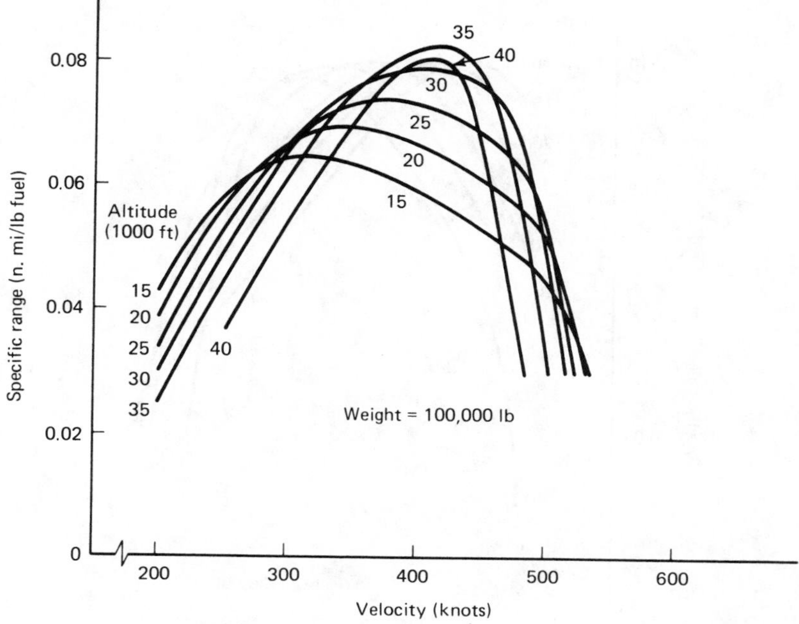

Figure 15.14 Specific range versus velocity for a small turbofan transport (with compressibility drag).

Equation 15.28 shows that with propellers, driven by constant-power engines, specific range depends on η, c, and L/D. Speed does not enter the problem. Assuming that η and c are approximately constant, the lift/drag ratio determines the range. Again, a C_L slightly lower than that for best L/D is chosen as discussed below. At each altitude, the speed appropriate to the weight must be used. The specific range at a particular weight is the same at all altitudes, but the higher cruise altitudes lead to higher speeds, as shown in Figure 15.15. Propellers start losing efficiency rapidly as the Mach number exceeds 0.65 to 0.70, a fact not considered in Figure 15.15. A new multibladed propeller called the propfan holds promise of having only a small loss up to $M = 0.8$.

Although the assumption of constant specific fuel consumption, c, is quite good near the design cruise power of a turboprop engine, the specific fuel consumption rises increasingly as power is reduced below 60% to 70% of the maximum cruise power. When a turboprop airplane, designed with the power to cruise at high altitude, perhaps 35,000 ft or higher, is operated at low altitude (e.g., 10,000 ft), the available power is greatly increased, and the power required to cruise at the slower best range speed is decreased. The engine then operates at a very low percentage of its design power and the specific fuel consumption is greatly increased. Such a design would show a large loss in specific range at low altitudes.

There are several reasons for choosing C_L slightly less than $C_{L_{(L/D)_{max}}}$. The loss in L/D is very small for a moderate decrease in C_L, as shown by the graph of L/D versus C_L in Figure 15.16. The decrease in C_L increases M_{cc}, tending to compensate for the drag increase in jets. The C_L margin with respect to high Mach number stall

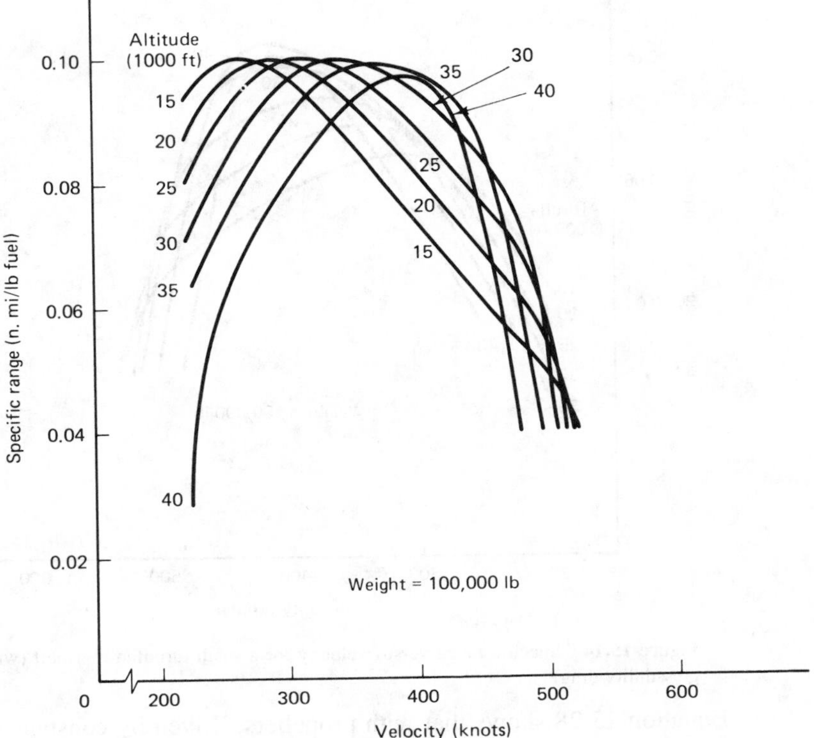

Figure 15.15 Specific range for a propeller-driven transport with the same drag as Figure 15.14 (assumes constant propeller efficiency).

or buffet is increased. The cruise altitude is decreased, at a given speed, and the higher-pressure air increases the thrust available from a given engine. If the engine is sized by the cruise thrust requirement, a smaller, lighter engine can be used. Furthermore, flying exactly at the minimum drag speed causes negative speed stability. If the airplane is flying at $V_{(L/D)_{\max}}$, a speed decrease due to rough air raises the drag and tends to slow the airplane further. Continual pilot throttle corrections are needed to maintain the cruise speed. Such an airplane lacks speed stability. If the airplane is cruising faster than $V_{(L/D)_{\max}}$, an inadvertent slowing decreases the drag. Then thrust exceeds drag and tends to accelerate the airplane automatically, returning it to the planned cruise speed.

Figure 15.16 also shows curves of $M(L/D)$ versus lift coefficient for several cruise Mach numbers. Since $V = Ma$, equation 15.27, the specific range equation for jets, can be written as

$$\text{mi/lb} = a\frac{M}{c}\frac{L}{D}\frac{1}{W}$$

The expression $M(L/D)$ is a measure of the specific range capability due to the aerodynamic characteristics of jet-powered airplanes. The curves of $M(L/D)$ versus lift coefficient show graphically the previously discussed fact that jet range is in-

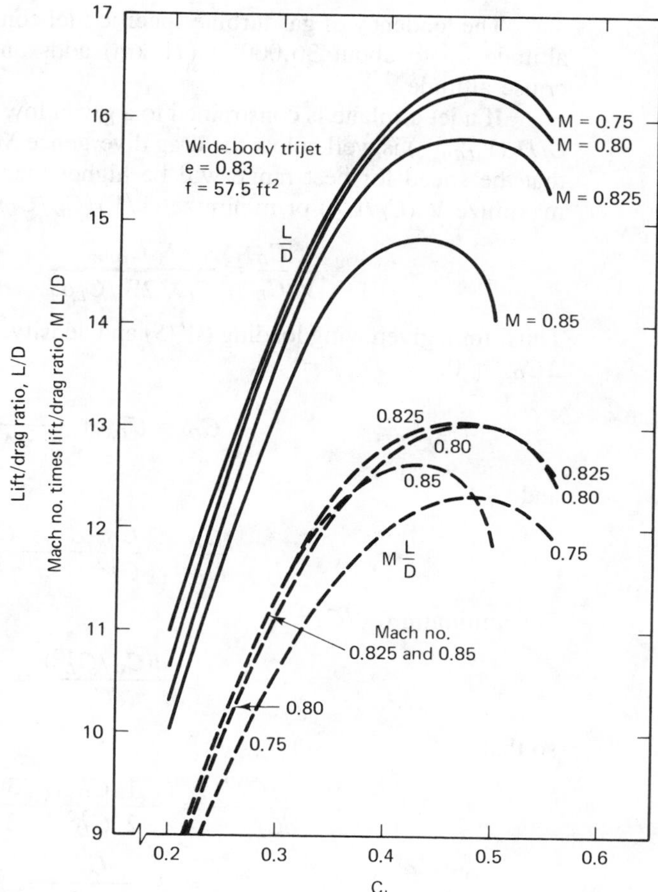

Figure 15.16 Lift-to-drag ratio and the product of Mach number and lift-to-drag ratio versus lift coefficient.

creased by high speed as well as by high L/D. Whereas the best L/D for the example in Figure 15.16 is obtained at a Mach number of 0.75 at a C_L of 0.48, the maximum $M(L/D)$ is obtained at a Mach number of 0.825 and a C_L of 0.46. Furthermore, the decrease in $M(L/D)$ is only 1% if the C_L is reduced to 0.405 at $M = 0.825$. This also illustrates that the jet airplane need not be flown at an exact altitude in order to obtain nearly maximum range. Note that at $M = 0.85$ the range [i.e., the $M(L/D)$] is decreased by only 3%, although the peak L/D is reduced by 9%. This discussion has assumed that the engine specific fuel consumption does not vary with Mach number, quite a good assumption with turbojets. However, turbofan sfc increases slowly with Mach number, as shown in Figure 17.14. Therefore, the optimum range will be obtained at a slightly lower Mach number than is indicated by Figure 15.16.

Because of speed stability, propeller-driven aircraft usually fly at speeds at least 5% to 10% faster than the best range speed. This significantly increases speed, the main purpose of aircraft, with only a small (1% to 2%) range loss. If the airplane flies 20% faster than the best range speed, the range loss is about 7%. Then C_L is $1/(1.2)^2 \times C_{L_{(L/D)\max}}$, or 30% below $C_{L_{(L/D)\max}}$.

The tendency of gas turbine specific fuel consumption c to decrease with higher altitude up to about 36,000 ft (11 km) adds another influence favoring a higher cruise altitude.

If a jet airplane is constrained to a given low altitude at which the speed for best L/D ($V_{(L/D)_{\max}}$) is well below the drag divergence Mach number, equation 15.27 shows that the speed for best range will be higher than $V_{(L/D)_{\max}}$. In this case, we want to maximize $V\,(C_L/C_D)$ or minimize $(1/V)(C_D/C_L)$. Now, since $V = \sqrt{2W/C_L\rho S}$,

$$\frac{1}{V}\frac{C_D}{C_L} = \frac{C_D}{C_L\sqrt{2W/C_L\rho S}} = \frac{C_D}{C_L^{1/2}\sqrt{2W/\rho S}}$$

Thus, for a given wing loading (W/S) and density, we must minimize $(C_D/C_L^{1/2})$. With $\Delta C_{D_C} = 0$,

$$C_D = C_{D_P} + \frac{C_L^2}{\pi \cdot \text{AR} \cdot e}$$

and

$$\frac{C_D}{C_L^{1/2}} = \frac{C_{D_P}}{C_L^{1/2}} + \frac{C_L^{3/2}}{\pi \cdot \text{AR} \cdot e}$$

For minimum $C_D/C_L^{1/2}$,

$$\frac{d(C_D/C_L^{1/2})}{dC_L} = 0$$

so that

$$\frac{d(C_D/C_L^{1/2})}{dC_L} = -\frac{1}{2}\frac{C_{D_P}}{C_L^{3/2}} + \frac{3}{2}\frac{C_L^{1/2}}{\pi \cdot \text{AR} \cdot e} = 0$$

$$C_{D_P} = 3\frac{C_L^2}{\pi \cdot \text{AR} \cdot e} \tag{15.29}$$

For best specific range of a jet at a given altitude, without compressibility drag, parasite drag equals three times induced drag and

$$C_{L_{\text{max range}}} = \sqrt{\frac{C_{D_P}\pi \cdot \text{AR} \cdot e}{3}} = \frac{1}{\sqrt{3}}\sqrt{C_{D_P}\pi \cdot \text{AR} \cdot e} \tag{15.30}$$

The C_L for maximum range for a jet airplane at constant altitude, $\Delta C_{D_C} = 0$, is $1/\sqrt{3}$ times the C_L for $(L/D)_{\max}$, the C_L for maximum range for a propeller-driven airplane. The corresponding best jet speed is 1.32 times $V_{(L/D)_{\max}}$. Actually, jets are seldom flown at such low altitudes or at this low C_L. When speed is constrained by compressibility drag, it is desirable to keep increasing altitude at almost constant M until the drag is minimized.

An example of a specific range curve at a particular altitude is given in Figure 15.17 for a DC-10. The 99% maximum miles per pound contour is used as the long-range condition. By operating 1% below the peak miles per pound, the speed is significantly improved. The 35,000-ft altitude shown is optimum for about

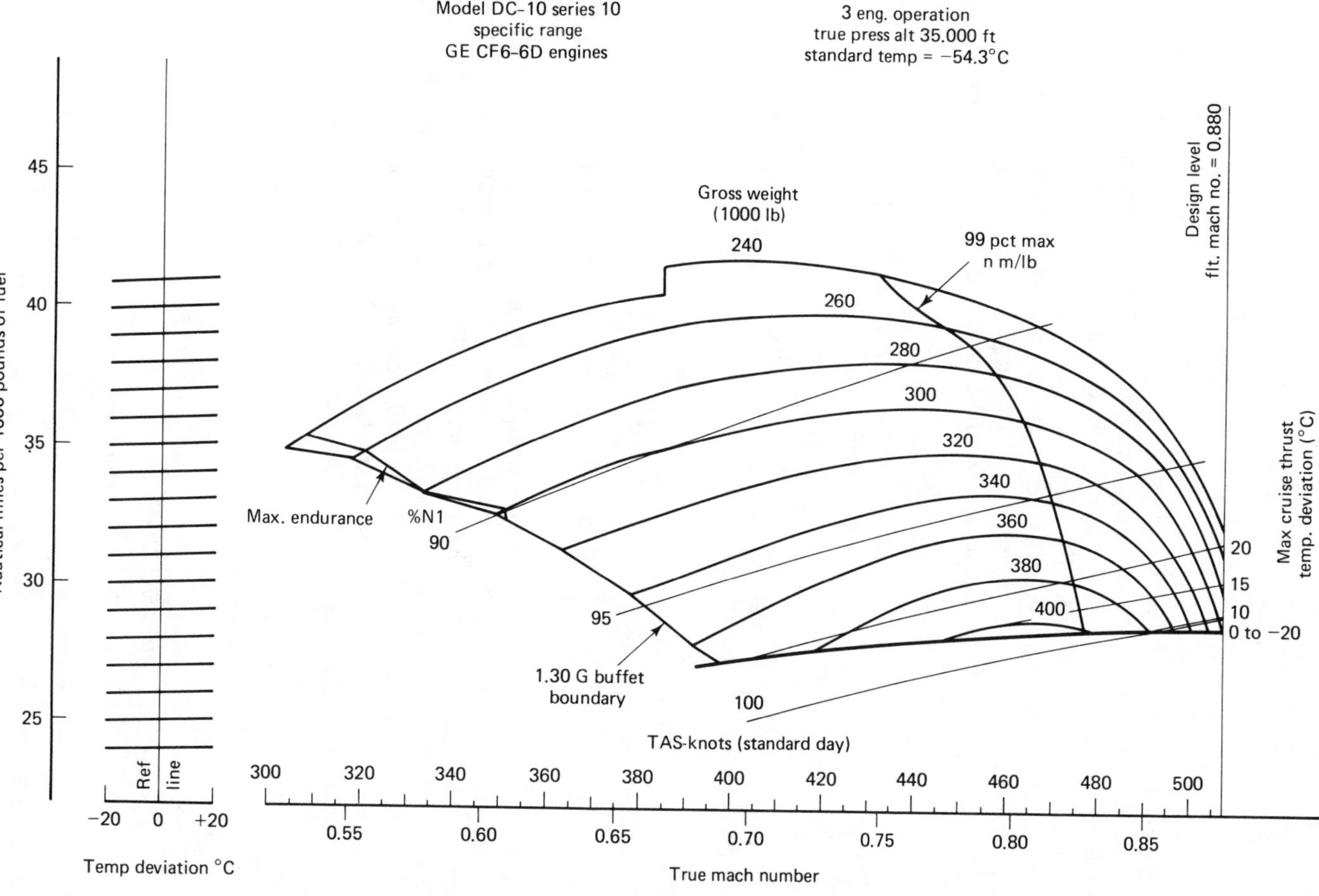

Figure 15.17 DC-10-10 specific range. (Courtesy of Douglas Aircraft Company.)

380,000 lb. The peak specific range for significantly lower weights is at too low a speed. Higher altitudes should be used for them.

The total cruise range is given by

$$\text{range} = \int_{W_f}^{W_i} \frac{\text{mi}}{\text{lb}} dW \tag{15.31}$$

A graphical solution is given in Figure 15.18, a plot of miles per pound versus airplane weight. W_i is the initial cruise weight and W_f the final cruise weight. The shaded area is the cruise range. This may be closely approximated by dividing the weight region between W_i and W_f into roughly equal divisions, ΔW. The range increment of each division will be the average miles per pound for that weight segment multiplied by ΔW. The total cruise range is the sum of the range increments.

If average values of η, c, and L/D can be chosen, the cruise range may be found analytically. For jets,

$$\begin{aligned} R \text{ (n. miles)} &= \int_{W_f}^{W_i} \frac{\text{mi}}{\text{lb}} dW = \int_{W_f}^{W_i} \frac{V}{c} \frac{L}{D} \frac{dW}{W} \\ &= \frac{V}{c} \frac{L}{D} \log_e \frac{W_i}{W_f} \end{aligned} \tag{15.32}$$

For propeller-driven planes,

$$\begin{aligned} R \text{ (n. miles)} &= \int_{W_f}^{W_i} 325 \frac{\eta}{c} \frac{L}{D} \frac{dW}{W} \\ &= 325 \frac{\eta}{c} \frac{L}{D} \log_e \frac{W_i}{W_f} \qquad \text{Breguet formula} \end{aligned} \tag{15.33}$$

where η, c, and L/D are assumed constant throughout the flight or, more realistically, are taken as effective averages; V is in knots; and c is in lb/lb-h or lb/bhp-h, as appropriate. The range formula for propeller-driven aircraft is called the Breguet formula, after its originator. The analogous jet formula is often similarly labeled.

The total range includes cruise range plus distance covered in climb and descent. It is conservative to assume that the fuel used in descent is about the same as in covering the same distance in cruise. Obviously, more fuel is used lifting the airplane,

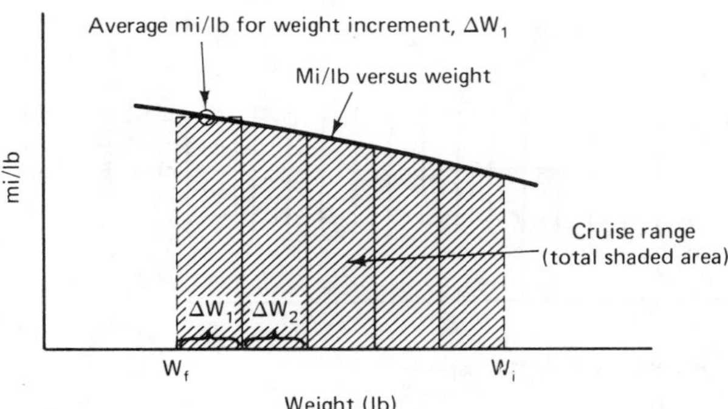

Figure 15.18 Specific range versus weight.

so the specific range in climb is reduced. In addition, fuel is used in taxi, takeoff, maneuvering around the airports, and landing. This is usually called *maneuver fuel*.

Detailed range studies require the calculation of the time, fuel, and distance to climb to various altitudes. A simple approximate method useful for preliminary design studies of turbofan-powered transports is shown in Figure 15.19. The chart shows the additional fuel required to climb to cruise altitude in terms of a percentage of the takeoff weight. The additional fuel is not the fuel used in climb but rather the increase in fuel required to climb compared to the fuel required to cruise the distance covered in climb. In using the chart, the fuel to cruise the entire distance from the departure airport to the destination is computed. The additional fuel to climb is then found from Figure 15.19. The additional climb fuel for propeller-driven aircraft will generally be lower than that given by Figure 15.19. The logic of the chart is that a certain energy is required simply to lift the weight to the cruise altitude. Other climb losses involve the effects of flying below the optimum cruise altitude during the climb and the effect of lower altitude on the engine specific fuel consumption. These are approximated by basing the curve on detailed climb results from several aircraft.

An approximation to maneuver fuel for modern high-bypass-ratio (about 5:1) turbofan engines is 0.7% of the takeoff weight. Low-bypass-ratio (about 1:1) engine-equipped aircraft require about 0.9%. Propeller-driven airplanes would need less maneuver fuel.

Use of the preceding approximations allows the range to be estimated from the Breguet equation using c, L/D, and η, where applicable, chosen at an average weight,

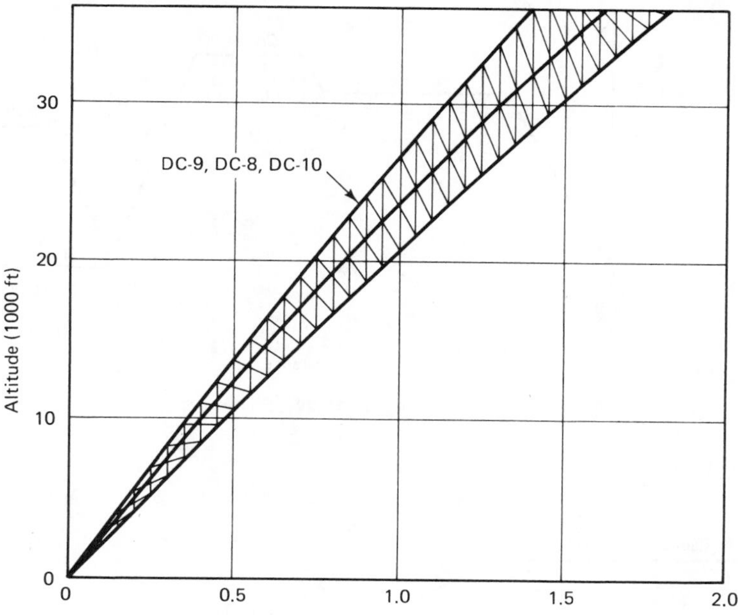

Figure 15.19 Additional fuel due to climb (to be added to cruise fuel calculated for total range).

or by the use of the method illustrated in Figure 15.18. The additional fuel due to climb and maneuver can then be added. For long flights, where the cruise portion is dominant, this approach is very satisfactory. For very short flights, where climb is a large portion of the total trip, the approximation would be subject to significant error.

To provide for errors in flight planning, winds, or navigation, and especially for diversions to alternate airports due to bad weather at the destination, reserve fuel is carried. This is usually based on assuming a balked landing at the scheduled destination, a climb to cruise altitude, and a flight to the alternate airport. Figure 15.20 shows a typical complete flight profile. The step climb is an altitude change to achieve the most efficient cruise altitude when sufficient fuel is consumed to permit the airplane to climb to the next higher altitude allowed by traffic control regulations.

Minimum reserve fuel requirements are established by the Federal Air Regulations. Reserve fuel is determined from a range calculation to the alternate airport plus a minimum required additional fuel quantity defined for domestic flights as 45 minutes of long-range cruise at normal cruise altitude. An approximate method of estimating the domestic reserve fuel for turbofan-powered aircraft is to assume reserve fuel to be equal to 8% of the zero fuel weight (i.e., the weight of the aircraft with payload but without any fuel) when the alternate airport is 200 n. mi from the original destination. If the distance to the alternate is 300 n. mi, 9% of the zero fuel weight is used. Reserve fuel for turboprops will be smaller by about 25%.

The summary of the payload–range characteristics of an airplane is usually shown in the form of Figure 15.21. As the range is increased, more fuel must be

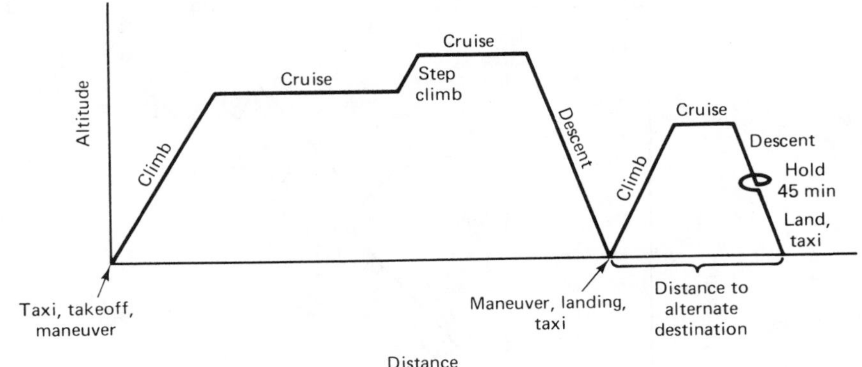

Figure 15.20 Flight profile.

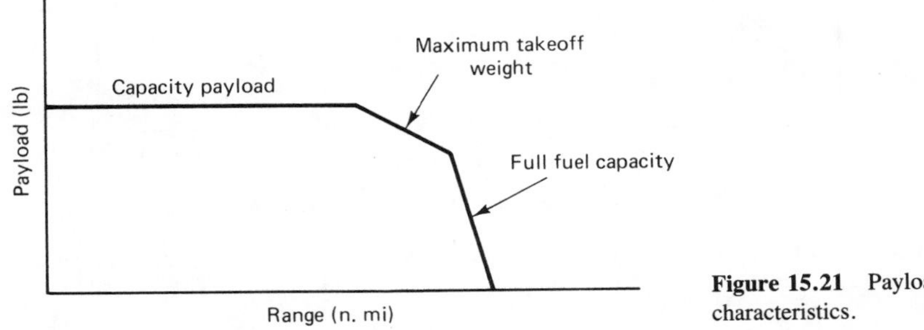

Figure 15.21 Payload-range characteristics.

carried. This continues until the takeoff weight reaches its maximum, limited by either performance or structural requirements. To fly farther requires more fuel, which can be carried only by reducing the payload. When the fuel tanks are full, range can be increased only by decreasing payload rapidly. The only range gain then comes from a lower average weight, as shown by equations 15.27 and 15.28. Aircraft seldom operate in this uneconomical mode.

ENDURANCE

The primary function of aircraft is the transportation of a payload over a prescribed distance. There are a few missions, however, in which the time of flight (i.e., the endurance) is the performance criterion to be maximized. Examples are military patrol missions, such as submarine hunting, and traffic spotting. In addition, all flights may require some holding period for traffic control or weather reasons. The performance objective in these flight segments is to consume as little fuel as possible during the holding period.

The endurance problem is similar to the range problem except that we are trying to determine how long the aircraft will fly rather than how far it will fly. The quantity analogous to specific range is specific endurance, the hours flown per unit quantity of fuel. In the usual units, specific endurance = hours per pound of fuel.

To maximize endurance, fuel flow per unit time must be minimized. Since the specific fuel consumption is nearly constant, the drag must be minimized for jet aircraft, while thrust horsepower must be as small as possible for propeller-driven aircraft. The endurance t_e for turbojet or turbofan aircraft is

$$t_e = \int_{W_f}^{W_i} (\text{h/lb fuel})\, dW = \int_{W_f}^{W_i} \frac{1}{Dc}\, dW$$

$$= \int_{W_f}^{W_i} \frac{1}{[W/(L/D)]c}\, dW = \int_{W_f}^{W_i} \frac{1}{c}\frac{L}{D}\frac{dW}{W} = \frac{1}{c}\frac{L}{D} \log_e \frac{W_i}{W_f} \qquad (15.34)$$

Again, c and L/D are assumed to be constant throughout the flight or taken as average values. Note the comparison between this equation and the jet range equation (15.32). Endurance is simply range divided by speed. For the greatest endurance, the aircraft should obviously fly at the speed for minimum drag. This, of course, assumes that c is a constant with speed, a good assumption for jets but not quite true for turbofans. In addition, at very low engine thrust levels, c tends to increase as the thrust is decreased. This may also influence the speed for best endurance.

For propeller-driven aircraft, endurance is

$$t_e = \int_{W_f}^{W_i} \frac{1}{\text{thp} \cdot c/\eta}\, dW = \int_{W_f}^{W_i} \frac{\eta}{c}\left(\frac{550}{DV \times 1.69}\right) dW$$

$$= \int_{W_f}^{W_i} 325 \frac{\eta}{c}\frac{L}{DV}\frac{1}{W}\, dW \qquad (15.35)$$

V is in knots and c is in pounds of fuel per brake horsepower per hour. We cannot assume that L/DV is a constant. L/DV is the ratio of lift to thrust power required, and we have shown in equation 15.12 that thp_{req} is a nonlinear function of weight. At any given lift coefficient, power required varies with $W^{3/2}$. However, it is shown in Ref. 15.1 that, if V is expressed as $\sqrt{2W/C_L\sigma\rho_S S}$, t_e is given by

$$t_e \text{ (hours)} = 37.9\left(\frac{\eta}{c}\right)\frac{C_L^{3/2}}{C_D}\sqrt{\frac{\sigma S}{W_i}}\left[\left(\frac{W_i}{W_f}\right)^{1/2} - 1\right] \qquad (15.36)$$

Here we see that, for best endurance, a propeller-driven airplane should be flown at the flight condition for maximum $C_L^{3/2}/C_D$. This expression is easily shown to be the inverse of the term in parentheses in equation 15.12. It is a maximum when the lift coefficient is $\sqrt{3}$ times the value for maximum lift/drag ratio. Since low speed as well as a particular lift coefficient is desirable for minimum power required, best endurance occurs at a high density (i.e., low altitude). Therefore, propeller aircraft endurance is best at low altitudes. The endurance of four-engine turboprop aircraft can often be improved by shutting down two engines and operating the other two engines at higher powers. This reduces the deterioration in specific fuel consumption associated with operating at low power settings.

Example 15.2

A DC-9-30 with a capacity of 100 passengers is cruising at a Mach number of 0.78 at a pressure altitude of 30,000 ft. Outside air temperature is −38°F. The initial cruise weight was 97,000 lb. According to the pilot's flight plan, he will start his descent at a weight of 82,000 lb. The DC-9-30 has a C_{D_P} of 0.0202, an e of 0.816, a wing area of 1000 ft^2, a wing span of 93.6 ft, and three lavatories. The compressibility drag coefficient ΔC_{D_C} is 0.0010. The JT8D-15 turbofan engines have an installed specific fuel consumption at cruise of 0.82 lb/lb-h. Determine:

(a) Distance covered at cruise altitude (assume that conditions at average weight can be considered as the average for the flight).
(b) Required engine thrust (total for two engines) at the average cruise weight.
(c) Fuel flow in gallons per hour (kerosene weighs 6.7 lb/gal).
(d) Seat-miles produced per gallon.
(e) Compare the DC-9 seat-miles/gallon with a five-passenger automobile having a fuel consumption of 20 mi/gal.

Solution: We are given h_p = 30,000 ft, M = 0.78, and T = −38°F = 422°R. Then

$$\text{speed of sound, } a = \sqrt{\gamma RT} = \sqrt{(1.4)(1718)(422)} = 1007.5 \text{ ft/s} = 686.9 \text{ mph}$$

$$q = \frac{\gamma}{2}pM^2 = \frac{1.4}{2}(629.7)(0.78)^2 = 268.18 \text{ lb/ft}^2$$

$$\text{average weight} = \frac{97{,}000 + 82{,}000}{2} = 89{,}500 \text{ lb}$$

$$C_L = \frac{W}{qS} = \frac{89{,}500}{(268.18)(1000)} = 0.3337$$

Since

$$\text{AR} = \frac{b^2}{S} = \frac{(93.6)^2}{1000} = 8.76$$

$$C_{D_i} = \frac{C_L^2}{\pi \cdot \text{AR} \cdot e} = \frac{(0.3337)^2}{\pi(8.76)(0.816)} = 0.00496$$

$$C_{D\,\text{Total}} = C_{D_P} + C_{D_i} + \Delta C_{D_c} = 0.0202 + 0.00496 + 0.0010 = 0.02616$$

$$\frac{L}{D} = \frac{C_L}{C_D} = \frac{0.3337}{0.02616} = 12.76$$

(a) Cruise distance $= \dfrac{V}{c}\dfrac{L}{D}\log\dfrac{W_i}{W_F} = \dfrac{(0.78)(686.9)}{0.82}\,12.76\log_e\dfrac{97{,}000}{82{,}000}$ (Mach no.: 0.78; Speed of sound: 686.9)

$= (8350)(0.168) = \underline{1403 \text{ statute miles}}$

(b) $T = D = C_{D\,\text{TOT}}qS = (0.02616)(268.18)(1000) = \underline{7015.6 \text{ lb}}$

or

$T = D = \dfrac{W}{L/D} = \dfrac{89{,}500}{12.76} = \underline{7014.1 \text{ lb}}$ (the difference is in the rounding of numbers)

(c) Fuel flow (lb/h) $= Tc = 7014.1(0.82) = 5752$ lb/h

Fuel flow (gal/h) $= \dfrac{5752}{6.7} = \underline{858.4 \text{ gal/h}}$

(d) Seat-miles/gal $= \dfrac{100 \times 0.78(686.9)}{858.4} = \underline{62.42 \text{ seat-miles/gal}}$

(e) Auto has 5 seats × 20 miles/gal = $\underline{100 \text{ seat-miles/gal}}$

$$\frac{\text{auto}}{\text{DC-9}} = \frac{100}{62.42} = \underline{1.6}$$

That is, the auto provides 60% more seat-miles/gal when all seats in both vehicles are filled.

THE ENERGY-STATE APPROACH TO AIRPLANE PERFORMANCE

We have discussed climb and cruise performance as quasi-steady problems in which all forces are in balance at any given instant. In the case of climb, we have recognized that some of the excess thrust may be used to accelerate the airplane along the flight path during the climb. A correction for this has been derived and simple expressions given for practical climb procedures, such as constant equivalent airspeed or constant Mach number operation.

The traditional method for determining the speed for best climb gradient, or best rate of climb, or best range climb is to calculate the rate or gradient of climb as a function of speed at various weights and altitudes. A plot of the results, shown for one weight and altitude in Figure 15.8, is then examined to establish the equivalent airspeed or Mach number that best matches the peak values of the desired quantity. The best range climb must be studied as a flight path that extends beyond reaching the cruise altitude and includes a portion of the cruise flight. The fuel consumed and distance traversed along this path must be evaluated for various climb speeds.

There is another approach to the climb problem that provides considerable insight and, in the case of supersonic aircraft, much more efficient climb flight paths. This approach recognizes that the basic problem is not just to climb but also to accelerate to cruising speed. The energy added to low-speed aircraft during the climb to cruise altitude and speed is overwhelmingly due to the potential energy of height. Therefore, when we focus on attaining an altitude as efficiently as possible, little error is introduced. As cruise speed increases, however, the kinetic energy becomes increasingly important. It then becomes useful to study the climb/accelerate problem as one in which the objective is to raise the total energy of the airplane in the most desirable manner. "Most desirable" may mean least time, least fuel, or greatest overall range with a given amount of fuel. In this "energy-state" analysis, the total energy becomes the fundamental variable.

The sum of the potential and kinetic energies is the total energy E of the aircraft. Thus

$$E = Wh + \frac{WV^2}{2g} = W\left(h + \frac{V^2}{2g}\right) \tag{15.37}$$

where W is the aircraft weight, h the altitude, and V the true speed.

It is convenient in analyzing performance using the energy method, and in comparing aircraft of different weights, to work with energy per unit aircraft weight. This is defined as specific energy h_e and is found by dividing equation 15.37 by W.

$$h_e = h + \frac{V^2}{2g} \tag{15.38}$$

h represents the energy per unit weight due to altitude, $V^2/2g$ is the kinetic energy per unit weight and has the same units as h, and h_e is the equivalent energy height at which the airplane would have the existing total energy if the speed were reduced to zero. Thus h_e is the height to which the airplane could zoom if its kinetic energy were converted to potential energy with no losses. Contours of constant specific energy h_e, determined from equation 15.38, are shown is Figures 15.22 and 15.23.

When an airplane gains height or speed, its specific energy must increase. To predict the rate at which an airplane increases its energy level, we determine the derivative of h_e with respect to time. From equation 15.38,

$$\frac{dh_e}{dt} = \frac{dh}{dt} + \frac{V}{g}\frac{dV}{dt} \tag{15.39}$$

The first term on the right side of equation 15.39 is the rate of climb; the second is the rate of energy change due to acceleration. From equation 15.20,

$$\sin\gamma = \frac{T - D}{W} - \frac{1}{g}\left(\frac{dV}{dt}\right)$$

$$\frac{dh}{dt} = V\sin\gamma = \frac{V(T - D)}{W} - \frac{V}{g}\left(\frac{dV}{dt}\right) \tag{15.40}$$

Combining with 15.39 yields

$$\frac{dh_e}{dt} = \frac{V(T - D)}{W} \tag{15.41}$$

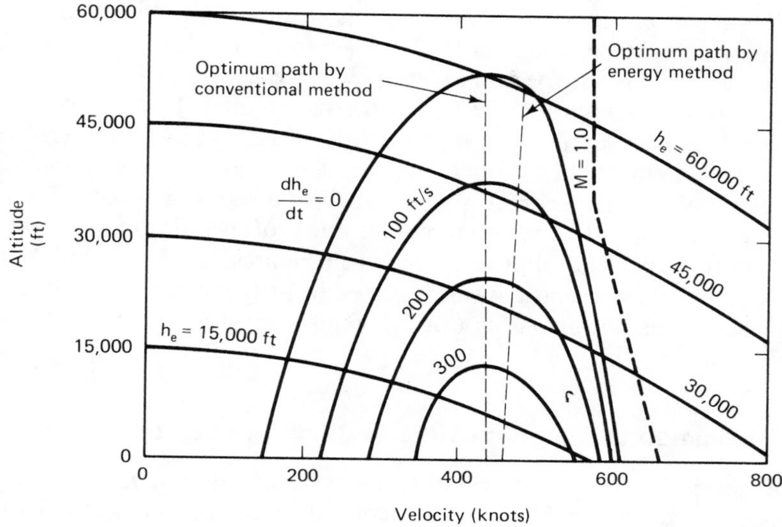

Figure 15.22 Contours of constant specific energy and constant specific excess power for a subsonic airplane. From E. S. Rutowski, "Energy Approach to the General Performance Problem," *Journal of the Aeronautical Sciences,* Vol. 21, No. 3, March 1954. Reprinted by permission of American Institute of Aeronautics and Astronautics.

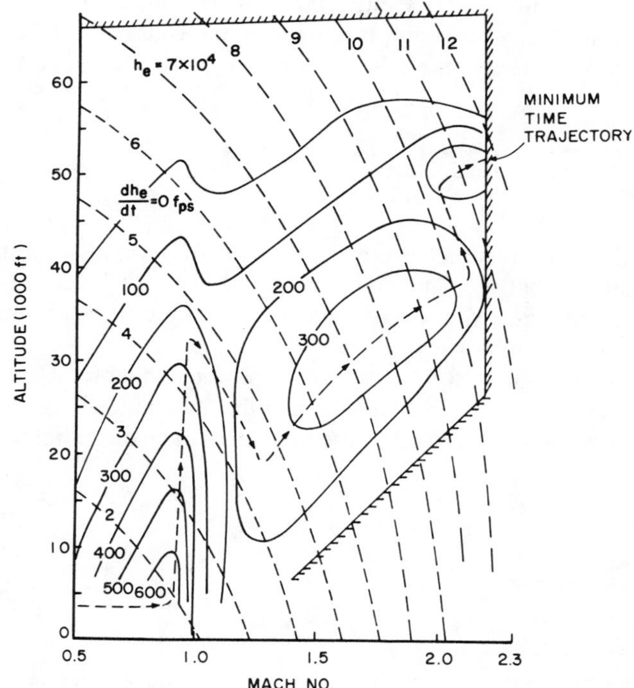

Figure 15.23 Contours of constant specific energy and constant excess specific power for a supersonic fighter. (From Leland M. Nicolai, *Fundamentals of Aircraft Design,* School of Engineering, University of Dayton, Dayton, Ohio, 1975, courtesy of Leland M. Nicolai.)

The rate of change of specific energy is equal to the excess power available per unit aircraft weight (see equation 15.19). Therefore, dh_e/dt is called *specific excess power*. Curves of constant specific excess power are shown in Figure 15.22 for a subsonic airplane and in Figure 15.23 for a supersonic fighter. These curves are constructed by calculating the specific excess power from equation 15.41 for a series of altitudes from sea level to the highest altitude of interest at a particular speed. dh_e/dt is then plotted against altitude. The process is repeated for a series of speeds. The combinations of altitude and speed for various constant values of specific excess power can be obtained from the chart and plotted as shown in Figures 15.22 and 15.23. The contour for $dh_e/dt = 0$ represents the steady-state level-flight performance boundary for the airplane for the power setting (e.g., maximum continuous thrust) for which the graph is drawn.

Application to Minimum Time to Climb/Accelerate

As stated previously, the climb problem is not only to attain a particular altitude but also to reach a particular cruise or combat speed at that altitude. To attain the energy level represented by the desired altitude/speed combination in the least time, the aircraft should be flown, at each level of energy, at the speed and altitude for which the rate of increase of energy is a maximum. From Figures 15.22 and 15.23, it can be seen that the highest specific excess power at each energy level occurs at the speed/altitude points at which the specific excess power curves are tangent to the specific energy curves. The locus of these tangent points describes the optimum energy climb schedule for minimum time to a given energy level and is shown on the figures.

The time to change from one energy level, h_{e_1}, to another, h_{e_2}, is found from the equation

$$\Delta t = \int_{h_{e_1}}^{h_{e_2}} \frac{dh_e}{dh_e/dt} \tag{15.42}$$

To evaluate this integral for Δt, we construct a graph of $1/(dh_e/dt)$ versus h_e for various points along the trajectory between the initial and final energy levels, as indicated in Figure 15.24. The area under the curve represents the time to reach the higher energy level, or the time to climb.

For the subsonic airplane in Figure 15.22, the excess power available at any altitude along the best rate of climb path is little different from that available along the best excess power path at the same altitude. Thus the time to climb will be only slightly

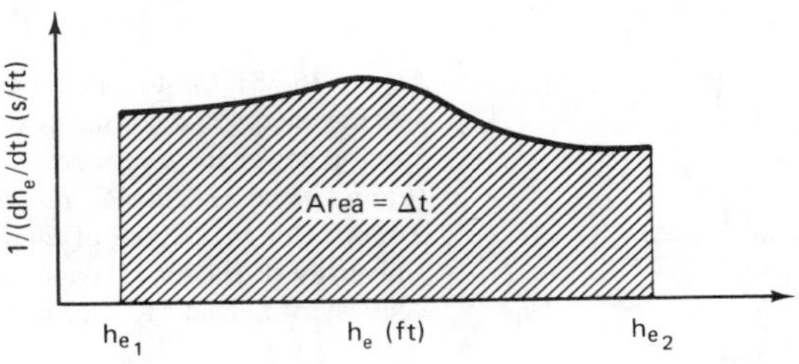

Figure 15.24 Determination of the time from an initial to a final energy level.

less. However, more speed will have been gained during the climb phase using the best energy path, so the time to a desired energy level defined by altitude and speed will be somewhat less using the best energy path. Furthermore, the average speed during the climb will be higher and more distance will be covered, both desirable characteristics toward which we would lean in choosing the climb operating procedure in the conventional approach. The energy method approximation assumes that, having attained an energy level, the desired proportion of potential and kinetic energy (i.e., altitude and speed) can easily be attained by a dive or a zoom in essentially zero time. This assumption is not true and leads to some error in the method. In any case, it is clear that for the subsonic airplane the energy method defines a path that is preferable to the best rate of climb speeds, but not very different from what would normally be done anyway.

It is in the supersonic case that the energy method offers dramatic advantages. Selecting, at each level of specific energy, the speed and altitude for maximum excess specific power defines flight paths entirely different from constant speed or nearly constant speed paths. As shown in Figure 15.23, the procedure leads to climbs with a large amount of acceleration and even accelerations in dives as part of the climb path. The physical reason for this lies in the variations of drag and thrust with velocity and altitude for supersonic airplanes (Figure 15.25). As the speed of an airplane approaches the speed of sound, the drag coefficient rises sharply. Above a Mach number of 1.0, the drag coefficient decreases. Since the dynamic pressure increases with speed at a given altitude, the actual drag continues to rise above Mach 1.0, but at a much slower rate than below Mach 1.0. The thrust of a supersonic engine also increases with speed but decreases with altitude. To increase the energy of a supersonic fighter as rapidly as possible, the excess specific power must be maintained at as high a level as possible. If the airplane climbs rapidly to a high altitude with little gain in speed, the loss in thrust due to altitude will reduce excess power so that low acceleration will occur through the transonic region. Acceleration to penetrate the transonic region may not even be possible. Therefore, there is a benefit to reducing the altitude, at constant specific energy, before penetrating the transonic region. The increased specific excess power helps the acceleration and the airplane acquires a high specific energy at a moderate altitude. The remaining altitude can then be quickly obtained by zooming at almost constant energy.

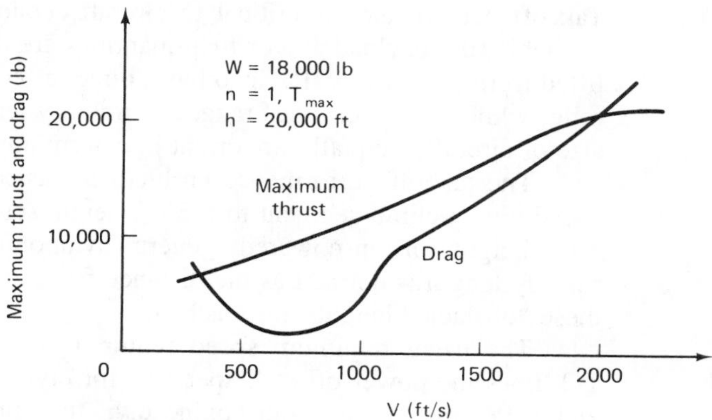

Figure 15.25 Drag and thrust of a supersonic fighter.

Application to Minimum Fuel to Climb/Accelerate

The problem of minimizing the fuel consumed to achieve a speed and altitude can be treated by the energy method in the same fashion as the time problem. For each increment of specific energy dh_e, we would like to use the least increment of fuel, dW_F. Thus we would like to minimize dW_F/dh_e or maximize dh_e/dW_F. The rate of change of specific energy per unit fuel consumption is

$$\frac{dh_e}{dW_F} = \frac{dh_e/dt}{dW_F/dt} = \frac{V(T - D)}{WTc} \tag{15.43}$$

where W_F is the fuel used and c is the specific fuel consumption.

To solve this problem, contours of constant values of dh_e/dW_F are constructed in a manner analogous to the procedure for finding the best time paths. The tangent points between contours of constant dh_e/dW_F and constant specific energy define the locus of points at each energy level that provide the next incremental gain of energy for the least fuel consumed. The fuel consumed is found by plotting $1/(dh_e/dW_F)$ versus energy level h_e along the locus. The area under the curve is the total fuel used, the procedure being similar to that shown in Figure 15.24 for the time-to-climb problem. This is, of course, a graphical solution to the equation

$$W_F = \int_{h_{e1}}^{h_{e2}} \frac{dh_e}{dh_e/dW_F} \tag{15.44}$$

Another use of the energy method is the study of fighter maneuverability. The available acceleration while maneuvering or the ability to maintain steady turns can be studied by constructing graphs similar to Figures 15.22 and 15.23 with load factors greater than 1.0. The increased wing lift required to turn or maneuver will increase the drag so that the excess specific power will decrease at every altitude–speed combination. The altitude–speed region for any level of excess power, including the level-flight contour, $dh_e/dt = 0$, will become smaller. Further discussion of turning performance is given in Chapter 16.

TAKEOFF PERFORMANCE

Takeoff performance is critical to aircraft economics. With limited runway length available, the payload and/or fuel quantities are dependent on the weight that can be lifted from the runway. Payload has a direct effect on cost per passenger mile or ton-mile, while fuel determines range. In military terms, the armament load for a given size of aircraft is equally important in determining the cost per unit of effectiveness.

The takeoff performance problem is basically an acceleration to the required speed plus a climb segment to a 35-ft height (civil turbine-powered transports) or a 50-ft height (piston-powered, general aviation or military aircraft). The required runway length is defined as the distance from the start of takeoff to the point where these "obstacle" heights are reached.

The usual minimum speed required at the obstacle height (35 or 50 ft) is 1.2 times the power-off stall speed V_S for civil jets and two-engine civil turboprops and 1.15 times V_S for four-engine civil turboprops. Military and general aviation requirements are often 5% to 10% lower.

For detailed calculations of takeoff distance, the acceleration is calculated as a function of speed from equation 15.45, which is derived as follows:

$$\text{acceleration } a = \frac{F}{M} = \frac{T - D}{W/g}$$

$$= \frac{T - [D_P + D_i + \mu'(W - L)]}{W/g} \tag{15.45}$$

$$= \frac{T - \left[C_{D_P}qS + K_L \dfrac{(C_{L_t})^2}{\pi \cdot \text{AR} \cdot e}qS + \mu'(W - C_{L_t}qS)\right]}{W/g}$$

where

T = total engine thrust (lb)
$C_{L_t} = C_{L_{\text{taxi}}}$ = lift coefficient in the taxi attitude with ground effect
K_L = ground effect correction to induced drag (Figure 9.10)
μ' = rolling coefficient of friction; μ' is 0.015 for large aircraft on a hard surface and may be twice that for small aircraft; on grass, μ' is about 0.05
C_{D_P} = parasite drag coefficient, including flaps, if deflected, and landing gear

The distance and speed at each instant during the takeoff run can be calculated using a step-by-step procedure. Some time interval Δt is assumed. In each time interval Δt_i, the speed increment ΔV_i is

$$\Delta V_i = a_i \Delta t_i$$

where a_i is the average acceleration during Δt_i and i is the number of the interval. The distance covered during the interval Δt_i is

$$\Delta d_i = V_i \Delta t_i$$

where V_i is the average speed during Δt_i. Then the acceleration distance to the lift-off speed over time $t = n\Delta t_i$ is

$$d_{\text{LO}} = \sum_1^n V_i \Delta t_i$$

If the acceleration a is a constant, $V = at$ and the distance to lift-off is

$$d_{\text{LO}} = \sum_1^n at_i \Delta t_i = \int_0^{t_{\text{LO}}} at\,dt = \frac{at_{\text{LO}}^2}{2}$$

Since $V_{\text{LO}} = at_{\text{LO}}$, $t_{\text{LO}} = V_{\text{LO}}/a$ and

$$d_{\text{LO}} = \frac{a}{2}\frac{V_{\text{LO}}^2}{a^2} = \frac{V_{\text{LO}}^2}{2a} \tag{15.46}$$

The acceleration is not actually constant since both T and D vary with speed.* It can be shown, however, that the effective average $(T - D)$ during a takeoff is

*The takeoff thrust of turbine engines decreases as the speed increases during the takeoff run. This variation is discussed in Chapter 17 and presented graphically for various engine bypass ratios in Figure 17.15.

the value at $V = (1/\sqrt{2})V_{LO}$, where V_{LO} is the lift-off speed. The acceleration distance is

$$d_{LO} = \frac{V_{LO}^2}{2a_{0.7V_{LO}}}$$

If we assume lift-off at $1.2V_S$,

$$V_{LO} = 1.2V_S = 1.2\sqrt{\frac{2}{\rho}\left(\frac{W}{S}\right)\frac{1}{C_{L_{MAX}}}}$$

Then

$$\begin{aligned} d_{LO} &= 1.44\left(\frac{2}{\rho}\right)\frac{W}{S}\left(\frac{1}{C_{L_{MAX}}}\right)\frac{1}{2\left(\dfrac{T - D}{W/g}\right)_{0.7V_{LO}}} \\ &= 1.44\frac{W^2}{g\rho SC_{L_{MAX}}(T - D)_{0.7V_{LO}}} \end{aligned} \tag{15.47}$$

The constant 1.44 in equation 15.47 is equal to $(V_{LO}/V_S)^2$. If we plan to attain a speed of $1.2V_S$ at 35 ft, we would lift off at a lower speed, say $1.16V_S$. Then the constant in equation 15.47 becomes $(1.16)^2 = 1.35$.

If the drag D is small compared to T, as is usually true during the takeoff run, we can assume that

$$d_{LO} = f\left(\frac{W^2}{\sigma SC_{L_{MAX}}T}\right) \tag{15.48}$$

where $\sigma = \rho/\rho_s$ and T = thrust at $0.7V_{LO}$. This suggests that if we plot the takeoff distances for many airplanes, determined accurately, over a range of weights and altitudes for each airplane, against the parameter $W^2/\sigma SC_{L_{MAX}}T$, the points will form a single curve. This is, in fact, very nearly the result of such a graph. The scatter is small and a single fairing gives an excellent approximation to the takeoff distance of any airplane.

The total required takeoff field length includes an air run that is a curved flight path until the steady-state climb angle is reached. The distance to reach a given height along such a path can be shown to be a function of the same variables used in the parameter in equation 15.48, although the exponents are not necessarily the same.

Since the ground run is about 80% of the total distance to a 35-ft height, $W^2/\sigma SC_{L_{MAX}}T$ is the dominant form. It has been shown that this parameter works very well in correlating the required runway-length results for many aircraft. Figure 15.26 shows the FAR (Federal Air Regulation) all-engines-operating takeoff field length to a 35-ft height, d_{35}, for jet or turbofan aircraft as a function of $W^2/\sigma SC_{L_{MAX}}T_{0.7V_{LO}}$. This chart applies to normal takeoff without engine failure and includes a 15% increase above the actual performance in accordance with the air transport requirements of FAR Part 25. The actual distance is the distance determined from Figure 15.26 divided by 1.15.

For propeller-driven aircraft, a comparable analysis shows that takeoff distance is a function of $W^2/\sigma SC_{L_{MAX}}P$, where P is the total brake horsepower. This is a less

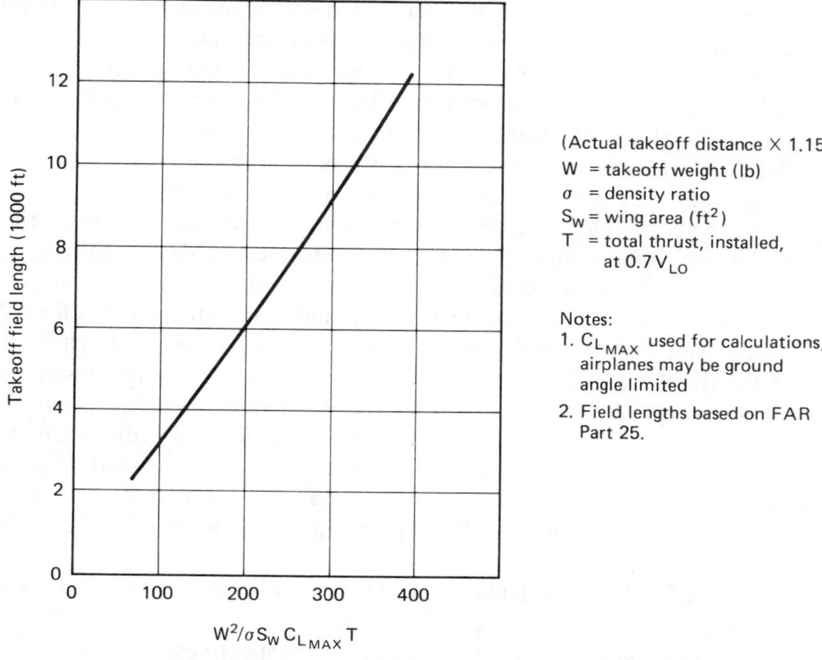

Figure 15.26 FAR takeoff field length to 35 ft (jet), all engines operating.

accurate approximation, since the effectiveness of the power depends on the propeller efficiency during takeoff. Use of this parameter assumes that all propellers are designed to attain a similar level of efficiency in takeoff, a fairly good but not perfect assumption.

Commercial takeoff runway lengths required by the FARs must be based on the assumption of an engine failure at the worst possible time, unless the all-engines-operating distance plus 15% is longer. If the engine fails sooner than the critical time, the airplane can be stopped on the runway in less distance, and if the engine fails at a higher speed, the airplane can continue the takeoff and reach a height of 35 ft in less distance. The airplane speed at which this worst possible time for engine failure occurs is called the *critical engine failure speed*. Allowing 1 to 2 seconds for the pilot to recognize the engine failure increases the speed at which the pilot actually decides whether or not to continue the takeoff. This slightly higher speed is called the *decision speed*, V_1. If an engine fails and the pilot recognizes it at a speed higher than V_1, the pilot must continue the takeoff because he cannot stop if the runway available is the minimum required field length. At a lower speed he must decide to stop. Field lengths based on the selection of V_1 such that the distance to accelerate to V_1 and stop (called the *accelerate–stop distance*) is equal to the distance to accelerate to engine failure, continue the acceleration with one engine inoperative, and then climb to a 35-ft height (called the *accelerate–climb distance*) are balanced field lengths. Field lengths shown in transport aircraft flight manuals are usually balanced field lengths. The critical engine failure speed is determined by calculating the accelerate–stop and the accelerate–climb distances as functions of assumed engine failure speeds and plotting them as in Figure 15.27. The intersection shows the critical engine failure speed and the balanced field length. In modern practice, computers may do all the work so that plots such as Figure 15.27 are not actually constructed, but the effect is the same. The calculation and later the confirming flight tests allow not only for the pilot recognition time, but also for reasonable time delays in applying brakes, spoilers, and other deceleration devices. Reverse thrust is not used in determining stopping distances but is an important reserve capability, especially on wet runways.

In addition to V_1, there are other speeds specified in the flight manuals that must be known to the pilot before takeoff. These include V_R, the proper speed for starting the rotation of the airplane to the lift-off attitude, and V_2, the proper climb speed to be reached at or before the 35-ft height and maintained in the climb if an engine has failed. Without engine failure, a higher climb speed is used. These speeds are given in the flight manuals as functions of takeoff weight, altitude, temperature, wind, and runway slope. V_1, V_R, and V_2 are in turn related to other defined speeds to assure safe margins above the stall speeds both in free air and in ground effect, and above the minimum speed V_{mc} at which the airplane can be controlled directionally with one engine inoperative on one side of the airplane. The takeoff procedure and the pertinent speeds are summarized in Figure 15.28. FAR Part 25 should be consulted for further details.

The FAR required takeoff field lengths with engine failure are given as functions of the generalized parameter $W^2/\sigma SC_{L_{\text{MAX}}}T$ in Figures 15.29, 15.30, and 15.31 for two-, three-, and four-engine jet aircraft, respectively. Figure 15.32 shows takeoff field length plotted against $W^2/\sigma SC_{L_{\text{MAX}}}P$ for older (1950s) aircraft with four piston engines with engine failure. This curve gives the distance to a 50-ft height, the applicable height with piston engines. Such aircraft are no longer built, but the curve is shown since modern turboprops would fit the same format. Turboprop airplane takeoff field lengths may also be estimated from the jet aircraft curves by determining the thrust available from the propellers at $0.7V_{\text{LO}}$. A curve for a propeller-driven STOL airplane is also shown in Figure 15.32. In using any of these generalized takeoff

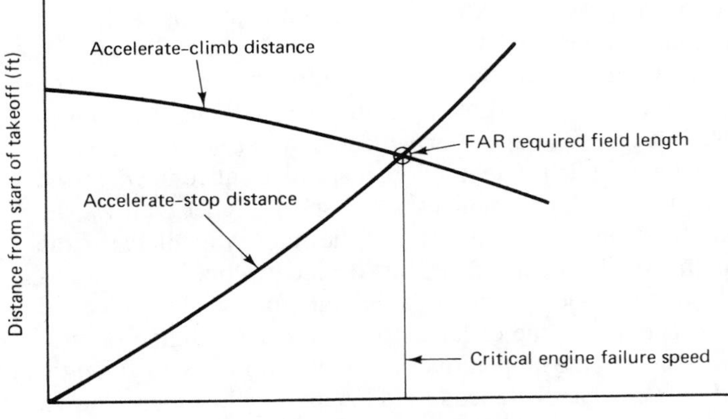

Figure 15.27 Determination of FAR takeoff field length with engine failure.

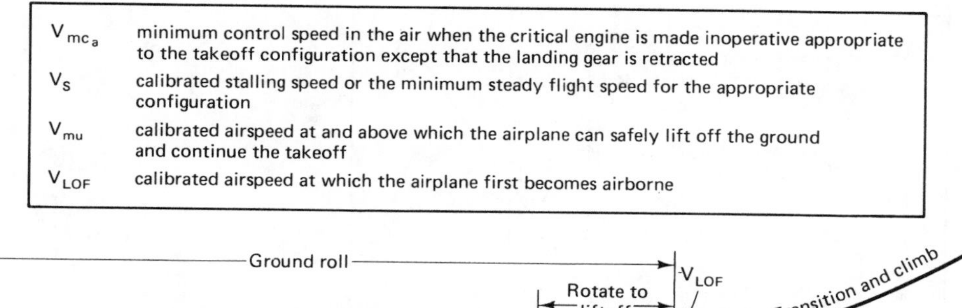

V_{mc_a}	minimum control speed in the air when the critical engine is made inoperative appropriate to the takeoff configuration except that the landing gear is retracted
V_S	calibrated stalling speed or the minimum steady flight speed for the appropriate configuration
V_{mu}	calibrated airspeed at and above which the airplane can safely lift off the ground and continue the takeoff
V_{LOF}	calibrated airspeed at which the airplane first becomes airborne

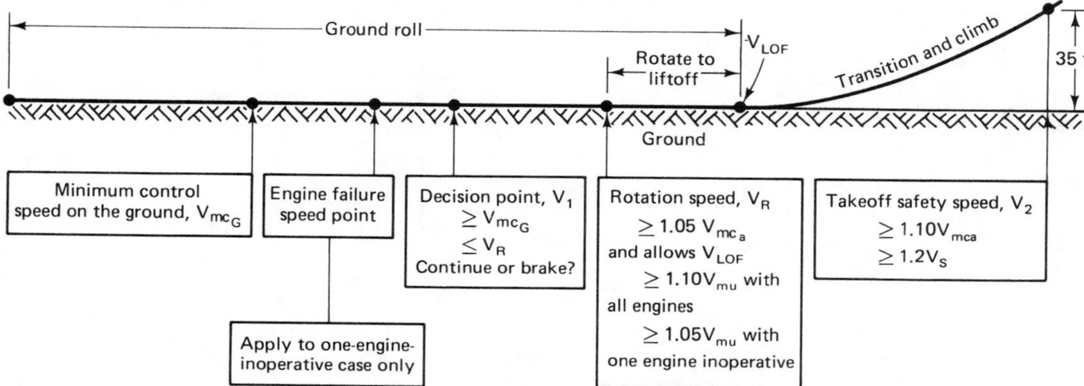

Figure 15.28 Diagram of takeoff path to 35 ft above the takeoff surface for multi-engine jet airplanes with engine failure.

performance curves, it is important to note that the thrust or power is the total installed value for all engines, not the thrust or power remaining after engine failure.

A final remark about this parametric method of predicting takeoff performance is in order. The reason the method works so well is that the distance required to accelerate a body to a speed V is dependent on the square of the speed and the reciprocal of the acceleration, as shown in equation 15.46. V^2 is dependent on the wing loading W/S and is inversely proportional to $C_{L_{MAX}}$ and the air density. The acceleration is directly proportional to the ratio of thrust to weight, T/W. We are neglecting the drag here. The parameter $W^2/\sigma S C_{L_{MAX}} T$ is simply the product of those terms. If the ratio of the takeoff distances of two aircraft are not quite close to the ratio of their respective values of this parameter, the calculations are wrong even if done by computer. Someone may have placed the wrong inputs into the computer. The only exceptions would be some strange configurations that used a flap angle with high drag for takeoffs, and even that discovery may have uncovered a design mistake. For takeoff with engine failure, braking effectiveness could also influence field length. Therefore, the parametric method can be used to check the reasonableness of detailed calculations of takeoff. Awareness of such fundamental relationships can be a great help to engineers by providing a simple confirmation of the reasonableness of solutions to various problems.

Example 15.3

At takeoff from San Francisco, the twin-engine DC-9-30 weighed 102,000 lb. The wing area is 1000 ft^2. The airport runways may be assumed to be at sea-level pressure altitude. The

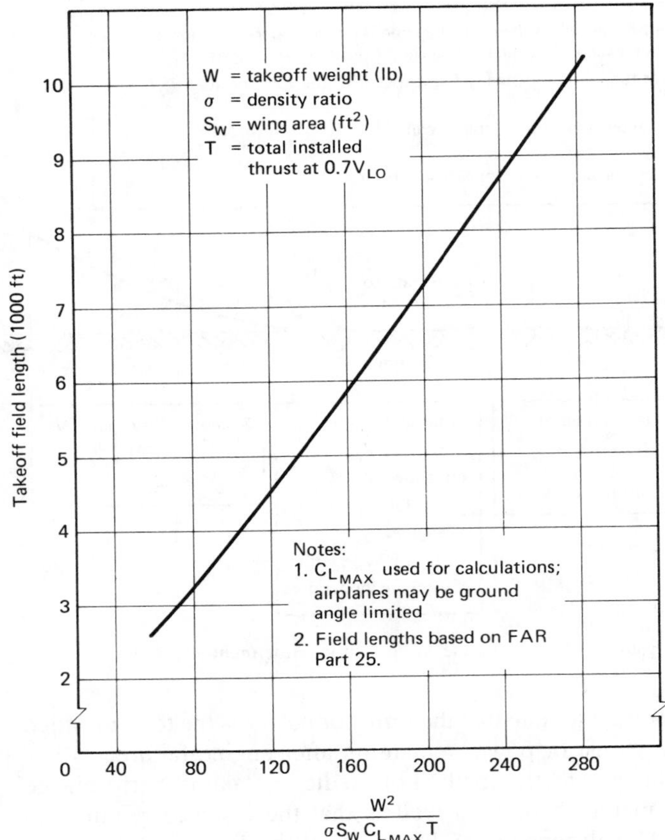

Figure 15.29 Two-engine jet aircraft FAR takeoff field length with engine failure.

temperature was 82°F. The engines have a static ($V = 0$) rating of 15,000 lb of thrust each, but lose 14% in thrust at the effective average takeoff speed ($0.7V_{LO}$) due to bleed and power extraction losses and the engine ram drag due to forward speed. For the takeoff, the flap angle is 15° and the slats are extended.

(a) Determine the FAR takeoff field length.
(b) What is the takeoff climb speed at $1.2V_S$?

Solution: At sea level,

$$T = 82°\text{F} = 542°\text{R}, \qquad \rho = \frac{p}{RT} = \frac{2116}{(1718)(542)} = 0.002272$$

$$W = 102{,}000 \text{ lb}, \qquad \sigma = \frac{\rho}{\rho_s} = \frac{0.002272}{0.002377} = 0.956$$

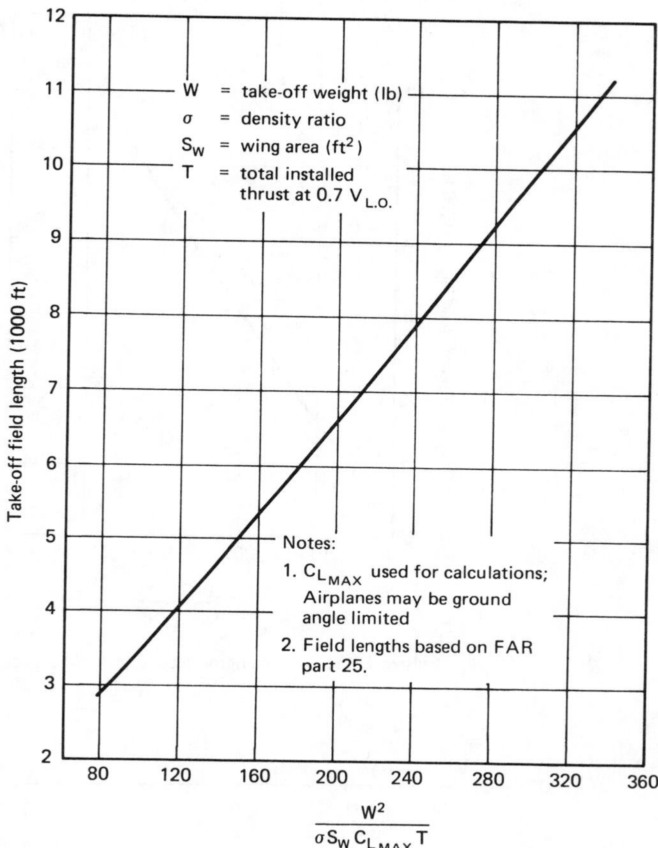

Figure 15.30 Three-engine jet aircraft FAR takeoff field length with engine failure.

At flap angle $\delta_F = 15°$, $C_{L_{\text{MAX}}} = 2.45$ (from Figure 14.15).

(a) Total thrust, $T = 15000 \times 0.86 \times 2 = 25{,}800$ lb.

$$\frac{W^2}{\sigma S C_{L_{\text{MAX}}} T} = \frac{(102{,}000)^2}{0.956(1000)\,(2.45)\,(25{,}800)} = 172.2$$

From Figure 15.29 for twin-engine aircraft, FAR takeoff field length = <u>6300 ft</u>.

(b) Climb speed $= 1.2V_S = 1.2\sqrt{\dfrac{2W}{C_{L_{\text{MAX}}}\rho S}}$

$$= 1.2\sqrt{\frac{2(102{,}000)}{(2.45)\,(0.002272)\,(1000)}} = \underline{229.73 \text{ ft/s}}$$

$$= \underline{156.6 \text{ mph}}$$

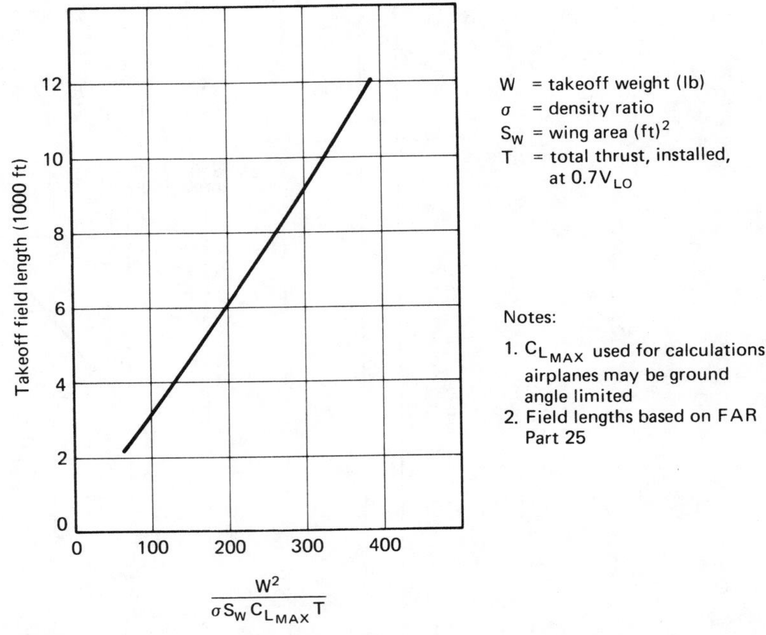

Figure 15.31 Four-engine jet aircraft FAR takeoff field length with engine failure.

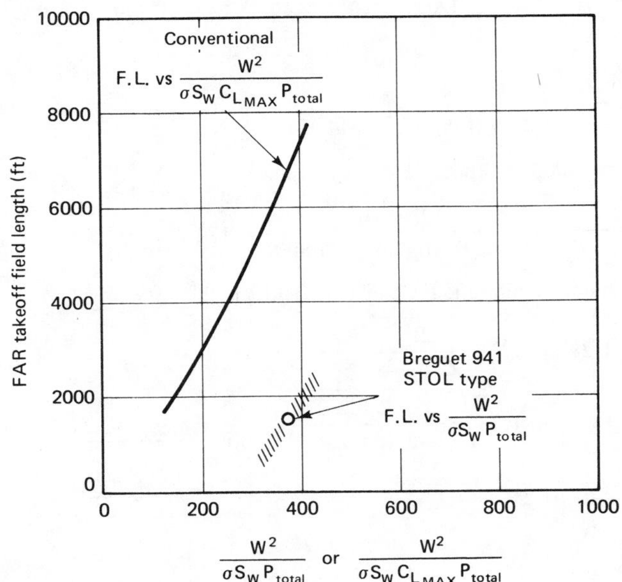

Figure 15.32 Four-engine propeller aircraft FAR takeoff field length with engine failure (based on DC-6).

LANDING PERFORMANCE

Landing distances consist basically of two segments, the air run from a height of 50 ft to the surface accompanied by a slight deceleration and flare, and the ground deceleration from the touchdown speed to a stop, as shown in Figure 15.33. Landing runway lengths required by FAR Part 25 for commercial aircraft require a safety factor of 1/0.6. The air distance d_{AIR} can be approximated by a steady-state glide distance d_{GL} plus an air deceleration distance d_{decel} at constant altitude, as shown in Figure 15.33.

V_{50} is the speed at the 50-ft height. In accordance with FAR Part 25, V_{50} must be at least $1.3V_S$. In practice, it is taken as equal to $1.3V_S$. V_L is the landing or touchdown speed and is usually about $1.25V_S$. The glide distance is

$$d_{\text{GL}} = 50\left(\frac{L}{D_{\text{eff}}}\right) \tag{15.49}$$

where $D_{\text{eff}} = D - T$.

$$d_{\text{decel}} = \frac{V_{50}^2}{2a} - \frac{V_L^2}{2a} = \frac{\frac{1}{2}\left(\frac{W}{g}\right)V_{50}^2 - \frac{1}{2}\left(\frac{W}{g}\right)V_L^2}{D_{\text{eff}}} \tag{15.50}$$

Since lift is essentially equal to the weight,

$$\begin{aligned} d_{\text{AIR}} &= 50\frac{L}{D_{\text{eff}}} + \frac{1}{2g}(V_{50}^2 - V_L^2)\frac{L}{D_{\text{eff}}} \\ &= \frac{L}{D_{\text{eff}}}\left[50 + \frac{1}{2g}(V_{50}^2 - V_L^2)\right] \end{aligned} \tag{15.51}$$

L/D_{eff} is the effective L/D ratio during the air run. It can be determined from flight test air runs by plotting flight test air run distances versus $(V_{50}^2 - V_L^2)$, as illustrated in Figure 15.34.

The ground deceleration distance is

$$d_G = \frac{V_L^2}{2a} = \frac{V_L^2}{2[R/(W/g)]} \tag{15.52}$$

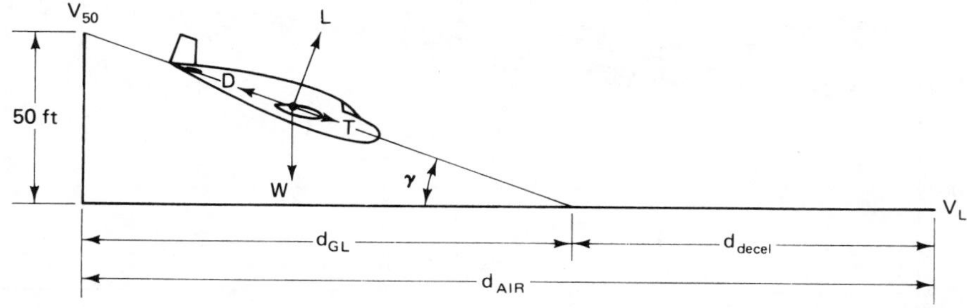

Figure 15.33 Two-segment approximation to the landing air run.

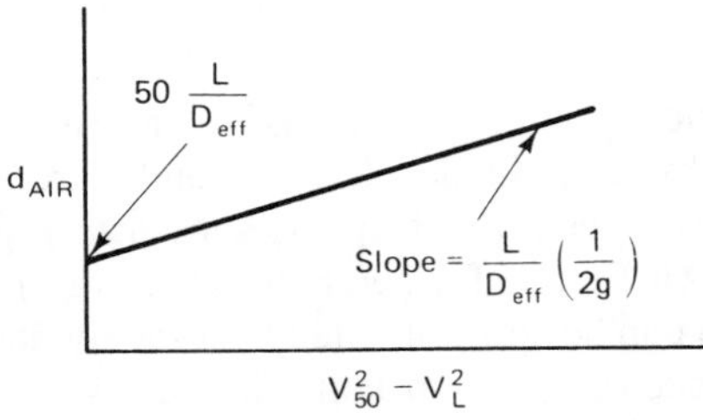

Figure 15.34 Format for plotting flight test landing air runs.

where

R = effective average resistance or total stopping force = $\mu(W - L) + D$

μ = braking coefficient of friction; μ = 0.4 to 0.6 on dry concrete, 0.2 to 0.3 on wet concrete, and 0.1 or less on ice.

D = drag, including drag of flaps, slats, landing gear, and spoilers.

Note that both d_{AIR}, air distance, and d_G, ground stopping distance, are directly proportional to V_{50}^2 and/or V_L^2. Both V_{50} and V_L are fixed percentages above V_S for safety reasons. Thus landing distance is linear in V_S^2 except for the glide distance from 50 ft, which depends only on the L/D in the landing configuration. Thus, for similar airplanes with similar L/D values and equivalent braking systems (i.e., similar μ), landing distances can be of the form

$$d_{\mathrm{land}} = A + BV_S^2 \tag{15.53}$$

where $A = 50(L/D)_{\mathrm{eff}}$. This suggests a means for predicting landing performance.

Figure 15.35 shows the FAR landing-field lengths on dry runways for seven transport airplanes plotted against the square of the stalling speed in the landing configuration. The landing configuration involves extended landing gear and usually

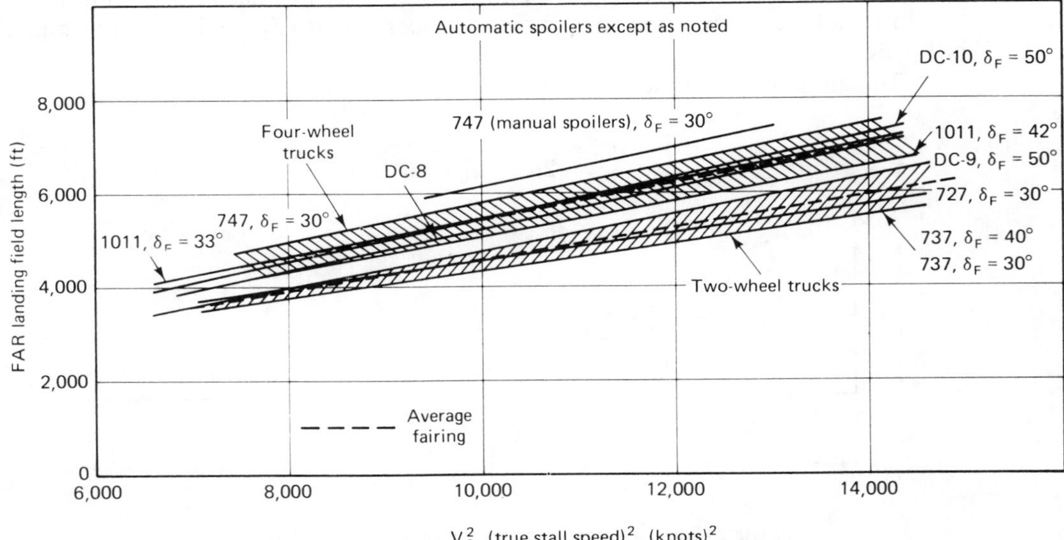

Figure 15.35 FAR landing-field length, dry runways.

full-flap deflection. In recent years, an alternate lower flap deflection has been provided to reduce the power required on the approach along the usual 3 degree ILS (instrument landing system) descent to the landing runway. The purpose of this lower flap angle is to reduce the community noise below the approach path. Figure 15.35 shows the landing data for two different flap angles for some airplanes.

The FAR landing field length is defined as the actual demonstrated distance from a 50-ft height to a full stop increased by the factor 1/0.60, a 67% increase. Although the individual points are omitted for clarity, the curves of landing field length versus V_S^2 are linear, as anticipated in equation 15.53. However, there are significant differences between the various airplanes. These differences are due to variations in the effective L/D in the air run (equation 15.51), in the effective coefficient of friction μ, and in the drag in the ground deceleration (equation 15.52).

Although flap drag plays a significant role in the air run, the pilot's control of the throttles is usually more important. If more power is maintained during the air run, the effect is the same as a higher effective L/D ratio. Furthermore, the touchdown speed is important since the wheel brakes are much more effective in retarding the airplane than the air drag during the air run. The sooner the airplane touches down and starts braking, the shorter the total distance will be. Thus the human factor plays a large role in landing distances. The official landing distance is partly a reflection of how hard the flight test pilot worked to optimize the landing. In practice, this is dependent on how important the landing-field length is to the usefulness of the airplane. If the landing distance is much shorter than the takeoff distance, a little longer flight test landing may not be detrimental.

Mechanical devices have a large influence on landing distances. Automatic spoilers are triggered by the rotation of the wheels at touchdown. The spoilers greatly decrease the lift, dump the weight on the wheels, and thereby make the brakes more effective. Manual spoilers, operated by the pilot, involve a delay. Even 2 s at a speed of 200 ft/s can increase the stopping distance by almost 400 ft. Including the safety factor of 67%, the effect on the field length can be close to 600 ft. With one exception, the curves on Figure 15.35 are for automatic spoilers. In the 747 example on the chart, manual spoilers are shown to cost 400 ft in field length. The adjustment of antiskid braking systems can also affect the average braking coefficient of friction during the deceleration.

These factors explain why all aircraft are not the same on Figure 15.35. In addition, there is a difference between the large aircraft with four-wheel landing trucks and those with two-wheel trucks. It appears that the effective coefficient of friction is less for wheels rolling immediately behind other wheels. Thus we have shown a scatter band for the four-wheel-truck aircraft and another band for the two-wheel-truck aircraft. The dashed fairings are in the center of the bands and are within 6% of the extremes of the scatter bands.

The dashed average fairings in Figure 15.35 represent a reasonable way to estimate landing-field lengths. The landing-field-length prediction is a function only of the square of the true stall speed.

Figure 15.35 is based on transport aircraft with highly developed antiskid braking systems. Aircraft with simple brakes and without spoilers will have considerably longer stopping distances than are built into the curves. However, quoted landing distances for small aircraft are sometimes based only on the ground run, without a safety factor. The total distance over a 50-ft height with the 1/0.6 factor is about 2.5 times as long.

FAR landing runway lengths for dry runways must be increased by 15% for operation on wet runways unless special wet-runway tests are run under airline operating conditions (e.g., 3 degree glide slope down to the 50-ft height, 80% worn tires, reverse thrust but with an engine inoperative, etc.). Even then a markup is required. The dry-runway tests do not permit reverse thrust.

Example 15.4

A twin-engine transport aircraft is scheduled to leave New York's La Guardia field on a 1200-km flight. The expected takeoff ambient temperature is 25°C, the pressure altitude is sea level, and the available runway length is 2135 m. The airplane has a wing area of 111.5 m^2, a static takeoff thrust rating of 82,288 N per engine up to 25°C, and a $C_{L_{\text{MAX}}}$ with a 15 degree flap angle of 2.5. With the anticipated 135 passengers, the payload will be 12,272 kg and the takeoff weight will be 59,800 kg. The effective average thrust at $0.7V_{\text{LO}}$ is 82% of the static thrust.

Determine whether this flight can be safely operated in accordance with the FAR takeoff field-length requirements. Assuming that a passenger weighs 91 kg, including baggage, how many more or fewer passengers can be carried to utilize the available runway fully.

Solution: Since the FAR takeoff performance charts are based on English units, the SI units must be converted. The required field length must be the longer of the all-engines operating field length (Figure 15.26), or the FAR field length with engine failure for two-engine airplanes (Figure 15.29).

First, evaluate the parameter $W^2/\sigma S C_{L_{\text{MAX}}} T$, where T is the total installed thrust at 0.7 V_{LO}.

$$W = 59{,}800 \times 2.2 = 131{,}560 \text{ lb}$$

$$S = 111.5 \times 10.76 = 1200 \text{ ft}^2$$

$$T = (82{,}288 \times 0.2248) \times 2(\text{engines}) \times 0.82 = 30{,}337 \text{ lb}$$

$$\rho = \frac{p}{RT} = \frac{101{,}325}{(287.05)(298)} = 1.1845 \text{ kg/m}^3$$

$$\sigma = \frac{\rho}{\rho_S} = \frac{1.1845}{1.2250} = 0.967$$

Then

$$\frac{W^2}{\sigma S C_{L_{\text{MAX}}} T} = \frac{(131{,}560)^2}{(0.967)(1200)(2.5)(30{,}337)} = 196.7$$

From Figure 15.26, all-engine field length = 6050 ft. From Figure 15.29, the field length with engine failure = 7200 ft. The engine-failure case is critical and exceeds the available runway length of 2135 m or 7000 ft.

From Figure 15.29, the value of $W^2/\sigma SC_{L_{MAX}}T$ is 191.0 for a field length of 7000 ft. To reduce the airplane value to 191.0, the square of the takeoff weight must be reduced in the ratio of 191.0/196.7. Thus

$$\left(\frac{W_2}{W_1}\right)^2 = \frac{191}{196.7}$$

$$W_2 = 131{,}560\sqrt{\frac{191}{196.7}} = (131{,}560)(0.9854) = 129{,}640 \text{ lb}$$

The weight reduction is (131,560 − 129,640) = 1920 lb. Each passenger, with baggage, weighs 91 kg or 200 lb. Thus 9.59 passengers (10 in practical terms) must be off-loaded for this flight or a refueling stop will be required.

Example 15.5

If a DC-8-63, which has four-wheel-landing-gear trucks, is landing at an altitude of 2000 ft on a standard day at a weight of 230,000 lb, determine:

(a) Stalling speed.
(b) Approach speed over the 50-ft height.
(c) Required FAR landing field length.

See Figure 14.15 for the DC-8 $C_{L_{MAX}}$ at the landing flap angle of 50 degrees. The DC-8-63 wing area is 2927 ft^2.

Solution: $C_{L_{MAX}}$ (at $\delta_F = 50°$) = 2.06. At 2000 ft, $\rho = 0.00224$ from Appendix A.

(a) $$V_S = \sqrt{\frac{2W}{\rho C_{L_{MAX}} S}} = \sqrt{\frac{2(230{,}000)}{(0.00224)(2.06)(2927)}} = 185 \text{ ft/s}$$

$$= 185 \times \frac{60}{88} = \underline{126.1 \text{ mph}}$$

$$= \frac{126.1}{1.152} = \underline{109.5 \text{ knots}}$$

(b) $V_{\text{approach}} = 1.3V_S = (1.3)(109.5) = \underline{142.4 \text{ knots}}$

(c) $V_S^2(\text{knots})^2 = (109.5)^2 = 11{,}990$

FAR landing field length = <u>6200 ft</u> (from the dashed fairing on Figure 15.35; the actual DC-8 curve shows 6100 ft).

PROBLEMS

15.1. A twin-engine, four-place airplane is flying at a pressure altitude of 6500 ft at a speed of 180 mph. Outside air temperature is 58°F. The gross weight is 4500 lb. The rectangular wing has an area of 179 ft^2 with a span of 31.9 ft. Wing thickness ratio is 0.13 and the sweep angle is zero; 90% of the wing is exposed. The wing parasite drag is 43% of the total parasite drag. Assuming a propeller (or propulsive) efficiency of 0.82, determine the required cruising brake horsepower per engine.

15.2. A two-place airplane is flying at a pressure altitude of 4000 ft at a speed of 120 mph. Outside air temperature is 50°F. The gross weight is 2000 lb. The rectangular wing has an area of 170 ft^2 with a span of 33.25 ft. Wing thickness is 14%. Wing parasite drag is 39% of the total parasite drag; 88% of the wing is exposed. Assuming a propeller (or propulsive) efficiency of 0.84, determine the required cruising brake horsepower.

15.3. A four-engine 747 with a capacity of 365 passengers is cruising at a Mach number of 0.82 at a pressure altitude of 37,000 ft. Outside air temperature is −50°F. The initial cruise weight was 630,000 lb. According to the pilot's flight plan, he will start his descent at a weight of 488,000 lb. The 747 may be assumed to have a C_{D_P} of 0.0145, an e of 0.86, a wing area of 5500 ft^2, and a wing span of 195.7 ft. The compressibility drag coefficient, ΔC_{DC}, at the average cruise weight is 0.0010. The JT9D-7 high bypass ratio turbofans have an installed specific fuel consumption at cruise of 0.65 lb/lb-h. Determine:

(a) Distance covered at cruise altitude (assume conditions at average cruise weight can be considered as the average for the flight).

(b) Required engine thrust, per engine, at the average cruise weight.

(c) Average cruise fuel flow in gallons per hour (kerosene fuel weighs 6.7 lb/gal).

(d) Seat-miles produced per gallon.

(e) Compare the 747 seat-miles/gal with a five-passenger automobile having a fuel consumption of 25 mi/gal.

15.4. A 747 with a capacity of 372 passengers is cruising at a Mach number of 0.84 at a pressure altitude of 35,000 ft. Outside air temperature is −60°F. According to the pilot's flight plan, he will start his descent at a weight of 450,000 lb. Assume that the 747 has a C_{D_P} of 0.0145, an e of 0.86, a wing area of 5500 ft^2, and a wing span of 195.7 ft. The compressibility drag coefficient, ΔC_{DC}, is approximately 0.0012. The JT9D-7 turbofan engines have an installed specific fuel consumption at cruise of 0.65 lb/lb-h. The cruise distance is 3100 nautical miles. Determine:

(a) Initial cruise weight. (Assume conditions at average cruise weight can be considered as the average for the flight. An iterative procedure may be used, i.e., assume a fuel weight. Then calculate the actual fuel weight, correct the average cruise weight, etc.)

(b) Required engine thrust, per engine, at the average cruise weight.

(c) Average cruise fuel flow in gallons per hour (kerosene weighs 6.7 lb/gal).

(d) Seat-miles produced per gallon.

(e) Compare the 747 seat-miles/gal with a 50-passenger bus having a fuel consumption of 5 mi/gal.

15.5. The airplane of Problem 15.1 has an initial cruise weight of 5100 lb and carries 960 lb of fuel usable in cruise. Engine specific fuel consumption is 0.46 lb/bhp-h. What is the cruise range?

15.6. The airplane of Problem 15.2 has an initial cruise weight of 2250 lb and will start its descent after consuming 430 lb of fuel in cruise. Specific fuel consumption is 0.48 lb/bhp-h. Determine the cruise range.

15.7. At takeoff from San Francisco, a twin-engine DC-9-30 required a takeoff runway length of 5700 ft. The wing area is 1000 ft^2. The airport runways may be assumed to be at sea-level pressure altitude. The temperature was 72°F. The engines have a static ($V = 0$) rating of 14,500 lb of thrust each, but lose 14% in thrust at the effective average takeoff speed ($0.7\ V_{LO}$) due to bleed and power extraction losses and the engine ram drag due to forward speed. For the takeoff, the flap angle is 20 degrees, and the slats are extended (see Figure 14.15).

(a) Determine the takeoff weight.

(b) What is the lift-off speed at $1.2V_S$?

15.8. At takeoff from Seattle, the three-engine B-727-200 transport weighed 168,000 lb. The wing area is 1700 ft^2. The airport runways may be assumed to be at sea-level pressure altitude. The temperature was 62°F. The engines have a static ($V = 0$) rating of 14,500 lb of thrust each, but lose 14% in thrust at the effective average takeoff speed ($0.7V_{LO}$) due to bleed and power extraction losses and the engine ram drag due to forward speed. For the takeoff, the flap angle is 20 degrees, and the slats are extended. The maximum lift coefficients are similar to the DC-9-30 (Figure 14.15).

(a) Determine the FAR takeoff field length.

(b) What is the lift-off speed at $1.2\ V_S$?

15.9. An airplane with 1500 ft^2 of wing area is taking off from a sea-level pressure altitude airport on a warm day with the temperature equal to 85°F. The three engines of 15,000-lb static rating each lose 18% of their thrust by the time they reach 70% of the lift-off speed. The takeoff maximum lift coefficient is the same as for a DC-9-30 with a flap angle of 20 degrees with slats extended. If a 6500-ft length is available, what is the maximum takeoff weight that can be flown from the runway and still meet the requirements of FAR 25?

15.10. If a DC-9-30 is landing at a pressure altitude of 4000 ft on a standard temperature day at a weight of 85,000 lb, what is the stalling speed, the approach speed over the 50-ft height, and the required FAR landing-field length? See Figure 14.15 for the DC-9 $C_{L_{MAX}}$ at the landing flap angle of 50 degrees with extended slats. Wing area is 1000 ft^2.

15.11. If a DC-9-30, landing at a pressure altitude of 4000 ft at a weight of 85,000 lb, requires a FAR landing-field length of 4900 ft, what is the stalling speed, the approach speed over the 50-ft height, and the ambient air temperature? See Figure 14.15 for the DC-9 $C_{L_{MAX}}$ at the landing flap angle of 50 degrees, slats extended. Wing area is 1000 ft^2.

REFERENCES

15.1. Perkins, Courtland D., and Hage, Robert E., *Airplane Performance, Stability and Control,* Wiley, New York, 1949.

15.2. McCormick, Barnes W., *Aerodynamics, Aeronautics, and Flight Mechanics,* Wiley, New York, 1979.

15.3. Rutowski, Edward S., "Energy Approach to the General Aircraft Performance Problem," *Journal of the Aeronautical Sciences,* Vol. 21, No. 3, March 1954.

15.4. Nicolai, Leland M., *Fundamentals of Aircraft Design,* School of Engineering, University of Dayton, Dayton, Ohio, 1975.

16

STABILITY AND CONTROL

We have discussed airplane lift, drag, and performance, characteristics that must be analyzed in the design of an efficient, useful flying machine. Having accomplished the considerable task of lifting the weight of the airframe, engine, payload, and enough fuel to fly the desired range, it is equally important to have a safe vehicle, one that responds as required to the controls and is not too demanding of the pilot's attention.

The study of the flying and handling characteristics is called stability and control. *Stability* refers to the ability of a vehicle to return to its original equilibrium position, without pilot assistance, after a disturbance has pitched, yawed, or rolled it to a different angle or has caused a speed change. *Control* describes the ability of the pilot, using the control surfaces such as the rudder, elevators, ailerons, and spoilers, to produce moments about the various axes that not only balance disturbing moments but also provide an adequate angular acceleration in the desired direction. Lift changes are usually produced by changing the angle of attack of the entire vehicle. Lift direction is changed by rolling. An example of a disturbing moment is the turning moment caused by a failed engine on a multiengine airplane.

Static stability refers to the initial tendency of a system to return to or move away from the equilibrium position after a disturbance. *Dynamic stability* concerns the entire history of the motion, in particular the rate at which the motion damps out.

In our discussion of stability, the airplane will be considered to be a rigid body. This assumption is well justified in many cases. At high dynamic pressures, however, deflections of fuselages and swept wings under the applied airloads change the relative angles of attack of the wing and tail and of the inner and outer wing panels, respectively. The result of the fuselage deflection is to change the effective angle of attack of the horizontal tail for a given wing angle, while the swept wing deflection reduces the wing lift curve slope and the negative (nose down) pitching moment about the aerodynamic center. Thus the $C_{M_{ac}}$ is made less negative. The study of the behavior of the structure under load and the interaction with the aerodynamic loads is called *aeroelasticity*. Aeroelastic effects usually tend to reduce static stability. In extreme cases, this interaction may lead to dangerous undamped structural oscillations called *flutter*. Flutter usually arises from deflections of structural elements such as control surfaces and wing sections in a more complex manner than the modes indicated above.

The rigid airplane we will study has 6 degrees of freedom (Figure 16.1). The airplane can move linearly along the flight path or x-axis, laterally along the y-axis, or vertically along the z-axis. It can rotate in pitch about the y-axis, in yaw about the z-axis, or in roll about the x-axis. In aircraft dynamics analyses, the origin of the coordinate system is usually fixed at the center of gravity of the airplane and moves with it. There are several alternative coordinate systems, including body axes, for which the x-axis is fixed in the fuselage, stability axes for which the x-axis is the perpendicular projection of the relative wind on the plane of symmetry, and wind axes for which the x-axis points into the relative wind. Positive moments or rotations are associated with nose up in pitch, nose to the right in yaw, and roll to the right.

The rigid body motions of aircraft may be conveniently divided into two classifications, longitudinal and lateral motions. *Longitudinal motions* occur in the plane of symmetry, which remains in its original position. *Lateral motions,* such as rolling, yawing, and sideslipping, displace the plane of symmetry. The technical significance of this distinction is that for normal symmetrical aircraft with small displacements these two types of motion are independent of each other. A pitching motion or a vertical velocity introduced by a pilot input or an atmospheric disturbance does not affect lateral motions. Therefore, it can be analyzed without considering lateral effects. On the other hand, rolling introduces side forces on the vertical tail and this, in turn, produces yawing moments. Similarly, an angle of yaw produces a rolling moment, so the roll and yaw modes are closely associated.

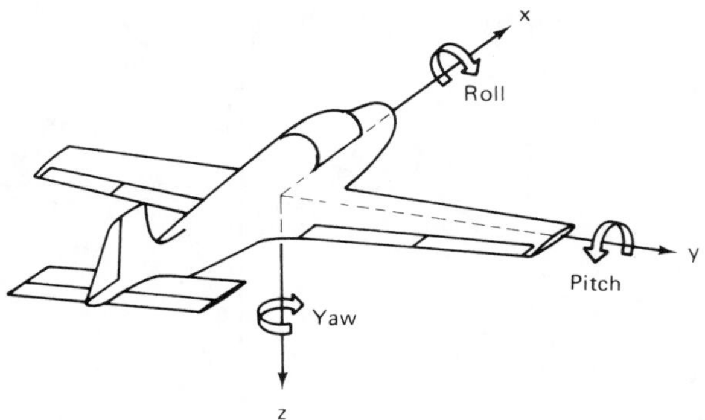

Figure 16.1 The 6 degrees of freedom of an airplane.

STATIC LONGITUDINAL STABILITY

Since longitudinal stability is concerned with motions in pitch, the pertinent aerodynamic characteristic is the variation of the pitching moment coefficient about the center of gravity with either angle of attack or lift coefficient (Figure 16.2). It will be remembered that the pitching moment coefficient is defined as

$$C_M = \frac{M}{(\rho V^2/2)S\overline{c}} \tag{16.1}$$

where $\overline{c}$ is the mean aerodynamic chord of the wing.

Since C_L is normally linear in angle of attack until the stall region is approached, either method of presentation is valid. We shall see that there is a special usefulness to the graph of C_M versus C_L, however. In Figure 16.2, the condition for equilibrium is the point where $C_M = 0$. This is called the *trim point*. The curves marked stable are those in which the pitching moment becomes more negative (i.e., nose down) as the angle of attack or the lift coefficient increases with respect to the trim point. If the airplane is momentarily pitched to a higher angle, the pitching moment change will tend to restore the aircraft to the original attitude. If the airplane is pitched nose down, the resulting moment change will tend to raise the nose. The unstable curves show the opposite behavior, tending to increase the deviation from equilibrium or trim. An aircraft is longitudinally statically stable if it has a negative value of dC_M/dC_L. In fact, the magnitude of the longitudinal stability is defined as the value of $-dC_M/dC_L$.

The elements of a normal airplane that primarily determine its longitudinal stability characteristics are the wing, horizontal tail, fuselage, and nacelles. The lift forces of fuselage and nacelles are small, but these bodies contribute moments in the form of couples. These are assumed to be moments about the center of gravity. The study of stability will therefore be based on a skeleton airplane, composed of a center of gravity, a wing, a horizontal tail, and the body moments about the center of gravity. Figure 16.3 shows such a skeleton airplane together with the aerodynamic forces acting on it. The wing lift may be assumed to act through the aerodynamic center if the wing couple due to camber, M_{ac}, is accounted for separately. Drag forces create

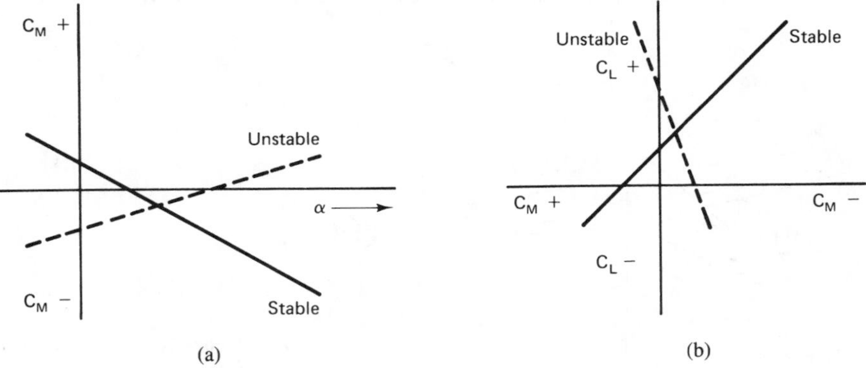

Figure 16.2 Stable and unstable pitching moment curves.

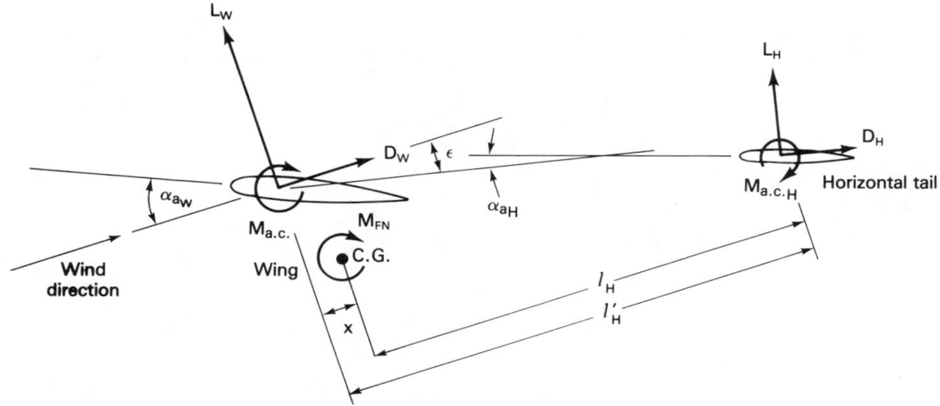

Figure 16.3 Schematic airplane showing primary forces and moments for longitudinal stability.

relatively small pitching moments in normal flight and will be neglected. By taking moments about the center of gravity and letting $(\cdot)_H$ denote "horizontal tail" and $(\cdot)_W$ denote "wing," the following equilibrium moment equations result:

$$M_{cg} = M_W + M_H + M_{F,N}$$
$$M_{cg} = M_{ac} + L_W x - L_H l_H + M_{F,N}$$

or

$$M_{cg} = C_{M_{ac}} q S_W \bar{c} + C_{L_W} q S_W x - C_{L_H} q S_H l_H \eta_H + C_{M_{F,N}} q S_W \bar{c} \tag{16.2}$$

where $M_{F,N}$ denotes the pitching moment due to fuselage and nacelles. Note that x is positive when the center of gravity is aft of the aerodynamic center. η_H, the *tail efficiency,* is the ratio of the effective dynamic pressure over the horizontal tail to the freestream dynamic pressure. A low tail position on the fuselage places the tail in the thick fuselage boundary layer, so the effective velocity is reduced. Typical η_H values are between 0.85 and 0.90. For T-tail arrangements, $\eta_H = 1.0$. Equation 16.2 assumes that drag moments are small compared to lift moments and that the cosine of the angles involved can be assumed to be 1.0. M_{ac_H} is usually zero since tail airfoils are normally symmetrical. Even if some tail camber is used, this term is very small and is neglected here.

Converting equation 16.2 to coefficient form by dividing $qS_W\bar{c}$, we obtain

$$C_{M_{cg}} = C_{M_{ac}} + C_{L_W}\frac{x}{\bar{c}} - C_{L_H}\frac{S_H}{S_W}\frac{l_H}{\bar{c}}\eta_H + C_{M_{F,N}} \tag{16.3}$$

Differentiating with respect to C_{L_W},

$$\frac{dC_{M_{cg}}}{dC_{L_W}} = \frac{x}{\bar{c}} - \frac{dC_{L_H}}{dC_{L_W}}\left(\frac{S_H}{S_W}\right)\frac{l_H}{\bar{c}}\eta_H + \left(\frac{dC_M}{dC_{L_W}}\right)_{F,N} \tag{16.4}$$

dC_{L_H}/dC_{L_W} depends on the rate of change of downwash at the tail with lift coefficient and the aspect ratios of the wing and tail. Thus it can be shown that

$$\frac{dC_{L_H}}{dC_{L_W}} = \frac{(dC_L/d\alpha)_H}{(dC_L/d\alpha)_W}\left(1 - \frac{d\epsilon_H}{d\alpha}\right)$$

so

$$\left(\frac{dC_M}{dC_{L_W}}\right)_{cg} = \frac{x}{\bar{c}} - \frac{(dC_L/d\alpha)_H}{(dC_L/d\alpha)_W} \cdot \left(1 - \frac{d\epsilon_H}{d\alpha}\right)\frac{S_H}{S_W}\frac{l_H}{\bar{c}}\eta_H + \left(\frac{dC_M}{dC_{L_W}}\right)_{F,N} \tag{16.5}$$

Longitudinal static stability has been defined as the value of $-dC_M/dC_L$, the rate of change of pitching moment coefficient with *total* airplane lift coefficient, C_L. The airplane is stable when this derivative is negative. Equation 16.5 differs from the static stability definition because it yields the rate of change of pitching moment coefficient with respect to *wing* lift coefficient. To obtain the equation for static stability, we note that

$$\left(\frac{dC_M}{dC_L}\right)_{cg} = \left(\frac{dC_M}{dC_{L_W}}\right)_{cg}\left[\frac{\left(\frac{dC_L}{d\alpha}\right)_W}{\frac{dC_L}{d\alpha}}\right]$$

$$= \left(\frac{dC_M}{dC_{L_W}}\right)_{cg}\left[\frac{\left(\frac{dC_L}{d\alpha}\right)_W}{\left(\frac{dC_L}{d\alpha}\right)_W + \left(\frac{dC_L}{d\alpha}\right)_H\left(1 - \frac{d\epsilon_H}{d\alpha}\right)\frac{S_H}{S_W}\eta_H}\right] \tag{16.6}$$

$$= \left(\frac{dC_M}{dC_{L_W}}\right)_{cg}\left[\frac{1}{\left[1 + \frac{(dC_L/d\alpha)_H}{(dC_L/d\alpha)_W}\left(1 - \frac{d\epsilon_H}{d\alpha}\right)\frac{S_H}{S_W}\eta_H\right]}\right]$$

so

$$\left(\frac{dC_M}{dC_L}\right)_{cg} = \left[\frac{x}{\bar{c}} - \frac{(dC_L/d\alpha)_H}{(dC_L/d\alpha)_W}\left(1 - \frac{d\epsilon_H}{d\alpha}\right)\frac{S_H}{S_W}\frac{l_H}{\bar{c}}\eta_H + \left(\frac{dC_M}{dC_{L_W}}\right)_{F,N}\right] \cdot \left[\frac{1}{\left[1 + \frac{(dC_L/d\alpha)_H}{(dC_L/d\alpha)_W}\left(1 - \frac{d\epsilon_H}{d\alpha}\right)\frac{S_H}{S_W}\eta_H\right]}\right] \tag{16.7}$$

There is a problem in using equation 16.7. l_H, the moment arm of the horizontal tail about the center of gravity, is itself dependent on the c.g. position and is, therefore, a function of $x/\bar{c}$. It is convenient to define as the tail length, l'_H, the longitudinal distance from the aerodynamic center of the wing to the aerodynamic center of the tail. l'_H is a constant for a defined airplane configuration. From Figure 16.3,

$$l'_H = l_H + x \quad \text{and} \quad l_H = l'_H - x$$

Substituting for l_H in equation 16.7, collecting the terms in $x/\bar{c}$, and simplifying leads to

$$\left(\frac{dC_M}{dC_L}\right)_{cg} = \frac{x}{\bar{c}} - \left[\frac{(dC_L/d\alpha)_H}{(dC_L/d\alpha)_W}\left(1 - \frac{d\epsilon_H}{d\alpha}\right)\frac{S_H}{S_W}\frac{l'_H}{\bar{c}}\eta_H - \left(\frac{dC_M}{dC_{L_W}}\right)_{F,N}\right] \cdot \left\{\frac{1}{\left[1 + \frac{(dC_L/d\alpha)_H}{(dC_L/d\alpha)_W}\left(1 - \frac{d\epsilon_H}{d\alpha}\right)\frac{S_H}{S_W}\eta_H\right]}\right\} \tag{16.8}$$

The first two terms are the most important to stability. $x/\bar{c}$ describes the position of the center of gravity with respect to the aerodynamic center as a fraction of the mean aerodynamic chord. The second term is the contribution of the tail to stability. The expression $S_H l'_H/S_W\bar{c}$ is a geometric ratio representing the effectiveness of the horizontal tail relative to the wing. It is called the *tail volume.*

Two very significant stability characteristics can be determined from equation 16.8. First, the *neutral point* is the center of gravity location for which neutral longitudinal stability occurs. From equation 16.8 with $(dC_M/dC_L)_{cg} = 0$,

$$\left(\frac{x}{\bar{c}}\right)_n = \left[\frac{(dC_L/d\alpha)_H}{(dC_L/d\alpha)_W}\left(1 - \frac{d\epsilon_H}{d\alpha}\right)\frac{S_H}{S_W}\frac{l'_H}{\bar{c}}\eta_H - \left(\frac{dC_M}{dC_{L_W}}\right)_{F,N}\right] \cdot \left[\frac{1}{\left[1 + \frac{(dC_L/d\alpha)_H}{(dC_L/d\alpha)_W}\left(1 - \frac{d\epsilon_H}{d\alpha}\right)\frac{S_H}{S_W}\eta_H\right]}\right] \tag{16.9}$$

where $(x/\bar{c})_n$ is the neutral point with respect to the wing aerodynamic center. The neutral point is usually expressed as the fraction of the m.a.c. from the leading edge of the m.a.c. [i.e., $(x/\bar{c})_n$ is added to the a.c. location expressed as a fraction of the m.a.c.].

Another important stability quality is the *static margin,* the distance that the center of gravity is ahead of the neutral point expressed as a fraction of the mean aerodynamic chord. Thus the static margin is

$$\text{static margin} = \left(\frac{x}{\bar{c}}\right)_n - \frac{x}{\bar{c}} \tag{16.10}$$

Solving for $x/\bar{c}$ in equation 16.8 and substituting this and equation 16.9 in equation 16.10 yields

$$\text{static margin} = -\left(\frac{dC_M}{dC_L}\right)_{cg} \tag{16.11}$$

The downwash at the tail comes almost entirely from the trailing vortex system of the wing, although there is a small contribution from the bound vortex. For an elliptical wing, the downwash angle in radians in the wing wake far behind the wing is $2C_L/(\pi \cdot \text{AR})$, exactly twice the amount calculated at the wing quarter-chord (equation 9.8). Viewed from a position far behind the wing, the trailing vortices extend to infinity both forward and aft, rather than just aft as is the condition at the wing. Thus the downwash is doubled.

In practice, one wing semispan behind the wing qualifies as "far behind," so that $2C_L/(\pi \cdot AR)$ gives a rather good first approximation to downwash at the tail. The downwash actually varies both with distance behind the wing and height of the tail above the wing wake.

The fuselage and nacelle contributions are almost always destabilizing. This follows from the nature of the flow around a body of revolution. The flow around a body of revolution at an angle of attack in a perfect fluid is shown in Figure 16.4. Pressures on the upper surface of the forward end and on the lower surface of the aft end are negative with respect to the freestream pressure. Remembering the nature of the pressure distribution in curved vortex flow, this result could be inferred from the increased curvature (i.e., the decreased radius of curvature) of the flow in these regions. On the opposite sides of the body the pressures are positive. The result is zero lift and a couple tending to pitch or yaw the body to a higher angle. In a real viscous fluid the moment is somewhat reduced and a small lift is created, but the primary effect is the unstable couple. Therefore, all bodies tend to turn crosswise to the wind unless they have tail surfaces that produce the necessary stabilizing moments.

Typical contributions of airplane components to the longitudinal stability are shown in Figure 16.5. If the moments are taken around the quarter-chord point of the mean aerodynamic chord of an unswept wing, the wing will have essentially neutral stability since the lift due to angle of attack acts through the quarter-chord, or very close to it. The moment due to camber at zero lift will give the wing a negative moment, however. The fuselage and nacelles will contribute unstable slopes. The separate contribution of the tail is large and stable, so the total configuration is stable (i.e., dC_M/dC_L is negative).

Figure 16.5 should be studied while referring to equations 16.3 and 16.8. The graph is a representation of the equations for the special case where x, the distance

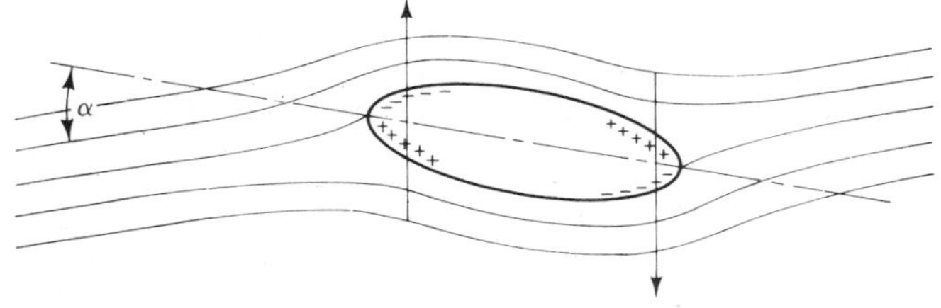

Figure 16.4 Potential flow about a body of revolution.

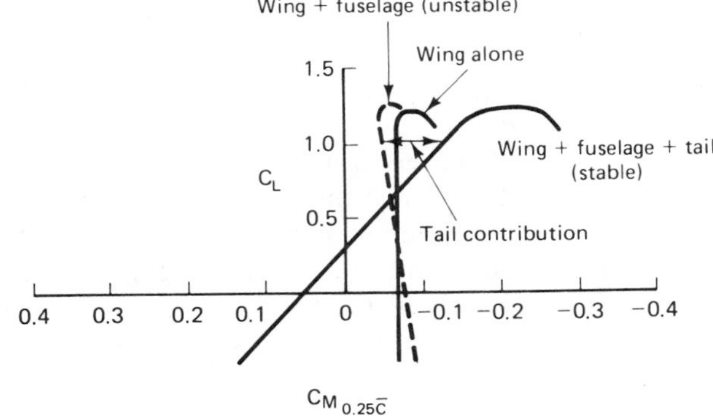

Figure 16.5 Typical contributions of airplane components to stability.

from the center of gravity, or the center of moments, to the wing a.c. is zero. Equation 16.8 shows that a change in $x/\bar{c}$ is directly reflected in an identical change in the stability, $-dC_M/dC_L$. Since $x/\bar{c}$ is a change in c.g. position in terms of a fraction of the m.a.c., it follows that the moment curves in Figure 16.5 drawn for the c.g. position of $0.25c$ can be converted to any other center-of-gravity position by changing the slope of the lift coefficient axis. Alternatively, the most aft center of gravity position for which the airplane will be stable can be determined by measuring the slope of the complete airplane moment curve. This slope, added to the c.g. position for which the graph is constructed, in this case 0.25, gives the "neutral point," the c.g. for neutral stability.

In Chapter 13, we discussed the aft movement of the wing aerodynamic center due to sweep. In Figure 16.5, moments are plotted for an unswept wing about the 25% m.a.c. point and the wing alone is neutrally stable. If Figure 16.5 were drawn for a swept-wing airplane, it might look unchanged provided that the moments were shown about a further aft point (e.g., 30% m.a.c.). It would also be acceptable to plot the moment coefficient for the swept-wing airplane with respect to the 25% m.a.c. point. In that case, the wing alone curve would be stable.

Often the wing alone stability is not available from a wind tunnel test, but a model of the complete airplane less tail surfaces has been tested. The available stability curve applies to the wing plus fuselage plus nacelles. *If this stability curve is based on moments about the* $\bar{c}/4$ *point, this value of* dC_M/dC_L *may be used in equation 16.8 to replace* $(dC_M/dC_{L_W})_{F,N}$ *and x may be taken as the distance between the wing* $\bar{c}/4$ *point and the center of gravity. The resulting error in this assumption is small.*

The general rule about the destabilizing effects of nacelles does not apply when the nacelles are mounted on the aft end of the fuselage on horizontal pylons, as in Figure 16.6. Here the pylon–nacelle combination acts as a low-aspect-ratio tail and actually contributes to stability at normal angles of attack. At very high angles of attack beyond the wing stall, the wake of the aft nacelles may blanket the tail and cause serious instability unless great care is taken in the design.

Another important function of the tail is damping, the creation of moments that oppose the pitching rate of the aircraft. If the airplane has a positive pitch rate with the nose rising, the tail is being moved downward. Relative to the tail, an upward velocity component is created leading to an effective increase in the angle of attack at the tail. The resulting increase in tail lift opposes the pitching motion. Since this angle of attack change is proportional to both the pitching rate and the tail length l_H, and the moment produced by a given tail lift is also proportional to the tail length, the damping moment varies as the square of the tail length. Therefore, a long fuselage that increases tail length, l_H, leads to better damped aircraft. For this reason, pilots always find a "stretched" fuselage version of an aircraft to be a "better flying" aircraft than the original.

In addition to the stability contribution of the tail, the tail must balance the airplane, primarily by opposing the moments resulting from the center of gravity not being at the aerodynamic center and from the $C_{M_{ac}}$, the wing couple about the aerodynamic center at zero lift. We have mentioned that when the sum of the moments is zero the airplane is trimmed. The lift coefficient at which the sum of the moments is zero is the *trim lift coefficient*. A stable airplane will fly steadily at the trim lift

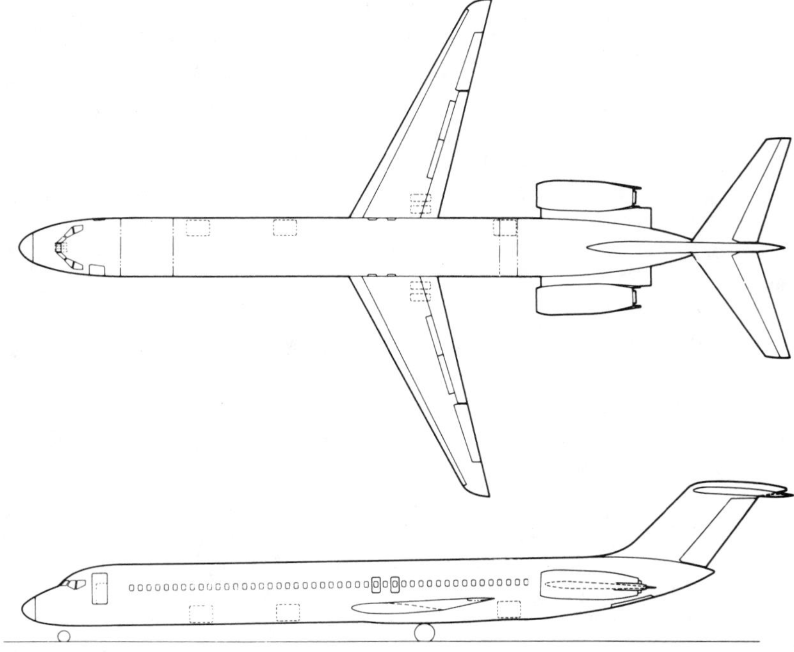

Figure 16.6 DC-9-50 aircraft with aft-mounted nacelles. (Courtesy of McDonnell Douglas Corp.)

coefficient and at the corresponding equivalent airspeed. The desired trim lift coefficient is obtained by varying the effective angle of attack of the horizontal tail. This is achieved by adjusting the stabilizer angle or the elevator angle. Figure 16.7 shows a series of pitching moment curves, C_M versus C_L, each for a different elevator angle.

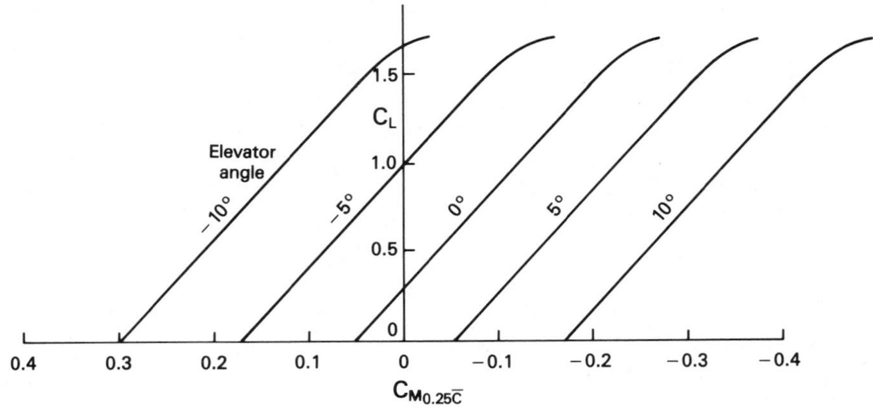

Figure 16.7 Effect of elevator deflection on the pitching moment curve.

A positive elevator angle produces positive lift on the tail. For conventional aft tails, a positive elevator deflection produces a negative, nose down, moment. Thus the effect of elevator angle change is to add an incremental C_M to the pitching moment curve, shifting the curve to the right or left for positive or negative elevator deflections, respectively. For each elevator angle, there is a specific trim C_L, the more positive, nose-down deflections trimming at lower lift coefficients that correspond to higher airspeeds.

Whether the airplane will fly level at the trim airspeed depends on the power or thrust settings. It is important to remember that in steady flight the elevator is a speed control, determining the trim C_L, while the throttle is a flight path angle control.

The static longitudinal stability we have been analyzing is known as *stick-fixed stability.* "Stick fixed" means that the pilot maintains the control column in a fixed position so that the control surfaces, specifically the elevators, are not permitted to move as the airplane changes its angle of attack. *Stick-free stability* describes the value of dC_M/dC_L with the elevators assuming the position, at each airplane angle of attack, for which the elevator aerodynamic *hinge moment* is zero. This will be the angle at which the elevators will float with no control force. Stick-free stability is important for airplanes with direct cable control of the surface and/or its control tabs mounted on the surface. Aircraft with full irreversible powered controls will have stick-fixed stability even when the pilot is flying "hands off" provided the pilot has trimmed the column to zero force at the initial flying speed.

Stick-free stability is normally less than stick-fixed stability because the direction of elevator float reduces the effectiveness of the horizontal tail in restoring an airplane to its original angle of attack. For example, if an airplane is pitched to a higher angle of attack than its original trim position, the elevator will tend to float trailing edge up, reducing the stabilizing increase in tail lift.

There is another form of longitudinal stability, *stick-force stability.* Stick-force stability is defined by the rate of change of stick force with flight speed. The Federal Air regulations, Part 25, for transport aircraft require that at any trim speed within the flight range a pull must be required to obtain and maintain speeds below the trim speed, and a push must be required to obtain and maintain speeds above the trim speed. The average gradient of the stable slope of the stick force versus speed curve may not be less than 1 lb for each 6 knots (Ref. 16.4). Similar but more detailed requirements are given for military aircraft in Ref. 16.5. To have a stable force gradient, an airplane must have stick-free stability since the force gradient results from the combined effects of stick-free stability and, for unpowered controls, the aerodynamic hinge moments on the control surfaces and the mechanics of the linkages. For fully powered control systems, the force gradients can be provided mechanically, but stick-fixed stability is usually a prerequisite.

There are exceptions to the preceding requirements. Some sophisticated modern aircraft are designed to be unstable aerodynamically. Very fast acting, full-time automatic controls, called *active controls,* provide simulated stability as far as the pilot is concerned. The advantages of such systems are the avoidance of a download on the tail at high lift coefficients, thereby increasing the total lift available for fighter maneuverability, and the possibility that smaller tail sizes can be used. Further discussion of active controls is given later in this chapter.

DYNAMIC LONGITUDINAL STABILITY

Dynamic stability of a vehicle denotes the complete study of the motion occurring after the vehicle has been disturbed from its equilibrium or trim condition. The equations governing such motions include the force and moment changes occurring both because of the displacement and because of the rate of change of displacement, the latter providing the damping that gradually reduces the motions in a dynamically stable case.

It is possible for a vehicle to be statically stable and dynamically unstable. If a statically stable vehicle acquires an angle of yaw from a crosswind disturbance, for example, it may tend to return to zero yaw angle but overshoot the equilibrium attitude. It then develops an opposite yawing moment, tending to return to the original heading but may overshoot again. If each succeeding deviation from equilibrium is of smaller magnitude, the vehicle is dynamically stable. If each succeeding overshoot is larger, the vehicle is dynamically unstable even though the static stability is positive. If the airplane returns to equilibrium without overshoot, the motion is a *simple subsidence*. If the yawing moment produced by the deviation tends to increase the yaw angle, the vehicle is, of course, statically and dynamically unstable and the motion is called *divergence*. These types of motion are illustrated in Figure 16.8.

Oscillating motions, as shown in Figure 16.8, can be described by two parameters, the period of time required for one complete oscillation, and the time to damp to half-amplitude, or the time to double the amplitude for a dynamically unstable motion.

It has been noted that longitudinal and lateral motions are quite independent of each other and can be considered separately. The longitudinal motion consists of two distinct oscillations, a long-period oscillation called the phugoid mode and a short-period oscillation referred to as the short-period mode.

The *phugoid, or long-period mode,* is one in which there is a large-amplitude variation of airspeed, pitch and flight path angle, and altitude, but almost no angle-of-attack variation. The phugoid oscillation is really a slow interchange of kinetic (velocity) energy and potential (height) energy about some equilibrium energy level as the airplane attempts to reestablish the equilibrium level-flight condition from which it has been disturbed. The motion is so slow that the effects of inertia forces and damping forces are very low. Although the damping is very weak, the period is so long that the pilot usually corrects for this motion without being aware that the oscillation even exists. Typically the period is 20 to 60 s.

The *short-period mode* is a very fast, usually heavily damped, oscillation with a period of a few seconds. The motion is a rapid pitching of the airplane about the center of gravity. The period is so short that the speed does not have time to change,

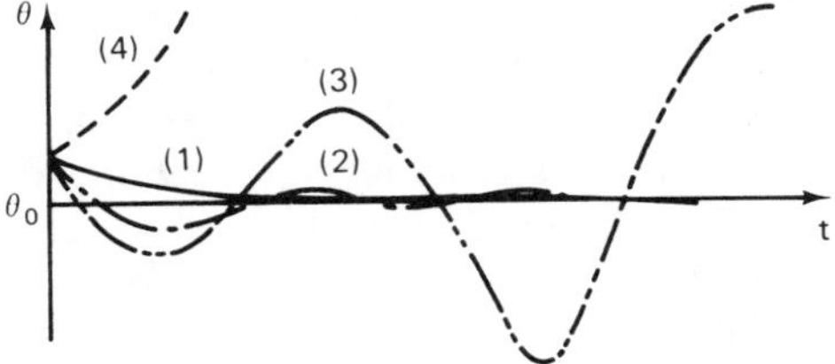

Figure 16.8 Types of dynamic motions: (1) simple subsidence, statically stable, dynamically stable; (2) damped oscillation, statically stable, dynamically stable; (3) divergent oscillation, statically stable, dynamically unstable; (4) divergence, statically unstable, dynamically unstable.

so the oscillation is essentially an angle-of-attack variation. The time to damp the amplitude to one-half of its value is usually on the order of 1 s.

STATIC LATERAL STABILITY

The lateral motions are of two types, assuming that the airplane possesses static stability about the directional, or yaw, axis and has roll or dihedral stability. *Directional stability* deals with the tendency of the airplane to point into the oncoming airstream and is provided by the vertical tail. Roll stability is the tendency of an airplane to naturally recover to a level-wing position after a bank or roll displacement has occurred. It is often called *dihedral stability,* since a wing dihedral angle provides the necessary restoring rolling moment. Actually, a pure roll does not immediately produce restoring moments even when an airplane has dihedral stability. A roll to the right, defined as a positive roll angle, inclines the lift vector to the right, as in Figure 16.9. The lateral component of the lift accelerates the airplane toward the right. The sidewise velocity adds to the forward velocity to produce an angle of yaw between the airplane centerline and the effective oncoming velocity. It is this angle of yaw, defined as negative when the nose is to the left of the freestream, that produces the rolling moment. An associated definition is the angle of sideslip, which is quantitatively equal to the angle of yaw but opposite in sign. If we think of the reference axis as parallel to the freestream the airplane yaws with respect to the freestream. If the reference axis is fixed in the airplane, the wind has an angle of sideslip β with respect to the airplane. When the wind approaches from the right of the airplane, the sideslip angle is positive.

Wing dihedral is positive when the wing tips are higher than the wing root. Dihedral produces rolling moments in a yawed or sideslip flight condition by increasing the effective angle of attack on the low wing panel that is moving into the wind and decreasing the angle of attack of the other side (Figure 16.10). The lateral wind component $V_0 \sin \beta$, can itself be resolved into components parallel to and normal to the wing plane. The component normal to the wing plane adds vectorially to the freestream velocity to provide a higher or lower angle of attack, depending on which wing panel is concerned. The change in angle of attack is $(\pm V_0 \sin \beta \sin \Gamma)/V_0 = \pm\beta\Gamma$, since for small angles the sine of an angle equals the angle in radians. The resulting moment acts to restore the airplane to the wing-level condition.

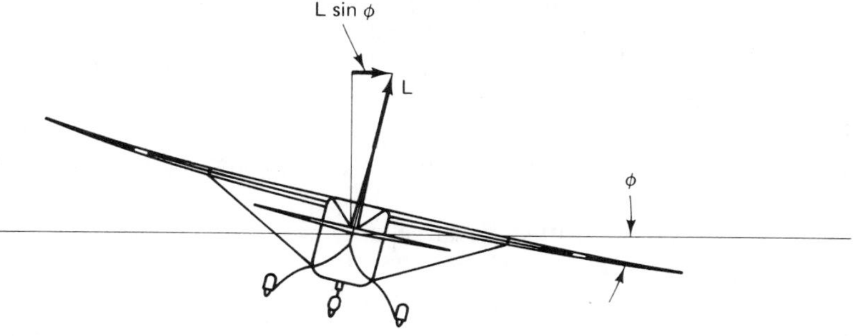

Figure 16.9 Side force produced by bank angle.

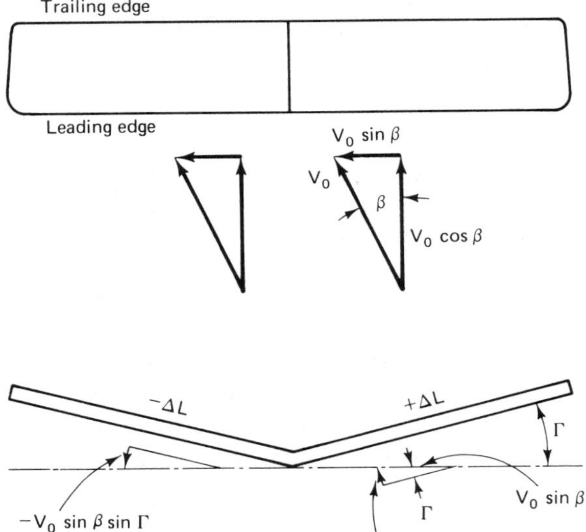

Figure 16.10 Dihedral effects on the rolling moment of a yawed wing.

DYNAMIC LATERAL STABILITY

There are two types of possible lateral dynamic motion. The first is called the *spiral mode*. If a spirally unstable airplane, through the action of a gust or other disturbance, gets a small initial roll angle to the right for example, a gentle sideslip to the right is produced. The sideslip causes a yawing moment to the right. If the dihedral stability is low, the directional stability keeps turning the airplane while the continuing bank angle maintains the sideslip and the yaw angle. As this process goes on, the spiral gets continuously tighter and steeper until finally, if the motion is not checked, a steep, high-speed, spiral dive results. The motion develops so gradually, however, that it is usually corrected unconsciously by the pilot, who may not be aware that the spiral instability exists. A combination of high directional stability and low dihedral, or roll, stability leads to a spirally unstable airplane.

The second lateral motion is an oscillatory combined roll and yaw motion called *Dutch roll* because of its similarity to an ice-skating figure of the same name. The Dutch roll may be described as a yaw and roll to the right, followed by a recovery toward the equilibrium condition, then an overshooting of this condition and a yaw and roll to the left, then back past the equilibrium attitude, and so on. The period is usually on the order of 3 to 15 s. Damping is increased by large directional stability and small dihedral and decreased by small directional stability and large dihedral. Although usually stable in a normal airplane, the motion may be so slightly damped that the effect is very unpleasant and undesirable.

Swept-wing aircraft have large dihedral stability even when they have no dihedral angle. This arises because the effect of an angle of yaw is to increase the sweepback angle of one wing panel and decrease it for the other side of the airplane. The change in sweep alters the effective dynamic pressure normal to the quarter-chord line of the wing panel, increasing the lift on one side of the wing, lowering it on the other side, and producing a restoring rolling moment. Nevertheless, many swept-wing airplanes have a considerable dihedral angle, not for aerodynamic reasons but to give adequate ground clearance for the wing tips and wing-mounted nacelles during landing and takeoff. These aircraft may then have too much dihedral effect for satisfactory Dutch roll damping. The problem is solved by installing a yaw damper, in effect a special-purpose automatic pilot that damps out any yawing oscillation by applying rudder corrections. Some swept wing aircraft have an unstable Dutch roll mode. If the Dutch roll is very lightly damped or unstable, the yaw damper becomes a safety requirement, rather than a pilot and passenger convenience. Dual yaw dampers are required and a failed yaw damper is cause for limiting flight to lower altitudes, and possibly lower Mach numbers, where the Dutch roll stability is improved.

CONTROL AND MANEUVERABILITY: TURNING PERFORMANCE

The primary aerodynamic controls available to the pilot are elevators, rudder, and ailerons. These are essentially plain flaps mounted on the rear portions of the horizontal tail, vertical tail, and the outer portions of the wing, respectively, as shown in Figures 2.1 and 2.2. Positive control surface deflection angles are those that produce positive lift on the control surface for the elevators and ailerons and a side force to the right for the rudder.

The flight paths of airplanes are controlled primarily by varying the magnitude and direction of the lift vector and by varying the thrust or power contributed by the power plants.

The magnitude of the lift vector is a direct function of the angle of attack or the lift coefficient. Since α and C_L are linear functions of each other over the useful flight region, we can think in terms of either one. The pilot varies α by controlling the pitching moment contributed by the horizontal tail so that the sum of the moments is zero at the desired C_L. The tail load is varied by changing the elevator angle.

Assume an airplane to be flying in steady level flight, as in Figure 16.11a. For this condition $L = W$. Then the pilot pulls back on the stick or wheel, depending on the type of control system, moving the elevator a few degrees in the negative, trailing-edge-up direction. The decrease in tail lift produces a positive, nose-up mo-

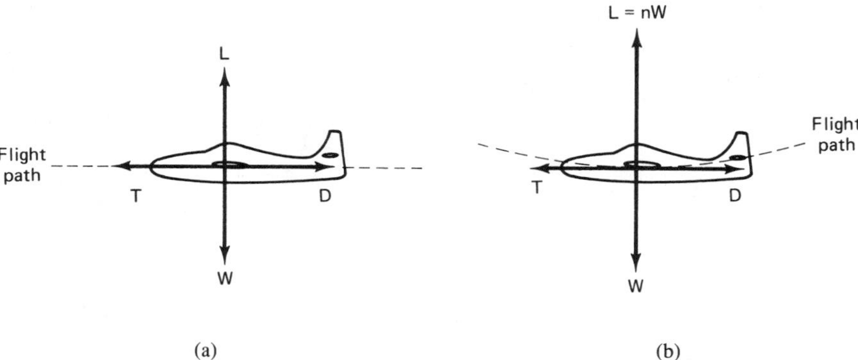

Figure 16.11 Forces in equilibrium and accelerated flight.

ment on the airplane, thereby starting a positive pitching motion. As the angle of attack increases, the basic stability of the airplane will cause an increasingly nose-down moment, as in Figure 16.2. When the increment in nose-down moment caused by the static stability equals the nose-up moment resulting from the elevator deflection, the sum of the moments will again be zero and the airplane will be trimmed at a new higher angle of attack. The lift will be greater at the higher α and now exceeds the weight, as in Figure 16.11b. The ratio of the lift to the weight is called n, the *load factor,* and

$$\frac{L}{W} = n \tag{16.12}$$

Since the weight is the force of gravity, an airplane with $L = W$ is said to have a lift of one g. If the lift is three times the weight, the airplane is subjected to 3 g's.

With $L > W$, assuming that the speed has not had time to change, the airplane will be given a vertical acceleration, which manifests itself as a curvature of the flight path (Figure 16.11b). The upward velocity component adds to the original horizontal velocity and inclines the flight path increasingly upward. The higher lift coefficient raises the induced drag coefficient. Furthermore, as the airplane starts to climb, a force of $W \sin \theta$ is developed in the drag direction as in Figure 15.7. The airplane speed then starts to decrease, decreasing the lift and drag. A stable airplane adjusts its speed and flight path angle quite rapidly, seemingly without all the complexity of following the steps we have just described, and settles down at the speed for which lift equals weight and at the flight path angle for which the forces are in climb equilibrium, as described by equation 15.17. The essence of the equilibrium is that the speed is determined by the $L = W$ condition (i.e., by the lift coefficient), and the flight path angle is determined by the thrust. Therefore, the speed is determined by the position of the elevators and the flight path angle by the throttle.

The direction of the lift vector is perpendicular to the wing plane. The wing angle of bank is controlled by the pilot through the use of ailerons. Ailerons deflect asymmetrically, one moving trailing edge down while the other moves in the trailing edge-up direction. The result is an increase in lift on one side of the airplane and a decrease on the other. The resulting rolling moment banks the airplane and tilts the lift vector to one side as in Figure 16.12a. The horizontal component of the lift vector accelerates the airplane laterally and curves the flight path as in Figure 16.12b. In a turn of radius R, the lateral force, $L \sin \phi$, where ϕ is the angle of bank, must balance the centrifugal force on the airplane. Thus

$$L \sin \phi = \frac{(W/g)V^2}{R} \tag{16.13}$$

For a level-flight turn, the weight W must be equal to the vertical component of lift, $L \cos \phi$. Substituting in equation 16.13, we obtain

$$L \sin \phi = \frac{[(L \cos \phi)/g]V^2}{R}$$

$$\tan \phi = \frac{V^2}{gR} \tag{16.14}$$

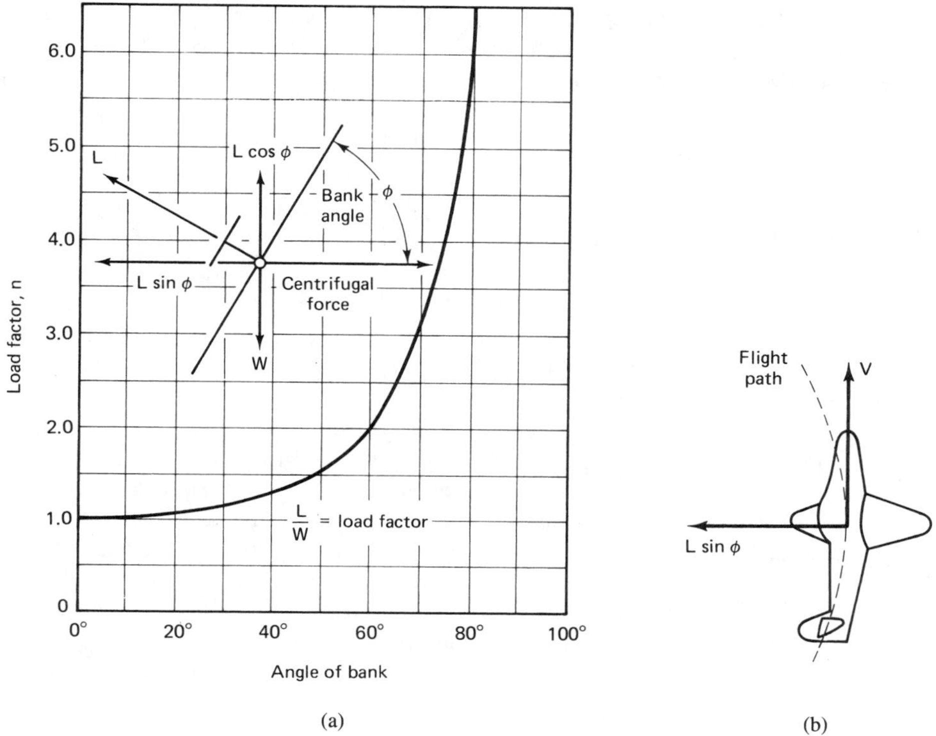

Figure 16.12 Turning flight at constant altitude.

Equation 16.14 specifies the angle of bank required for any speed and radius of turn. Conversely, the radius of turn is given by

$$R = \frac{V^2}{g \tan \phi} \tag{16.15}$$

Also, since for a level-flight turn

$$W = L \cos \phi \tag{16.16}$$

it follows that the lift for such a turn must be given by $L = W/\cos \phi$ and

$$\frac{L}{W} = \frac{1}{\cos \phi} = n \tag{16.17}$$

In a level-flight turn, the lift required is greater than the weight by the ratio $1/\cos \phi$. For a 45 degree bank, for example, the lift is 41% greater than the weight and the airplane is subjected to 1.41 g's (i.e., $n = 1.41$). This is illustrated in Figure 16.12a. Therefore, in a turn the stalling speed is increased by the square root of the ratio $1/\cos \phi$, 19% in our 45 degree example. This is one of the reasons for maintaining a margin of 20% to 30% above the level flight stall speed for takeoff and landing. This margin is required for maneuvering that may become necessary to avoid an obstacle.

The turning performance of an airplane in level flight can be developed by starting with equation 16.15. Since the load factor n normal to the wing surface is $1/\cos\phi$, it can be seen from Figure 16.13 that $\tan\phi = \sqrt{n^2 - 1}$. Substituting in equation 16.15 gives the radius of turn as

$$R = \frac{V^2}{g\sqrt{n^2 - 1}} \tag{16.18}$$

The rate of turning in radians per second is

$$\frac{d\gamma_1}{dt} = \frac{V}{R} = \frac{g\sqrt{n^2 - 1}}{V} \tag{16.19}$$

where γ_1 is the flight path angle in a horizontal plane measured with respect to an earth reference such as north.

The radius of turn is reduced and the rate of turn is increased by increasing the load factor for level-flight turns. There are three limitations on the maximum achievable load factor. The first is the highest load factor permitted by the structure. As noted in Chapter 18, the design maneuver load factor is 2.5 for transports, up to 3.8 for small general aviation aircraft, and up to 7 to 8 for combat aircraft. Second, the lift coefficient increases with load factor. When the maximum lift coefficient is reached, no higher load factor can be attained and the turning radius is the minimum. Actually, a practical minimum radius occurs earlier. Because of the danger of a stall and a spin following a turning stall, some angle-of-attack margin must be retained. From equation 16.17, we can write

$$n_{\text{MAX}} = \frac{C_{L_{\text{MAX}}} qS}{W} \tag{16.20}$$

Thus the minimum turn radius at a given speed is obtained with high $C_{L_{\text{MAX}}}$, high q, and low wing loading. With a fixed speed, the q decreases at high altitude. Major aerodynamic design objectives for fighters are the highest possible $C_{L_{\text{MAX}}}$ and lowest wing loading compatible with other requirements. Wing leading edge strakes and small canard surfaces placed just ahead of and above the wing are used to create vortices at a high angle of attack. These vortices postpone the stall and increase usable wing lift. Figure 16.14 illustrates a maneuver V–n diagram. The maximum maneuvering structural load factor and the maximum load factor obtainable at $C_{L_{\text{MAX}}}$ are shown as functions of V_E at a given altitude. From equations 16.18 and 16.19, we

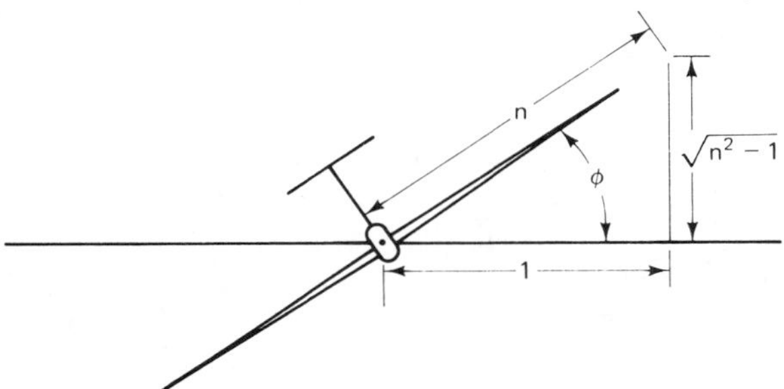

Figure 16.13 Geometry of bank angle ϕ. From equation 16.17, $\cos\phi = 1/n$; then $\tan\phi = \sqrt{n^2 - 1}$.

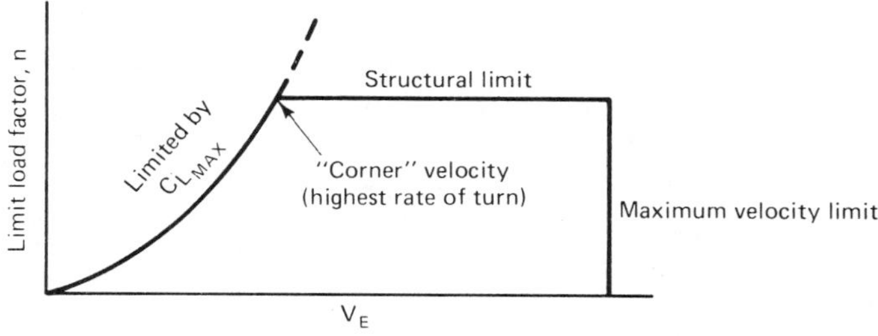

Figure 16.14 Maneuver *V-n* diagram.

see that the turn rate is maximized with the highest n and the lowest V. Therefore, the fastest turning will occur at the intersection of the $C_{L_{MAX}}$ and the structural limits. V–n diagrams are discussed further in Chapter 18.

The third limitation on load factor in a turn is the additional drag created by the high lift coefficient producing the turning force. When the drag exceeds the available thrust, a level turn cannot be sustained. Therefore, obtaining the highest permissible C_L with as little drag as possible is another fighter design objective. The cruise drag obtained using the usual methods applies, in general, only to lift coefficients between about 0.2 and 0.7. At higher C_L, the effective e decreases. The decrease is about 5% to 10% at $0.7C_{L_{MAX}}$, typical of the climb immediately after takeoff. At still higher lift coefficients, the drag increase is even greater. Normally, this is not a flight region of importance for performance. For fighters, however, it is crucial. The problem is particularly difficult at high Mach numbers.

In a thrust-limited turn,

$$T = D = C_D \frac{\rho}{2} V^2 S \tag{16.21}$$

At a given speed and altitude, the available thrust will be known. The allowable drag coefficient can be found from equation 16.21. If the speed is below the Mach number at which compressibility drag is involved, the allowable turning lift coefficient can be determined from the equation for C_D.

$$C_D = C_{D_P} + \frac{C_L^2}{\pi \cdot \text{AR} \cdot e} \tag{16.22}$$

If compressibility drag is present, the lift coefficient must be determined from a curve of C_D versus C_L for the turning flight Mach number. The usable load factor is then found from equation 16.20, in which the maximum lift coefficient is not the aerodynamic $C_{L_{MAX}}$ but rather the maximum permitted by the available thrust. The turning rate and radius can then be found from equations 16.18 and 16.19.

When the airplane is accelerated sidewise in a turn, the directional or weathercock stability will rotate the airplane about its vertical axis automatically to keep the airplane pointing into the relative wind. Therefore, turns can be made with ailerons only. The airplane always has an angle of *sideslip* in such a turn to cause the directional correction. This increases the drag, a condition that can be avoided by use of the rudder to yaw the airplane at the rate that keeps the airplane always headed into the relative wind during the "coordinated" turn. When turns are made with ailerons only, the aileron which increases the lift of the rising wing also increases the drag on that side of the airplane. The result is an initial yaw in the direction opposite to the desired turn direction. This is called *adverse yaw*. After the angle of bank has developed and the sideslip started, the correct direction of yaw follows. The rudder is also essential to correct the asymmetric thrust caused by failure of an engine on a multiengine airplane and to balance yawing moments introduced by the rotation of the propeller slipstream. Turns can be made with rudder only because the yaw angle caused by a rudder deflection induces a roll if the airplane had dihedral stability. The roll inclines the lift vector and curves the flight path. In a turn in which the relative wind approaches from the outside of the turn, the airplane is said to be *skidding*.

High-speed aircraft often have wings that are sufficiently flexible so that aileron effectiveness is seriously reduced due to wing twist under load. The aileron lift at the rear of the outer panel may twist the wing to a lower angle of attack, which counteracts much of the aileron lift. In extreme cases, the net effect may even be reversed, so that a positive aileron angle might actually reduce the outer panel wing lift. In such designs the outboard ailerons are utilized only during the low-speed regime, and high-speed roll control is obtained by small inboard ailerons for gentle maneuvers and by spoilers for higher rates of roll. Spoilers are essentially flat plates of perhaps 5% to 10% chord located just ahead of the flaps. When raised, as shown in Figure 2.3, they cause a flow separation and a loss of lift. Spoilers are actuated only on the wing panel to be lowered in the turn, since they are capable only of reducing lift. Spoilers are raised symmetrically after landing to sharply reduce the lift and dump the airplane's weight on the wheels. This greatly increases the stopping effectiveness of the wheel brakes. Spoilers are also used as dive brakes to permit high rates of descent without attaining excessive speeds and for deceleration when rough air is encountered.

CONTROL SYSTEMS

Control systems vary from a simple cable or push rod driving the elevator, rudder, or aileron to triple or quadruple full-power systems with artificial feel for the pilot.

Although aerodynamic control surfaces are generally similar to plain flaps, unless full power is used a great deal of engineering effort is required to properly shape their leading edges and to locate their hinge lines. These design features are not usually important to the surface effectiveness, but they dominate the surface hinge moments, which in turn determine the forces felt by pilots. Obtaining smoothly varying and stable pilot forces for all flight conditions is an important design requirement for both military and commercial aircraft. There are many specific requirements in the Federal Air Regulations and in the military specifications for the maximum and minimum forces permitted in various maneuvers.

For small aircraft, a simple directly driven hinged surface is adequate (Figure 16.15). As the airplane size grows, the forces required to move the control surfaces against the aerodynamic moments tending to return the surfaces to zero deflection become large. To reduce this restoring moment, the hinge line is moved aft, closer to the surface center of pressure. The surface is said to be aerodynamically balanced using nose overhang (Figure 16.16b). The nose contour is shaped so that the nose does

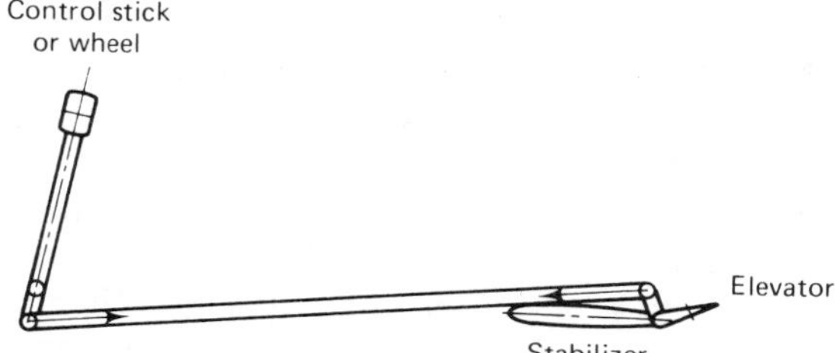

Figure 16.15 Simple pushrod-driven elevator control system.

not protrude into the airstream even at high control surface deflections. A highly curved protrusion, which would generate large suctions on the nose, would overbalance the surface, tending to move it to a higher angle. The pilot then would have to hold the surface from going too far instead of having a stable force curve in which more force is required to obtain a larger surface deflection.

A paddle or horn aerodynamic balance used on older aircraft is shown in Figure 16.16c. Figure 16.16d illustrates another way to obtain nose overhang balance. Called a *sealed internal overhang,* it utilizes an internal paddle to feel the pressure difference between the top and bottom surfaces and provide the balancing moment. Its advantages are that the seal prevents the energy loss from flow from one side of the surface to the other through the control surface gap and that the contour can be

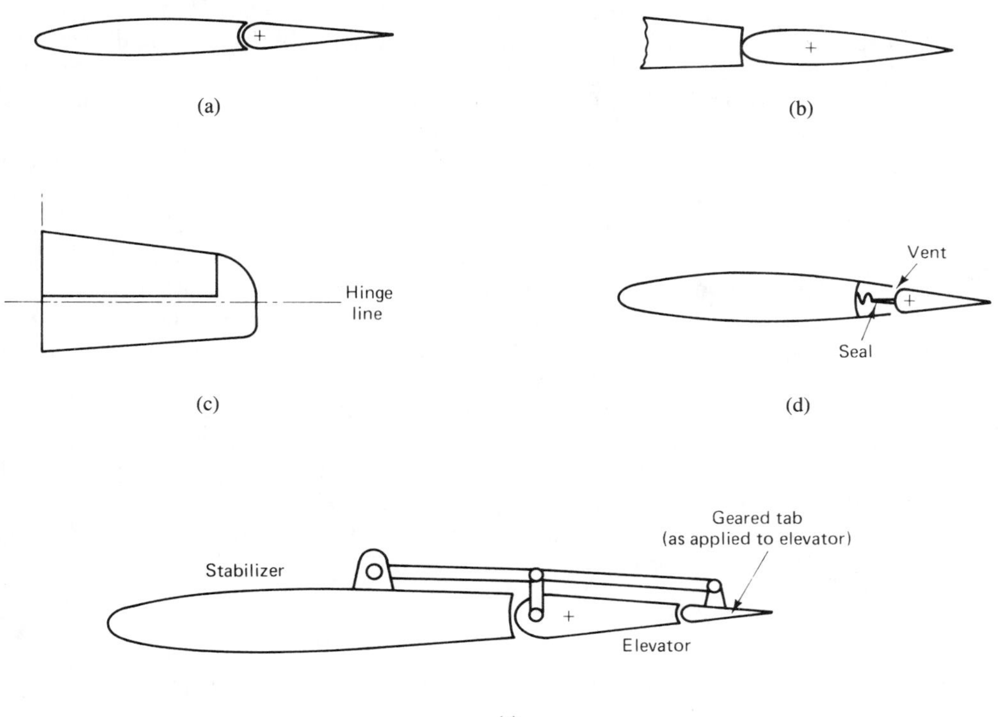

Figure 16.16 Control surfaces with various types of aerodynamic balance: (a) unbalanced control surface; (b) nose overhang balance; (c) paddle or horn balance; (d) internal sealed overhang balance; (e) geared tab.

smooth without the discontinuity of the nose overhang in Figure 16.16b. The result is that the parasite drag may be reduced by as much as 4% to 5% of the total parasite drag.

Control surfaces can be aerodynamically balanced without nose treatment by using a *geared tab,* a small flap on the surface trailing edge (Figure 16.16e). The tab applies a moment tending to increase the surface angle, its deflection being proportional to the control surface deflection. The geared tab lifts in the direction opposite to that of the basic control surface and thereby reduces the effectiveness of the surface. It is easily adjusted to obtain the desired amount of balance, however.

As control surfaces become larger, the design of control surface shapes to obtain comfortable pilot forces without the risk of overbalance at extreme flight conditions becomes impossible. An alternative is to drive the tab directly and let the tab provide the complete moment required to move the elevator, rudder, or aileron. This arrangement is called a *flying tab.* An intermediate design, called the *linked tab,* is shown schematically in Figure 16.17. It is a system in which the pilot force is directed partly to the control surface and partly to a control tab on the trailing edge. By adjusting the point of connection between the input rod from the pilot and the link between the control surface and the tab (point P in the diagram), the proportion of the force going to the tab can be varied to obtain the desired pilot forces. This system is in wide use on large aircraft that can still be controlled by pilot force.

As aircraft become larger, the point is finally reached at which full power must be used to drive the control surfaces. This may also be true of small, very high performance aircraft operating over a wide speed range. Since a complete failure of these systems is intolerable, they must be very reliable and multiply redundant. All wide-bodied transport aircraft have full-power systems. They must be designed with three or four completely independent hydraulic systems, any one of which can provide the pilot with adequate control. Spoilers that have large and nonlinear hinge moments are always fully powered on large aircraft.

One other use of tabs should be mentioned. Tabs are used on aircraft of all sizes to provide a zero control surface hinge moment at the control surface angle for steady flight. The purpose is to reduce the required pilot force to zero. The airplane is said to be trimmed and will fly "hands off" at the flight condition for which the trim tab angle is set. A *trim tab* is illustrated in Figure 16.18.

Trimming may also be accomplished by varying the incidence angle of the horizontal stabilizer. Aircraft with a high angle of attack for stall and powerful flaps with a large negative pitching moment require large moment capability from the horizontal tail. If the elevator is not sufficiently powerful, the tail power can be increased by providing the ability to vary the stabilizer angle for different flight regimes. All jet transports have this capability.

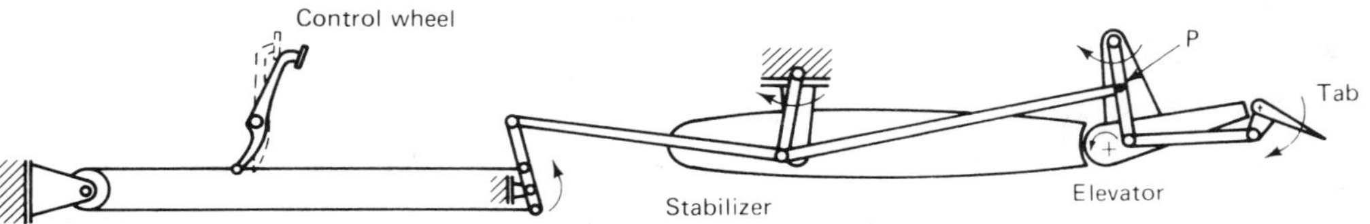

Figure 16.17 Linked tab control system (simplified). (Courtesy Douglas Aircraft Company.)

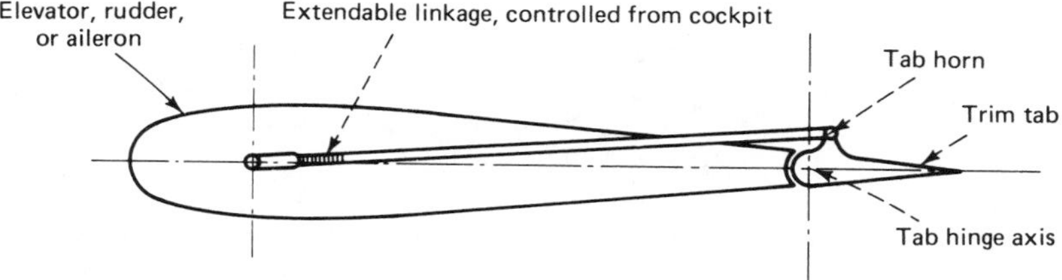

Figure 16.18 Schematic drawing of a trim tab.

ACTIVE CONTROLS

The three traditional disciplines of aeronautics are aerodynamics, structures, and propulsion. In the last decades the increasing use of automatic controls for stability augmentation and navigation has made guidance and control an important fourth element. This is particularly so with the introduction of active controls that provide full-time artificial stability and load alleviation. Active controls are fast-acting control systems that reduce required tail sizes by providing full-time artificial stability, reduce gust-induced loads by compensating for the gust by control movements, and can even prevent dangerous high-speed wing flutter by damping potentially unstable wing motions by rapid aileron or flap deflections. The development of light, highly reliable microcomputers has been the key to this technology. Even then, relying on artificial controls to provide stability and to limit critical structural loads requires multi-redundant systems so that no failure or remotely likely combination of failures will cause a dangerous flight condition. Thus active control systems will probably require three- or four-way duplication of the entire system, including sensors such as accelerometers and gyros, computers, and control surface actuators. Considerable active control capability has been built into the General Dynamics F-16 fighter. The long-range version of the Lockheed L-1011, the L-1011-500, employs limited active control functions to reduce the wing loads due to gusts and maneuvering. The advantage to the L-1011 is a reduction in wing structural weight.

Example 16.1

A large jet transport with a low horizontal tail has the following characteristics:

$$\begin{aligned}
\text{wing area, } S_W &= 2927 \text{ ft}^2 \\
\text{wing span, } b_W &= 148.4 \text{ ft} \\
\text{mean aerodynamic chord, } \bar{c} &= 22.73 \text{ ft} \\
\text{horizontal tail area, } S_H &= 559.1 \text{ ft}^2 \\
\text{horizontal tail span, } b_H &= 47.5 \text{ ft} \\
\text{tail length, } l'_H &= 71.2 \text{ ft}
\end{aligned}$$

The total of the swept wing, fuselage, and nacelle contribution to stability is slightly stable about the 25% $\bar{c}$ point.

$$\left(\frac{dC_M}{dC_L}\right)_{\substack{\text{wing+} \\ \text{fuselage+nacelle}}} = -0.016; \quad \text{also} \quad \frac{d\epsilon_H}{d\alpha} = 0.43$$

What is the most aft allowable center of gravity position at which stability, dC_M/dC_L, is at least -0.10? (Note that $dC_M/dC_L = -0.10$ corresponds to the center of gravity being 10% of the mean aerodynamic chord ahead of the "neutral" point.)

This problem involves the use of equation 16.8. The lift curve slopes of the wing and tail appear in this equation, but only in the form of the ratio of $(dC_L/d\alpha)_H$ to $(dC_L/d\alpha)_W$. For strict engineering analysis, the lift curve slopes should be calculated from equation 13.4. For this application involving ratios, however, the effects of sweep and Mach number are mostly canceled, so for preliminary work we can assume that

$$\frac{dC_L}{d\alpha} = \frac{a_0}{1 + 57.3a_0/(\pi \cdot \text{AR})} \quad \text{per degree}$$

Solution:

Wing $\begin{cases} S_W = 2927 \text{ ft}^2 \\ b_W = 148.4 \text{ ft} \\ \bar{c} = 22.73 \text{ ft} \end{cases}$ $\quad \text{AR}_W = \frac{b^2}{S} = \frac{(148.4)^2}{2927} = 7.52$

Horizontal tail $\begin{cases} S_H = 559.1 \text{ ft}^2 \\ b_H = 47.5 \text{ ft} \\ l_H' = 71.2 \text{ ft} \end{cases}$ $\quad \text{AR}_H = \frac{(47.5)^2}{559.1} = 4.04$, Assume $\eta_H = 0.90$

$$\left(\frac{dC_M}{dC_L}\right)_{\text{F,N}} = -0.016, \qquad \frac{d\epsilon_H}{d\alpha} = 0.43$$

The static stability equation (16.8) is

$$\left(\frac{dC_M}{dC_L}\right)_{\text{cg}} = \frac{x}{\bar{c}} - \left[\frac{(dC_L/d\alpha)_H}{(dC_L/d\alpha)_W}\left(1 - \frac{d\epsilon_H}{d\alpha}\right)\frac{S_H}{S_W}\frac{l_H'}{\bar{c}}\eta_H - \left(\frac{dC_M}{dC_{L_W}}\right)_{\text{F,N}}\right] \left\{\frac{1}{\left[1 + \frac{(dC_L/d\alpha)_H}{(dC_L/d\alpha)_W}\left(1 - \frac{d\epsilon_H}{d\alpha}\right)\frac{S_H}{S_W}\eta_H\right]}\right\}$$

Note that

$$a_0 = \frac{2\pi(0.95)}{57.3} = 0.104 \quad \text{per degree}$$

Then

$$\left(\frac{dC_L}{d\alpha}\right)_H = \frac{a_0}{1 + 57.3a_0/(\pi \cdot \text{AR})} = \frac{0.104}{1 + \frac{(57.3)(0.104)}{\pi(4.04)}} = 0.0708$$

and

$$\left(\frac{dC_L}{d\alpha}\right)_W = \frac{0.104}{1 + \frac{(57.3)(0.104)}{\pi(7.52)}} = 0.0831$$

Then

$$\left(\frac{dC_M}{dC_L}\right)_{cg} = \frac{x}{\bar{c}} - \left[\frac{0.0708}{0.0831}(1 - 0.43)\frac{559.1}{2927}\left(\frac{71.2}{22.73}\right)0.90 - (-0.016)\right]$$

$$\cdot \left[\frac{1}{1 + \dfrac{0.0708}{0.0831}(1 - 0.43)\dfrac{559.1}{2927}(0.90)}\right]$$

$$= \frac{x}{\bar{c}} - [0.2615 + 0.016]\left[\frac{1}{1.0835}\right] = \frac{x}{\bar{c}} - 0.2561$$

But $(dC_M/dC_L)_{cg}$ must be -0.10. Then

$$-0.10 = \frac{x}{\bar{c}} - 0.2561$$

$$\frac{x}{\bar{c}} = 0.2561 - 0.10 = 0.1561$$

quarter chord of $\bar{c}$

Thus the aft c.g. position $= 0.25 + 0.1561 = 0.406$, or 40.6 aft of the leading edge of the m.a.c.

Example 16.2

A fighter is flying at a Mach number of 0.85 at 26,000 ft. To avoid an attacker, the pilot enters a $5g$ level-flight turn. What is the radius of turn and the distance covered in reversing direction? What C_L is required if the wing loading is 50 lb/ft^2?

Solution: At 26,000 ft, the speed of sound is 1011.9 ft/s. The pressure is 752.71 lb/ft^2.

$$V = (0.85)(1011.9) = 860.12 \text{ ft/s}$$

For a level-flight turn, from equation 16.17,

$$\cos\phi = \frac{1}{n} = \frac{1}{5} = 0.2$$

Then bank angle $\phi = 78.46$ degrees; $\tan\phi = 4.90$. From equation 16.15,

$$\text{radius } R = \frac{V^2}{g\tan\phi} = \frac{(860.12)^2}{(32.17)(4.90)} = \underline{4693.17 \text{ ft}}$$

To turn 180°, the distance is half a complete circle:

$$180° \text{ turn distance} = \frac{2\pi R}{2} = \pi(4693.17) = 14{,}744 \text{ ft}$$

$$= \underline{2.79 \text{ statute miles}}$$

Now, $L = 5W$(i.e., $n = 5$) and $W/S = 50$ lb/ft^2.

$$C_L = \frac{5W}{(0.7)(752.71)(0.85)^2S} = \frac{5(50)}{380.78} = \underline{0.66}$$

PROBLEMS

16.1. A large turboprop transport has the following characteristics:

$$\text{wing area, } S_W = 3520 \text{ ft}^2$$
$$\text{wing span, } b_W = 155 \text{ ft}$$
$$\text{mean aerodynamic chord, } \bar{c} = 25.0 \text{ ft}$$
$$\text{horizontal tail aspect ratio} = 4.2$$
$$\text{tail length, } l_H^1 = 83.0 \text{ ft.}$$

The airplane has a T-tail, the horizontal being mounted on top of the vertical tail. The wing is unswept. The fuselage and nacelle contribution to stability about the quarter-chord is unstable; that is,

$$\left(\frac{dC_M}{dC_L}\right)_{\substack{\text{fuselage}\\ +\text{nacelle}}} = 0.06; \quad \text{also} \quad \frac{d\epsilon_H}{d\alpha}\text{(at the tail)} = 0.50$$

Assume that the wing and tail have elliptical lift distributions so that $dC_L/d\alpha = a_0/(1 + 57.3\, a_0/(\pi AR))$ (per degree). The most aft allowable center of gravity position, at which stability, dC_M/dC_L, is at least -0.10, is 35% of the m.a.c. (Note that $dC_M/dC_L = -0.10$ corresponds to the c.g. being 10% of the mean aerodynamic chord ahead of the 'neutral' point.) Determine the horizontal tail area.

16.2. A fighter airplane is pursuing a target at 22,000 ft on a standard day at $M = 0.87$. The pilot has a pressure suit so that he can withstand a maximim load factor of $6g$. What is the fighter's turn radius? What distance must be covered to complete a 180 degree turn? If the fighter has a wing loading of 58 lb/ft^2, what is the lift coefficient?

16.3. A transport flying at 39,000-ft pressure altitude at $M = 0.83$ makes a 180 degree turn. Temperature is standard. The pilot limits the bank angle to 25 degrees to avoid alarming the passengers. What is the turn radius and the time to complete the turn?

16.4. A DC-8-50, the original turbofan-powered version of the DC-8, has the following characteristics:

$$\text{wing area, } S_W = 2883 \text{ ft}^2$$
$$\text{wing span, } b_W = 148.4 \text{ ft}$$
$$\text{mean aerodynamic chord, } \bar{c} = 22.98 \text{ ft}$$
$$\text{horizontal tail area, } S_H = 559.1 \text{ ft}^2$$
$$\text{horizontal tail span, } b_H = 47.5 \text{ ft}$$
$$\text{tail length, } l_H^1 = 68.4 \text{ ft.}$$

The total of the wing, fuselage, and nacelle contributions to stability is slightly stable about the $\bar{c}/4$ point; that is,

$$\left(\frac{dC_M}{dC_L}\right)_{\substack{\text{wing}+\\ \text{fuselage}\\ +\text{nacelle}}} = -0.04$$

At the tail,

$$\frac{d\epsilon_H}{d\alpha} = 0.45$$

Assuming that the wing and tail have elliptical lift distributions so that $dC_L/d\alpha = a_0/(1 + 57.3\, a_0/(\pi \text{AR}))$ (per degree), what is the most aft allowable center of gravity position at which stability, dC_M/dC_L, is at least -0.10?

REFERENCES

16.1. Perkins, Courtland D., and Hage, Robert E., *Airplane Performance, Stability and Control,* Wiley, New York, 1949.

16.2. McCormick, Barnes W., *Aerodynamics, Aeronautics, and Flight Mechanics,* Wiley, New York, 1979.

16.3. Abramson, Norman, *An Introduction to the Dynamics of Airplanes,* Dover, New York, 1958.

16.4. United States Federal Air Regulations Part 25, paragraph 25.171.

16.5. United States Air Force specification MIL-F-8785C, 5 November 1980.

17

PROPULSION

All propulsion is based on *Newton's third law:* To every action there is an equal opposed reaction. If body A exerts a force on body B, body B exerts an equal but oppositely directed force on body A.

In a fluid the propulsion unit pushes on a fluid (e.g., air), and the fluid pushes back on the propulsion unit, creating thust. This is equally true for a propeller, a jet engine, a turbofan, the oar of a rowboat, the paddle of a canoe, or a rocket. Quantitatively, the thrust produced is equal to the time rate of change of momentum of the fluid.

A propeller is merely a collection of rotating wings using airfoils similar to those used on wings. Turbojet or turbofan engines use large numbers of small "wings" called *blades* attached to a rotating core. A turbojet passes all the incoming air through the combustion chambers. A turbofan directs only part of the air it initially compresses through the burners; the remainder is ducted around the higher stages of the compressor and the burners and reenters the freestream. A rocket is a jet engine that carries its own oxidizer, as well as the fuel, and so produces thrust without requiring air.

PISTON ENGINES

From the 1903 flight of the Wright Brothers to the mid-1940s, the piston engine was the only power plant used for aircraft. Originally, the engines were water cooled, involving high-weight, high-drag radiators. In 1908, an air-cooled rotary design, the Gnome, was built in France. The cylinders of this engine were arranged in a circle around a crankcase. The propeller was fastened to the front of the crankcase and the entire engine rotated. This was done to increase the air velocity over the cylinders for cooling purposes. The weight/power ratio was significantly lower for this engine, although the gyroscopic forces were large during maneuvering. A development of this engine powered the famous Nieuport fighter during World War I. Later air-cooled engines were radials with a similar arrangement of cylinders, but with the engine fixed and only the propeller rotating.

One of the major aircraft design debates during the next 25 years concerned the relative drag, weight, and maximum power capabilities of liquid-cooled versus air-cooled engines. Many of the best fighters of World War II were powered by liquid-cooled engines such as the Rolls-Royce Merlin and the Allison V-1710 (Figures 17.1 and 17.2a). By the mid-1940s, however, the debate was over and air-cooled engines

Figure 17.1 British World War II Spitfire fighter powered by the Rolls-Royce Merlin liquid-cooled engine. (Courtesy of Smithsonian Institution.)

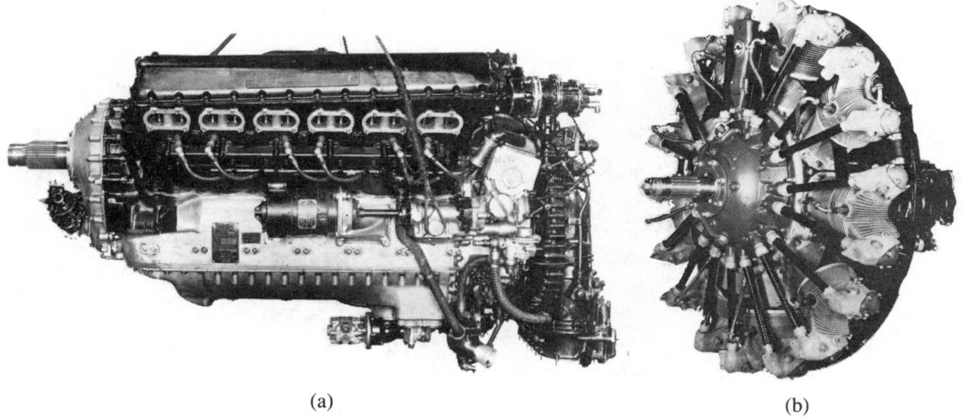

(a) (b)

Figure 17.2 Types of piston aircraft engines: (a) Rolls-Royce Merlin liquid-cooled engine (Courtesy of National Air and Space Museum, Smithsonian Institution); (b) Pratt & Whitney Wasp air-cooled radial engine. (Courtesy of Pratt & Whitney.)

were the victors. All major transport and military aircraft of the postwar period were powered by radial air-cooled engines (Figure 17.2b).

One major development assisting the air-cooled engines was the development of the NACA cowling (Figure 17.3), an enclosure for the engine that limited the flow of air over the engine cylinders to the air actually in contact with the cooling fins of the cylinders. Since the air flowing over the cylinders suffered a large pressure loss and contributed a large drag, this advance was very important in improving the efficiency of aircraft.

Weight characteristics of piston engines are illustrated in Figure 17.4. The data shown are the specific weights, the ratio of weight to power. The open symbols show the later technology. The major specific weight reductions were the result of greater engine size. Also shown are the much lower specific weights for turboprop engines. This explains the trend in the larger general aviation aircraft toward turboprop

Figure 17.3 Ten-passenger Boeing 247 transport showing the first step toward controlling cooling air with a Townend ring and the 247D with a full NACA cowling. (Courtesy of United Airlines.)

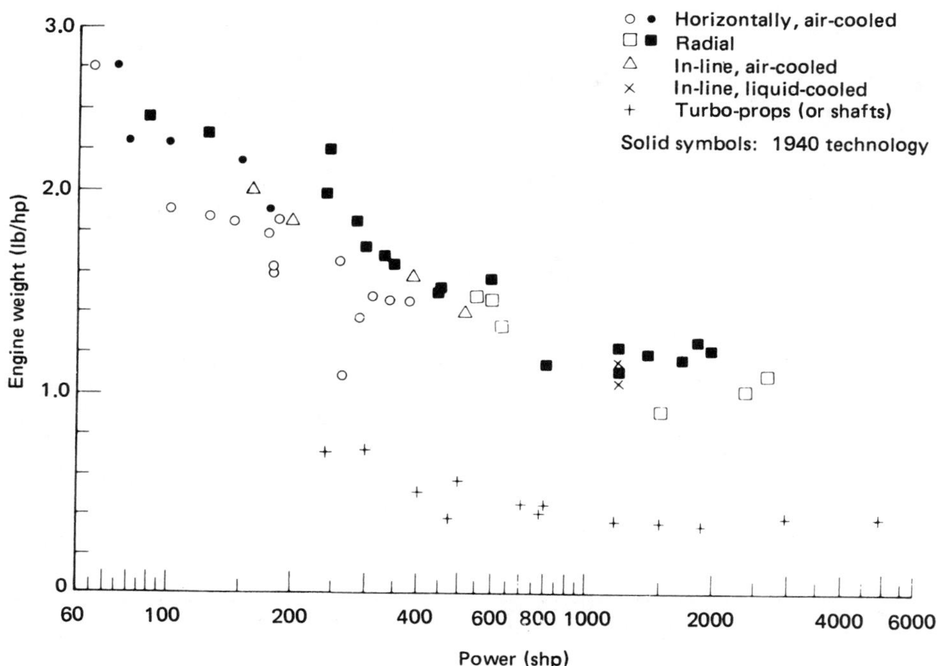

Figure 17.4 Engine weight-to-power ratios for piston and turboprop engines. From B. McCormick, *Aerodynamics, Aeronautics, and Flight Mechanics,* 1979. Reprinted by permission of John Wiley & Sons, New York.

engines. Smaller general aviation aircraft still use piston engines because of the much greater cost of the turbine engines. Turboprop weight data must be used with caution in a comparison with turbofans, since the complete turboprop power plant weight must include the propeller, a major factor not contained in the definition of turboprop weights. Very often the weight data for larger turboprop engines do not include the gear box weight, while piston engine weights always do. The gear box may weigh more than half as much as the basic turboprop engine.

Figure 17.5 shows the brake specific fuel consumption (BSFC) characteristics for existing piston and turboprop engines. BSFC is expressed in terms of pounds of fuel used per hour per shaft horsepower. Turboprop engines usually have a significant amount of gross jet exhaust thrust. The net jet thrust is the gross jet thrust minus the ram drag, as shown in equation 17.1. The incremental *equivalent shaft horsepower* due to the jet thrust varies with both the thrust and the speed in accordance with equation 15.9, being zero at zero speed. The net jet thrust may be negative at high speeds or it may add as much as 10% or more to the effective power. The turboprop BSFC data in Figure 17.5 apply to static sea-level conditions. It applies reasonably well for helicopters. For high-speed turboprops, favorable speed and altitude effects reduce the BSFC by 10% to 15%, and the net jet thrust provides a further decrease, or increase, depending on airplane speed and the engine design. Most present turbo-

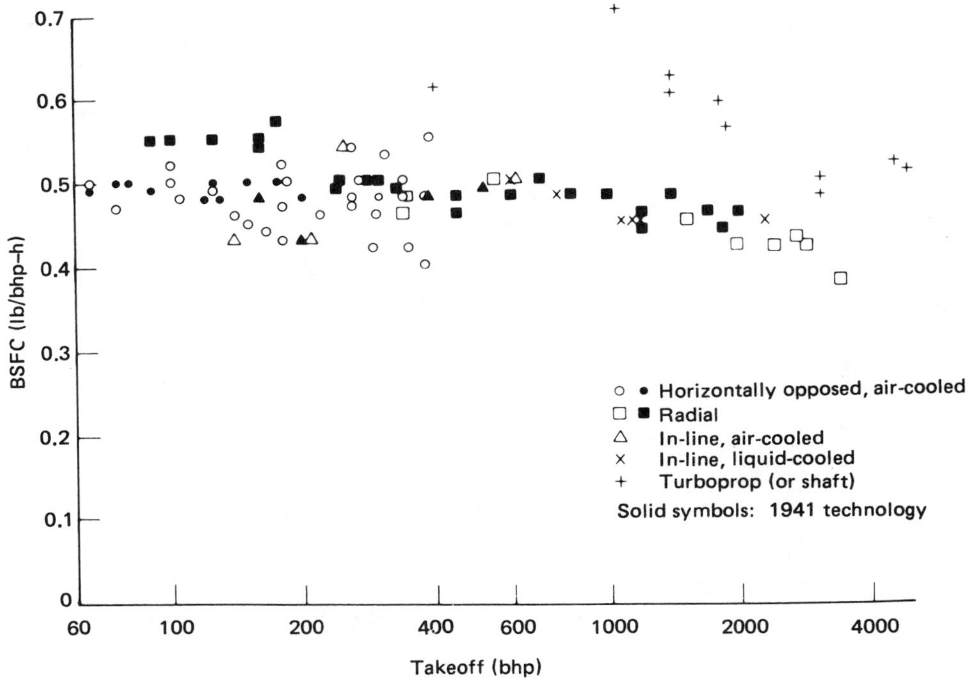

Figure 17.5 Brake specific fuel consumption for piston and turboshaft engines. From B. McCormick, *Aerodynamics, Aeronautics, and Flight Mechanics,* 1979. Reprinted by permission of John Wiley & Sons, New York.

props are derivatives of much older engines. Advanced turbine technology applied to future turboprop engines may bring the *equivalent* BSFC at high speed and altitude down to about 0.3.

In the SI system of units, the BSFC is expressed in newtons per kilowatt-hour. A value of 0.5 lb/bhp-h in the English system becomes 2.98 N/kW-h in the SI system.

The power of piston engines depends primarily on the density of the intake air. Therefore, full throttle power decreases with altitude. To maintain power to higher altitudes, some engines are equipped with either a gear-driven or an exhaust-turbine-driven supercharger. All piston engines used in fighters and transports after about 1935 were supercharged.

Piston engine power is set using two parameters, the engine speed in revolutions per minute (rpm) and the intake manifold pressure (MP) in inches of mercury. The power obtained depends on these settings, the pressure altitude, which affects the standard temperature of the intake air and the back pressure on the engine, and the deviation of the temperature from the standard for that altitude. A representative piston engine performance chart is shown in Figure 17.6.

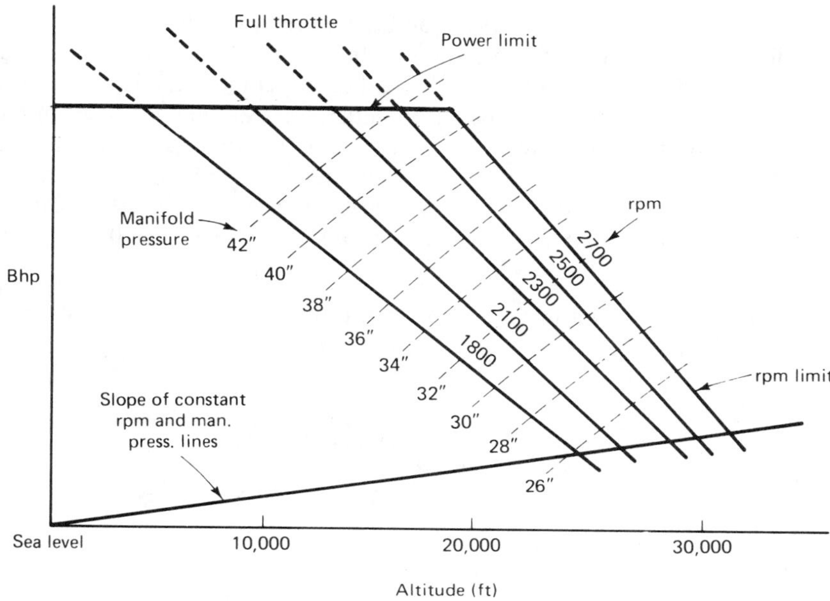

Figure 17.6 Full-throttle performance chart for a supercharged piston engine.

GAS TURBINES

An aircraft gas turbine is a device in which freestream air is taken in through a carefully designed inlet, compressed in a rotating compressor, heated in a combustion chamber, and expanded through a turbine. The gas then leaves through a nozzle at a velocity greater than freestream. The power output of the turbine is utilized to drive the compressor and any mechanical load, such as electrical generators and hydraulic pumps, connected to the drive shaft. In the case of turboprops, most of the gas energy is extracted by the turbine to drive the propeller shaft. A wide variety of cycle arrangements are possible.

A *turbojet* is a gas turbine in which no excess power (above that required by the compressor) is supplied to the shaft by the turbine. The available energy in the exhaust gases is converted to the kinetic energy of the jet.

A *turboprop* is a gas turbine in which the turbine absorbs power in excess of that required to drive the compressor. The excess power is used to drive a propeller. Although most of the energy in the hot gases is absorbed by the turbine, turboprops still have appreciable jet thrust under most conditions.

A *turbofan* is similar to a turboprop except that the excess power is used to drive a fan or low-pressure compressor in an auxiliary duct, usually annular around the primary duct. Both the turbofan and the turboprop impart momentum to greater volumes of air than a turbojet, but the velocity added is less.

Reheat, or *afterburning,* can be used in any of these devices to give additional thrust at the expense of fuel economy. In reheat, additional fuel is added to the exhaust gases and burned, thereby increasing the temperature, the jet velocity, and the thrust.

The thrust equation for a turbojet engine can be derived from the general form of Newton's second law (i.e., force equals the time rate of change of momentum), $F = d(MV)/dt$. Figure 17.7 illustrates the inlet and exhaust flows of the turbojet. The negative thrust due to bringing the freestream air almost to rest just ahead of the engine is called *momentum drag* or *ram drag*. The net thrust is given by equation 17.1.

$$\begin{aligned}\text{Net thrust } F_N &= \text{gross thrust } F_G - \text{momentum drag}\\ &= \underbrace{\frac{W}{g}V_e}_{\text{momentum thrust}} + \underbrace{(p_e - p_0)S_e}_{\text{pressure thrust}} - \underbrace{\frac{W}{g}V_0}_{\text{momentum drag}}\\ &= \frac{W}{g}(V_e - V_0) + (p_e - p_0)S_e \qquad (17.1)\end{aligned}$$

where

W = weight of air passing through engine, N/s (lb/s)
V_e = jet stream velocity, m/s (ft/s)
p_e = static pressure across propelling nozzle, N/m^2 (lb/ft^2)
p_0 = atmospheric pressure, N/m^2 (lb/ft^2)
S_e = propelling nozzle area, m^2 (ft^2)
V_0 = aircraft speed, m/s (ft/s)

Since p_e is usually close to p_0 for subsonic engines, the second term in equation 17.1 is small, on the order of 1% to 2%.

In deriving equation 17.1, we have omitted the fact that the exhaust gas flow consists of the mass of air flowing through the engine plus the fuel added to the air in the combustion chambers. Thus the momentum thrust should be larger by the ratio of the mass of air plus fuel to the mass flow of the air alone. Since the fuel/air ratio is only about 0.01, the error is small and the simplification helps the physical understanding of equation 17.1.

The equations for a turbofan are similar except that there are two airstreams, one passing through the fan, compressor, burner, and turbine stages and the other

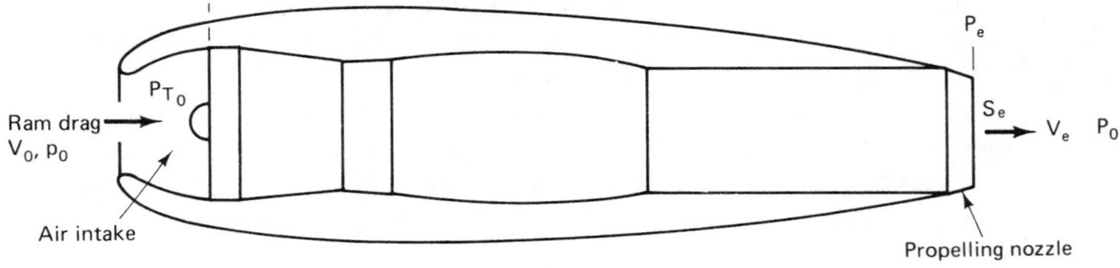

Figure 17.7 Flow through a turbojet engine.

passing only through the fan. The gross thrust is the sum of the gross thrusts of the separate gas streams. The ratio of the weight of air bypassing the compressor and burners (i.e., the fan air) to the weight of primary air flowing through them is called the *bypass ratio*.

Fan air is compressed to a much lower pressure than primary air and exits at a lower velocity. A given thrust requires imparting a certain momentum to the engine air so that, assuming that $p_e = p_0$, $F_N = (W/g)[V_e - V_0]$. The kinetic energy left in the airstream by the engine exhaust gases is $(W/g)[(V_e - V_0)^2/2]$ and is a measure of the efficiency of the engine. The less the kinetic energy per unit time left in the exhaust gases, the greater the efficiency will be. It can be seen that, for a specified momentum imparted to the air each second, the kinetic energy will be less for a higher weight flow and a lower velocity. Turbofans do exactly that, adding a lower velocity increment to a larger weight of air. As bypass ratio is increased (i.e., as more energy is extracted by the turbine to drive the fan), the exit velocity of the primary gas is reduced. Higher bypass ratios lead to lower exit velocities and higher efficiency. Figure 17.8 shows the variation of specific fuel consumption (sfc) with bypass ratio. Although sfc decreases as bypass ratio increases, the net gains to an airplane are less than indicated because the higher the engine bypass ratio, the greater the nacelle diameter, weight, and drag will be. Furthermore, the engine cost will increase with bypass ratio. Whether bypass ratios beyond about 8 will be economically viable is not yet clear. Figure 17.8 is based on the engine technology of the 1970s. The most advanced engines in the 1980s show a similar trend with bypass ratio, but the entire level of sfc is lower by about 10%. The improvements are due to higher component aerodynamic efficiencies, higher pressure ratio, and reduced leakage around the tips of the compressor and turbine blades.

Another important gain from high-bypass-ratio turbofans is reduced exhaust jet noise. Jet noise energy varies as the eighth power of the relative exhaust velocity (i.e., the velocity of the exhaust gases with respect to the freestream). Figure 17.9 shows the variation of jet noise with relative exhaust velocity. Representative aircraft are also indicated. A reduction of 10 decibels corresponds to reducing the perceived noise level

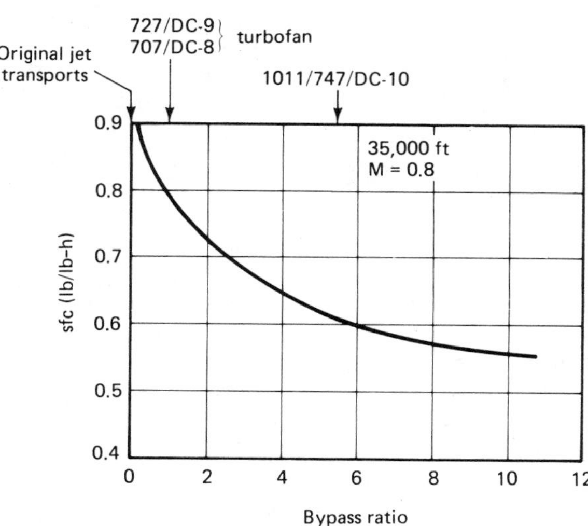

Figure 17.8 Typical variation of specific fuel consumption with bypass ratio.

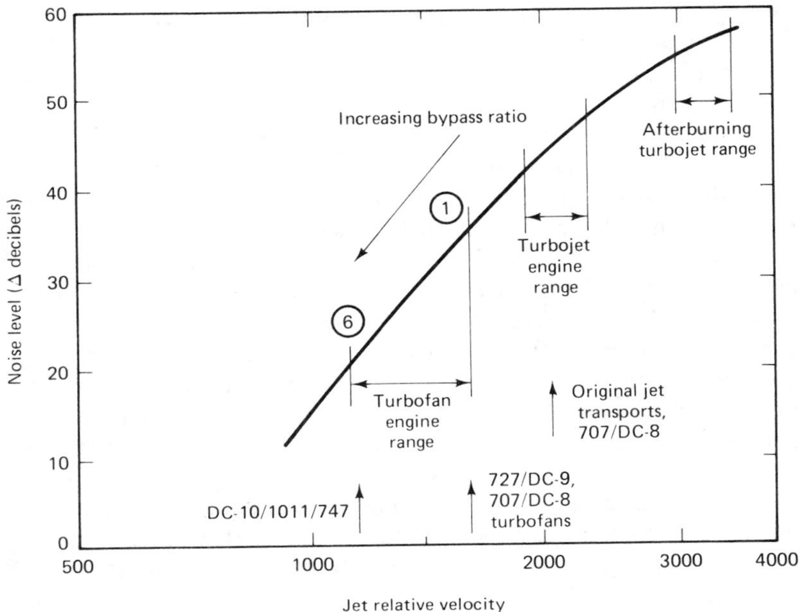

Figure 17.9 Relative jet exhaust noise level versus jet relative velocity for equal engine thrust.

by about 50%. Total engine noise arises from many sources, however, as shown in Figure 17.10. When the originally dominant jet noise is reduced, other sources become responsible for the noise! Therefore, high-bypass-ratio engines are designed to minimize other noise sources also. Careful engine design to minimize simultaneous interaction between fan and compressor blades and the boundary layer wakes of preceding stages, larger spacing between stages, lower blade tip Mach numbers, and the use of sound-absorbing linings in inlet and exhaust ducts are all necessary to attain the lower noise levels achieved by modern high-bypass-ratio engines.

Various types of gas turbine engines are shown in the following figures. Figure 17.11 shows a cross-section of a turbojet engine and compares its cycle to that of

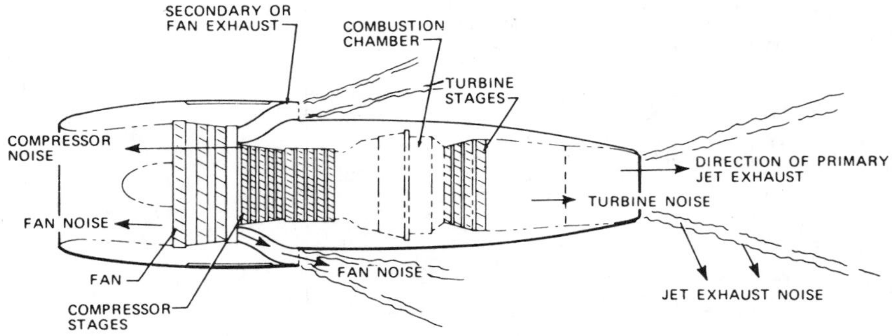

Figure 17.10 Turbofan engine noise sources. (Courtesy of Douglas Aircraft Company.)

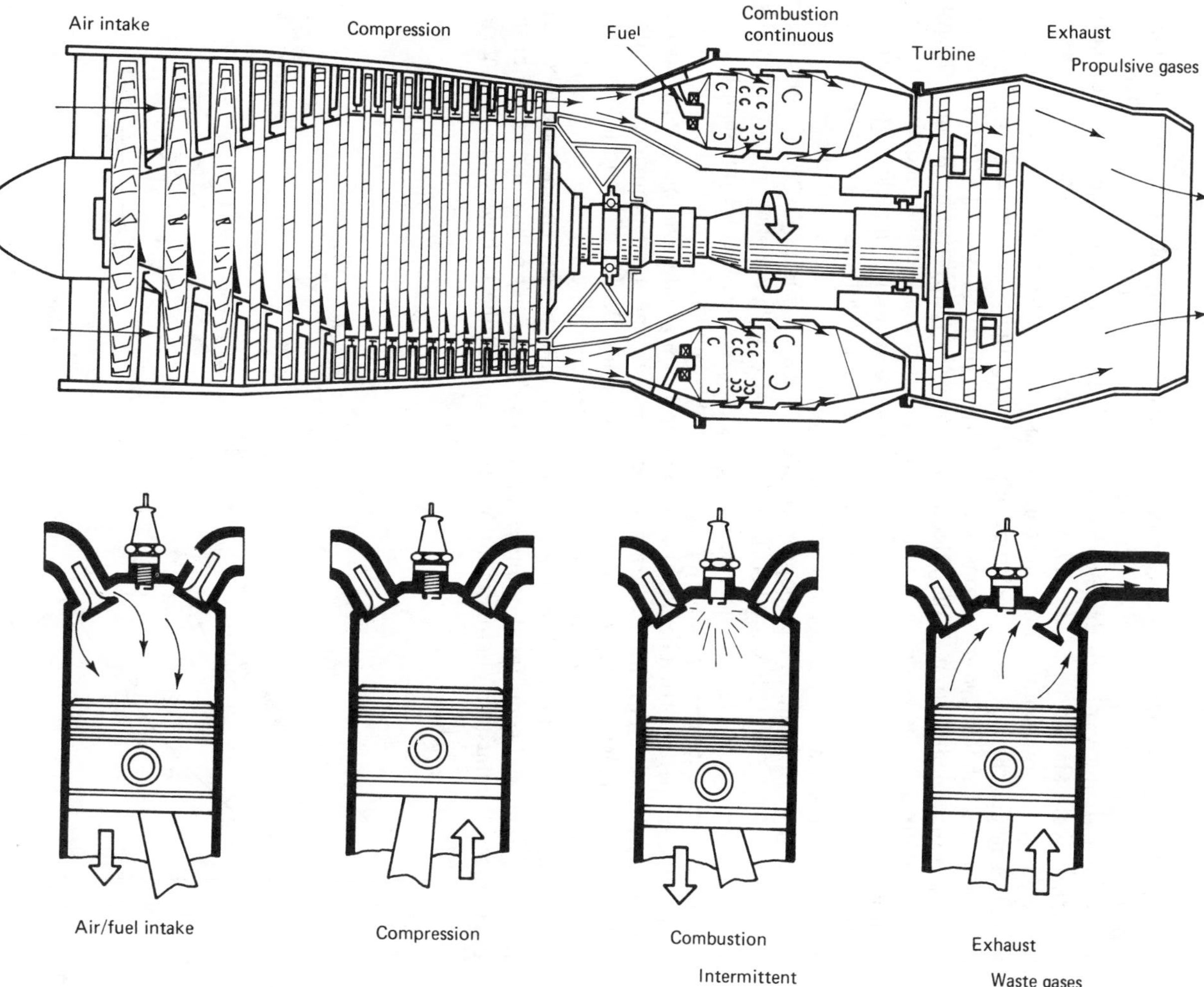

Figure 17.11 Comparison between the working cycle of a piston engine and a turbojet engine. (Courtesy of Rolls Royce, Limited.)

a four-cycle piston engine. The engine shown is a single-shaft, axial-flow engine. Large modern engines usually have two concentric shafts, with portions of the compressor and turbine mounted on each shaft. This is done to permit the various stages to operate at their best speeds.

While the comparison between the four-cycle piston engine and the turbojet shown in Figure 17.11 is generally correct, there is a significant difference between the thermodynamic cycles of these two engines. The piston engine cycle is essentially the *Otto* cycle shown in Figure 17.12a. An isentropic compression of the gas is followed by a rapid combustion at nearly constant volume at the top of the piston stroke. The gas then expands isentropically, pushing the piston downward. The gas turbine follows the *Brayton* or constant pressure cycle (Figure 17.12b). Air is compressed isentropically, first in the inlet as the air is slowed from freestream velocity to the relatively low velocity at which it enters the compressor, and then in the compressor. The air then enters the combustors where fuel is injected and burning occurs at constant pressure. The gas total temperature and volume is greatly increased. The gas then expands isentropically through the turbine stages and the jet nozzle.

A cutaway picture of a turboprop engine is shown in Figure 17.13. This engine, the Garrett TPE 331-3, has a takeoff rating of 904 equivalent shaft horsepower. As noted previously, the term "equivalent" means that the residual jet thrust in the exhaust has been converted to an equivalent increment in shaft power after consideration of the propeller efficiency. This conversion has no real meaning at zero speed, but to show some credit for jet thrust in turboprop engine data summaries, an equivalent shaft horsepower is often determined assuming 2.0 lb of static ($V_0 = 0$) jet thrust to be equivalent to 1 hp. This engine uses centrifugal compressor stages to reduce the length and cost of the engine. Some small turboprop engines use a combination of axial and centrifugal compressor stages. To further reduce the length of the engine, the combustion chambers are wrapped around the turbine stage. The gases

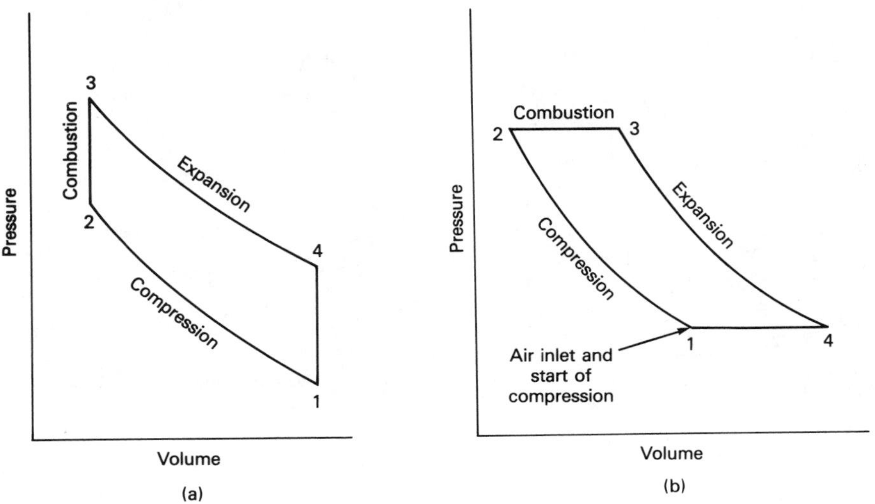

Figure 17.12 Comparison between (a) the piston engine Otto cycle and (b) the gas turbine Brayton cycle.

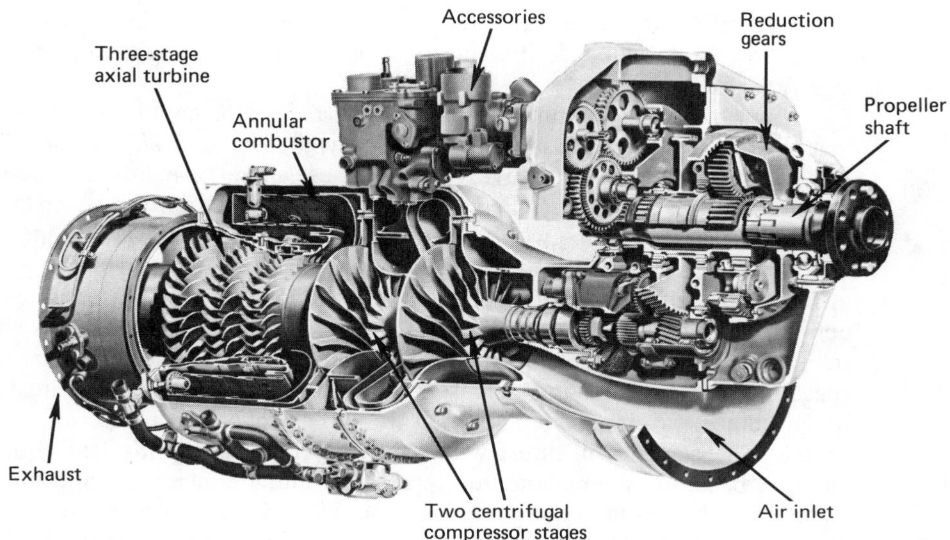

Figure 17.13 Cutaway view of the Garrett TPE 331-3 turboprop engine. (Courtesy of Garrett Turbine Engine Co.)

flow through the burners and then forward before turning 180 degrees to flow through the turbine stages. The engine weight is only 355 lb, including the integral gear box.

Figure 17.14 shows a modern high-bypass-ratio engine. This engine features a large fan, two concentric shafts, and adjustable stator blades in the compressor. Stators

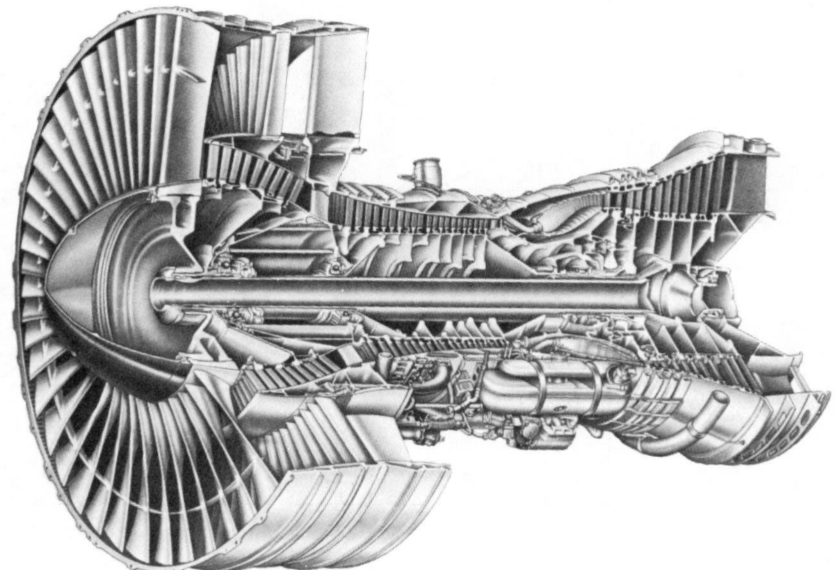

Figure 17.14 Cutaway drawing of the Pratt & Whitney JT9D-7R4 high-bypass-ratio turbofan engine. (Courtesy of Pratt & Whitney Aircraft Group.)

are stages of fixed blades between the rotating stages. They serve to adjust the swirl in the flow so that the effective angles of attack of the following rotating stages are as efficient as possible.

Figure 17.15 is a summary of the performance of the high-bypass-ratio turbofans that power wide-body transport aircraft such as the Boeing 747, McDonnell-Douglas DC-10, and the Lockheed L-1011. Although these aircraft use different engines manufactured by Pratt & Whitney, General Electric, and Rolls-Royce and each basic engine type has several versions with sea-level static (i.e., zero forward airspeed) takeoff thrust ratings from 40,000 to over 52,000 lb, Figure 17.15 provides a good overall understanding of the variation of maximum available cruise thrust and specific fuel consumption (sfc) with Mach number and altitude. The cruise thrust is shown as a fraction of the sea-level static takeoff thrust. The large decrease in available thrust as speed increases is due to the increase in ram or momentum drag. The thrust at a given Mach number is directly proportional to ambient pressure in the isothermal atmosphere. The decrease in thrust with altitude increase is slightly less rapid at lower altitudes because the temperature drops as the altitude increases. Specific fuel consumption improves at higher altitudes, at a given Mach number, until the isothermal region is reached. Then the sfc is generally unaffected by further altitude increase. At very high altitudes, the decreasing Reynolds number on the gas turbine engine blades may cause a slight increase in sfc. The sfc is raised moderately by higher cruise speeds.

Although Figure 17.15 is not exact for any engine, it is a good approximation to all the engines in this class. Engine types entering service in the late 1980s can be expected to have cruise fuel consumptions lower by about 6% to 10%. Note that the

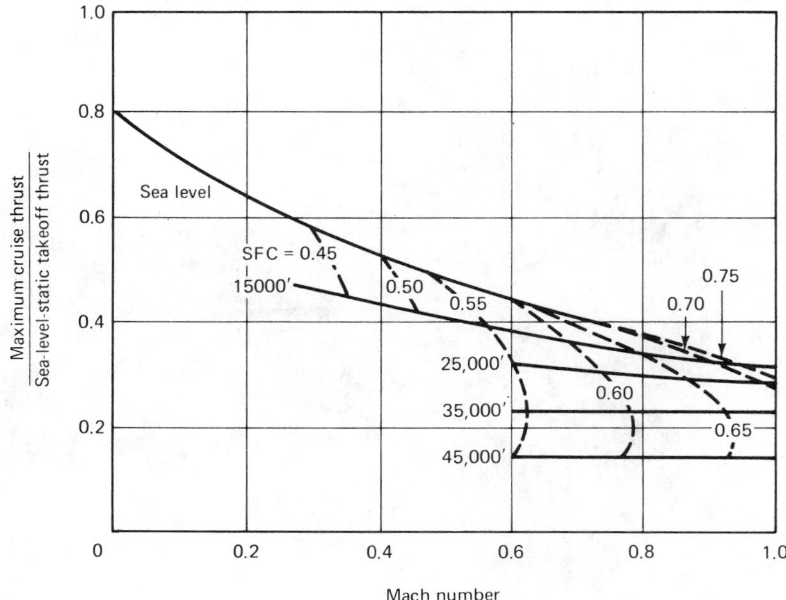

Figure 17.15 Maximum cruise thrust and specific fuel consumption variation with altitude and Mach number for typical high-bypass-ratio (5 to 6) turbofans.

cruise sfc at a representative cruise speed and altitude, such as a Mach number of 0.8 and 35,000 ft, is the important value. The sea-level static value is often quoted and guaranteed by engine manufacturers because it is easy to measure on a ground test stand. Because of the large effect on sfc of very large speed changes, the static value is useful only as a reference. The engines in Figure 17.15 have bypass ratios of 5 to 6. Engines with lower bypass ratios have a smaller variation of sfc and thrust with speed, so only cruise comparisons are meaningful at all.

Figure 17.16 shows the variation in installed takeoff thrust with Mach number for pure turbojets and for low- and high-bypass-ratio turbofans. The data are shown in a generalized form, the ratio of installed thrust at a given Mach number to the engine specification static (zero speed) thrust. The ratio is shown as 0.97 at zero Mach number to account for typical engine installation losses, such as inlet total pressure loss, nozzle loss, engine air bleed for cabin pressurization, and power extraction for hydraulic pumps and electrical generators. Because the lower-bypass-ratio engines have less airflow per pound of thrust, they suffer less ram drag as speed increases. Therefore, the rate of thrust decrease with Mach number is less. Although the thrust decrease with speed for individual engine types may vary slightly from the curves in Figure 17.16, the effects of bypass ratio are generally correct as shown.

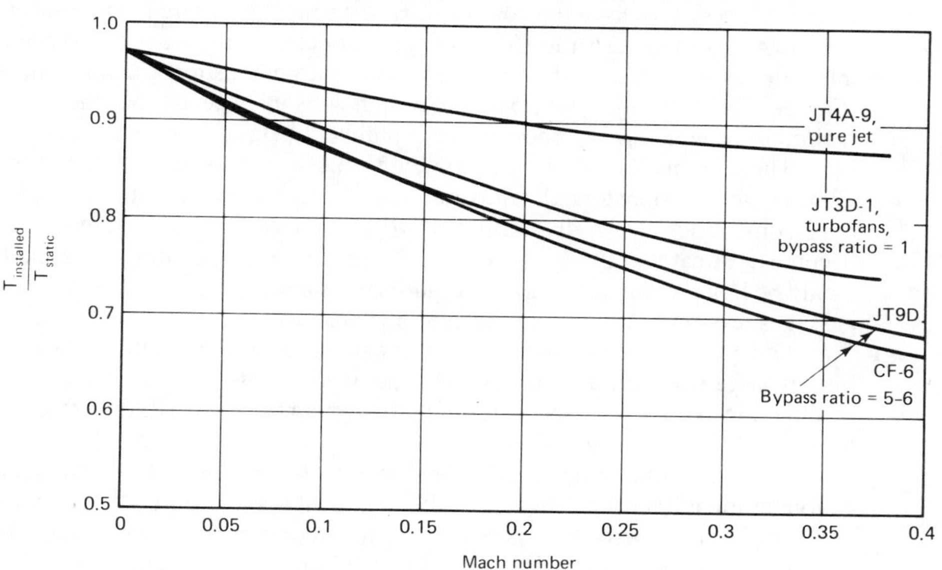

Figure 17.16 Variation of takeoff thrust with Mach number.

SPEED LIMITATIONS OF GAS TURBINES: RAMJETS

The maximum thrust of a turbojet or turbofan with a given airflow is limited by the maximum allowable turbine inlet temperature. Excessive temperature weakens the turbine materials and can cause blade or disc failure. Most of the wear on turbine engines is due to fatigue from stress and temperature cycles, with the greatest damage occurring during high-thrust takeoff operation. The requirement for long engine life is an important factor in setting the maximum turbine inlet temperature. To increase the allowable temperature, cooling air is often circulated through passages in the turbine blades.

We have mentioned that some of the incoming air compression occurs in the inlet diffuser, which greatly slows the air velocity. Neglecting the low remaining velocity at the engine face, the total temperature and pressure can be assumed to be representative of the static values. The total temperature and pressure at the engine compressor entrance can be determined by equations 7.18 and 7.19, respectively. As the flight Mach number increases, the inlet total temperature rises (see Figure 19.2). Then the air temperature after passing through the compressor is also increased, and the amount of fuel that can be added and burned to reach the limiting temperature is reduced. At a high enough Mach number, the fuel allowed becomes so small that no net thrust can be developed. Even at a lower Mach number than this, the engine becomes hopelessly inefficient. Turbojet engines are limited to Mach numbers between 3.0 and 3.5.

As Mach number increases and the total inlet temperature rises, so does the total inlet pressure. As the Mach number approaches 3, the inlet pressure rise is sufficient to permit the compressor to be omitted. The engine will still develop thrust. Of course, without the need to drive the compressor, no turbine is required. The resulting engine is known as a *ramjet*. A ramjet is basically a duct with the front end shaped to be the inlet, the aft end designed as a nozzle, and the combustion chamber in the middle (Figure 17.17). Ramjets will operate at high subsonic speeds, but the thrust specific fuel consumption is very poor at Mach numbers below 3 to 4.

The elimination of the turbine eases the materials temperature problem and allows higher combustor exit temperatures. As Mach number increases further into the hypersonic range, generally defined as $M > 5.0$ (see Chapter 19), temperature again becomes critical and the walls of a ramjet require active cooling. The likely coolant would be liquid hydrogen, which would serve as a heat sink as it flows through cooling passages around the ramjet to the fuel injectors where it serves as a fuel.

One problem with a ramjet is that it cannot function until the pressure rise in the inlet is large (i.e., until a high speed is attained). Thus a ramjet can only be used in conjunction with a low-speed propulsion system such as a turbojet or a rocket to provide the initial acceleration.

At its high operating speeds, the ramjet provides a large gross thrust and suffers a large momentum drag. Therefore, the net thrust is the difference between two large numbers. A small percentage loss in gross thrust or increase in inlet drag would cause a large loss in net thrust. It is essential to minimize the losses in inlet compression in

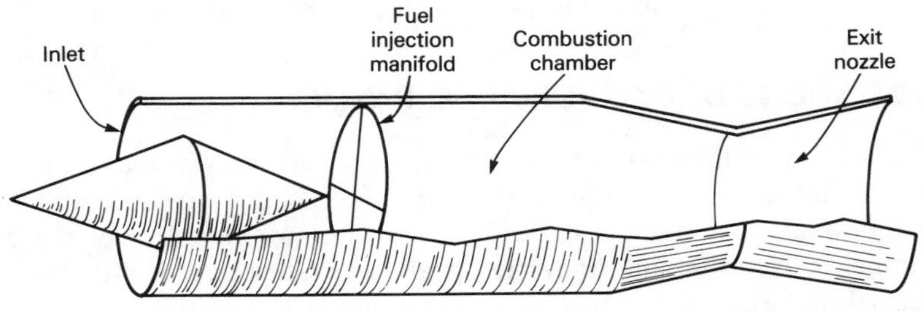

Figure 17.17 Schematic drawing of a ramjet.

order to maximize the gross thrust. One way to achieve this is to place the ramjet so that the intake air has already been partly compressed by the inevitable shock waves on the wing or body. This also saves weight by shortening the required diffuser length. Similarly, the aft end of the body may provide a surface that acts as part of the nozzle, with the exhaust gas flowing over the carefully shaped body surface. In effect, the wing and body surfaces become part of the propulsion system. Such a design is so integrated that the whole vehicle must be designed together, even more than for turbine-powered aircraft. Propulsion system and airframe can no longer be separated from each other.

As Mach number increases in the hypersonic region, the temperature rise due to inlet compression becomes excessive. The high static temperature can cause dissociation and ionization of the air within the combustor, a process that absorbs energy and reduces the temperature increase sought from the burning of the fuel. To solve this problem, the *supersonic combustion ramjet,* or *scramjet,* has been developed. The inlet flow is decelerated only as much as required to obtain the necessary pressure rise, reducing the static temperature increase. The flow remains supersonic passing through the burner. This requires very rapid mixing and burning of the fuel, a problem still requiring intensive research. In addition to avoiding dissociation, the scramjet reduces the total pressure losses suffered in the inlet, and permits shorter, lighter inlets and nozzles. Combustor length is increased, however.

Practical use of ramjets has been limited to a few missile applications. Much research was focused on ramjets in the 1960s when a hypersonic airplane was seriously proposed. After this project was discontinued, hypersonic research was reduced to a low level until the 1980s. At that time, the National Aerospace Plane (NASP) was initiated. This program seeks a one-stage-to-orbit aircraft that can takeoff from a runway, accelerate to hypersonic and orbital speeds ($M = 25$), and then go into orbit. A commercial aircraft cruising at Mach numbers of about 7 was also discussed. The high-temperature problems of the airframe and the engine, the efficiency of the ramjet, the complex multiple propulsion system, and the hypersonic aerodynamics make an economically viable commercial version extremely unlikely and the military version difficult. Much useful knowledge will be obtained, however, and the military version may be useful even if two stages are required.

One problem with hypersonic ramjets is that available ground test facilities are limited to Mach numbers of about 7 or 8. Operation at higher Mach numbers will have to be estimated from theory and confirmed in flight test. Such tests are planned for the X-30, an experimental version of an aerospace plane. The X-30 is scheduled for flight in the early 1990s, but the optimistic schedule will almost certainly be extended.

PROPELLERS

A propeller is a group of rotating blades, or "wings," so oriented that the direction of the resultant "lift" is primarily forward. The lift produced at each airfoil section is perpendicular to the effective resultant air velocity approaching the blade at that point (Figure 17.18). This velocity is the vector sum of the velocity due to the forward speed of the airplane, the induced flow in the plane of the propeller, and the velocity due to the rotation of the blade. The lift vector is therefore inclined at an angle to the flight direction and it is the component perpendicular to the flight direction that creates most

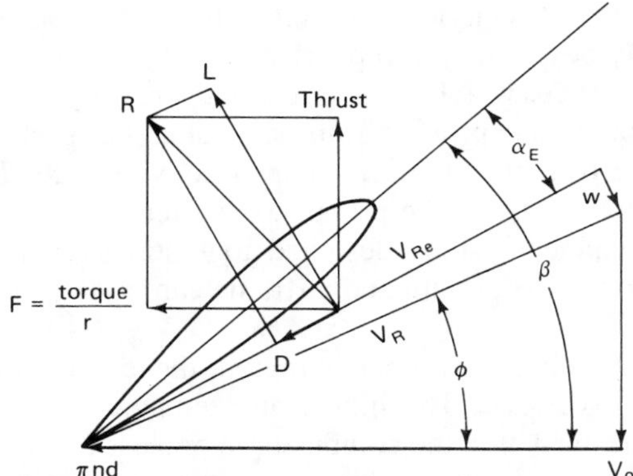

n, propeller revolutions per second;
d, diameter at the airfoil station;
β, blade angle at airfoil station;
w, induced velocity;
V_R, resultant air velocity before considering induced flow;
V_{Re}, effective resultant air velocity including induced flow.

Figure 17.18 Propeller blade element.

of the torque that the engine must overcome. In addition, the skin friction drag of the blades adds to the power the engine must supply. The flow induced into the propeller plane is analogous to the downwash on a wing.

The detailed analysis of a propeller is very complex and is based on blade element theory. In this theory, each element of a blade is regarded as an airfoil moving in its appropriate manner. The lift on each airfoil is associated with circulation around the blade. Due to the variation of this circulation along the blade from root to tip, trailing vortices will spring from the blade and pass downstream with the fluid in approximately helical paths. Each blade experiences a flow field modified by the passage of the previous blade and the vortex pattern it left in the stream.

One interesting result can be seen from Figure 17.18. At a given V_0, n, and d, a particular value of airfoil angle β is required to produce the effective angle of attack α_E that produces the best value of section lift/drag ratio L/D. Since the d varies at every airfoil station from the root to the tip of the blade, $V_0/\pi nd$ will also vary, and a different blade angle will be required at every radial station. For this reason, propeller blades are twisted to match the best effective angle of attack at every station along the blade for some flight condition. For other flight conditions, such as different airplane speeds V_0 or different propeller rotational speeds n, the blade angle will not be optimum. Therefore, on all but the most simple aircraft, the propeller blade angle is adjustable. The adjustment may be accomplished by a simple two-position design or by a continuously varying system operated by a governor that varies blade angle to maintain a constant engine speed. The blade angle is usually defined by the airfoil angle at the station three-fourths of the distance from the propeller axis to the tip.

Simple Momentum Theory

A simple method of considering the operation of a propeller depends on studying the momentum and kinetic energy of the system. The propeller is assumed to have a large number of blades, so that it becomes effectively an "actuator" disk with the thrust uniformly distributed over the disk. Slipstream rotation is ignored. The axial velocity of the fluid is continuous in passing through the propeller disk to maintain continuity of the flow. The pressure of the fluid receives a sudden increment, Δp, at the propeller disk. Δp is equal to the thrust on a unit area of the disk, and a slipstream of increased axial velocity is formed behind the propeller.

Consider an actuator disk of area A in a stream with velocity V_0, as in Figure 17.19. On approaching the disk, the axial velocity rises to $V_0 + aV_0$ and the pressure falls from p_0 to p_1. The velocity is constant through the disk, but rises to $V_0 + bV_0$ in the final slipstream. The pressure rises to $(p_1 + \Delta p) = p_2$ immediately behind the disk and then falls to its original value, p_0. It is legitimate to apply Bernoulli's equation to the flow before and behind the disk separately, but not through the disk.

$$\text{original total pressure } p_{T_1} = \underbrace{p_0 + \frac{\rho}{2}V_0^2}_{\text{freestream}} = \underbrace{p_1 + \frac{\rho}{2}(V_0 + aV_0)^2}_{\substack{\text{just ahead of the disk} \\ \text{(plane of the propeller)}}} \tag{17.2}$$

$$\text{final total pressure } p_{T_2} = \underbrace{p_0 + \frac{\rho}{2}(V_0 + bV_0)^2}_{\text{far behind the propeller}}$$
$$= \underbrace{(p_1 + \Delta p) + \frac{\rho}{2}(V_0 + aV_0)^2}_{\text{just behind the disk}} \tag{17.3}$$

Using the freestream definition for p_{T_1} and the far-field value for p_{T_2},

$$\Delta p = p_{T_2} - p_{T_1} = \frac{\rho}{2}(2bV_0^2 + b^2V_0^2) = \rho V_0^2\left(b + \frac{b^2}{2}\right) = \rho V_0^2\left(1 + \frac{b}{2}\right)b \tag{17.4}$$

Also, the thrust is the rate of change of momentum from far ahead of the disk to far behind it:

$$T = \Delta pA = \rho(V_0 + aV_0)AbV_0 \tag{17.5}$$

so

$$\Delta p = \rho(V_0 + aV_0)bV_0 = \rho V_0^2(1 + a)b \tag{17.6}$$

Comparing 17.4 with 17.6, it is seen that

$$a = \frac{b}{2} \tag{17.7}$$

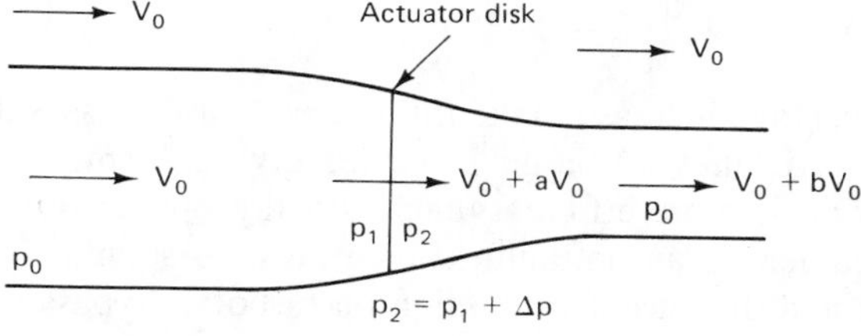

Figure 17.19 Actuator disk flow.

Thus half of the added velocity in the final slipstream occurs ahead of the propeller and half behind it. Substituting equation 17.7 in 17.5 yields

$$T = 2A\rho V_0^2(1 + a)a \tag{17.8}$$

The increase of kinetic energy of the fluid per unit time in the slipstream is the difference between the kinetic energy in the final slipstream and the kinetic energy of the same mass of air far ahead of the propeller. Letting M be the mass flow through the actuator disk per unit time,

$$\begin{aligned}\Delta\text{K.E.} &= \frac{M[V_0(1 + b)]^2}{2} - \frac{MV_0^2}{2} \\ &= \frac{A\rho V_0(1 + a)}{2}\{[V_0(1 + b)]^2 - V_0^2\} \\ &= \frac{A\rho V_0(1 + a)}{2}\left[V_0^2\left(1 + \frac{b}{2}\right)2b\right]\end{aligned} \tag{17.9}$$

Substituting $b = 2a$, we obtain

$$\begin{aligned}\Delta\text{K.E.} &= \frac{A\rho V_0^3(1 + a)^2(4a)}{2} \\ &= 2A\rho V_0^3(1 + a)^2 a\end{aligned} \tag{17.10}$$

Substituting equation 17.8 for thrust in 17.10, we obtain

$$\Delta\text{K.E.} = TV_0(1 + a) \tag{17.11}$$

This increase in fluid kinetic energy is the work done on the fluid by the propeller. The useful work done is thrust times distance. In unit time, this is TV_0. The ideal efficiency η is then

$$\begin{aligned}\eta &= \frac{\text{output}}{\text{input}} = \frac{TV_0}{\Delta\text{K.E.}} \\ &= \frac{TV_0}{TV_0(1 + a)} \\ &= \frac{1}{1 + a}\end{aligned} \tag{17.12}$$

Thus the greater the percentage increase in the fluid velocity as it passes through the propeller, the lower the efficiency. A large propeller giving a small velocity increase to a large amount of air is more efficient than a small propeller. Of course, the large propeller will be heavier, so an optimum choice must be sought. Note the similarity between this result and the earlier discussion on turbofan bypass ratio.

Total Propeller Efficiency

The momentum theory of the propeller gives an ideal efficiency that can never be realized in practice since the theory ignores:

1. The friction drag of the blades.
2. The kinetic energy of the rotation of the slipstream.
3. The fact that thrust is not uniformly distributed over the blades.

The maximum actual propeller efficiency is usually about 90%. When the propeller is installed on an airplane and the slipstream drag on the nacelle or fuselage and wing is subtracted from the thrust, the propulsive efficiency does not exceed 87%.

The thrust horsepower provided by the propeller is

$$\text{thp} = \frac{TV_0}{550} = \text{bhp} \cdot \eta \tag{17.13}$$

where

T = thrust (lb)
V_0 = velocity (ft/s)
bhp = engine brake horsepower
550 = conversion factor for horsepower (i.e., ft-lb/s per horsepower)

In practice, propeller efficiency can be determined from a chart, the basis of which can be a great many detailed calculations or many test points. Such a propeller chart can be presented in several forms. Figure 17.20 shows a typical chart in the form of C_p versus J for contours of equal propulsive efficiency η. Contours of equal blade angle are also shown. C_p is a nondimensional *power coefficient*, and J is known as the *advance ratio*. They are defined as

$$C_p = \frac{\text{bhp} \times 550}{\rho n^3 D^5} \quad \text{and} \quad J = \frac{V_0}{nD}$$

D is the propeller diameter and n is propeller revolutions per second. From Figure 17.18 it is seen that $J = \pi \tan \phi$ at the propeller tip and is related to the angle at which the resultant airflow approaches the propeller blade. It can be shown that the effective average lift coefficient of a propeller blade is proportional to the ratio of C_p to J. Thus these two parameters define the geometry of the blade–airflow system and the blade lift coefficient. They therefore determine the efficiency, including the effects of inflow, rotational energy in the slipstream, and viscous losses. Charts such as Figure 17.20 are valid only for the blade planform and number of blades for which the chart is constructed. The data of Figure 17.20 apply to a four-bladed propeller with a blade activity factor of 135. The activity factor is a parameter dependent on blade width weighted for the greater effectiveness of faster-moving sections near the tip.

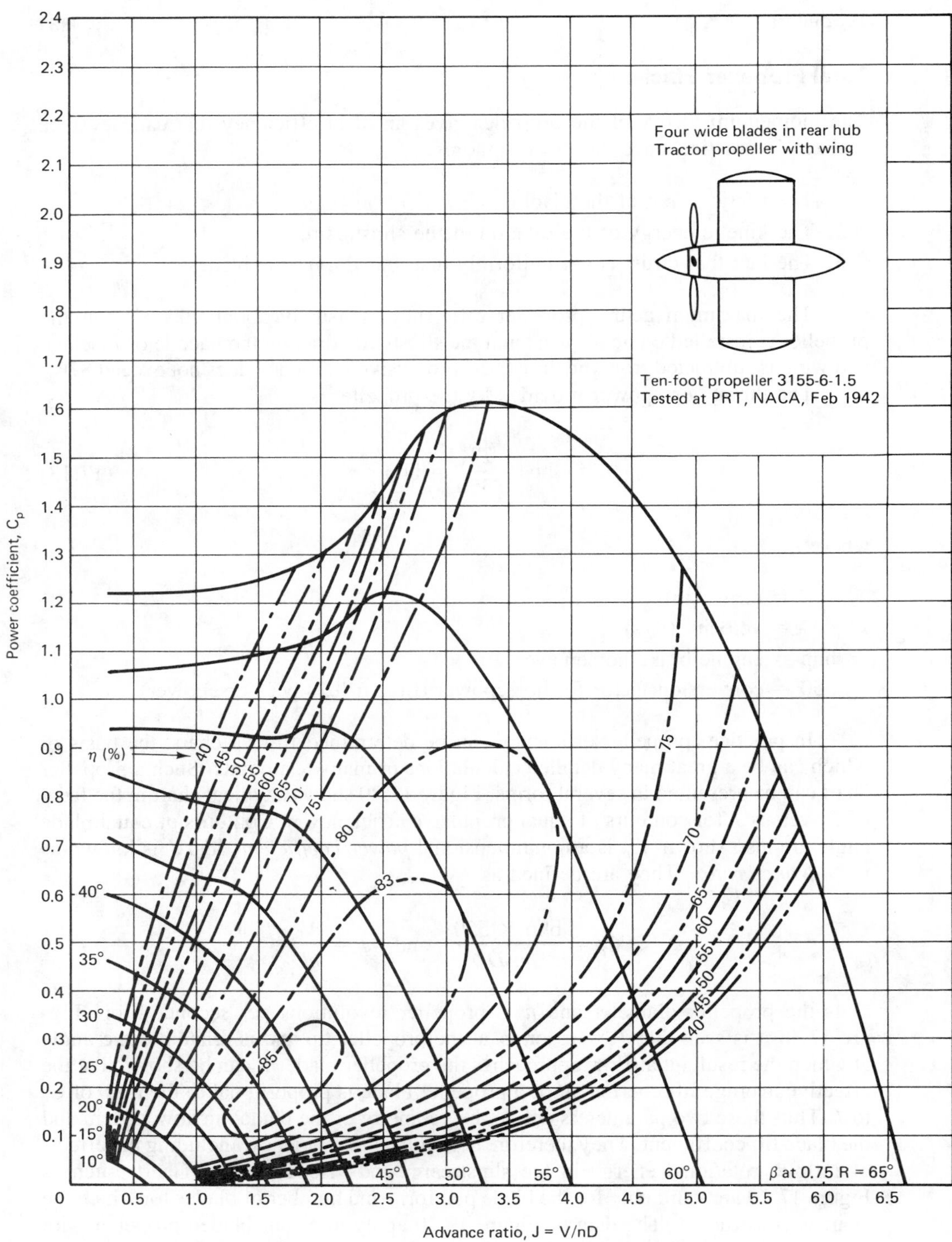

Figure 17.20 Propeller chart for a four-bladed tractor propeller with a blade activity factor of 135. Reprinted with permission from Bierman, Gray, and Maynard, "Wind-Tunnel Tests of Single and Dual-Rotating Tractor Propellers of Large Blade Width," NACA Wartime Rep. L-286, Sept. 1942.

There are simple methods of correcting the data from a particular propeller chart to apply to propellers of different blade planforms and numbers of blades.

In addition to the basic propulsive efficiency described here, it may be necessary to apply a compressibility correction to account for higher blade drag if the blade tip Mach numbers approach 1.0. The blade lift coefficient determines the critical tip Mach number just as it does for wings. Recent developments in propeller technology have shown the possibility of delaying this tip loss by the use of sweptback blades. For large, high-speed turboprops, a special form of the propeller known as a propfan shows promise of maintaining good efficiency up to airplane cruise Mach numbers of 0.8, whereas conventional propellers have a serious efficiency decrease above Mach numbers of 0.65 to 0.70. Propfans feature 8 to 10 very wide blades of sweptback planform (Figure 17.21).

Another form of the high-solidity propeller (i.e., many wide blades) is the General Electric unducted fan (UDF). The UDF features an aft-mounted fan running on the same shaft as the turbine. The gear box between the engine and propeller is eliminated, resulting in substantial savings in weight and cost. On the other hand, the fan and the turbine cannot operate at their individual optimum speeds. Propfans in either form offer potential fuel saving of 15% to 25%. Since the efficiency advantage is greater at lower speeds, short-range aircraft that fly a larger percentage of their mission at lower speeds will have larger fuel savings. Figure 17.22 shows a flight demonstration version of the General Electric GE-36 unducted fan mounted on a McDonnell Douglas MD-80.

Figure 17.21 Proposed propfan. (Courtesy of Hamilton Standard.)

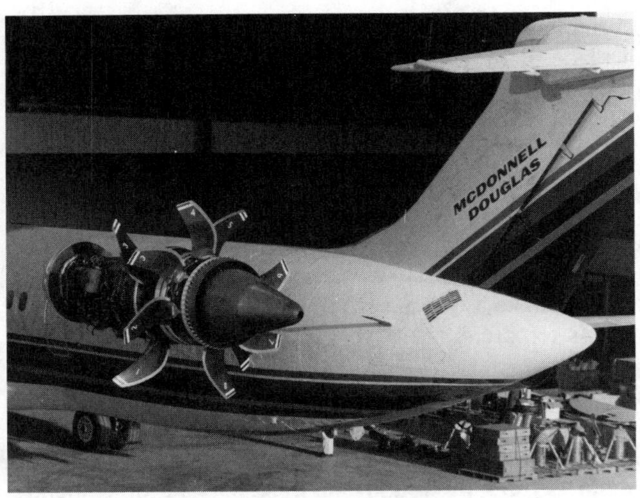

Figure 17.22 General Electric unducted fan during installation on a McDonnell Douglas MD-80. The nacelle cowling and some of the blades on the counterrotating eight-bladed fans have not yet been installed. (Courtesy of McDonnell Douglas Corporation.)

OVERALL PROPULSION EFFICIENCY

We have discussed the improvement in specific fuel consumption of a turbojet or turbofan engine that results from imparting a lower additional velocity to a larger amount of air. The ideal efficiency of a propeller has been found to vary in the same manner. We can now address a more general definition of propulsion efficiency. The *overall efficiency*, η_{overall}, of a propulsion system is equal to the ratio of the useful work performed by the system to the heat energy available from perfect combustion of the fuel. The useful work per unit time is the product of the net thrust, T, and the vehicle velocity through the atmosphere. The heat energy available is equal to the product of the fuel flow per unit time, that is, Tc for turbojets and turbofans, and the *heat energy available per unit weight of fuel, h_f,* expressed in mechanical-energy units. Then

$$\eta_{\text{overall}} = \frac{\text{useful work}}{\text{heat energy input}} = \frac{TV_0}{Tch_f} = \frac{V_0}{ch_f} \qquad \text{jets} \tag{17.14}$$

The overall efficiency is itself composed of two parts, a propusive efficiency and a thermal efficiency so that $\eta_{\text{overall}} = \eta_{\text{prop}} \times \eta_{\text{th}}$. The propulsive efficiency, η_{prop}, is the ratio of the useful work to the mechanical energy produced in the fluid. This mechanical energy is the increase in kinetic energy of the fluid in unit time (i.e., the difference between the kinetic energy of the jet exhaust and the free stream). Then

$$\eta_{\text{prop}} = \frac{\text{useful work}}{\substack{\text{mechanical energy}\\ \text{produced in system}}} = \frac{TV_0}{(W/g)[(V_e^2 - V_0^2)/2]} \tag{17.15}$$

From equation 17.1, assuming the ideal condition for which $p_e = p_0$, $T = W/g(V_e - V_0)$. Substituting in equation 17.15 and defining $a = 1/2[(V_e - V_0)/V_0]$, that is a is half of the fractional increase in gas velocity between V_0 and V_e,

$$\eta_{\text{prop}} = \frac{TV_0}{T(V_e + V_0)/2} = \frac{2}{V_e/V_0 + 1} = \frac{1}{1 + a} \tag{17.16}$$

This result agrees with the ideal efficiency for the propeller shown in equation 17.12.

The thermal efficiency, η_{th}, is the ratio of the mechanical energy produced in the system to the heat energy input.

$$\eta_{\text{th}} = \frac{\text{mechanical energy produced in system}}{\text{heat energy input}} = \frac{(W/g)[(V_e^2 - V_0^2)/2]}{Tch_f}$$

Making the same substitution for thrust, T, as used before,

$$\eta_{\text{th}} = \frac{T(V_e + V_0)/2}{Tch_f} = \frac{(V_e + V_0)/2}{ch_f} \tag{17.17}$$

Returning now to η_{overall}, equation 17.14 shows that $V_0/c = \eta_{\text{overall}}h_f$. Substituting in the range equation for jet aircraft (equation 15.32),

$$R, \text{ft} = \frac{V}{c}\frac{L}{D}\log_e\frac{W_i}{W_f} = \eta_{\text{overall}}h_f\frac{L}{D}\log_e\frac{W_i}{W_f} \qquad \text{jets} \tag{17.18}$$

In this form, it is seen clearly that range depends on overall propulsion efficiency, fuel heating value, and the lift-drag ratio.

As always, it is very important to use consistent units. Since specific fuel consumption, c, is usually expressed as (lb fuel)/(lb thrust-h) and velocity is usually a distance flown per hour, using hours as the time unit and pounds as the force unit is probably best.

The heating value of JP-4 jet fuel is about 18,400 Btu/lb. One Btu = 778 ft-lb, so h_f = 18,400 × 778 = 14,315,200 ft-lb/lb. Using feet as the distance measure and hour as the time unit requires that V be expressed as feet per hour. Thus the range in equation 17.18 is in feet.

To determine the overall propulsion efficiency of a jet transport equipped with the most efficient high-bypass-ratio turbofan engines available in the late 1980s, assume a cruise velocity of 475 knots (M = 0.83) and a specific fuel consumption of 0.57. Then, from equation 17.14,

$$\eta_{\text{overall}} = \frac{V_0}{ch_f} = \frac{475 \times 6083\ (\text{ft/n. mi.})}{(0.57)(14{,}315{,}200)} = 0.354$$

The corresponding equation for the overall propulsion efficiency of a propeller-driven aircraft is

$$\eta_{\text{overall}} = \frac{\text{useful work}}{\text{heat energy input}} = \frac{TV_0}{\dfrac{T(V_0/3600)}{\eta(550)}ch_f} = \frac{(3600)(550)\eta}{ch_f}$$

$$= 1{,}980{,}000\frac{\eta}{ch_f} \qquad \text{propellers} \tag{17.19}$$

In this derivation for the propeller case, V_0 is in feet per hour and c is the usual specific fuel consumption in pounds of fuel per horsepower per hour. The factor 3600 is introduced to change the speed to feet per second so that the familiar factor 550 properly calculates horsepower upon which c is based. Note that speed no longer appears in the equation as we would expect from the propeller range equation, 15.33. As an example of the overall propulsion efficiency of a propeller-driven system, assume a propeller efficiency, η, of 0.85. h_f is still 14,315,200 ft-lb/lb. First, assume a value of $c = 0.48$, typical of existing turboprop engines. Then

$$\eta_{\text{overall}} = 1{,}980{,}000 \frac{0.85}{0.48(14{,}315{,}200)} = 0.245$$

This value is lower than the most advanced turbofan efficiencies. However, advanced-technology turboprop engines are expected to have cruise specific fuel consumption values of about 0.3. Then

$$\eta_{\text{overall}} = 1{,}980{,}000 \frac{0.85}{0.30(14{,}315{,}200)} = 0.392$$

and the turboprop is superior in cruise overall propulsion efficiency by about 10%. One advantage of the propeller, which in its high-speed form is really the propfan previously discussed, is that its overall propulsion efficiency is little changed by lower speed flight. The turbofan deteriorates rapidly with speed. Thus, at reduced speeds in takeoff and landing operations, holding around airports and climb, the gains due to the propfan are much larger. This is very important for shorter-range flights, but less important on long flights.

Substituting the value for η/c from equation 17.19 in the range equation for propeller-driven aircraft, equation 15.33, rederived to obtain range in feet, yields

$$R, \text{ft} = \eta_{\text{overall}} h_f \frac{L}{D} \log \frac{W_i}{W_f} \qquad \text{propellers} \qquad (17.20)$$

The range equations for jets and propellers are the same when expressed in terms of overall propulsion efficiency. It should be remembered, however, that the definition of η_{overall} contains velocity for the jet case but not for the propeller aircraft.

Figure 17.23 shows the variation of η_{overall} with Mach number for several types of power plant. Efficiency is improved about 65% from the original transport airplane turbojet engines to the best types flying in the last half of the decade of the 1980s. Another 15% to 25%, depending on the speed, can be expected from the propfans.

Example 17.1

A turboprop-powered airplane is flying at 304 knots at 20,000 ft on a standard day. Each of the four 14-ft-diameter propellers is driven by an engine delivering 1920 shaft horsepower.

The propeller speed is 1050 rpm. Assume that the propellers are four bladed with a blade activity factor of 135, so that the propeller chart (Figure 17.20) can be used directly. Determine:

(a) Propulsive efficiency.
(b) Thrust per propeller.
(c) Ideal efficiency from simple momentum theory.

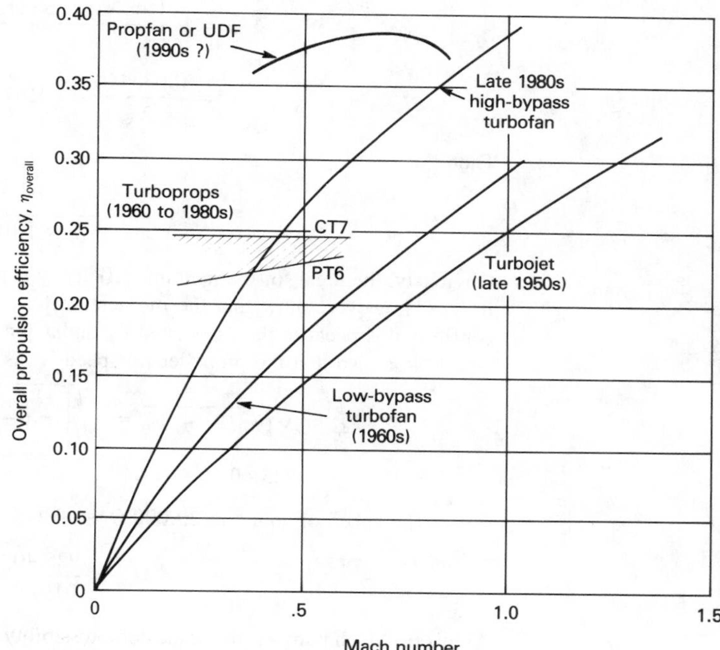

Figure 17.23 Variation of overall propulsion efficiency with Mach number for various types of engines.

Solution:

(a) To use the propeller chart, we must calculate the power coefficient C_p and J. At 20,000-ft altitude, $\rho = 0.0012673$ slug/ft^3.

$$C_p = \frac{\text{bhp} \times 550}{\rho n^3 D^5} = \frac{(1920)(550)}{(0.0012673)(1050/60)^3(14)^5} = 0.289$$

Power is converted to ft-lb/s. Therefore, n must be in revolutions per second and V in ft/s.

$$J = \frac{V}{nD} = \frac{(304)(1.69)}{(1050/60)(14)} = 2.10$$

From Figure 17.20, $\underline{\eta = 0.851}$.

$$\text{(b) } T = \frac{550 \text{ bhp} \cdot \eta}{V} = \frac{(550)(1920)(0.851)}{(304)(1.69)} = \underline{1751 \text{ lb}}$$

(c) The ideal efficiency from simple momentum theory is given by equation 17.12. To find a, use equation 17.8, which expresses thrust in terms of a.

$$T = 2A\rho V_0^2(1 + a)a$$

$$1751 = 2\left[\frac{\pi}{4}(14)^2\right](0.0012673)(304 \times 1.69)^2(a + a^2)$$

$$= 102{,}985(a + a^2)$$

$$a^2 + a - 0.017 = 0$$

$$a = \frac{-1 \pm \sqrt{1 - 4(1)(-0.017)}}{2} = \frac{-1 \pm 1.0334}{2}$$

$$= \frac{0.03344}{2} = 0.0167$$

Then

$$\eta_{\text{ideal}} = \frac{1}{1 + a} = \frac{1}{1.0167} = \underline{0.984}$$

Obviously, the ideal, or momentum, efficiency is not very useful for performance purposes. It does, however, show that the momentum loss is small in this case. The propeller chart confirms this because the associated C_p and J put this point near peak efficiency.

Note also that the propeller tip speed is, as shown in Figure 17.18,

$$V_{\text{tip}} = \sqrt{(\pi n D)^2 + V_0^2} = \sqrt{\left[\pi\left(\frac{1050}{60}\right)14\right]^2 + [304 \times 1.69]^2}$$

$$= 925.40 \text{ ft/s}$$

The speed of sound at 20,000 ft is 1036.9 ft/s, so the propeller tip Mach number is

$$M_{\text{tip}} = \frac{925.40}{1036.9} = 0.89$$

At this tip Mach number, the blade compressibility loss is small. If the tip Mach number were much higher, a compressibility correction to η would be required.

ROCKET ENGINES

The gas turbine and reciprocating internal combustion engines are both air-breathing power plants. They ingest air from the surrounding atmosphere and use the oxygen in the air as the oxidizer for the chemical burning process that extracts the heat energy from the fuel. Only the fuel is carried aboard the vehicle. A *rocket* is a device that burns fuel and an oxidizer, both of which are carried by the vehicle. The forward thrust is obtained by applying a rearward momentum to the products of combustion, the mass of which is clearly limited by the weight-carrying capacity of the vehicle. Therefore, to obtain as large a thrust as possible with a given mass flow, the rearward velocity must be as large as possible. A rocket is the only means of obtaining thrust in a vacuum or near vacuum such as exists outside or near the outer edge of the atmosphere.

There are two basic types of rockets, liquid propellant rockets and solid propellant rockets. *Liquid propellant rockets* employ liquid propellants that are fed under pressure from tanks into the combustion chamber. A schematic diagram of a liquid propellant system is shown in Figure 17.24. The propellants consist of a liquid oxidizer, such as liquid oxygen, red fuming nitric acid, or hydrogen peroxide, and a liquid fuel (e.g., gasoline, ammonia, or liquid hydrogen).

In the combustion chamber the propellants react to form hot gases at high pressure, which in turn are accelerated and ejected at a high velocity through a nozzle. The momentum imparted to the gases per unit time is equal to the thrust developed

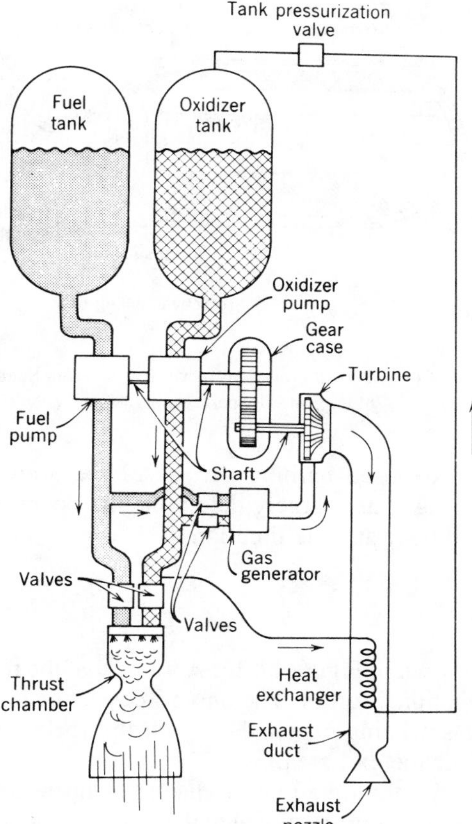

Figure 17.24 Simplified schematic diagram of a liquid propellant rocket system. From Sutton and Ross, *Rocket Propulsion Elements,* 1976. Reprinted by permission of John Wiley & Sons, New York.

by the rocket. A liquid rocket propulsion system is relatively complicated since it requires several precision valves, a complex feed mechanism with propellant pumps and turbines, or a propellant pressurizing device.

Solid propellant rockets contain all the propellant within the combustion chamber. This type of rocket is simple since a feed system, valves, or pumps are not required. Solid rockets are usually limited to short-duration firing (1⁄10 to 25 s). Long-duration solid rockets require excessively large and heavy combustion chambers. Solid propellant rockets have been widely used for jet-assisted takeoff (JATO) purposes for aircraft with marginal takeoff performance, as well as for the initial launch and acceleration of missiles and spacecraft. Figure 17.25 shows a cross section of a solid propellant rocket motor.

The solid propellant charge contains all the chemical elements necessary for complete combustion. The propellants usually have a plasticlike caked appearance and burn on their exposed surfaces to form hot exhaust gases at a nearly constant rate. The body of the propellant is called the *grain*. The grain may be a heterogeneous mixture of several chemicals, for example a mixture of oxidizing crystals of perchlorate in a matrix of an organic, plasticlike fuel such as asphalt. It may be a homogeneous charge of special chemicals, such as modified nitrocellulose-type gun-

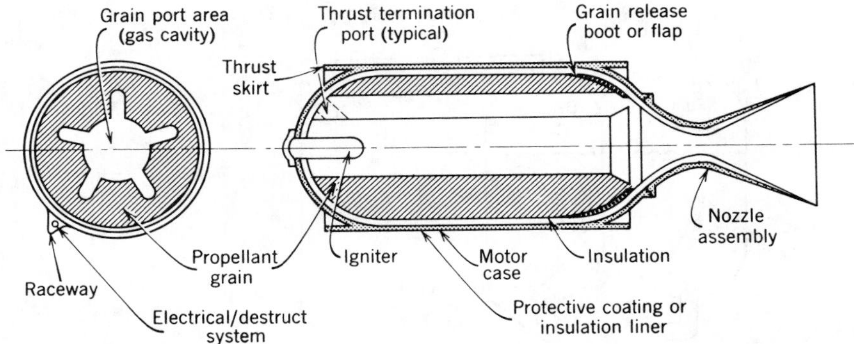

Figure 17.25 Sectional view of typical solid propellant rocket motor. From Sutton and Ross, *Rocket Propulsion Elements,* 1976. Reprinted by permission of John Wiley & Sons, New York.

powder. The shape, size, and exposed burning surface of the grain influence the burning characteristics of the rocket and largely determine the operating pressure in the combustion chamber, the thrust, and the duration.

ROCKET MOTOR PERFORMANCE

Consider a rocket in flight as shown in Figure 17.26. Assume that the rocket is operating in a vacuum without gravitational forces. The only forces acting on the rocket are the reaction to the exhaust gases being expelled through the rocket nozzle and the nozzle exit pressure acting over the exit area, S_e.

From Newton's second law, the time rate of change of linear momentum of a body, in this case the exhaust gases, is proportional to the force acting on it. Therefore, the force on the exhaust gases is

$$F_1 = V_e \frac{dm}{dt}$$

where

F_1 = force, N (lb)
dm/dt = mass flow rate of the exhaust gases, kg/s (slugs/s)
V_e = exhaust gas velocity relative to the rocket, m/s (ft/s)

Since a rocket is designed to maintain essentially constant temperature and pressure in the combustion chamber or reservoir, the flow is similar to the supersonic wind

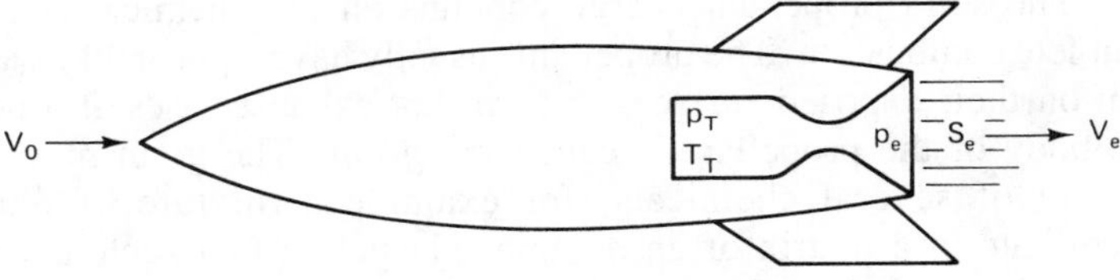

Figure 17.26 Rocket in flight.

tunnel flow discussed in Chapter 7. The combustion chamber pressure is always high enough to obtain sonic velocity at the throat or minimum section of the nozzle. The equations of Chapter 7 apply, and for a given exit area the exit velocity is constant.

From Newton's third law, the force imposed on the exhaust gas by the rocket engine is equal to the force exerted on the engine by the exhaust gases. The latter is the major part of the thrust. The other component is the pressure term given by $p_e S_e$, so the total thrust, in a vacuum, is

$$F = V_e \frac{dm}{dt} + p_e S_e \tag{17.21}$$

If the rocket motor were immersed in the atmosphere at $V_0 = 0$ and with zero thrust ($V_e = 0$), atmospheric pressure p_0 would act on all surfaces and the net pressure effect would be zero. If the rocket is then ignited, all the pressures remain the same except at the nozzle exit, where the pressure will be determined by the throat area, the exit area S_e, and the combustion chamber (reservoir) pressure in accordance with the one-dimensional compressible fluid equations in Chapter 7. The additional force due to pressure is then

$$\Delta F = (p_e - p_0)S_e$$

and the total thrust is

$$F = V_e \frac{dm}{dt} + (p_e - p_0)S_e \tag{17.22}$$

Because of the relationship between S_e, V_e, and p_e, maximum thrust is obtained when the exhaust pressure is equal to the atmospheric pressure. A rocket nozzle design that permits the expansion of the propellant products to the pressure of the surrounding fluid (see Figures 7.5 and 7.6) is said to have an *optimum expansion ratio*. An exit area differing from the optimum area will result in less rocket thrust, although the loss is small for quite large deviations from the optimum area. Since most rockets experience very large variations in atmospheric pressure along their flight path, exit area is designed for an intermediate altitude that produces the most efficient overall performance for the rocket powered vehicle.

The *effective exhaust velocity* is defined by the equation

$$c_e = \frac{F}{dm/dt} \tag{17.23}$$

The effective exhaust velocity is a fictitious velocity equal to the actual exhaust velocity plus the increment in exhaust velocity that would produce the thrust increment actually contributed by the pressure term $(p_e - p_0)S_e$. c_e is equal to the actual exhaust velocity when $p_e = p_0$.

An important performance parameter for rockets is *specific impulse* or *specific thrust*. It can be defined as the thrust that can be obtained with a propellant weight flow of 1 unit per second. It is the reciprocal of specific fuel consumption. Thus

$$\text{specific impulse } I_{\text{sp}} = \frac{F}{(dm/dt)g} \tag{17.24}$$

Since, from equation 17.23, $F = c_e(dm/dt)$,

$$I_{sp} = \frac{c_e}{g} \tag{17.25}$$

The total impulse I_t is the i: tegral of thrust F over the operating duration t.

$$I_t = \int_0^t F\,dt = \int_0^t I_{sp}\left(\frac{dm}{dt}\right)g\,dt$$

For constant thrust,

$$I_t = Ft = I_{sp}\left(\frac{dm}{dt}\right)gt = I_{sp}W_p \tag{17.26}$$

where W_p is the total weight of propellant, N (lb).

Thus the performance of a rocket depends primarily on specific impulse, which, in turn, is proportional to the effective exhaust velocity c_e. The magnitude of c_e for chemical rockets ranges from 2000 m/s (6562 ft/s) to 4000 m/s (13,123 ft/s), with a typical value of about 3048 m/s (10,000 ft/s).

The exhaust velocity for a particular rocket can be determined from equation 7.15:

$$c_p T_e + \frac{V_e^2}{2} = c_p T_T \tag{17.27}$$

so

$$V_e = \sqrt{2c_p(T_T - T_e)} = \sqrt{2c_p T_T\left(1 - \frac{T_e}{T_T}\right)}$$

Since $T_e/T_T = (p_e/p_T)^{(\gamma-1)/\gamma}$ (equation 7.2), and $c_p = \gamma R/(\gamma - 1)$ (equation 7.17),

$$V_e = \sqrt{\frac{2\gamma R T_T}{\gamma - 1}\left[1 - \left(\frac{p_e}{p_T}\right)^{(\gamma-1)/\gamma}\right]} \tag{17.28}$$

From equation 17.28, we can see that the exhaust velocity is a function of T_T, p_e/p_T, and the constants R and γ. T_T is a function of the chemical reaction of the fuel and the oxidizer. Any fuel–oxidizer combination at a particular pressure will burn at a particular temperature determined by the heat of reaction and called the *adiabatic flame temperature*. Thus T_T depends primarily on the propellant mixture. p_T is dependent on the nozzle throat area and the mass flow rate at which the rocket fuel and oxidizer are consumed. This, in turn, is determined by the rate at which the pumps drive fuel into a liquid rocket engine combustion chamber or by the burning surface area in a solid propellant rocket combustion chamber. The exit pressure p_e is determined by the rocket exit area, which is usually designed to bring the exit pressure equal to the ambient pressure at the average height during the burning phase of the rocket flight path.

The other two factors in equation 17.28 are the constants R and γ. Unfortunately, the values we have been using so far for R and γ are applicable only to a particular gas, air. The gas constant R is more generally defined as the universal gas constant $\overline{R}$ divided by the molecular weight of the gas, $\overline{M}$. Thus

$$R = \frac{\overline{R}}{\overline{M}}$$

where $\overline{R}$, the universal gas constant = 8314 J/K-kg mole.

The molecular weight of air is 28.96. γ is not really a constant at all, because the composition and temperature of the gas are changing as the gas flows through the rocket motor. However, for preliminary design of rockets, γ is often taken as some average between 1.2 and 1.35.

Table 17.1 shows the combustion chamber temperature, the molecular weight of the products of combustion, the exhaust velocity, the resulting specific impulse, and γ for several rocket fuel–oxidizer combinations. We can see from equation 17.28 that a fuel–oxidizer combination with a high value of R, which results from a low-molecular-weight $\overline{M}$, and a high T_T will increase the exhaust velocity and therefore will be more efficient. Other factors must also be considered. Hydrogen–fluorine, for example, suffers from being extremely corrosive and toxic. The choice of a rocket fuel must be made not only on the basis of its performance but also after consideration of the difficulty of designing the storage tanks, the pumps and piping that bring the fuel to the motor, the motor itself, and the threat to personnel handling the equipment.

Note that the specific impulse as calculated for Table 17.1 is simply V_e/g. This is because the assumed rocket has been designed so that the exit pressure is equal to the ambient pressure (i.e., $p_e = p_0$ in equation 17.22). Also, the ratio of p_T to p_e has been taken as 68.03. This corresponds to a combustion chamber pressure of 1000 lb/in.^{2} at sea level, a combination widely used to compare rocket engine performance.

The use of equation 17.28 to determine rocket motor performance is shown in the following example.

Example 17.2

A rocket motor using liquid hydrogen and liquid oxygen as the fuel and oxidizer has a combustion chamber temperature and pressure of 2700 K and 25 atm, respectively. The rocket motor throat area is 0.07 m^2. The exit area is designed for a standard pressure altitude of 17 km. γ may be assumed as 1.26. The molecular weight of the combustion products is 9.5.

TABLE 17.1 TYPICAL PROPERTIES OF SEVERAL LIQUID ROCKET PROPELLANTS[a]

Fuel–oxidizer combination	Combustion chamber temperature, K	Molecular weight of combustion products	V_e, m/s	I_{sp}, s	γ
Kerosene–oxygen	3555	21.9	2788	285	1.24
Hydrogen–fluorine	3869	11.8	3774	385	1.33
Hydrogen–oxygen	2689	8.9	3760	384	1.26

[a] $p_T/p_e = 68.03$; sea level, $p_e = p_0$.

At the design altitude of 17 km, calculate the exit velocity, specific impulse, and the thrust of the engine. Also determine the Mach number and the area at the exit.

Solution: Since the rocket is designed for an altitude of 17 km, the nozzle exit area will be designed to give an exit pressure equal to the ambient pressure at 17 km. From Table A.1, this pressure is, by interpolation, 8852 N/m^2.

The chamber pressure is 25 atm (i.e., 25 times the sea-level standard pressure). Thus $p_T = 25 \times 101{,}325 = 2{,}533{,}125$ N/m^2. The given problem is then

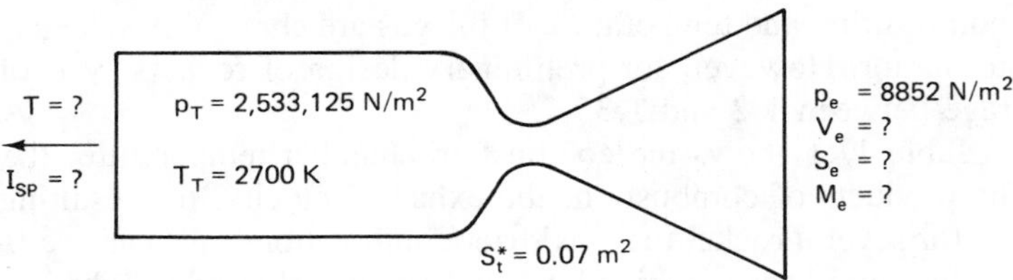

From equation 17.28,

$$V_e = \sqrt{2\frac{\gamma}{\gamma - 1}RT_T\left[1 - \left(\frac{p_e}{p_T}\right)^{(\gamma-1)/\gamma}\right]}$$

$$R = \frac{\overline{R}}{\overline{M}} = \frac{8314}{9.5} = 875.2 \quad \text{and} \quad \gamma = 1.26$$

Then

$$V_e = \sqrt{2\frac{1.26}{1.26 - 1}(875.2)(2700)\left[1 - \left(\frac{8852}{2{,}533{,}125}\right)^{(1.26-1)/1.26}\right]}$$
$$= \underline{3971.79 \text{ m/s}}$$

Since the nozzle is designed with $p_e = p_0$, I_{sp} is found from equation 17.25, where $c_e = V_e$.

$$I_{sp} = \frac{V_e}{g} = \frac{3971.79}{9.8} = \underline{405.29 \text{ s}}$$

From equation 7.2,

$$\frac{p_e}{p_T} = \left(\frac{T_e}{T_T}\right)^{\gamma/(\gamma-1)} \quad \text{so that} \quad \frac{T_e}{T_T} = \left(\frac{p_e}{p_T}\right)^{(\gamma-1)/\gamma}$$

and

$$T_e = 2700\left(\frac{8852}{2{,}533{,}125}\right)^{(1.26-1)/1.26} = 840.3 \text{ K}$$

Then the speed of sound at the nozzle exit is

$$a_e = \sqrt{\gamma RT_e} = \sqrt{1.26(875.2)(840.3)} = 962.62 \text{ m/s}$$

and

$$M_e = \frac{V_e}{a_e} = \frac{3971.79}{962.62} = \underline{4.126}$$

The area of the exit is found from equation 7.41. Note that Figure 7.5 is not applicable since we are not working with air and γ is not 1.4.

$$\left(\frac{S_e}{S_t^*}\right)^2 = \frac{1}{M_e^2}\left[\frac{2}{\gamma+1}\left(1+\frac{\gamma-1}{2}M_e^2\right)\right]^{(\gamma+1)/(\gamma-1)}$$

$$= \frac{1}{(4.126)^2}\left[\frac{2}{2.26}\left(1+\frac{0.26}{2}(4.126)^2\right)\right]^{2.26/0.26} = 517.5$$

and

$$S_e = S_t^*\sqrt{517.5} = 0.07(22.75) = \underline{1.592 \text{ m}^2}$$

Thrust $= V_e(dm/dt)$, so we must find dm/dt, the mass flow rate.

$$\frac{dm}{dt} = \rho_e S_e V_e$$

$$\rho_e = \frac{p_e}{RT_e} = \frac{8852}{(875.2)(840.3)} = 0.0121 \text{ kg/m}^3$$

Then

$$\frac{dm}{dt} = 0.0121(1.592)(3971.79)$$

$$= 76.51 \text{ kg/s}$$

and

$$\text{thrust} = (3971.79)(76.51)$$

$$= \underline{303{,}879 \text{ N}}$$

$$= \underline{68{,}318 \text{ lb}}$$

PROPULSION–AIRFRAME INTEGRATION

Propulsion–airframe integration is the process of locating the power plants and designing their installation to meet many operating requirements while minimizing drag and weight penalties. The arrangement of the propulsive units influences aircraft safety, structural weight, flutter, drag, control, maximum lift, propulsive efficiency, maintainability, and aircraft growth potential.

For prop-driven aircraft, the propeller requirements almost always place the engines on the wing or, for single-engine airplanes, at the fuselage nose. An unusual design used in converting a small piston-engine-powered commuter airplane from a twin-engine configuration to a trimotor was to place the center engine on the vertical tail. A recent trend in turboprop executive aircraft has been to place the engines on the rear portion of the wing. The pusher propellers, at the rear of the engines, are behind the aft bulkhead of the passenger cabin. The latter is actually the fundamental design objective because it reduces the noise and vibration in the cabin caused by the propellers. Because the engines are so far aft, the center of gravity is also far aft, and the moment arm of the horizontal tail is small. For this reason, longitudinal control provided by the tail is augmented by a canard, or, as in the case of the Beech Starship (Figure 1.60), the aft horizontal tail is replaced by a canard.

For jet or turbofan engine aircraft, propulsion–airframe integration is a complicated subject. Jet or turbofan engines may be placed in the wing, on the wing, above the wing, or suspended on pylons below the wings. Engines may be mounted in the rear of the fuselage, on top of the rear fuselage, or on the sides of the fuselage. Wherever the nacelles are placed, the detailed spacing with respect to wing, fuselage, or other nacelles is crucial.

Engines buried in the wing root have minimum parasite drag and probably minimum weight. Their inboard location minimizes the yawing moment due to asymmetric thrust after engine failure. However, they pose a threat to basic wing structure in the event of a blade or turbine disk failure, have much more difficulty in optimizing inlet efficiency, and more difficult accessibility for maintenance. If a larger-diameter engine is desired in a later version of the airplane, the entire wing may have to be redesigned. Such installations also eliminate the flap in the region of the engine exhaust, thereby reducing $C_{L_{MAX}}$. For all these reasons, this approach is no longer used, although the first commercial jet, the DeHaviland Comet, had wing-root-mounted engines. Figure 17.27 shows the Hawker-Siddeley Nimrod, developed from the Comet design.

Wing-mounted nacelles can be placed so that the basic engine is forward of the front spar to minimize wing structural damage in the event of a disk or blade failure. This automatically locates the inlet well ahead of the wing leading edge (Figure 17.28) and away from the high upwash flow near the leading edge. It is relatively simple to obtain high ram recovery in the inlet, since the angle of attack at the inlet is minimized and no wakes can be ingested.

Nacelles must be placed laterally to avoid superposition of induced velocities from the fuselage and nacelle or from adjoining nacelles. The interference problem is even greater with respect to wing–pylon–nacelle interference and is most easily solved by nacelle locations sufficiently forward and low with respect to the wing to avoid drag increases from high local velocities, especially premature occurrence of local supersonic velocities. In some designs the nacelle may be close to the wing but with the curved portions (i.e., the inlet and the aft end) located well forward and aft of the wing, respectively. Structurally, outboard nacelle locations are desirable to reduce wing bending moments in flight, but flutter requirements are complex and may show more inboard locations to be more favorable. The latter also favors directional control after engine failure.

Another influence of wing-mounted nacelles is the effect on flaps. The high-temperature, high-q exhaust impinging on the flap increases flap loads and weight and may require more expensive titanium structure. The impingement also increases drag, a significant factor in takeoff climb performance after engine failure. Eliminating the flap behind the engines reduces $C_{L_{MAX}}$. On the Boeing 707 and 747 and the McDonnel-Douglas DC-10, the flap behind the inboard engine is eliminated and this area is used for inboard all-speed ailerons.

Pylon–wing interference can and does cause serious adverse effects by increasing local velocities near the wing leading edge. Drag increases and $C_{L_{MAX}}$ losses result. A pylon that goes over the top of the leading edge is much more harmful in this

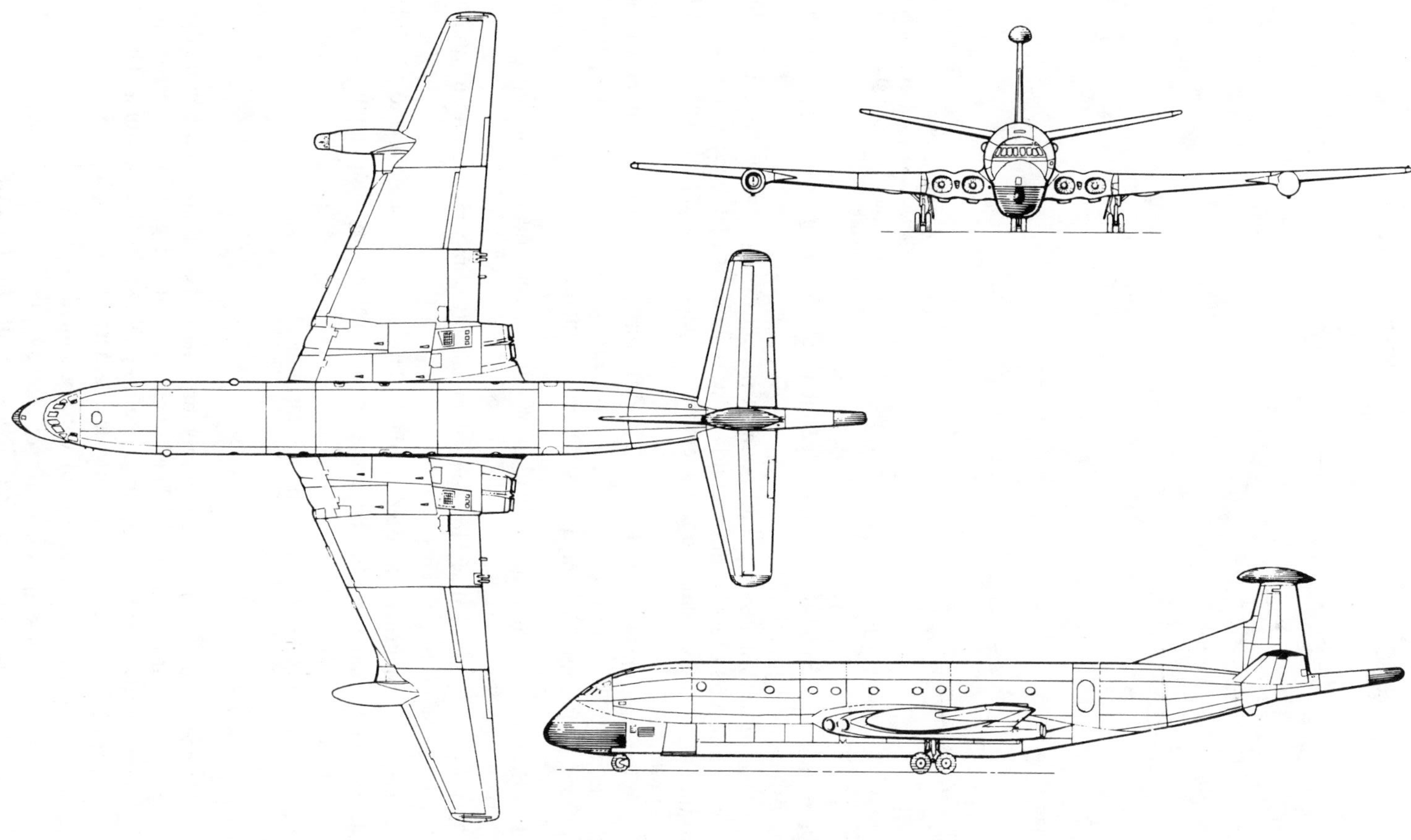

Figure 17.27 Hawker-Siddeley Nimrod, a reconnaissance airplane with jet engines installed in the wing root. (Courtesy of British Aerospace.)

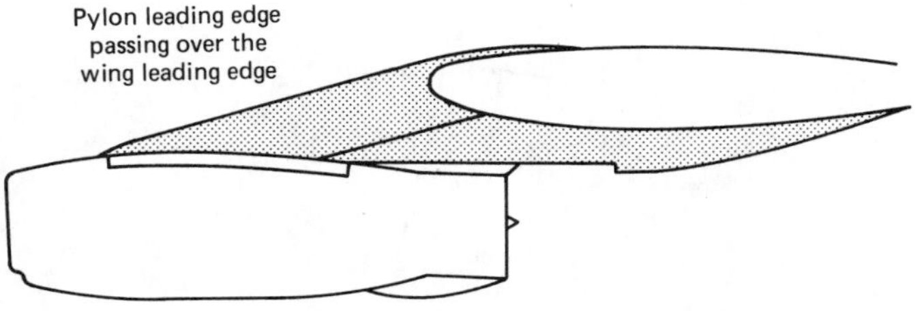

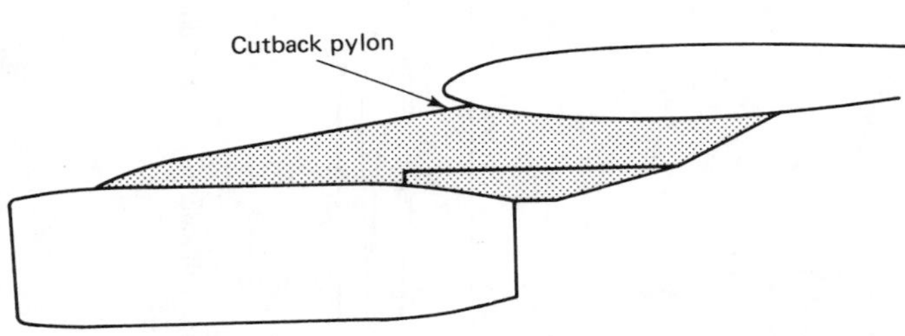

Figure 17.28 Over-the-wing and cutback pylons for wing-mounted jet or turbofan engines. (Courtesy of Douglas Aircraft Company.)

regard than a pylon whose leading edge intersects the wing lower surface behind the leading edge (Figure 17.28).

One disadvantage of pylon-mounted nacelles on low-wing aircraft is that the engines, mounted close to the ground, tend to suck dirt, pebbles, rocks, and so on, into the inlet. Serious damage to the engine blades can result. It is known as *foreign object damage*.

When aircraft become smaller, it is difficult to place engines under a wing and still maintain adequate wing–nacelle and nacelle–ground clearances. This is one reason for the aft-engine arrangements. Other advantages are:

1. Greater $C_{L_{MAX}}$ due to elimination of wing–pylon and exhaust-flap interference (i.e., no flap cutouts).
2. Less drag, particularly in the critical takeoff climb phase, due to eliminating wing–pylon interference.
3. Less asymmetric yaw after engine failure with engines close to the fuselage.
4. Lower fuselage height above the ground, permitting shorter landing gear and airstair lengths.
5. Last, but not least, it may be the fashion.

Disadvantages are:

1. The center of gravity of the empty airplane is moved aft, well behind the center of gravity of the payload. Thus a greater center of gravity range is required to accommodate both full-payload and no-payload flight conditions. This leads to more difficult balance problems and generally a larger tail.
2. The wing weight advantage of wing-mounted engines is lost.
3. The wheels kick up water on wet runways, and special deflectors on the gear may be needed to avoid water ingestion into the engines.

4. At very high angles of attack, the nacelle wake blankets the T-tail, necessary with aft-fuselage-mounted engines, and may cause an unstable stall. This requires a large tail span that puts part of the horizontal tail well outboard of the nacelles.
5. Vibration and noise isolation for fuselage-mounted engines are difficult problems.
6. Aft-fuselage-mounted engines reduce the rolling moment of inertia. This can be a disadvantage if there is a significant aerodynamic rolling moment created by asymmetric stalling. The result can be an excessive roll rate at the stall.
7. Last, but not least, it may not be the fashion.

It appears that in a DC-9-size aircraft the aft engine arrangement is to be slightly preferred, but the difference is small. Smaller aircraft benefit from an aft engine arrangement.

A center turbofan engine is always a difficult problem. Early DC-10 studies examined two engines on one wing and one on the other, and two engines on one side of the aft fuselage and one on the other, in an effort to avoid a center engine. Neither of these proved desirable. Four center engine possibilities are shown in Figure 17.29.

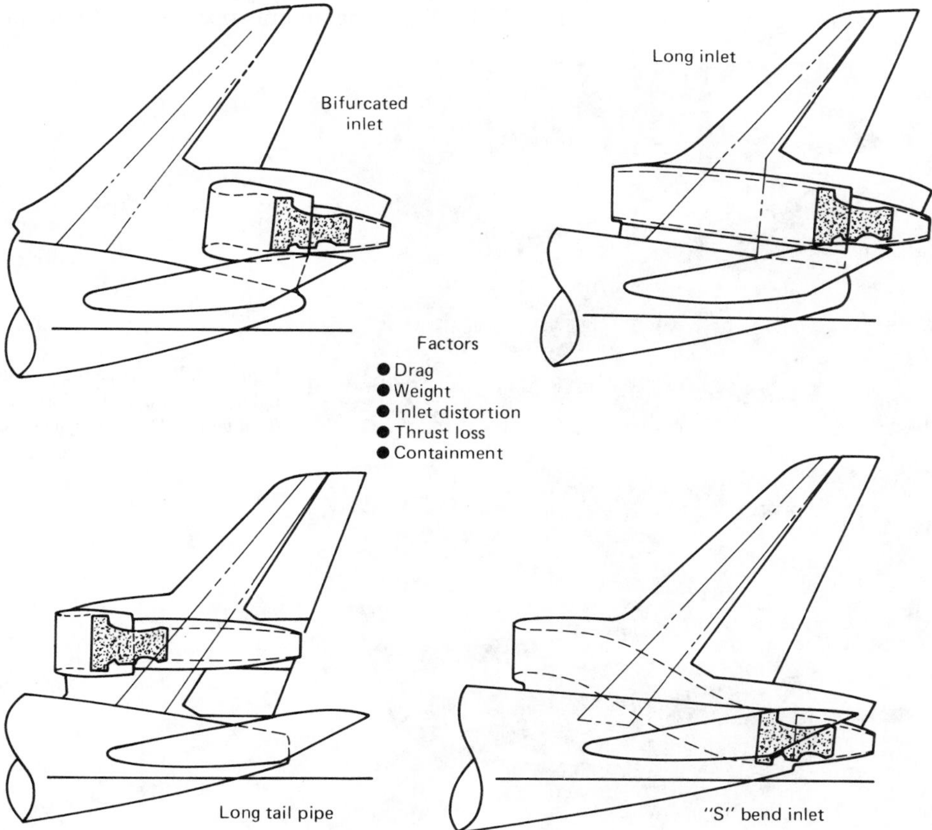

Figure 17.29 Aft engine locations studied for the DC-10 transport aircraft. (Courtesy of Douglas Aircraft Company.)

Each possibility entails compromises of weight, inlet total pressure loss, inlet flow distortion, drag, reverser effectiveness, and maintenance accessibility. The two usually used are the S-bend, which has a lower engine location and uses the engine exhaust to replace part of the fuselage boat-tail (saves drag) but has more inlet loss, a flow distortion risk due to inlet duct curvature, a drag from fairing behind the inlet duct, and cuts a huge hole in the upper fuselage structure, and the straight-through inlet with the engine mounted on the fin. This location has an ideal aerodynamic inlet free of distortion, but does have an inlet loss due to the length of the inlet and an increase in fin structural weight to support the engine. Such are the complexities of propulsion integration!

Inlet design presents the problem of designing the smallest-diameter entrance, to keep the nacelle diameter small, that will pass the airflow required by the engine without having excessive Mach numbers, preferably $M < 0.75$, at the minimum inlet cross section. In addition, the inlet ahead of the minimum, or inlet area, must be shaped to avoid separation from the inlet lips over the entire range of airplane speeds, angles of attack, angles of yaw, and engine intake airflows.

Fighter aircraft turbojet power plants are usually installed in the aft end of the fuselage (Figures 17.30 and 17.31). Since maximum speed is an important objective

Figure 17.30 General Dynamics F-16 fighter, showing engine installation. (Courtesy of General Dynamics.)

Figure 17.31 McDonnell Douglas F-15 fighter aircraft, showing engine installation. (Courtesy of McDonnell Aircraft Company.)

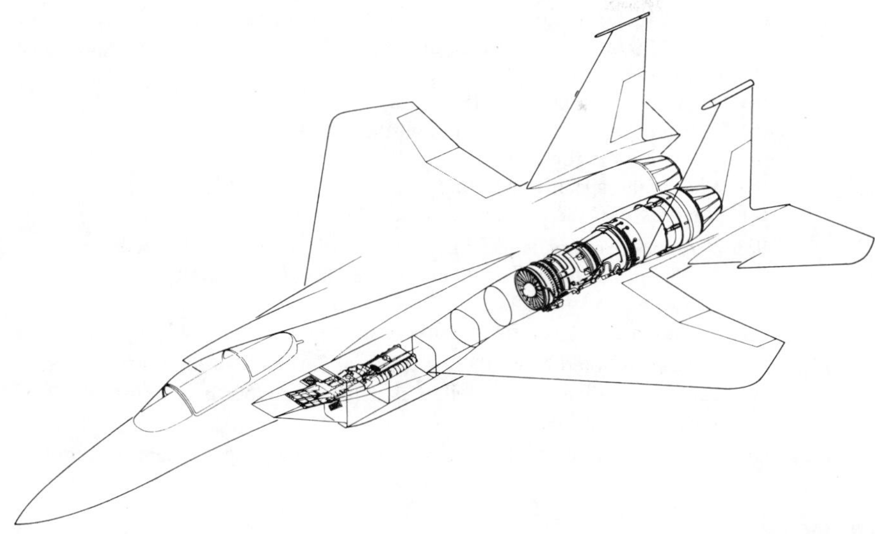

Figure 17.31 continued

for these aircraft, obtaining the smallest size to minimize the drag is a primary consideration. Placing the engines in the fuselage achieves this goal. The engine intake air must be brought to the front face of the engine, or engines, through a duct with minimum curvature to avoid total pressure losses in the duct. Such losses reduce engine thrust and efficiency. The inlet ducts extend forward with the inlets on the bottom or sides of the fuselage (Figures 17.30 and 17.31). Twin-engine fighters have a particular problem fairing the aft end of the fuselage around twin exhausts. It is an aerodynamic challenge to avoid excessively steep boat-tail angles that cause flow separation around the exhaust nozzles. As noted in Chapter 11, fighter aircraft drag is difficult to estimate because the wing, nacelles—or the portion of the fuselage enclosing the inlet ducts that serve the same purpose as nacelles—and the fuselage are so integrated that it is difficult to decide where one ends and the other begins.

PROBLEMS

17.1. A proposed twin-engine turboprop transport airplane will cruise at 28,000 ft at a true speed of 400 mph. Air temperature is 430°R. The airplane will be powered by engines with a takeoff rating of 1,500 shp and a maximum cruise rating at 28,000 ft of 800 shp. At cruise, propeller speed is 1200 revolutions per minute.

(a) Assuming four-bladed propellers, with a blade activity factor of 135, so that the propeller chart in Figure 17.20 is directly applicable, find the variation of propulsive efficiency with propeller diameter. What propeller diameter would you choose to obtain the highest cruise efficiency at maximum cruise power? (Assume several propeller diameters and determine their efficiencies. Plot η versus diameter. Check additional values of diameter as necessary to locate the optimum.)

(b) What is the ideal efficiency of the propeller you selected?

17.2. Consider a rocket motor with a combustion chamber (reservoir) temperature of 3517 K, a pressure of 20 atm, a throat area of 0.11 m^2, and an exit pressure equal to the pressure at a standard-day altitude of 18 km.

(a) Determine the exit velocity.

(b) Determine the mass flow through the motor.

(c) Determine the rocket thrust.

(d) What is the specific impulse?

(e) What justifies the c_p value in note 3 below?

Notes: A rocket motor is just like a supersonic wind tunnel except:

1. The gas is not air, so R and γ are changed. Assume $R = 519.1$ J/kg K.
2. Assume $\gamma = 1.22$.
3. Assume c_p (specific heat at constant pressure) = 2879 J/kg K.

17.3. A turboprop-powered Navy P3V is flying at 350 mph at 20,000 ft on a standard day. Each of the four 14-ft-diameter propellers is delivering 1750 lb of thrust. What is the *ideal* efficiency of this propeller?

17.4. Discuss the factors that influence a decision about aft placement of turbofan engines.

REFERENCES

17.1. McCormick, Barnes W., *Aerodynamics, Aeronautics and Flight Mechanics,* Wiley, New York, 1979.

17.2. Sutton, George, P., and Ross, Donald M., *Rocket Propulsion Elements,* Wiley, New York, 1976.

18

STRUCTURES

IMPORTANCE OF STRUCTURAL WEIGHT AND INTEGRITY

Although the science of aerodynamics is fundamental to the understanding and determination of lift, drag, performance, stability, and control of the airplane, structural technology makes an equally vital contribution to a useful flying machine. The structure of an airplane must withstand the applied aerodynamic and inertial loads not only for normal flight but also for extreme conditions that may be encountered very rarely. Typical of these are encounters with high-velocity vertical gusts or inadvertent overspeeds to speeds well above the normal flight regime. The essential character of an aircraft structure is light weight, because weight plays such an important role in the performance and economics of an airplane. In fact, this crucial significance of the structural weight is the difference between aircraft structures and other forms of structural engineering.

The importance of empty weight should be clear from the limitation placed on maximum takeoff weight by the available runway length. A pound more of structural weight is a pound less of payload. Furthermore, the specific range is inversely proportional to the average airplane cruising weight, so an increase in structural weight raises the fuel consumption and the fuel cost. The first cost of the airframe (i.e., the airplane less engines) is generally found to be proportional to the weight empty. If the payload and range cannot be reduced, a higher structural weight requires a larger engine and/or wing area to meet the takeoff and landing requirements, thereby raising the structural weight even further.

For all these reasons, aircraft structural design has always sought to meet the load requirements with a minimum acceptable margin of safety and the least possible weight. Nevertheless, the potentially disastrous effect of an airplane structural failure means that the structures must be designed for long life either with safe life criteria or with fail-safe design. *Safe life* means that the stresses in a component are so low that fatigue failure is not possible over the anticipated life of the airplane, or at least until some period has passed after which a part replacement is required. *Fail-safe* means that the structure has alternative load paths so that no single failure will be hazardous to the aircraft. This can be achieved by designing so that no one component carries a large part of a load. Therefore, if one part fails, the remainder of the structure can still carry most of the maximum load. Since the maximum load is rarely encountered and the structure has a safety factor of 1.5, the structure remains safe until the failure is found and repaired.

DEVELOPMENT OF AIRCRAFT STRUCTURES

Early aircraft were built from very light materials such as bamboo and wood assembled using the basic concepts of bridge building, beam, and truss construction. A *truss* consists of an arrangement of straight bars connected in such a manner as to divide the area enclosed by the perimeter of the truss into a system of triangular spaces. As shown in Figure 18.1, the Bleriot XI fuselage was built in this manner, with the diagonal members being tensioned wires. The World War I Sopwith Camel wing shows similar diagonal wire bracing (Figure 18.2). For wings, however, the main structural members were the *spars,* beams that extended from wing tip to wing tip. The spars were essentially solid rectangular section beams with the spar caps at the

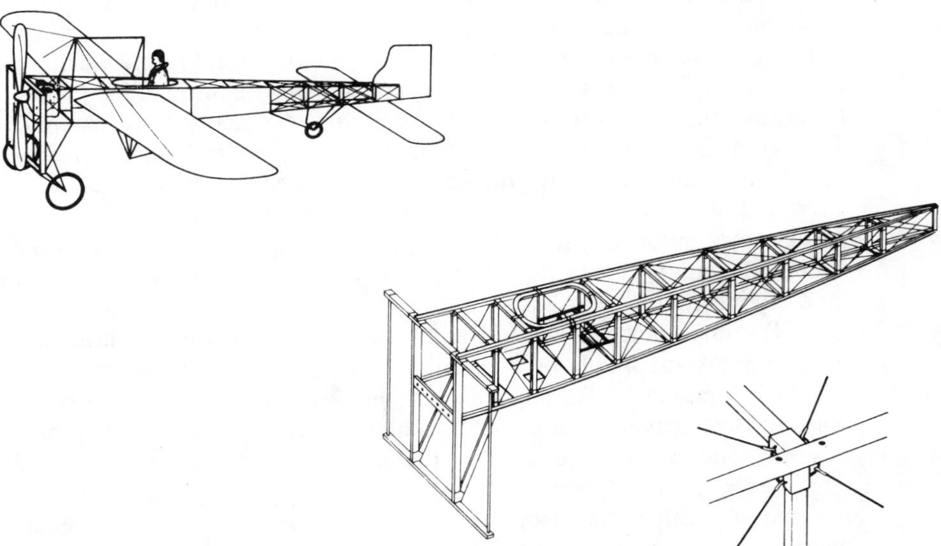

Figure 18.1 Bleriot XI monoplane, showing fuselage construction.

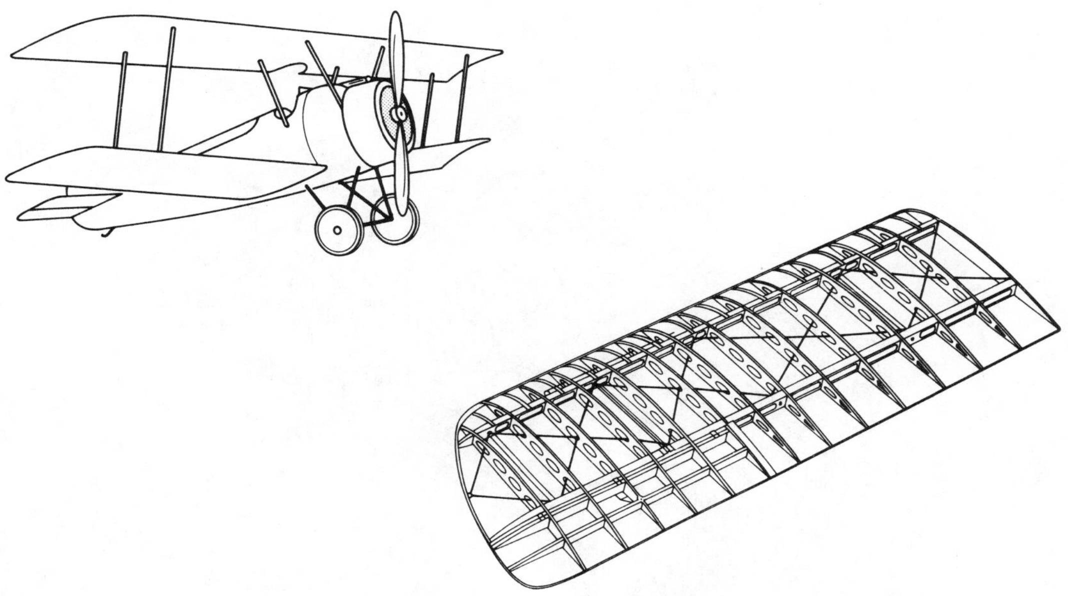

Figure 18.2 Sopwith Camel World War I fighter.

top and bottom of the spars carrying the bending load, while the web connecting them carried the shear loads. The ribs maintained the airfoil shape and transferred the local air loads on the fabric surface covering to the spars (Figure 18.2).

The next step in aircraft structure was the substitution of metal for wood. The structural technique was little different except for the substitution of thin steel tubing for the wood members in the fuselage. Ribs were assembled by riveting many pieces of metal together or by stamping the rib from aluminum sheet when aluminum alloys became available starting in the late 1920s.

The next major development, and one of the most important, was the use of metal skins designed to carry much of the load. Originally, metal skins were very thin and employed only because they were less subject to weathering than was the cloth covering. When structural engineers began to use the skins to carry some of the load, the weight of the underlying structure could be lightened. In the extreme, for example, a fuselage could be a shell or monocoque structure, eliminating everything but the skin. In practice, however, the skin must be stiffened to prevent buckling at a relatively low compressive stress. This is accomplished by attaching small lengthwise metal strips to the fuselage skin. These strips have various possible cross-sectional shapes designed to provide stiffness with the least weight. They are called *stiffeners, stringers,* or *longerons*. In addition, frames are added at intervals along the length of the fuselage to transfer the internal loads to the skin and to maintain the cross-sectional shape, just as ribs do in the wing (Figure 18.3). On wings, the use of load-carrying skin permits a reduction in the weight of the spar caps. As on the fuselage, the wing skins are stiffened by stiffeners or stringers running spanwise on the inside surfaces of the skin (Figure 18.4). The stiffeners not only carry their own load but, by pre-

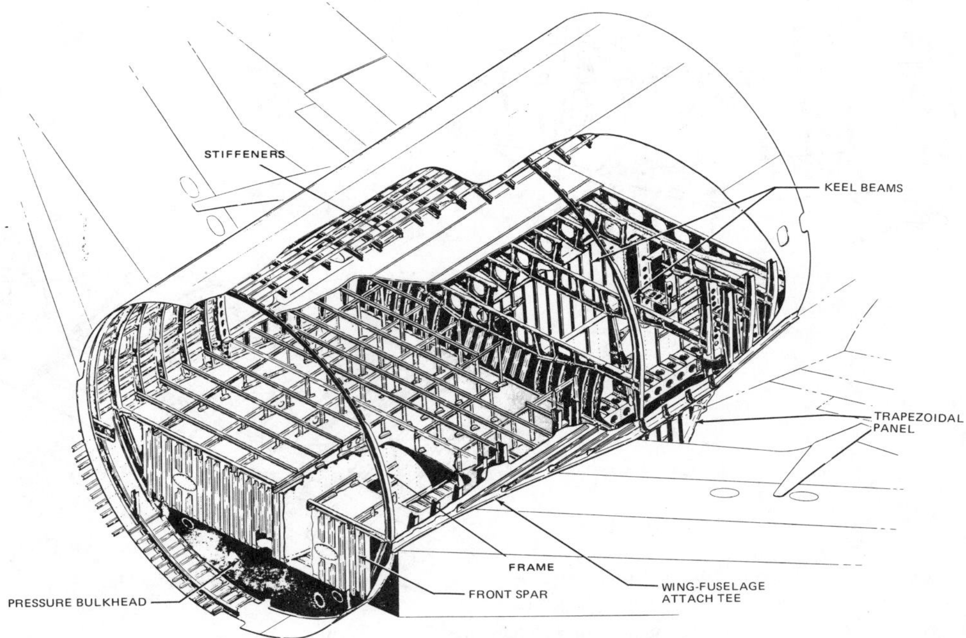

Figure 18.3 Fuselage structure. (Courtesy of Douglas Aircraft Company.)

venting early buckling of the skin, they increase the stress that can be supported in compression by the skin before buckling. Furthermore, the skin close to the stiffeners continues to carry increased loads even after some of the skin is buckled. This condition is permitted in the design only at loads above the normal flight loads, since any skin buckling in cruise would be aerodynamically unacceptable due to effects on drag.

Wing skins that started as minimum-thickness aluminum alloy stiffened by stringers became heavier as their load-carrying ability was used more intensely. With higher wing loadings and larger wing spans, skin thicknesses increased until they are close to an inch at the root of the largest aircraft. To minimize weight, the thickness is tapered with spanwise location to maintain approximately a constant stress as the bending loads reduce toward the tip. On the outer panel of large aircraft the minimum thickness is about 0.06 in., a limitation set by lightning-strike requirements. Minimum skin thickness is 0.05 in. on other parts of the aircraft such as the fuselage to permit countersinking rivets. Executive jet minimum skin thickness may be 0.04 on wings, with rib thicknesses as low as 0.025 in. Light, low-speed airplane skins are as thin as 0.02 to 0.03 in.

The increased use of metals was accompanied by the evolution from the biplanes with many struts and bracing wires (Figure 18.5), to monoplanes with wing struts, and finally to the cantilever monoplanes (Figure 18.6), free of these drag-producing protuberances. The use of two wings connected by struts and wires in the biplane configuration gave the assembly a large effective beam depth that contributed to achieving light but stiff structure, even though the materials were wood and fabric. The biplane also allowed more wing area within a given span at a time when low wing loadings

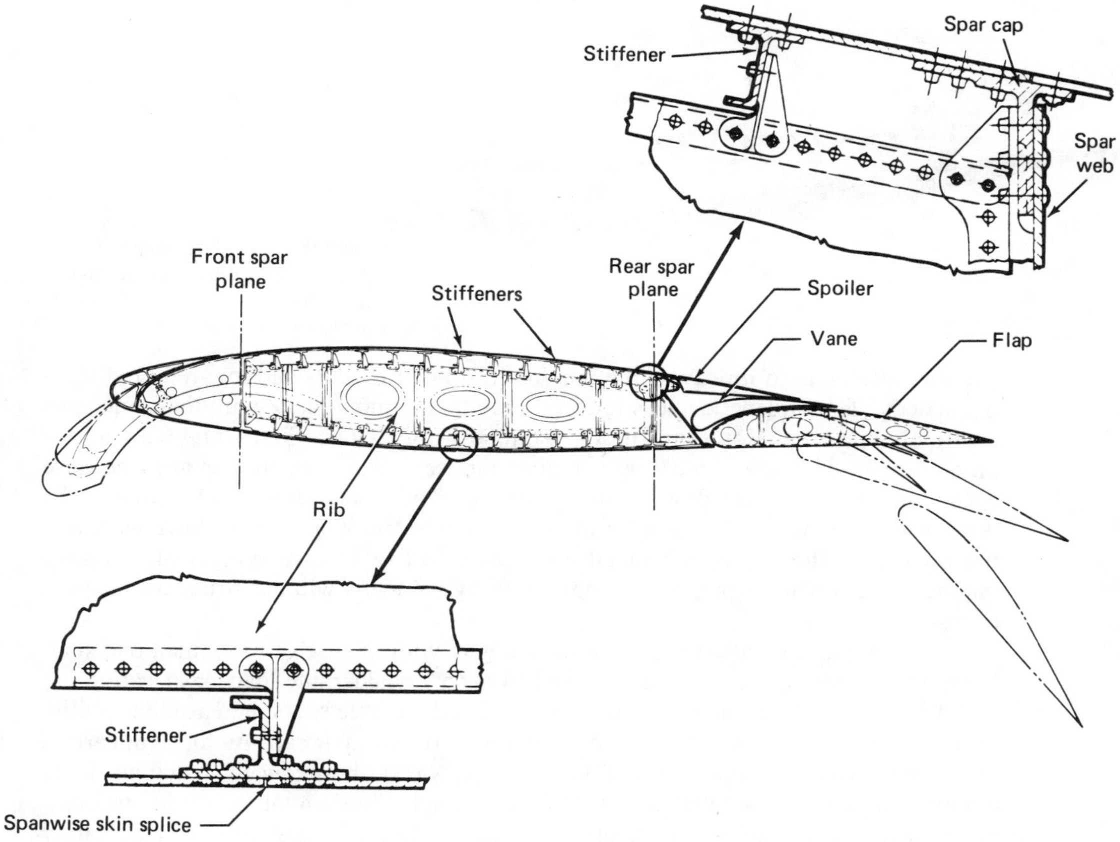

Figure 18.4 Wing structure: skin, stiffeners, spars, and ribs. (Courtesy of Douglas Aircraft Company.)

Figure 18.5 Curtiss Condor transport, 1927. (Courtesy of Trans World Airlines.)

Figure 18.6 The Douglas DC-3 represented one of the greatest single steps in transport aircraft development (Courtesy of Western Airlines.)

were required due to low available maximum lift coefficients and low required stalling speeds. The biplane has less induced drag than a monoplane with the same span and total wing area. However, for a given total wing area and individual wing aspect ratio, biplanes have a significantly higher induced drag than monoplanes because each wing operates in the downwash of the other. The greater downwash increases the drag of each wing, just as drag is increased due to the wing's own downwash acting on itself. The use of the metal cover as effective load-carrying material gave unbraced monoplane wings the strength to carry the loads without struts and without excessive weight.

Achieving fail-safe design in a wing began with the use of more than one spar. Even after the wing skins were designed to carry considerable load, spar caps were still primary structural members. Use of two and then three spars increased the ability of the airplane to survive after a complete failure of one spar cap. As the skin carried increasing loads, the importance of the spar caps was further reduced so that no single component carried more than about 20% of the load. Thus a total failure in one component would leave sufficient strength to tolerate all normal loads with a full 1.5 safety factor or more. Even the loads associated with an extreme gust could be carried, but with a reduced safety factor. At worst, the result would be a permanent deformation in the structure because the yield point might be exceeded. To prevent one sheet of skin from becoming critical, wing skins are often divided into spanwise sections so that no one skin section is excessively important (i.e., carries too large a percentage of the load).

ELEMENTS OF AIRCRAFT STRUCTURES

Modern aircraft structures are very complex. One advantage of the simple truss structure is that the loads in each member can be easily determined. The stiffened shell of a fuselage or the stiffened sheet covering and spar caps on the top and bottom of a wing have many members, and the load in each depends not only on the applied loads but also on the deflection of each component. The analysis of this redundant structure is difficult. The theories of elasticity and elastic stability must be applied to many points in the structure simultaneously to obtain a solution. This is far beyond the scope of our discussion, but some understanding of the nature of such structures can be obtained by indulging in some gross yet useful simplifications.

Aircraft structures are basically composed of three types of structural elements, stiffened shells such as the fuselage (Figure 18.3) or nacelles, stiffened plates such as the top and bottom of the wing, and beams (Figure 18.7). A wing may be thought of as an I-beam, with the upper and lower flanges of the I-beam consisting of the upper and lower wing skins together with the wing stiffeners and the spar caps. The spar webs serve as the beam web, carrying the shear loads. The stresses in tension are simple to determine, basically just by dividing the tension load by the cross section of the resisting material. The difficulties arise mostly in compression, where the thin skins of the shells or plates will generally buckle before reaching the compressive failure stress of the material. The word "thin" here refers to the ratio of thickness to width.

The simplest form of buckling occurs in a long slender rod or column (Figure 18.8). In a short section under increasing compressive load, the material would eventually fail, in the sense of being permanently deformed, when the compressive stress passes the compressive yield point of the material. The yield point is the stress beyond which the material will not return to its original form (i.e., the material is permanently deformed). We know from experience that a long slender column will suffer a lateral bending deformation and will be unable to carry significantly greater

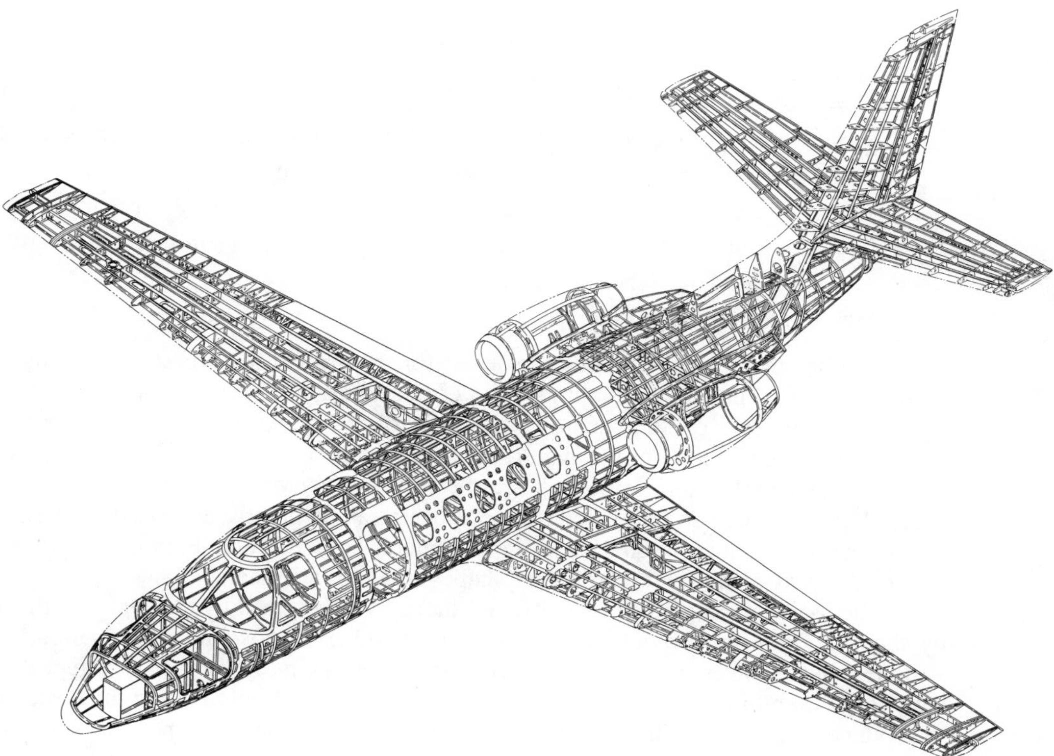

Figure 18.7 The structure of the Cessna Citation II executive jet airplane. (Courtesy of Cessna Aircraft Company.)

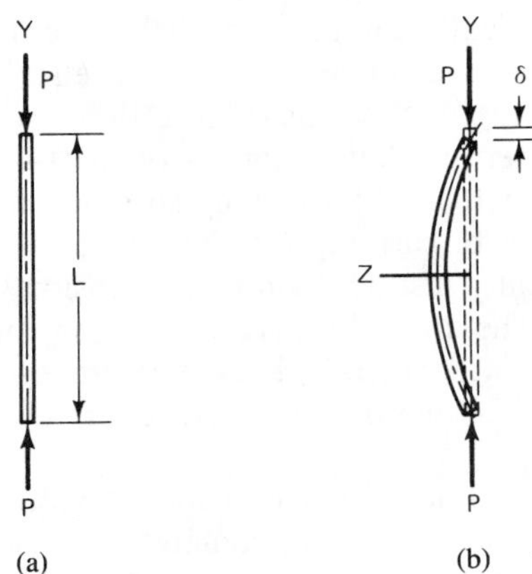

Figure 18.8 Simple column under axial load: (a) load less than critical load; (b) buckled column, load greater than critical load.

loads at a stress much below the yield stress of the material. The critical load beyond which a column will buckle was derived by Euler and is given by the Euler column formula as

$$P_E = \frac{\pi^2 EI}{L^2}$$

where

P_E = critical load

E = modulus of elasticity of the material, the ratio of the applied stress to the strain; the strain is the deformation per unit length of a material

I = moment of inertia of the cross section of the column, a measure of the bending resistance of the cross-sectional shape; for a solid circular section $I = \pi d^4/64$, where d is the column diameter

L = column length

Similar, more complex equations exist for the critical load or stress for sheets or shells with compressive loadings. The critical stress in a sheet, for example, is dependent primarily on the square of the ratio of the thickness to the unsupported width of the sheet, the E of the material, and a complex function of the ratio of the width to the length. Also important are the restraint conditions at the edges of the sheet. The load that can be carried by such sheets can be greatly increased by the use of stiffeners, which postpone buckling to a much higher stress. Figure 18.9 shows some examples of the wide variety of shapes used for stiffeners.

The load-carrying ability of stiffened sheets is determined by test and calculated by theory. There are obviously an enormous number of possible combinations of stiffener shapes, sizes, and spacings used with various skin thicknesses. An appropriately large literature exists giving the results of many tests, which allow the designer to produce a structure with a fairly good degree of confidence. Samples of new skin and stringer configurations are always tested, however, to confirm strength characteristics. The goal of the designer is to achieve the required strength with the lightest possible structure.

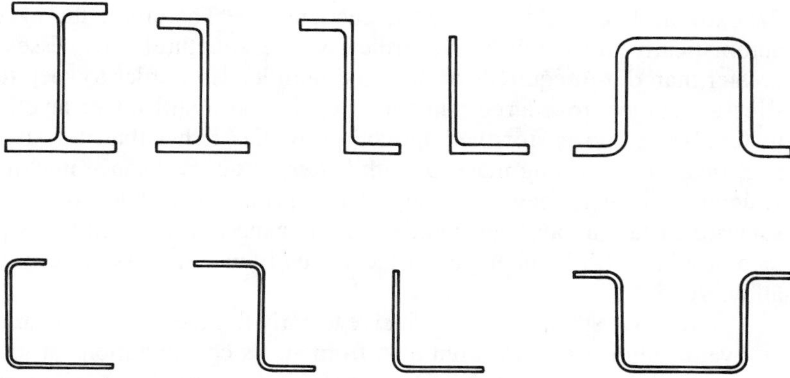

Figure 18.9 Stiffener cross-sectional shapes.

IMPORTANCE OF FATIGUE

An important aspect of aircraft structures is the need to design for a long fatigue life. Metals suffer a gradual deterioration under repeated application and removal of loads, even though the loads may be much smaller than those under which failure may occur with a single application. An airplane wing goes through a major load cycle when at takeoff a large upward air load is applied to the wing, followed by a landing when the air load is removed and a downward load from the weight of the wing is imposed. In addition, the various gusts encountered in flight produce a great many load cycles, usually of smaller magnitude. Fuselages receive similar cyclic loadings not only due to the flight loads on the airplane but because of the pressurization cycle. Aircraft used to be designed for relatively short lives and many military aircraft still are. Modern transports may fly for over 70,000 hours, however. To accommodate this requirement, fatigue life rather than strength requirements often dominates a structural design.

The number of load cycles a material can tolerate without failure depends on the stress level. The lower the stress, the greater the number of cycles the part can withstand. A typical fatigue test (Figure 18.10) shows the enormous life increase obtainable by limiting a material to cyclic stress levels much below the ultimate strength. This is not as difficult as it might at first seem, since the airplane is designed

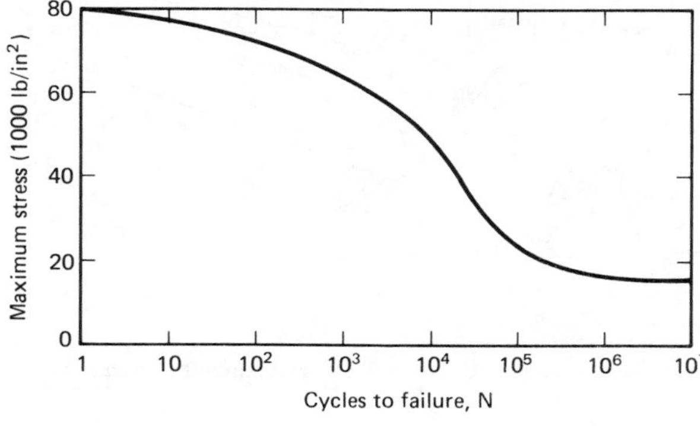

Figure 18.10 Fatigue test of joints and splices, 7075-76 aluminum alloy.

for extreme loads seldom, if ever, encountered. Therefore, the everyday loads are automatically much lower. Nevertheless, the structural thicknesses used are often greater than those required for the maximum loads in order to keep the normal flight stresses, subject to a large number of cycles, at a still lower level. In some cases, materials are chosen for their fatigue properties rather than for strength. If fatigue is critical, a less strong material with better fatigue resistance may result in a lighter structure. Since fatigue is of importance primarily in tension, an aluminum alloy superior in fatigue and designated 2024 is generally used on the lower tension surface of wings, while the upper surface, usually in compression, can utilize a stronger alloy, 7075.

Excessive stress levels conducive to early fatigue failure may arise not only from an overall high stress level but also from stress concentrations at local points in the structure. Holes, particularly of small radius, can cause large increases in stress level. Fittings and joints, which serve to carry load from one structural component to another, may, if not designed very carefully, introduce increased local stresses leading to fatigue failure long before any problem occurs in the basic structure.

The load flowing through a structure may be considered analogous to a subsonic fluid flowing through a duct. Whenever the material cross-sectional area is reduced, the stresses increase, since the same total load, analogous to the quantity of fluid per unit time, must be carried by the smaller available cross section. Just as a small radius of curvature along the contour of a wing or body leads to high local velocities in a

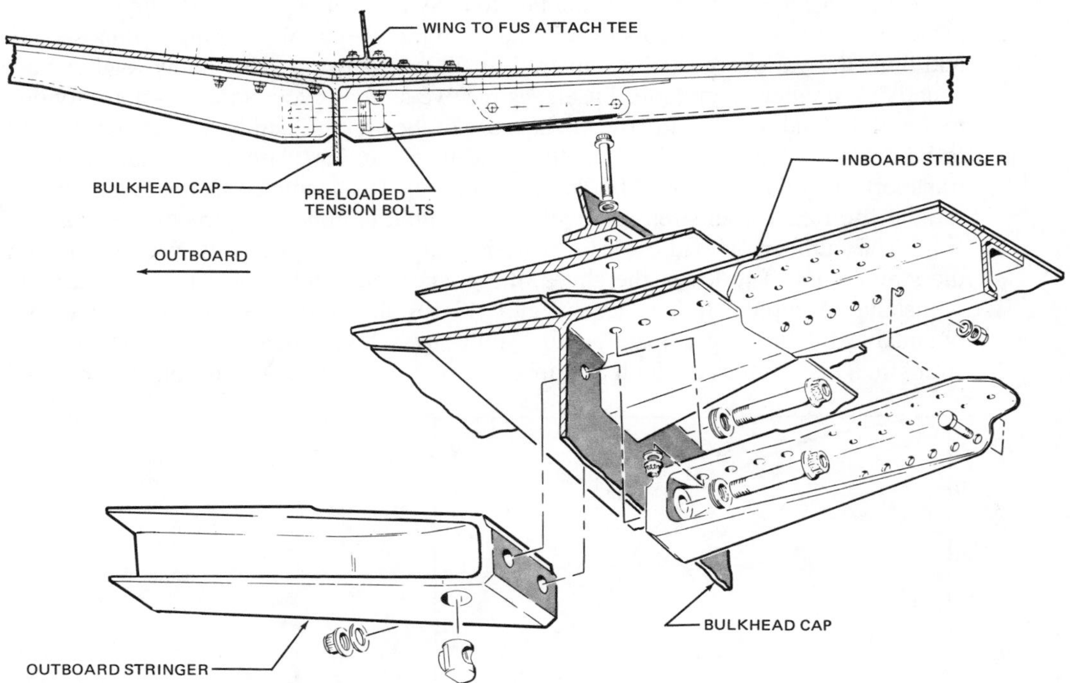

Figure 18.11 DC-10 wing panel joint at side of fuselage. (Courtesy of Douglas Aircraft Company.)

small portion of the fluid close to the surface, so a small radius along the load path in the material leads to a high local stress. A small crack that may be created by this local fatigue failure creates its own stress concentration and accelerates the process of further fatigue failure. Thus the crack will grow and may eventually lead to a total failure of the component if not detected and repaired. A major part of the design of airplane structure is the avoidance of stress concentrations by careful detail design. Examples are avoiding small radius holes and lowering the basic stress level approaching a hole or fitting by thickening the material. This is usually done by attaching an extra sheet of metal called a *doubler* to the inside surface of the wing skin around an inspection opening or to the fuselage skin around a window. Special attention is also given to fitting design (Figure 18.11). In spite of this care in detailed structural design, a vital aspect of airframe maintenance is a never-ending search on a scheduled basis for structural cracks.

MATERIALS

Figure 18.12 shows the distribution of metallic structural materials in modern Boeing transport aircraft. Eighty percent of the material is aluminum alloy, which is the lightest for most purposes, especially where buckling resistance is involved. This follows because for a given weight the thickness of aluminum is greater than other metals, and the critical buckling stress varies with the square of the thickness. Steel comprises 17% of the structural weight and is used for highly loaded parts such as landing gears, engine fittings, and flap tracks. Titanium, 3% of the total, is a more expensive material but is particularly suited to high-temperature applications such as nacelle structure, firewalls, and pneumatic ducting because it retains its strength at higher temperatures.

Graphite and boron composite materials are being used extensively in secondary, noncritical structures such as control surfaces, flaps, and spoilers. In military aircraft, a few primary applications, such as horizontal stabilizers and wings, have been made. These materials are very light and stiff and may in future years become the major aircraft structural material. Their use is limited now because of cost, which is expected to continue to decline, and lack of experience with composites in primary structure. Composite material is made of fibers of graphite or boron embedded in a matrix or

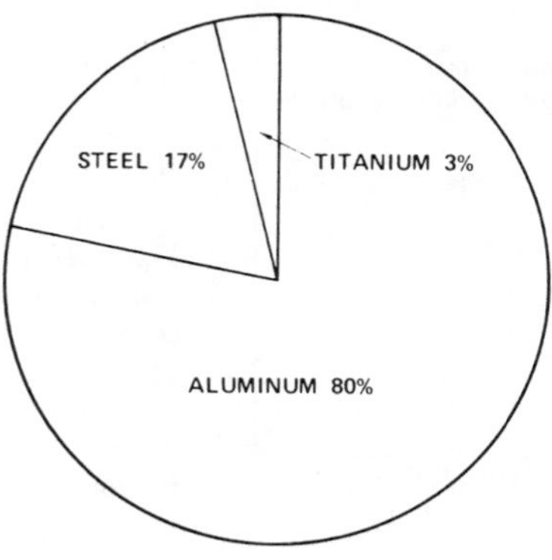

Figure 18.12 Distribution of metallic structural alloy in Boeing aircraft. From Lovell and Disotell, AIAA Paper 78-1552, "Structural Material Trends in Commercial Aircraft," 1978. Reprinted by permission of American Institute of Aeronautics and Astronautics.

binding of epoxy. The strength is enormous parallel to the fibers but much less at right angles to them. Therefore, composite materials are usually composed of layers in which the fibers are oriented in different directions to achieve the desired directional strength characteristics (Figure 18.13). Graphite epoxy composite properties in terms of the ratio of ultimate tensile stress to material density and modulus of elasticity to density are compared to aluminum, titanium, and steel in Figure 18.14. The composite is pseudo-isotropic, that is, approximately equal strength in all directions. Graphite composite is about 50% better than aluminum. Because of lower cost compared to boron composite, graphite composite is becoming the dominant composite material.

Figure 18.14 indicates potential weight savings of over 30% for pseudo-isotropic graphite epoxy. In actual use, however, compromises have to be made at fittings and joints in order to maintain the integrity of the material. Some parts will have to be made of metal both at fittings and to provide current flowpaths in the event of lightning strikes. Because composites are brittle, they may not be usable in a pressurized fuselage shell; or, if they are, a composite/metal mix may be necessary. It has been estimated in the past that up to 25% structural weight savings would someday be possible with composites. Weight savings of this magnitude are achieved in relatively simple components such as control surfaces. Larger, more complex primary structures such as fuselages and wing structural boxes have been more difficult to design, and the weight savings have generally been much less. At this time, it seems that overall structural weight savings of 10% to 15%, compared to conventional aluminum structures, are a better estimate for complete aircraft built of composites.

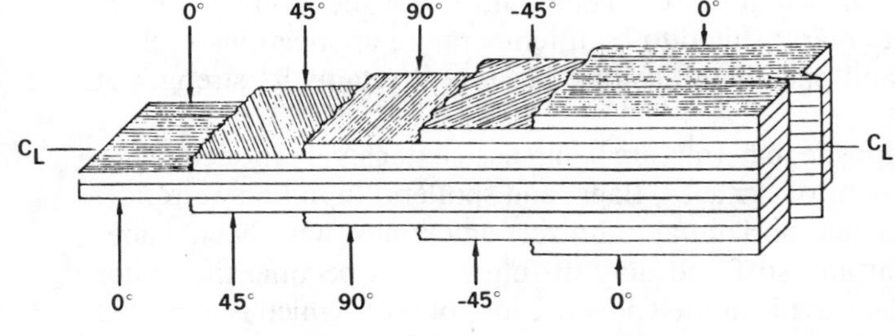

Figure 18.13 Symmetrical balanced lay-up of composite material. From AIAA Paper 78-1530, Aug. 1978. Reprinted by permission of American Institute of Aeronautics and Astronautics.

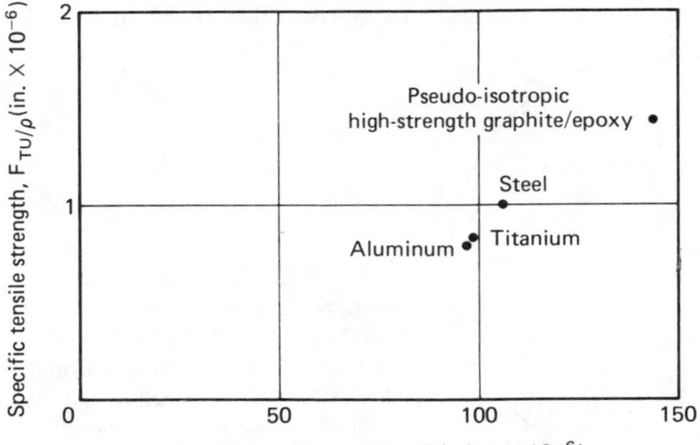

Figure 18.14 Graphite–epoxy composite material specific strength and modulus comparison. From R. S. Shevell, "Technological Development of Transport Aircraft—Past and Future," *Journal of Aircraft,* Vol. 17, No. 2, February 1980. Reprinted by permission of American Institute of Aeronautics and Astronautics.

Advanced metals are also candidates for future aircraft. In particular, aluminum–lithium, an aluminum alloy containing lithium, is expected to show a weight reduction of about 10% for the same strength and stiffness. Such an alloy would achieve much of the weight reduction anticipated with graphite–epoxy. Composite material requires a complete factory retooling since the processes of lay-up and curing are entirely different. The investment in production facilities is very large. Aluminum–lithium would be used just as present aluminum alloys are used, so this large startup cost would not be incurred. For these reasons, the introduction of all composite structures may occur very slowly, if at all. An advanced metal structure with some composite components may prove to be the most efficient solution.

Another characteristic of structural materials is loss of strength at high temperatures. As discussed in Chapter 19, temperatures in boundary layers increase with Mach number and transmit heat to the vehicle skin. Aircraft skin temperature is not a serious problem for flight below a Mach number of about 2.0. As cruise Mach number rises above 2.0, loss of strength of the usual aluminum alloy structural material becomes significant, so allowable design stresses must be reduced. The result is heavier structure. To ameliorate this problem, more and more of the structure must be built of other materials, such as titanium alloys, which have improved high-temperature characteristics. Figure 18.15 shows typical variations with temperature of

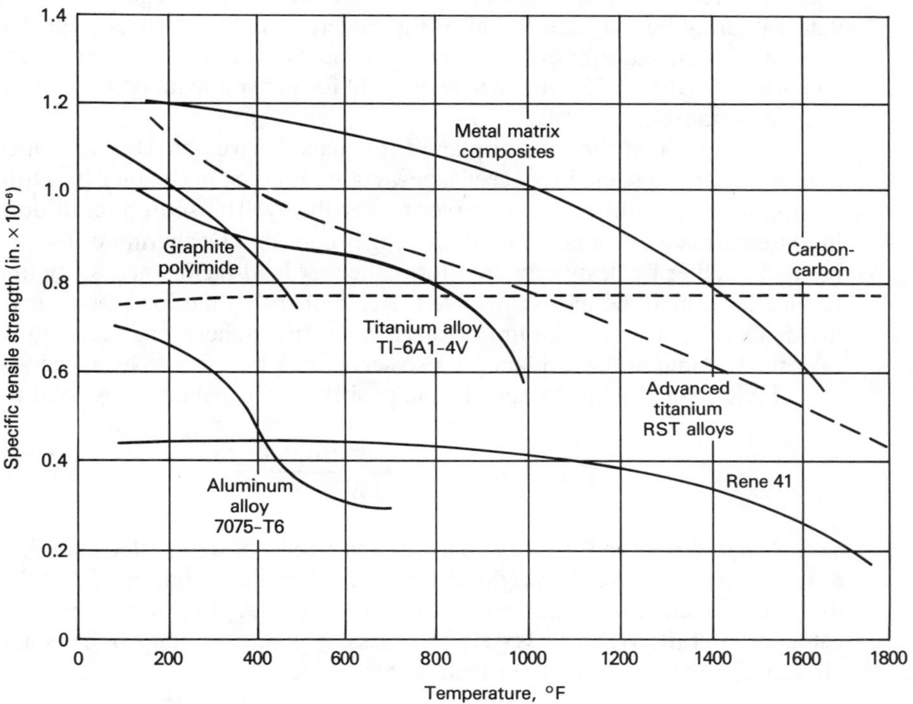

Figure 18.15 Variation of specific strength with temperature for present and possible future materials.

specific strength, the ratio of ultimate tensile stress to density, for several structural materials. It should be noted that each alloy of a metal has a different characteristic. Thus the curve shown for the particular aluminum alloy 7075-T6 will not be identical with other aluminum alloys, but the trend will be similar. Aluminum alloy fades rapidly above about 300° F. René 41, a stainless steel alloy, is much superior to aluminum above 400° F but has a lower specific strength than aluminum at lower temperatures. Titanium alloys, present and advanced, are much better than either of the others but are very expensive. Also shown are curves for future metal matrix composites, graphite polyimide composite, and carbon–carbon, a composite material made of carbon fibers in a carbon matrix or binder. Of course, there is more to materials than specific strength. Fatigue properties, failure modes, cost, and ease of manufacture are a few of the characteristics that must be acceptable to permit use of new materials. Almost always, improved material properties come with significant cost increases.

LOADS

The structural design of an airplane must begin with the determination of the applied loads. Design load criteria have been developed both by rational analysis and experience acquired ever since the development of the airplane began. The results of this data-gathering activity are found in the Federal Air Regulations, which specify the required design loads for every major component of the aircraft for every critical flight condition. FAR Part 23 provides the criteria for general aviation aircraft, and Part 25 addresses transport aircraft.

In level flight the total airplane lift equals the weight. During maneuvers such as turns or pull-ups, the lift exceeds the weight, as noted in Chapter 16. Different types of aircraft require different maneuvering capability. Transport aircraft do not engage in extreme maneuvers because of passenger comfort. Furthermore, the structure can be made lighter by designing only for maneuver load factors necessary in the type of service for which the airplane is used. Remember that the load factor is the total lift divided by the weight. The limit load factor is the highest load factor to be expected over the lifetime of the airplane. It also serves as a limit not to be exceeded by pilots.

FAR 23 and FAR 25 specify the positive limit maneuvering load factor, n, as

$$n = 2.1 + \frac{24{,}000}{W + 10{,}000} \tag{18.1}$$

except that n may not be less than 2.5 and need not be greater than 3.8. W is the design maximum takeoff weight in pounds. When the weight is 50,000 lb or higher, the limit maneuvering load factor is 2.5, the value applicable to almost all transport category aircraft. When W is 4118 lb or less, n is 3.8. Military fighters and aerobatic airplanes are stressed for limit load factors up to 7 or 8.

Airplanes must be designed to withstand the limit load without a permanent deformation in any part of the structure. In other words, the yield stress must not be exceeded at the limit load. In addition, the structure must withstand the ultimate load factors, defined as 1.5 times the limit load factors, without failure. Thus the ultimate allowable stresses of the material must not be exceeded at loads 50% greater than the highest expected load over the life of the airplane.

A structural design must consider not only the maneuvering load factors, but also the load factors that might occur as the result of flying into a sharp edge vertical gust. Such a vertical gust changes the effective angle of attack on the wing from the original angle α to $\alpha + U_g/V_0$, where U_g is the velocity of the vertical gust (Figure 18.16). At high speed, the increased lift from such a gust, together with the original $1g$ lift existing in smooth level flight, may result in a higher limit load factor than the required maneuvering load factor. In that case, the airplane is said to be *gust critical* and the applied loads are based on the maximum gust load factor. The increased lift from the gust is

$$\Delta L = \frac{dC_L}{d\alpha}\frac{U_{gE}}{V_E}\frac{1}{2}\rho_s V_E^2 S \tag{18.2}$$

Equivalent velocities V_E are used here because the gust velocities defined in the FARs are given as equivalent velocities. ρ_s is the standard sea-level air density.

The gusts used in load analysis are specified in the FARs and are based on measurements taken over a long period on many aircraft. The basic gust used for design at the maximum allowable level flight speed V_C is specified as having a vertical velocity of 50 ft/s, defined as an equivalent velocity. This value applies from sea level to 20,000 ft altitude. Above 20,000 ft, the gust can be reduced linearly to 25 ft/s at 50,000 ft. V_C is selected by the airplane designer as a speed that should never be deliberately exceeded. It is a legal maximum speed. V_C is chosen sufficiently beyond the maximum cruising speed so that the pilot will not normally exceed it. A higher gust of 66 ft/s equivalent speed, reducing to 38 ft/s at 50,000 ft, must also be considered. This is called the *rough air gust* and is assumed to be the highest gust velocity to be anticipated even in very turbulent air. It is assumed that, before such a gust might be

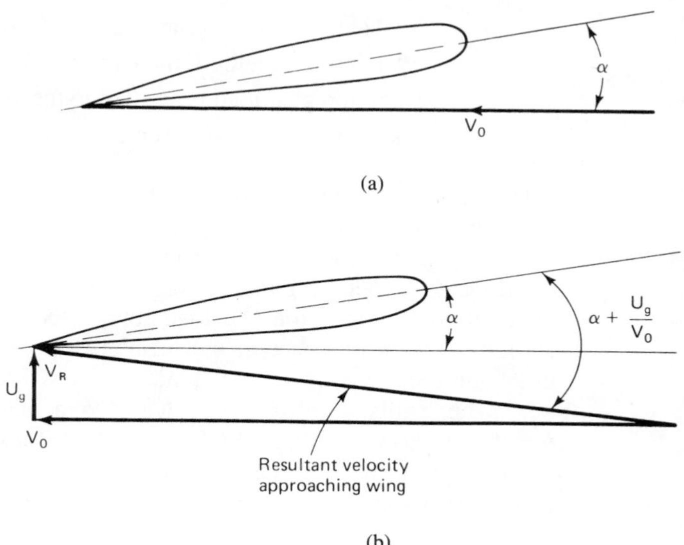

Figure 18.16 Effect of a positive vertical gust on angle of attack: (a) smooth air, level flight; (b) vertical gust encountered.

encountered, the pilot would have slowed to a safe speed known as the *rough air speed*. The rough air speed V_B is the speed at which the increased angle of attack caused by the rough air gust would also cause the airplane to stall. This places an automatic maximum limit on the load that might be generated on the wing. A third defined gust is studied at the maximum airplane dive speed V_D, a speed sufficiently beyond V_C that it is highly improbable that it will ever be exceeded in an inadvertent overspeed. Since this would happen so seldom, hopefully never, it is assumed that the maximum likely coincident gust velocity is only half as large as for the gust considered for normal flight up to V_C.

The gust load factor formula is given in the FAR as

$$n = 1 + \frac{K_g U_{g_E} V_E a}{498(W/S)} \tag{18.3}$$

where

$$K_g = \frac{0.88\mu_g}{5.3 + \mu_g} = \text{gust alleviation factor}$$

$$\mu_g = \frac{2(W/S)}{\rho \overline{C} a g} = \text{airplane mass ratio}$$

U_{g_E} = equivalent gust velocities (ft/s)
ρ = density of air (slugs/ft^3)
W/S = wing loading (lb/ft^2)
$\overline{C}$ = mean geometric chord (ft)
g = acceleration due to gravity (ft/s^2)
V_E = airplane equivalent speed (knots)
a = slope of the airplane normal force coefficient curve per radian if the gust loads are applied to the wings and horizontal tail surfaces simultaneously by a rational method; the wing lift curve slope, C_L per radian, may be used when the gust load is applied to the wings only and the horizontal tail gust loads are treated as a separate condition

The notation used here for U_{g_E} and V_E differs from the FARs to be consistent with the notation of this book.

Equation 18.3 is derived from equation 18.2 with the addition of the gust alleviation factor, which accounts for the fact that the airplane tends to rise away from the gust, giving some automatic load relief, and that a true sharp edge gust does not actually exist. The constant 498 results from the constants converting to consistent units and the value of ρ_S, the standard-day sea-level value of ρ.

The results of a maneuver and gust load factor analysis at a particular altitude and weight are shown in Figure 18.17. The charts show the maneuver and gust load factors versus equivalent airspeed. For some aircraft the gust loads are critical, while for others the maneuver loads predominate. Although negative load factors must also be

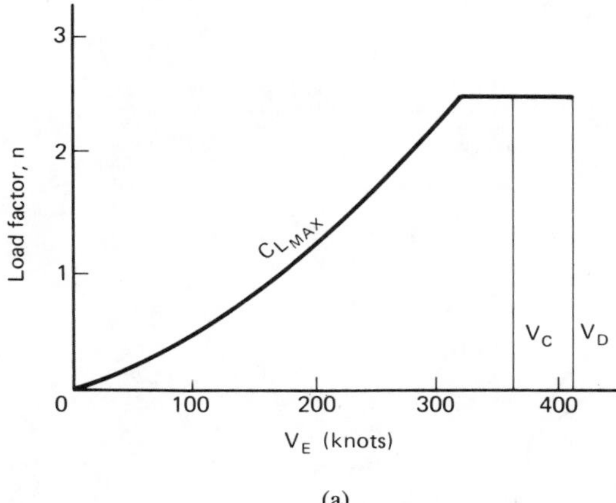

(a)

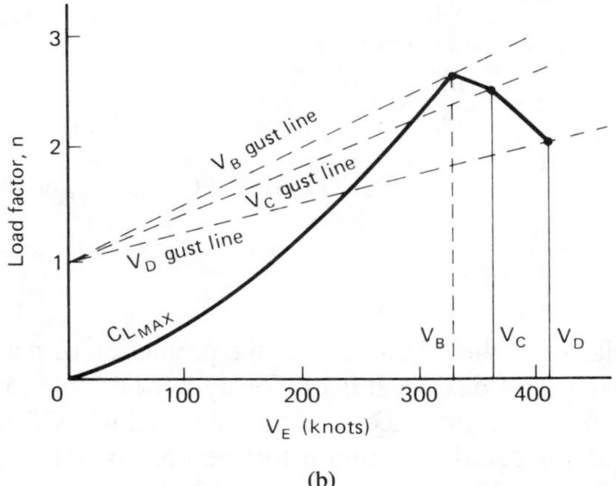

(b)

Figure 18.17 Representative *V–n* diagrams for transport aircraft: (a) maneuver diagram; (b) gust diagram.

considered, they are smaller than the positive, upward load factors and are omitted in Figure 18.17 for clarity. As an example, the negative maneuvering load factor for transports is −1.0. The charts in Figure 18.17 are called *V–n* diagrams and are constructed over the entire altitude range of an airplane to determine the highest load factor at each equivalent speed.

Once the critical load factors are determined, the wing spanwise distribution of the appropriate load is calculated. This aerodynamic loading is adjusted by the opposing downward load of the wing structure, the power plants, if they are located on the wing, and the wing fuel to obtain a total load that must be borne by the wing structure. Note that the effects of dead weight distributed over the wing span, such as fuel, engines, and structural weight, are to reduce structural loads. Such a loading curve is shown in Figure 18.18. From this loading, the shear loads and bending moments can be determined at each wing station. Similar loading analyses must be carried out for all parts of the airplane.

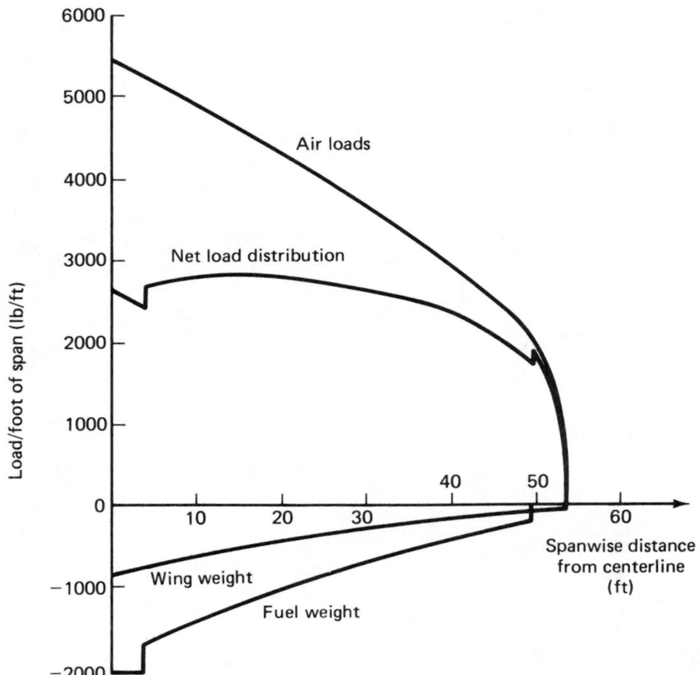

Figure 18.18 Wing load distribution.

WEIGHT ESTIMATION

The weight empty of an airplane is a vital parameter in the prediction of performance and economics. In the early stages of design, it is obviously not possible to calculate the weight of every piece of the structure since no detailed drawings will have been made. In fact, it takes months of detailed design effort before enough information begins to be available to do that. Furthermore, there are many other contributions to weight empty about which little will be known. Examples are the hydraulic system, the interior furnishings such as insulation, wall linings, food equipment and seats, electrical system, and so on. On the other hand, there is no point in spending the money and effort on detailed structural and system design on an airplane that may never be built. Clearly, a method is required to estimate the weight empty of an airplane with sufficient accuracy to appraise the capability of the airplane before any detailed design is started.

The solution lies in the use of statistical methods. The weights of major components and systems of past aircraft are analyzed to determine the parameters upon which these weights depend and the manner in which the weights vary with these parameters. Then the new proposed aircraft need be defined only well enough to define these parameters. To be useful, these parameters must be physically logical. They must also depend on characteristics of the airplane that are significant variables in the design. In other words, it is important to be able to predict the effect on weight empty of design variables such as wing area, span, thickness ratio and taper ratio or fuselage depth, width and floor loadings. A simple example of such statistical weight estimation is that landing gear weight is 4% of the maximum takeoff weight. Although this is not exact, a study of many airplanes shows surprisingly good agreement between this simple estimate and the actual weight of landing gears. A more complex example is given in Figure 18.19, which shows the variation of wing weight per square foot of wing planform area with a wing weight index. The wing weight index can be shown to be proportional to the skin and spar cap compression and tension loads due to the bending moments at the root of a wing, with the exception of the taper ratio terms, which are empirical. Of course, the graph is entirely empirical; but if the index were not fundamentally sensible, there would be much more scatter and the graph would be useless. This curve is applicable to high-speed transports and is of the form

$$\text{wing weight/ft}^2 = A + BI_W$$

where I_W is the wing weight index. The A term takes care of the portion of the structure not dependent on the bending loads.

Similar statistical curves exist for other components of the airplane. The sum of the component weight estimates lead to a total weight-empty estimate. In spite of the difficulty of weight estimation, most aircraft weights come within a few percent of the estimate. This partially results from a process of redesign undertaken if the detailed weights, determined from the actual drawings near the end of the engineering process, are excessive. Then the structure is analyzed to find areas with excessive strength margins. Such parts are redesigned to remove the unnecessary weight. In rare, desperate cases, a total change in structural design concept may be undertaken to meet the original weight estimates. These estimates are so important because they have probably been guaranteed to the customer with a small tolerance of about 2% to 3%.

Although structures are very complex, Figure 18.20 shows that the structural weight is roughly a constant fraction of an effective maximum airplane weight for a wide range of high-speed transport aircraft. The effective maximum weight is taken in Figure 18.20 as the square root of the product of the maximum takeoff weight, TOW, and the maximum zero fuel weight, ZFW. The latter is the maximum allowable airplane weight without fuel. It is significant because the added weight of fuel, almost always carried in the wing, does not add to the wing structural weight. We have mentioned before that the fuel load actually reduces the wing bending moment under the critical flight load conditions. The higher total weight due to the fuel does, of course, affect the load on some parts of the structure, such as the landing gear. Some average weight between the TOW and ZFW is most significant. Although the form of Figure 18.20 could lead to only a first approximation, the relatively small scatter in the structural fraction is interesting. It is also noteworthy that the heavier side of the scatter band are the earlier versions of the aircraft, while the lower values generally apply to the later developments of each aircraft model. All aircraft designs are done under a fairly tight schedule. Only in the later versions has time permitted a refinement of the design to reduce the excess strength in many parts of the first version. Remember that every part must be checked to meet the strength requirements. None can be understrength, but an overstrength condition is safe, although a waste of weight. Only when an airplane is "grown" to a greater capacity, such as lengthening the fuselage and/or increasing the takeoff weight, is full use made of all the original overstrength

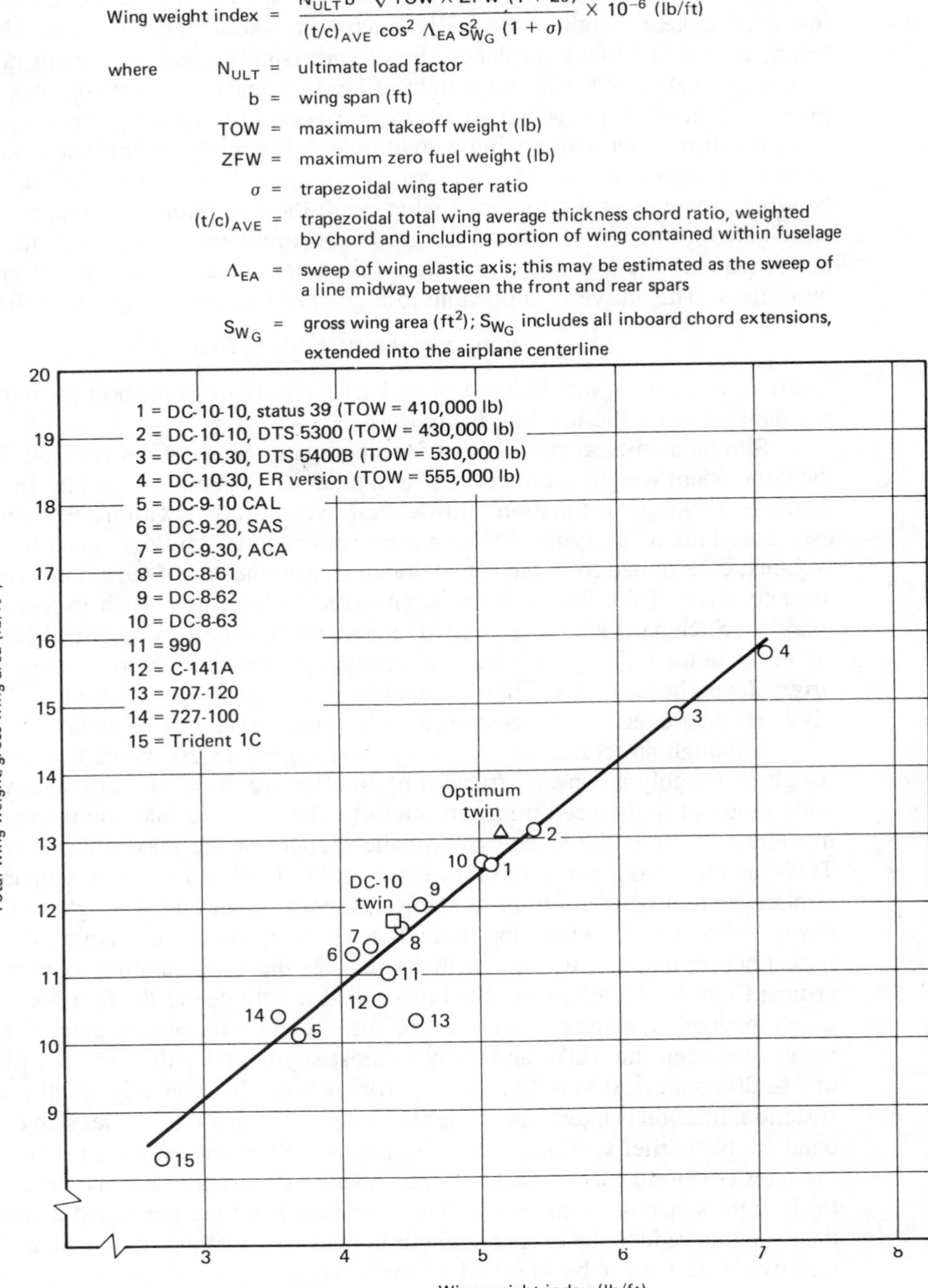

Figure 18.19 Statistical wing weight correlation. (Courtesy of Douglas Aircraft Company.)

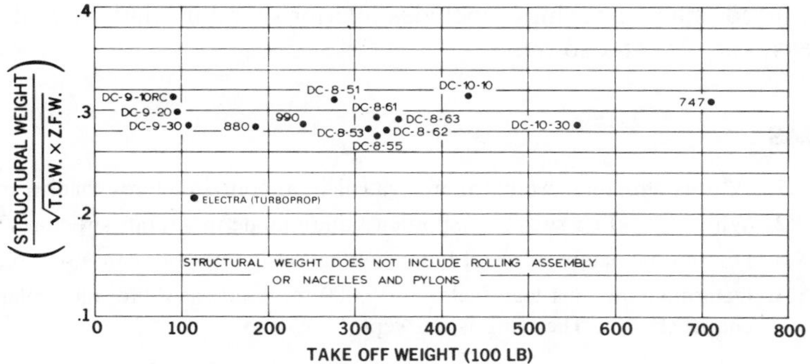

Figure 18.20 Structural weight trends of transport aircraft.

parts. The weight fraction is then improved and falls near the lower edge of the scatter band in Figure 18.20. Note that the slower Lockheed Electra turboprop had a lower structural weight fraction, probably due to much lower design speeds and lower fuselage pressurization requirements. The lower speeds reduce loads and permit the use of higher wing and tail thickness ratios.

Table 18.1 presents the weight breakdowns of several transport-type aircraft varying in size from a Cessna Citation executive jet to the Boeing 747. A weight breakdown shows the weights of the various components and systems of an aircraft. The structural weight is the sum of the wing, tail, body, nacelle system, and the alighting gear except for the wheels and rolling assembly. Study of these weight breakdowns will yield considerable insight into the sources of empty weight in aircraft. If these airplanes were cargo versions, the furnishings weight would be greatly

TABLE 18.1. TYPICAL WEIGHT BREAKDOWNS, LB (REF. 18.3)

Aircraft system	Citation-500	DC-9-30	L-1011	747
Wing	1,020	11,391	47,401	88,741
Tail	288	2,790	8,570	11,958
Body	930	11,118	49,432	68,452
Alighting gear	425	4,182	19,923	32,220
Nacelle system	241	1,462	8,916	10,830
Propulsion system (less dry engine)	340	2,190	8,279	9,605
Dry engine weight	1,002	6,160	30,046	35,700
Flight controls (less auto pilot)	196	1,434	5,068	6,886
Auxiliary power system	0	817	1,202	1,797
Instrument system	76	575	1,016	1,486
Hydraulic and pneumatic system	94	753	4,401	5,067
Electrical system	361	1,715	5,490	5,305
Avionics system (including auto pilot)	321	1,108	2,801	4,134
Furnishings and equipment system	794	8,594	32,829	48,007
Air-conditioning system	188	1,110	3,344	3,634
Anti-icing system	101	474	296	413
Empty weight (M.E.W.)	6,377	55,873	229,014	333,339
Takeoff gross weight	11,650	108,000	430,000	775,000

reduced, since furnishings includes interior sound insulation, wall linings, seats, galleys, carpets, and so on.

PROBLEMS

18.1. Why is structural weight more critical in aircraft structures than in most other structures?

18.2. Why is buckling such an important failure mode in aircraft structures?

18.3. Discuss the nature of metal fatigue. What design practices are used to avoid fatigue failure?

18.4. Determine the gust load factor at 25,000 ft altitude at V_C for an airplane with the following characteristics. The wing is unswept.

$$\text{maximum takeoff weight} = 30{,}000 \text{ lb}$$
$$\text{wing area} = 700 \text{ ft}^2$$
$$\text{wing span} = 76.0 \text{ ft}$$
$$\text{structural design speed, } V_C = 260 \text{ knots, equivalent}$$

REFERENCES

18.1. Lovell, D. T., and Disotell, M. A., *Structural Material Trends in Commercial Aircraft,* AIAA Paper 781552, Aug. 1978.

18.2. Shevell, R. S., "Technological Development of Transport Aircraft—Past and Future," *Journal of Aircraft,* Vol. 17, No. 2, Feb. 1980.

18.3. Beltramo, M. N., Trapp, D. L., Kimoto, B. W., and Marsh, D. P., *Parametric Study of Transport Aircraft Systems Cost and Weight,* NASA CR 151970, April 1977.

18.4. Sechler, Ernest E., and Dunn, Louis G., *Airplane Structural Analysis and Design,* Dover, New York, 1963.

19

HYPERSONIC FLOW

Several flow regimes have previously been discussed. *Incompressible flow* is based on the assumption that density is constant as the pressure changes, an assumption found to introduce negligible error at Mach numbers below about 0.4 and only small errors for most flight characteristics up to at least $M = 0.5$. This assumption was originally used because the analysis is greatly simplified when density is constant. *Subsonic compressible flow* requires accounting for effects of variable density. The importance of density changes gradually increases as Mach number rises so that there is a smooth transition from the speed region in which incompressible methods are accurate to the high-subsonic-speed region. When local supersonic flow occurs on the bodies or surfaces, local shock waves and supersonic expansion waves cause more abrupt changes. *Supersonic flow* was found to be an entirely different regime in which effects of shock waves caused large pressure changes. The mechanisms producing supersonic lift and drag are entirely different from the subsonic physics, so the equations used for analysis are radically altered. The two-dimensional supersonic equations for lift and drag given in Chapter 12 are *linearized* results in which higher-order, nonlinear terms have been considered negligibly small. This assumption requires that the induced velocities be small compared to the freestream velocity (i.e., the disturbances are small). When the Mach number gets sufficiently above 1.0 or the angle of attack becomes too large, the neglected terms begin to be significant and the linearized theory deviates increasingly from experimental results.

Hypersonic flow is flow at velocities much higher than the speed of sound. The term *hypervelocity* is sometimes used. There is no strict definition that delineates the hypersonic Mach number range, although Mach numbers above 5.0 are generally defined as hypersonic. The basic physical definition is that the hypersonic region begins when the oblique shock wave begins to lie very close to the surface of a slender body. This is illustrated in Figure 19.1. The drawing in Figure 19.1a represents supersonic flight at a Mach number of 2.0. The oblique shock is far from the surface of the 4.0% thick diamond airfoil at 2.3 degrees angle of attack. At a Mach number of 7.0 (Figure 19.1b), the shock is close to the airfoil surface. This characteristic can be verified from Figure 12.20, which shows that for any flow deflection angle the shock wave angle decreases continuously to a limiting value as the Mach number increases to infinity. The *shock layer,* the region between the shock wave and the surface, becomes very thin for slender bodies. The linear theory becomes increasingly inadequate because the disturbances created by the body are no longer small.

A second hypersonic characteristic is an increased *viscous interaction* between the boundary layer and the oblique shock wave. At the high Mach numbers, very high temperatures will be developed in regions where the flow is decelerated, such as in the boundary layer. The high temperature in the boundary layer increases the viscosity and decreases the density. The result is that the boundary layer thickness is much greater than at the same Reynolds numbers in subsonic or supersonic flow and is no longer small compared to the shock layer thickness. The *viscous interaction* may alter the external flow significantly.

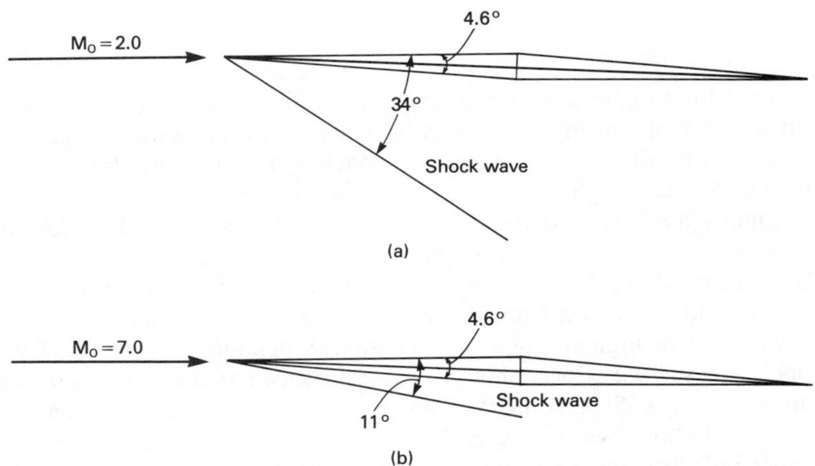

Figure 19.1 Effect of freestream Mach number on shock wave angle, $\alpha = 2.3$ degrees.

TEMPERATURE EFFECTS

Another important effect of very high Mach number is the high surface temperatures encountered. We have developed an equation for stagnation temperatures in Chapter 7. Equation 7.18 states that

$$\frac{T_T}{T_0} = 1 + \frac{\gamma - 1}{2} M_0^2$$

There are limitations to this equation, however, because as the air temperature rises above about 550 K (1000°R) the ratio of specific heats γ begins to vary instead of retaining the constant value of 1.4. The equations that account for the effect of variable γ on the ratio of stagnation or total temperature to the ambient temperature become extremely complex. Charts showing the results of these complex calculations can be found in Ref. 19.1.

Figure 19.2 shows the variation of T_T with Mach number at an altitude of 18 km (59,000 ft). Temperatures are shown based on both equation 7.18 with $\gamma = 1.4$ and on the modification involving variable γ. The effect of variable γ is significant only above a Mach number of 4, with the ambient temperature of 216.66 K existing at this altitude. At Mach 5, variable γ reduces T_T by about 6%. At Mach 8, the reduction is about 15%.

From Figure 19.2, at $M = 6$, $T_T = 1620$ K (2916°R), while at $M = 9$, $T_T = 3270$ K (5885°R). Although these temperatures exist only at the stagnation point, the boundary layer temperatures are only about 10% lower over much of the vehicle surface. These temperatures are destructive to materials, so a hypersonic vehicle must be protected by insulation, by the use of ablative materials on the affected surfaces to absorb the heat, or by active cooling with liquid hydrogen or equivalent. In addition, special structural materials that retain their strength to high temperatures must be used as discussed in Chapter 18. The reader should refer to Figure 18.15.

We have noted that at high gas temperatures the value of γ varies. This is due at first, at about 500 K, to the increase in specific heat arising from the excitation of the vibrational modes of the molecules. Above 2000 K, molecular dissociation of the oxygen commences, and above 4000 K the nitrogen begins to dissociate. At temperatures of about 9000 K, the gas is mostly dissociated and ionization begins. Obviously, the nature of the gas is severely modified. The ratio of specific heats for air decreases from 1.4 at normal temperatures to about 1.2 at 3000 K. Flow calculations for the high-temperature, chemically reacting shock layers become enormously complex, and the equations must be solved numerically by computer.

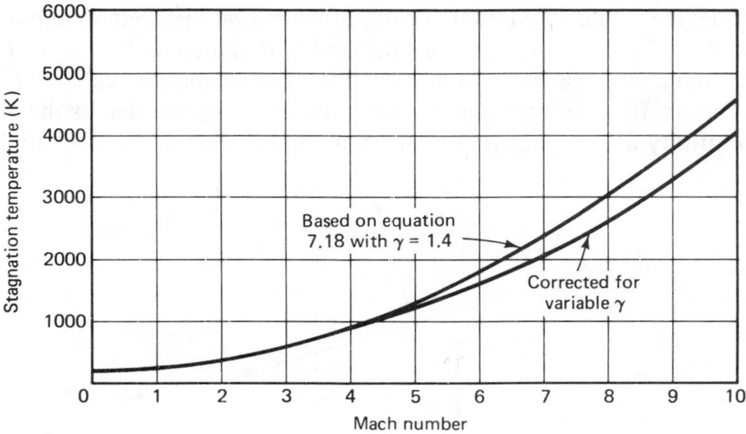

Figure 19.2 Variation of stagnation (total) temperature with Mach number at an altitude of 18 km (59,000 ft).

NEWTONIAN THEORY

Although the detailed theory is very complex, there are some simpler theories that can be used for fairly satisfactory first approximations to the pressure distribution on bodies in hypersonic flow. Surprisingly, the simplest theory in practical use was developed from Isaac Newton's theory of fluid forces, which was given in his *Philosophiae Naturalis Principia Mathematica* published in 1687. Newton, of course, knew nothing of supersonic or hypersonic flow and intended his derivation to apply to all fluid flows. Yet we shall see that the Newtonian theory's only possibility of application is at hypersonic speeds.

Newton assumed that a uniform flow consisted of a steady stream of particles moving along straight parallel paths and that, upon encountering a surface, such as the inclined plane in Figure 19.3, the particles lose their momentum perpendicular to the plate and flow parallel to the plate. The force exerted on the plate is the time rate of change of fluid momentum normal to the plate. From Figure 19.3, the mass of air affected in unit time is the plate area, S, times the sine of the plate angle of inclination to the fluid stream times the velocity, V_0, times the density. Thus, the air mass flow rate is $\rho V_0 S \sin \alpha$. The velocity component normal to the plate is $V_0 \sin \alpha$, and the force on the plate is $F = \rho V_0^2 S \sin^2\alpha$. The net force is the difference between the force with the stream striking the plate and the ambient force when the velocity is zero. In terms of pressure, the force per unit area is $p - p_0$ and the pressure coefficient is

$$C_p = \frac{p - p_0}{\frac{1}{2}\rho V_0^2} = \frac{F}{\frac{1}{2}\rho V_0^2 S} = \frac{\rho V_0^2 \sin^2\alpha}{\frac{1}{2}\rho V_0^2}$$

$$= 2 \sin^2\alpha \qquad \text{Newtonian sine}^2 \text{ law} \tag{19.1}$$

Equation 19.1 shows a maximum pressure coefficient, $C_{P_{max}}$, of 2.0 at $\alpha =$ 90 degrees. This maximum occurs, of course, at the stagnation point. We know from Bernoulli's equation, equations 6.7 and 6.9, and the definition of C_P, equation 8.10, that in incompressible flow the maximum, or stagnation, value of C_P is 1.0. From the theory of flow through shock waves, it can be shown that as the Mach number goes to infinity and γ goes to 1.0, the maximum value of C_P goes to 2.0, agreeing with

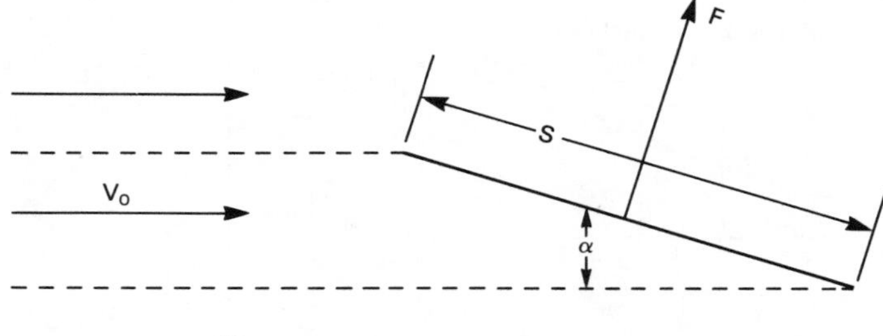

Figure 19.3 Newton's concept of force on an inclined surface in a fluid stream.

the Newtonian theory. The student can easily show that an intermediate stagnation value at $M = 1.0$ is 1.28 (combine equations 8.10 and 7.19).

At high but finite Mach numbers, $C_{P_{max}}$ values will be somewhat less than 2.0. As noted above, shock-wave theory provides a means of evaluating the maximum stagnation pressure behind a shock wave. Reference 19.2 discusses shock wave theory and provides tables of the loss in stagnation pressure across normal shock waves based on $\gamma = 1.4$. At a Mach number of 10, for example, $C_{P_{max}}$ is 1.84 with γ assumed to be 1.4. Actually, γ decreases as temperature rises, so $C_{P_{max}}$ may be somewhat closer to 2.0.

The significance of the reduced $C_{P_{max}}$ is that, if we know that the maximum pressure coefficient is less than 2.0, the Newtonian equation can be improved by replacing the constant 2 by the $C_{P_{max}}$ applicable to the Mach number being studied. Thus the modified Newtonian equation is

$$C_P = C_{P_{max}} \sin^2\alpha \qquad \text{modified Newtonian law} \qquad (19.2)$$

Because of its simplicity, the modified Newtonian law is frequently used in the preliminary design of hypersonic vehicles. Agreement with computational fluid dynamic solutions is generally better at high Mach numbers and with three-dimensional bodies.

The reason why the Newtonian solution works is because at hypersonic speeds, where the shock waves lie very close to the body, the Newtonian assumption of straight particle flow until the particles impact the body is nearly true. At supersonic speeds, the shock wave angle is much greater than the body angle, so the fluid particles are deflected before reaching the body surface. At subsonic speeds, the fluid feels the presence of the body far ahead, so the streamlines start curving around the body even before the leading edge of the body is reached.

There are intermediate methods of hypersonic analysis that are more complex than the Newtonian, but much simpler than detailed numerical solutions. Among these are the tangent wedge and the tangent cone methods. There are exact inviscid supersonic solutions for simple wedges and for cones at zero angle of attack. These results are available in tables or graphs. The tangent methods assume that the local pressure on a body is equal to the exact value of the pressure on a wedge or cone surface having the same inclination to the flow direction as the body surface and with the same Mach number.

SUMMARY

Figure 19.4 illustrates schematically how pressure coefficient on a wedge varies with Mach number over the entire speed range based on the applicable linear theories. Starting with a zero Mach number $C_{P_{inc}}$, the C_P rises with Mach number subsonically in accordance with equation 8.13, $C_P = C_{P_{inc}}/\sqrt{1 - M^2}$. At $M = 1.0$, this equation provides an infinite value of C_P and is obviously useless. Above $M = 1.0$, the

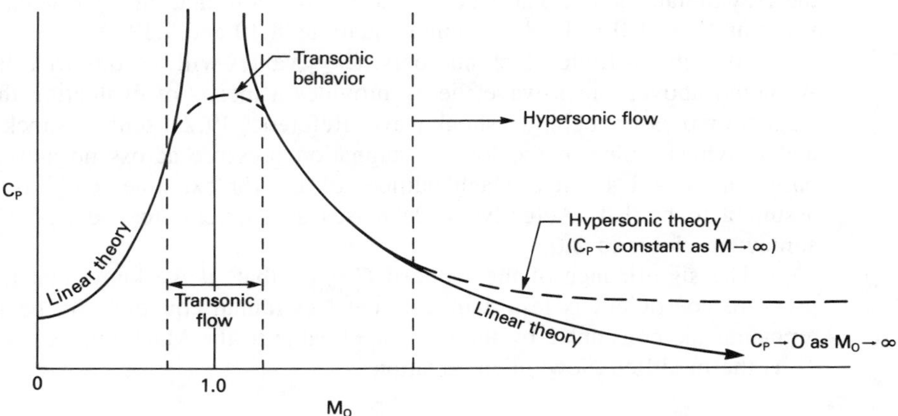

Figure 19.4 Pressure coefficient trends over the entire Mach number range.

linearized supersonic equation, 12.11, $C_P = 2\theta/\sqrt{M^2 - 1}$ applies. Again, as M approaches 1.0, the equation goes to infinity. In the transonic speed range, from approximately $M = 0.8$ to $M = 1.2$, both subsonic and supersonic flows are present. Transonic theory provides some understanding to permit correlation of experimental results (see Figure 12.13), but numerical results were lacking until the advent of numerical solutions using computational fluid dynamics methods and high-speed computers. The line labeled transonic behavior is representative of such experimental data. At very high Mach numbers, the supersonic C_P from equation 12.11 goes to zero, while hypersonic theory, including the Newtonian equation, yields a constant value independent of Mach number.

We can summarize the hypersonic discussion by noting that, while the simple Newtonian results are useful, most detailed calculations will involve powerful computers running computational fluid dynamic algorithms of great complexity. When the temperatures in the shock layer rise to the point where dissociation and ionization occur, the accuracy of the results is in doubt. One problem is that no experimental facilities exist to obtain pressures and forces at Mach numbers above about 10 with long enough run times for proper measurement. It is also difficult to match both Mach numbers and the appropriate temperatures. This makes hypersonic vehicle development a difficult business. Existing hypersonic vehicles are rockets, missiles, and the space shuttle, which share the characteristic of short flight times in the atmosphere (see Chapter 20). Present hypersonic research focuses on the national Aerospace Plane (NASP) which would take off and land like an airplane and have the capability of going into orbit. Hypersonic cruise aircraft for interceptor and transport purposes are also envisioned. These missions involve much longer flight times in the atmosphere at high temperatures and the use of scramjet propulsion (see Chapter 17). The problems are much more complex. Nevertheless, the next decades will see significant advances in this field. A conceptual drawing of a hypersonic aircraft is shown in Figure 19.5.

Figure 19.5 Conceptual drawing of a hypersonic aircraft. (Courtesy of McDonnell-Douglas Corporation.)

REFERENCES

19.1. Ames Research Staff, *Equations, Tables, and Charts for Compressible Flow,* NACA Report 1135, 1953.

19.2. Anderson, John D., Jr., *Fundamentals of Aerodynamics,* McGraw-Hill, New York, 1984.

19.3. Cox, R. N., and Crabtree, L. F., *Elements of Hypersonic Aerodynamics,* English Universities Press Ltd., London, 1965.

20

ROCKET TRAJECTORIES AND ORBITS

Satellites and space vehicles differ in many fundamental ways from aircraft. Most of an airplane's flight is in a steady-state condition for which the lift is equal to or very nearly equal to the weight. Except in descent, the thrust is equal to or greater than the aerodynamic drag. Satellites and space vehicles operate without an atmosphere or in regions for which the air density is so low that its effect is very small. Launch vehicles, of course, must pass through the atmosphere, so the drag in rocket trajectories is significant. The efficiency of a rocket is enhanced by a rapid passage through the atmosphere, however, so the acceleration and velocity must be high. As a result, the thrust is large compared to the drag, and the percentage effect on rocket performance of a small drag error is relatively small.

Airplane aerodynamics is deeply involved with the creation of lift and drag and the means of evaluating them. Because of the differences we have just described, satellites and space vehicles are not concerned with lift and drag except during the early stages of launch and during reentry to the earth's atmosphere. Flight of launching rockets is mostly a balance between thrust and gravity forces, while satellite flight paths involve the interaction of gravity and centrifugal force. A spacecraft flight path depends on the gravitation fields of various celestial bodies and flight path corrections provided by small rockets.

A reentry vehicle depends on lift and drag at hypersonic speeds ($M > 5$), for which, as discussed in Chapter 19, the physics are quite different from the physics at subsonic and supersonic speeds. Of course, a vehicle that lands like an airplane (e.g., the space shuttle) must be capable of controllable flight in all speed regimes. The largest problems for reentry vehicles are caused by the high surface temperatures generated at high Mach numbers.

ROCKET TRAJECTORIES

Consider the velocity history of a rocket in field-free space (i.e., no gravity or drag). From the discussion of rocket propulsion we have seen that the thrust is

$$F = c_e \frac{dm}{dt}$$

where

F = thrust, N (lb)

c_e = effective exhaust velocity, m/s (ft/s) (assumed constant)

$\frac{dm}{dt}$ = mass flow rate of the propellant exhaust gases, kg/s (slugs/s)

The thrust must equal the time rate of change of momentum of the rocket, so

$$F = M\frac{dV}{dt} = c_e \frac{dm}{dt}$$

where

M = instantaneous mass of the rocket, kg (slugs)

V = velocity of the rocket, m/s (ft/s)

Then

$$dV = \frac{c_e}{M} dm$$

Starting at $V = 0$ and letting M_1 equal the mass of the rocket without fuel, we obtain

$$V_b = \int_0^{V_b} dV = c_e \int_{m_t}^{0} \frac{dm}{M} = -c_e \int_{M_1+m_t}^{M_1} \frac{dM}{M} \tag{20.1}$$

where

V_b = burnout velocity when all propellant is consumed

M_1 = rocket mass without propellant

m_t = total propellant mass

Note that we have changed the variable from the mass of propellant consumed, m, to the instantaneous mass of the rocket, M. $dm/dt = -dM/dt$, since a positive amount of propellant burned reduces the total vehicle mass. Then

$$V_b = \left(-c_e \log_e M\right)_{M_1+m_t}^{M_1} = c_e \log_e \frac{M_1 + m_t}{M_1} \tag{20.2}$$

Thus

$$V_b \text{ (final speed)} = c_e \log_e \frac{\text{initial weight}}{\text{final weight}} \tag{20.3}$$

Now introduce a gravity field such as that which exists near the earth's surface. There will be a component of gravity opposing the rocket thrust, as shown in Figure 20.1. Equating forces parallel to the flight path,

$$F_{\text{net}} = M\frac{dV}{dt} = c_e\frac{dm}{dt} - Mg \sin\theta \tag{20.4}$$

$$dV = \frac{c_e}{M}dm - g \sin\theta\, dt \tag{20.5}$$

If we assume a constant value of g, such as g_0 at the surface of the earth, we can easily integrate from $t = 0$ to t_b, the burnout time, and obtain

$$V_b = \int_0^{V_b} dV = c_e \log_e\left(\frac{M_1 + m_t}{M_1}\right) - g_0 t_b \sin\theta \tag{20.6}$$

Thus the velocity gained by the rocket will be reduced by the gravity-loss term $-g_0 t_b \sin\theta$. For a vertical flight path, θ is 90 degrees and $\sin\theta = 1.0$. The burning time has now become important. In the ideal no-gravity case, burning time did not have any effect.

Actually, as noted in Chapter 5, g is not constant but varies inversely as the square of the distance from the center of the earth. For antiaircraft rockets, whose targets are usually below 16 km (52,500 ft) and almost never over 20 km, g differs from the surface value by only about 0.5% at the top of the trajectory. The average error is approximately half of that. For space and orbital vehicles, however, the variable effect of g is large and must be considered. At a height of 50 km (31 statute miles or 164,000 ft), g is 1.5% less than at the surface of the earth. Similarly, at

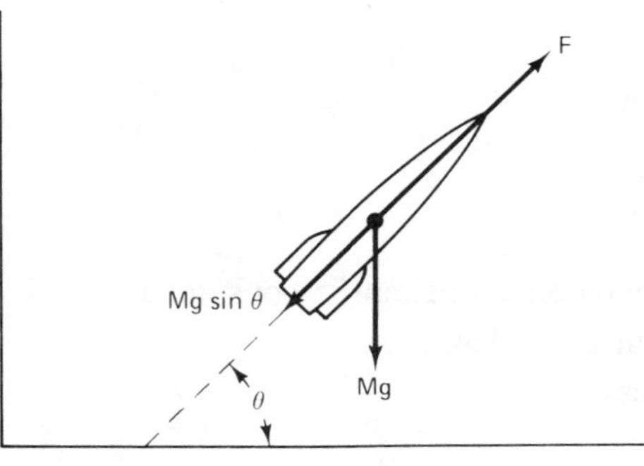

Figure 20.1 Rocket accelerating under the influence of gravity.

heights of 100 km (62.1 st. mi), 300 km (186.4 st. mi), and 600 km (372.8 st. mi), g is less than the surface value by 3.1%, 8.8%, and 16.5%, respectively. For such cases, equation 5.14,

$$g = g_0\left(\frac{r_0}{r_0 + h}\right)^2$$

must be substituted in equation 20.5. The gravity-loss term $-g_0 t_b \sin\theta$ in equation 20.6 then becomes

$$-g_0 \int_0^{t_b} \left(\frac{r_0}{r_0 + h}\right)^2 \sin\theta\, dt \tag{20.7}$$

In general, the flight path will be curved rather a straight line. A rocket with a flight path that is not vertical will have forces acting perpendicular to the flight path unless the rocket is flown at an angle of attack for which the lift of the body and fins balances the weight component perpendicular to the flight path. Figure 15.7 for a climbing aircraft illustrates these forces. For the case with zero angle of attack (i.e., no lift), the acceleration in the perpendicular direction is

$$\frac{dV_\perp}{dt} = \frac{-mg\cos\theta}{m} = -g\cos\theta \tag{20.8}$$

The change in the flight path angle per unit time will be

$$\frac{d\theta}{dt} = \frac{dV_\perp}{dt} \times \frac{d\theta}{dV_\perp} = \frac{dV_\perp}{dt} \times \frac{1}{V} = -\frac{1}{V} g\cos\theta \tag{20.9}$$

Equations 20.5, 20.7, and 20.9 can only be solved by numerical integration, usually done by computer.

In a typical vertical booster flight, with $t_b = 100$ s, velocity losses due to gravity are of the order of 975 m/s (3200 ft/s), about 9% to 13% of the velocity required for escape from the earth or orbital flight, respectively. In addition, drag must be taken into account. The velocity losses due to drag for an average rocket booster vertically are about 10% of those due to gravity or about 100 m/s (328 ft/s). On the other hand, decreasing atmospheric pressure increases effective exhaust velocity and thrust and contributes to an increase in vehicle velocity of 5% to 10%.

The difficulty of designing space launching systems can be quickly understood from a study of equation 20.3. Choosing a range of final speeds, assuming a value of effective rocket exhaust velocity of 12,300 ft/s, typical of liquid hydrogen and oxygen, and solving for the ratio of initial weight to final weight yields the following results:

Final velocity, ft/s	Initial weight / Final weight	Fuel weight / Initial weight
10,000	2.25	.56
20,000	5.08	.80
25,000 (orbital)	7.63	.87
36,700 (escape)	19.76	.95

To obtain orbital or escape velocity with a single rocket requires that the fuel weight be 87% or 95%, respectively, of the total weight of the vehicle. The final weight, including the structure, the engine, the fuel system, controls, communications, and the payload must be only 13% of the initial weight for the orbital case or 5% for the escape velocity mission. The requirements for skillful design of all parts of the rocket are obviously very stringent. In fact, the problem is so difficult that space vehicles are not generally built as a single vehicle but consist of several stages as discussed in the next section. Note that we have been optimistic in this analysis because equation 20.3 does not include the velocity losses due to gravity and drag.

MULTISTAGE ROCKETS

We have seen from equation 20.2 that the velocity obtained by a rocket is very much dependent on the ratio of the initial weight to the weight after the fuel is burned (i.e., the final weight). Accelerating the structure of a large rocket to high speeds takes a lot of energy. Elimination of fuel tanks and their supporting structure after their fuel is used permits more efficient use of the remaining fuel. In practice, this is achieved by *staging,* the carrying of fully loaded small rockets as payload on larger ones. When the large rocket tanks are emptied, this whole rocket is dropped. The engine is started on the smaller rocket and only its smaller tanks and structure need be accelerated to higher speeds. Multistage rockets may have two, three, or even more stages.

If the thrust of an upper-stage rocket begins as soon as the thrust of the stage beneath it has terminated, the net velocity increment is

$$V_b = \Delta V_1 + \Delta V_2 + \cdots$$

or

$$V_b = c_{e_1} \log_e\left(\frac{M_1 + m_t}{M_1}\right)_1 + c_{e_2} \log_e\left(\frac{M_1 + m_t}{M_1}\right)_2 + \cdots \\ -g(t_1 \sin \theta_1 + t_2 \sin \theta_2 + \cdots) \tag{20.10}$$

c_{e_i}, $[(M_1 + m_t)/M_1]_i$, and t_i are the effective exhaust velocities, mass ratios, and flight times of each stage. The stages are not independent since each constitutes the payload of the one beneath. Thus, for each pound added to the terminal payload of a three-stage rocket, it may be necessary to add 100 lb to the first-stage booster if a specified terminal velocity is to be achieved.

With a given total mass of structure and propellant, and assuming that the exhaust velocity c_e is the same for all stages, the optimum terminal velocity is obtained when all stages have the same mass ratio, $R = (M_1 + m_t)/M_1$. For this case with n stages, assuming no gravity,

$$V_b \text{ (optimum)} = nc_e \log_e \frac{M_1 + m_t}{M_1} = nc_e \log_e R = c_e \log_e R^n \tag{20.11}$$

$$= n(\Delta V_b \text{ for one stage}) \tag{20.12}$$

Example 20.1

An instructive example is to calculate the velocity achieved by two identical rockets in tandem (Figure 20.2). If each has mass ratio $R = 6$, total weight = 1800 lb, and exhaust velocity $c_e = 11{,}520$ ft/s, we may calculate individual stage velocities from equation 20.2. In Figure 20.2, W_1 is the empty weight and W_p is the propellant weight. Drag will be neglected.

Solution: In field-free space, considering the upper stage alone,

$$\Delta V_2 = c_{e_2} \log_e R_2 = 11{,}520 \ln\left(\frac{1800}{300}\right)$$
$$= 20{,}640 \text{ ft/s}$$

For the first or booster stage,

$$\Delta V = c_{e_1} \log_e R_1 = 11{,}520 \ln\left(\frac{1800 + 1800}{1800 + 300}\right)$$
$$= 6210 \text{ ft/s}$$
$$\Delta V_{\text{TOT}} = 20{,}640 + 6210 = 26{,}850 \text{ ft/s}$$

Thus, when one rocket is loaded by a duplicate of itself, it achieves only about one-third of its normal velocity. In the earth's field, both stages are slowed down still further. If we assume that burning time $t_b = 100$ s, then each stage loses $-gt_b = 3200$ ft/s, and the net velocity is

$$\Delta V_{\text{net}} = 26{,}850 - (2 \times 3200) = 20{,}450 \text{ ft/s}$$

The actual attained speed of the pair is less than what one rocket alone would achieve in a field-free vacuum. If the masses were redistributed to give equal R's to each stage, about 10% improvement in ΔV_{TOT} would result. On the other hand, if t_b were increased to 200 s, the rocket would not be able to leave the ground.

The trajectory equations, including both drag and gravity terms, become rather complex and defy a closed-form solution. A numerical integration (i.e., a step-by-step calculation) is used.

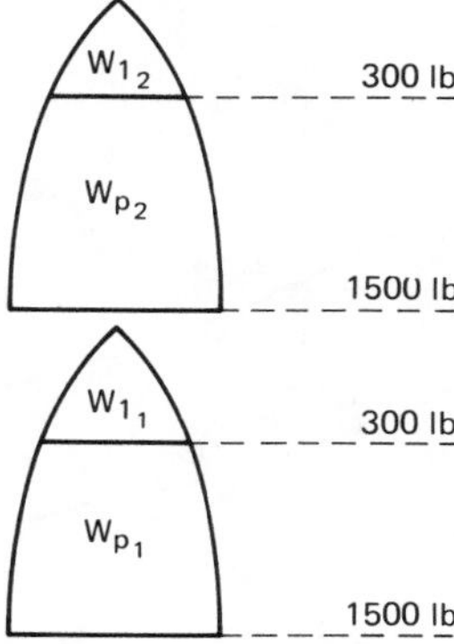

Figure 20.2 Two-stage rocket, both stages having the same total weight and mass ratio.

ESCAPE VELOCITY

One interesting requirement for space travel is the need to attain the escape velocity, the velocity that gives the vehicle sufficient kinetic energy to provide the work necessary to overcome the gravitational force of the earth. This velocity can be found by equating the kinetic energy of a body to the gravity-caused work. Neglecting the rotation of the earth and the relatively small attraction of other celestial bodies,

$$\frac{MV^2}{2} = \int_{r_0+h_b}^{\infty} Mg\,dr \tag{20.13}$$

where

r_0 = radius of the earth, 6.378×10^6 m (20.92×10^6 ft)
h = instantaneous altitude above sea level, m (ft)
h_b = altitude at end of fuel burn

From equation 5.14,

$$g = g_0\left(\frac{r_0}{r_0 + h}\right)^2$$

Substituting this expression for g in equation 20.13 and integrating yields

$$V_e = r_0\sqrt{\frac{2g_0}{r_0 + h_b}} \tag{20.14}$$

where V_e is the escape velocity in m/s (ft/s).

The velocity of escape at the earth's surface is 11,200 m/s (36,700 ft/s) and does not vary appreciably within the earth's atmosphere. The variation of V_e with altitude is shown in Figure 20.3. Since it is physically impossible to travel this fast in the atmosphere without burning up from the excessively high skin temperature, a space rocket must pass through the atmosphere at relatively low velocities and then accelerate to V_e outside the atmosphere.

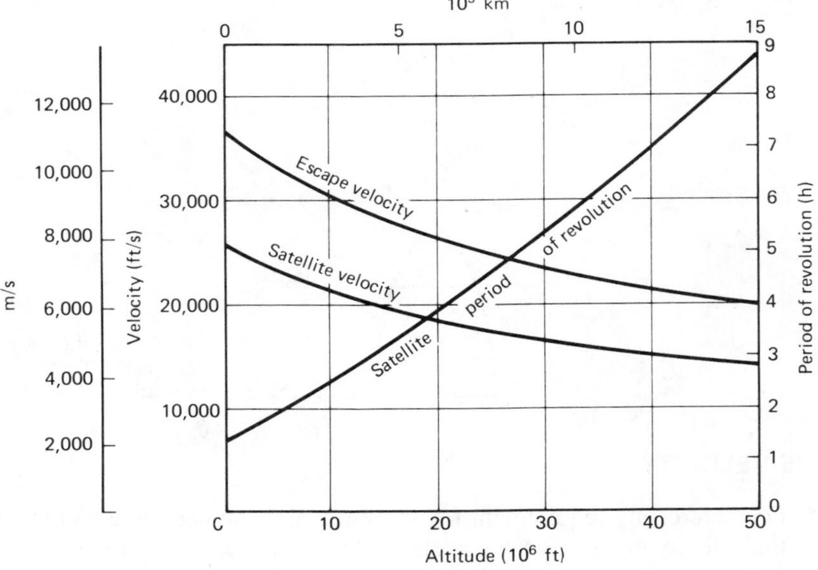

Figure 20.3 Satellite velocity, period of revolution, and escape velocity as functions of altitude. A circular satellite orbit is assumed.

CIRCULAR ORBITAL OR SATELLITE VELOCITY

A satellite or orbital vehicle could remain in orbit outside the earth's atmosphere for an indefinite time without the addition of further energy. The altitude of the orbit must be above the earth's atmosphere to prevent expenditure of energy in the form of drag. Any drag would momentarily reduce the speed and allow the satellite path to bend below the circular. The pull of gravity then accelerates the satellite along a steepening flight path. The higher air density together with the increasing speed raises the drag and consumes increasing amounts of the satellite energy. Eventually, the satellite descends into the lower atmosphere at high speeds and burns up.

For a circular trajectory, the velocity of the satellite must be sufficiently high so that its centrifugal force will balance the earth's gravitational attraction. Thus, if V_s is the satellite velocity, M the satellite mass, and r_0 the radius of the earth,

$$\frac{MV_s^2}{r_0 + h} = Mg$$

Again substituting equation 5.14 and solving for V_s yields

$$V_s = r_0\sqrt{\frac{g_0}{r_0 + h}} \tag{20.15}$$

Thus the circular satellite velocity is lower than the escape velocity from the same altitude by factor of $1/\sqrt{2}$. The period, in seconds, of one revolution for a circular orbit relative to a stationary earth is

$$\begin{aligned}\tau &= \frac{\text{circumference at altitude, } h}{V_s} \\ &= \frac{2\pi(r_0 + h)}{V_s}\end{aligned} \tag{20.16}$$

Substituting equation 20.15, we obtain

$$\tau = \frac{2\pi(r_0 + h)}{r_0\sqrt{g_0/(r_0 + h)}} = \frac{2\pi(r_0 + h)^{3/2}}{r_0\sqrt{g_0}} \tag{20.17}$$

The satellite velocity V_s and the period τ are shown in Figure 20.3 as a function of altitude. A satellite circling the earth at an altitude of 483 km (300 statute miles) would have a velocity of 7582 m/s (24,874 ft/s) and would circle a stationary earth in 1.570 h.

ELLIPTICAL ORBITS

The circular orbit is a special case of the more general elliptic orbit. An ellipse is defined as a closed path enclosing two fixed points, called the *foci,* such that the sum of the distances from any point on the circumference of the ellipse to each of the foci is a constant. Thus, from Figure 20.4, $r + r' = \text{constant} = c_2$. In the case of an earth satellite, the center of the earth is located at one of the foci, called the *occupied or true focus*. The other focus is the *vacant focus*.

The line segment connecting the foci and extending out to the circumference of the ellipse is known as the *major axis*. The end points of the major axis are the *vertices*. The vertex closer to the earth (i.e., the occupied focus) is called the *perigee*. The vertex farther from the earth is the *apogee*. The distance between the vertices is the length of the major axis and is designated as $2a$. The midpoint of the major axis is the *origin* of the ellipse. Each vertex is located at a distance a from the origin, half the length of the major axis. Thus a is known as the *semimajor axis*.

For the special case where the satellite is at perigee, the radius from the occupied focus, F (see Figure 20.4), is equal to r measured to the perigee point, and r', the radius from the unoccupied focus, is equal to $2a$ less the distance from V' to F'. From symmetry, it is clear that $V'F'$ is equal to r. Then

$$r + r' = r + (2a - r)$$

$$r + r' = 2a \tag{20.18}$$

By definition of an ellipse, $r + r' =$ constant everywhere along the circumference. Thus we have proved that the constant is $2a$ (i.e., the length of the major axis) and equation 20.18 applies at all points along the ellipse.

The distance from the origin to either focus in Figure 20.4 is called the *linear eccentricity, c,* of the ellipse. It is a measure of the "flattening" or elongation of the ellipse. When $c = 0$, the ellipse is a circle. The effect of linear eccentricity on the shape of an ellipse depends on the size of the ellipse. A nondimensional shape parameter is produced by dividing c by a. This parameter is the *eccentricity, e,* of the ellipse. Thus

$$e = \frac{c}{a} \tag{20.19}$$

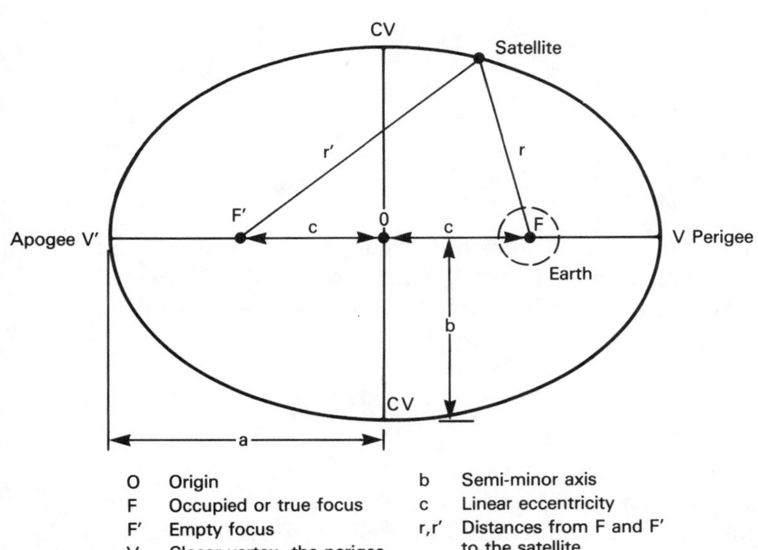

Figure 20.4 Geometry of the elliptical orbit.

When $e = 0$, the ellipse is a circle. When $e = 1.0$, the ellipse becomes a straight line on the major axis.

The *minor axis* of an ellipse is a line segment perpendicular to the major axis at the origin, O, and extending out to the ellipse. The intersections of the minor axis with the ellipse are called the *covertices*. The length of the minor axis is $2b$, so b is the length of the *semiminor axis*. Studying the geometry of an ellipse when the satellite is located on the minor axis at a covertex makes clear that for this symmetrical configuration

$$r' = r$$

Substituting in equation 20.18,

$$2r = 2a \quad \text{so} \quad r = a$$

By the Pythagorean theorem,

$$b = \sqrt{a^2 - c^2} \tag{20.20}$$

Recognizing that $c = ae$ and simplifying,

$$\frac{b}{a} = \sqrt{1 - e^2} \tag{20.21}$$

Also

$$e = \sqrt{1 - \frac{b^2}{a^2}} \tag{20.22}$$

Thus the ratio of b to a, which defines the shape of the ellipse, depends on the eccentricity e, or conversely, given the shape of the ellipse, e is defined by b/a. The ratio b/a and the dimension of either b or a, or e and the length of either b or a completely define the ellipse. Notice that when $b/a = 1$ the ellipse becomes a circle and $e = 0$.

Two other interesting expressions, easily derived, are that the radius at perigee, r_p, is

$$r_p = a(1 - e) \tag{20.23}$$

and the radius at apogee, r_a, is

$$r_a = a(1 + e) \tag{20.24}$$

Note that if any two of the parameters a, b, c, e, r_p, and r_a are given the geometry of the elliptical orbit can be completely described.

The total energy of a satellite in orbit remains constant if there is no friction or other braking force, a situation that exists in orbits outside the atmosphere. The total energy consists of kinetic energy, which equals one-half the mass times the square of the velocity, and the potential energy, which represents the work done in lifting the body to the radius, r, from the center of the earth. Note that r is the same quantity we have designated as $r_0 + h$ (i.e., the radius of the earth plus the altitude).

An increase in potential energy shows up as a loss in kinetic energy, and vice versa. A satellite at perigee or minimum radius of its orbit possesses minimum potential energy and maximum kinetic energy and speed. At apogee, a satellite has maximum potential energy and minimum speed. For a circular orbit, the perigee height equals the apogee height, and the kinetic energy is constant throughout the orbit.

In the early part of the seventeenth century, Johannes Kepler, a German astronomer and mathematician, induced some fundamental relationships of planetary motion from a careful study of astronomical observations. These laws of planetary motion also apply to artificial satellites such as earth satellites. They are as applicable today as they were almost four centuries ago.

Kepler's first law states that a satellite travels in an elliptical path around its center of attraction, the latter being located at one of the foci of the ellipse. It follows that the orbit must lie in a plane containing the center of the attracting body (e.g., the earth).

Kepler's second law states that the radius vector from the center of the earth (we shall assume that the earth is the attracting body) sweeps out equal areas in equal times. This law defines the nature of the speed variation around the orbit. In Figure 20.5, a satellite's travel time between adjacent points is the same although the distances between adjacent points vary, so the area swept by the radius vector will be constant. It can be seen that the second law requires a higher speed when the radius vector is small near perigee and a lower speed near apogee where the radius is the maximum in order to cover equal areas in equal time. This conclusion was already intuitive from energy considerations, which required the lowest speed at apogee.

The variation of velocity along the path of an elliptical orbit around the earth, or any other body, can be further visualized in Figure 20.6. The velocity vectors, which are always tangent to the ellipse, are shown at various points around the ellipse. Also shown are the components of the velocity vectors in the radial direction, from the satellite toward the earth, and in the angular direction, perpendicular to the radius. The lengths of the vectors show the relative speed.

Figure 20.6 shows that the velocity is only in the angular direction at perigee and apogee. At all other points, there is a velocity component toward or away from the earth. The speed at apogee is much lower than at perigee. When traveling from perigee to apogee, the radial velocity component is away from the earth, but with ever decreasing speed due to the attraction of the earth. When moving from the apogee to perigee, the radial velocity component is toward the earth, and the pull of gravity increases the speed.

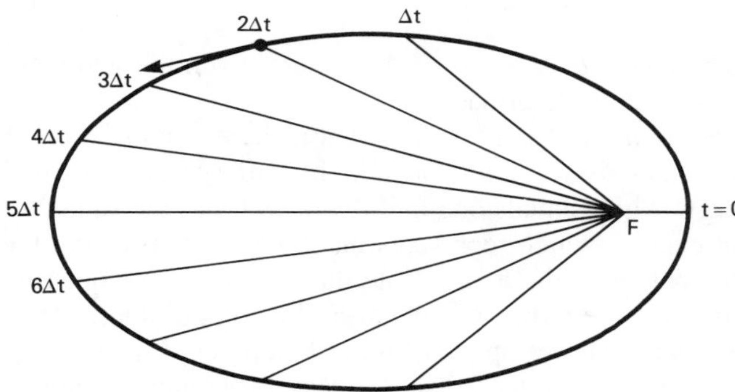

Figure 20.5 Illustration of Kepler's second law showing the position of the radius vector from the center of the earth to an earth satellite in equal time intervals. The area swept by the radius vector in equal time intervals is constant.

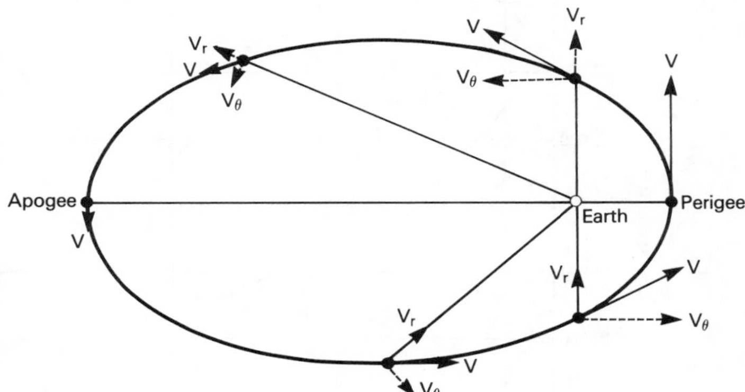

Figure 20.6 Velocity variation around an elliptical orbit. Velocity components are shown along and perpendicular to the radius vectors.

Kepler's third law specifies that the ratio of the squares of the orbital periods of any two satellites about the same body equals the ratio of the cubes of the semimajor axes of the respective orbits. Thus

$$\frac{\tau_1^2}{\tau_2^2} = \frac{a_1^3}{a_2^3} \tag{20.25}$$

The equations of motion of a satellite, derived from Kepler's laws, can be used to determine the speed at any point along the elliptical orbit as

$$V = \sqrt{g_0 r_0^2\left(\frac{2}{r} - \frac{1}{a}\right)} \tag{20.26}$$

Study of equation 20.26 gives further confirmation that the highest velocity is obtained when the orbit comes closest to its attracting focal center at perigee, and a minimum velocity occurs at apogee, where r is as large as possible. It can also be shown that the velocity at apogee, V_a, and the velocity at perigee, V_p, are given by

$$V_a = \sqrt{\frac{g_0 r_0^2(1 - e)}{a(1 + e)}} \tag{20.27}$$

and

$$V_p = \sqrt{\frac{g_0 r_0^2(1 + e)}{a(1 - e)}} \tag{20.28}$$

For a circular orbit, $r = a = b$, $e = 0$, and equations 20.26, 20.27, and 20.28 become identical with equation 20.15.

Figure 20.7 shows the variation in speed with radius around an elliptical orbit with a selected value of the length of the semimajor axis, a, of 13,185 n. mi. These speeds are compared with the satellite speeds for circular orbits. When the radius of the elliptical orbit equals a, at 13,185 n. mi., i.e. at a covertex, the speed is identical with the speed for the circular orbit with the same radius. At lower radii, nearer perigee, the speed around the ellipse exceeds the speed for a circular orbit with the same radius. At higher radii, the elliptical speeds are lower then the circular at the same radii.

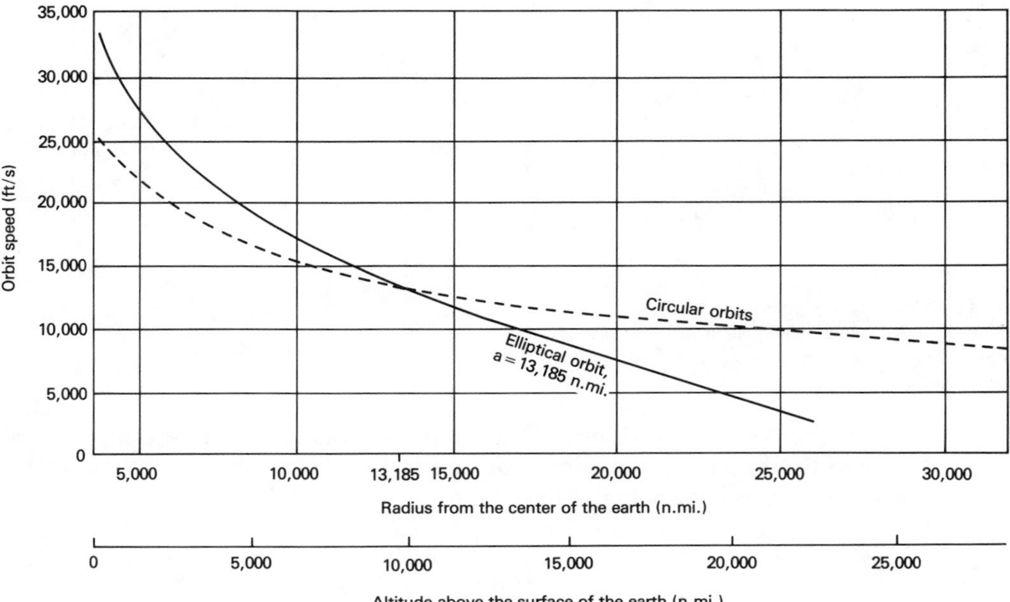

Figure 20.7 Variation of speed with radius for circular orbits and for an elliptical orbit with semi-major axis = 13,185 n. mi. (From Reference 20.2)

The time to complete a circuit of an ellipse (i.e., the period) can be shown to be

$$\tau = \frac{2\pi\sqrt{a^3}}{r_0\sqrt{g_0}} \tag{20.29}$$

Again, for a circular orbit, $a = (r_0 + h)$, and equation 20.29 reduces to equation 20.17.

Another useful speed relation can be derived from Kepler's second law. At apogee and perigee, the flight paths are perpendicular to the radii. Thus at these points the elliptical paths match circular paths. The area swept by the radius of a circle moving through the angle θ is $\theta r^2/2$. In unit time, θ is V/r, so the area is $Vr/2$. Since the areas swept are equal in equal times at both the perigee and the apogee,

$$V_p r_p/2 = V_a r_a/2$$

and

$$V_p r_p = V_a r_a \tag{20.30}$$

Equation (20.30) can also be proved by dividing equation 20.27 by equation 20.28 and inserting equations 20.23 and 20.24. However, it is more instructive to derive it from the second law.

ORBITAL MANEUVERS

Spacecraft are not usually launched directly into their *target* orbits. Therefore, it is necessary to maneuver the spacecraft from their initial or *parking* orbit into the desired orbit. This procedure is called *orbit acquisition*.

The correction of an orbit can change the size, shape, or orientation in space of a satellite's orbit. The latter involves a change in the orbit plane, which we shall not discuss in depth in this introductory discussion.

Orbital parameters are changed by applying a velocity increment using the spacecraft's rockets. It should be remembered that, although for aircraft purposes we have often used the terms *speed* and *velocity* interchangeably, speed is the magnitude of velocity. Velocity is a vector comprising both magnitude and direction. In the aircraft discussions, we have generally dealt with steady-state conditions for which the velocity direction was clearly understood. In the case of orbital vehicles, thrust may be applied in any required direction to obtain the desired *velocity* increment. However, when orbits are to be changed within the same plane so that only the size and shape of an orbit is to be modified, only the speed or velocity magnitude need be corrected, provided that the speed increment is applied at the perigee or the apogee. If the orbital plane is to be altered, then the velocity direction must be changed.

Figure 20.8 shows an elliptical orbit and three reference circular orbits. One circular orbit is tangent to the elliptical orbit at perigee, while another is tangent at apogee. At these points of tangency, the velocity vectors point in the same direction. However, the magnitude of the velocity vectors (i.e., the speeds) are not equal. A third

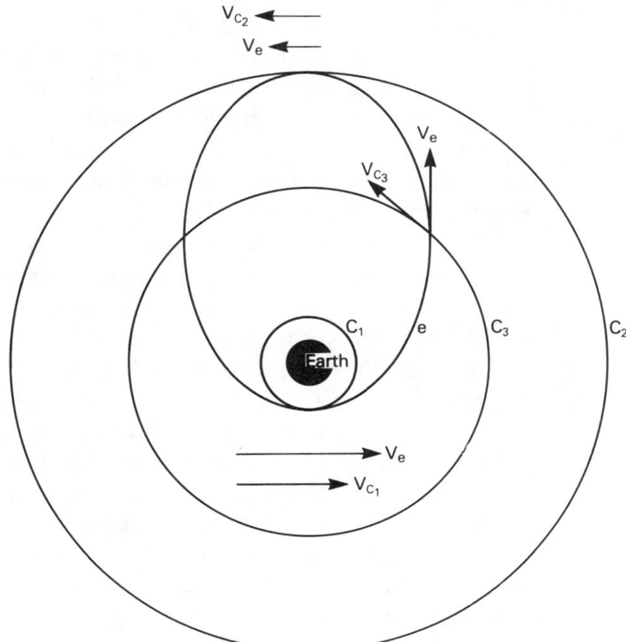

Figure 20.8 Comparison of satellite velocities in an elliptical orbit and several circular orbits.

circle with the same semimajor axis as the ellipse is drawn intersecting the ellipse at the covertices (i.e., the ends of the minor axis). We have noted that the speeds of satellites in circular and elliptical orbits of the same semimajor axis are equal when the elliptical orbit satellite is at radius $r = a$. This occurs at the covertices. The directions of the two satellites are different, however, and the flight paths are different. It is easier to change the orbit at tangent points where the directions of the original and the changed orbit are the same.

A velocity increment applied to a satellite is referred to as a *delta-V* since the symbol delta, Δ, denotes a change. A delta-V applied in the flight direction is known as an *in-track delta-V.* A *radial delta-V* produces a change in velocity directly toward or away from the center of the earth and changes the flight direction. In-track and radial velocity increments do not change the orbital plane. A change in the orbital plane requires a *cross-track* delta-V, a correction applied in a direction perpendicular to the orbital plane. Maneuvers performed without cross-track delta-V's are known as *coplanar transfers*.

We shall assume that the delta-V is applied instantaneously so that the spacecraft does not have time to move while the speed is being changed. This type of velocity change is called an *impulsive* delta-V. The assumption is usually valid for small velocity increments and for large delta-V's produced by solid rocket motors or high-thrust liquid rockets. Low-thrust corrections are much more difficult to evaluate since the distance covered during the maneuver may be considerable and the orbital direction is constantly changing.

In general, a satellite's initial or parking orbit will not be tangent to the final or target orbit. A third orbit that is tangent to both is frequently used; it is called a *transfer* orbit. A typical example is the requirement to transfer a spacecraft from a low-altitude circular orbit to a high-altitude circular orbit. The optimum method is to use a transfer orbit whose perigee radius equals the semimajor axis of the parking orbit and whose radius at apogee equals the semimajor radius of the target orbit. This procedure is called a *Hohmann transfer*. It is illustrated in Figure 20.9. Coplanar Hohmann transfers are used when maneuvering between coplanar circular orbits and can be accomplished using in-track delta-V's.

The required delta-V for an in-track correction is the difference in magnitude between the orbital velocities of the initial and final orbits. If the delta-V is applied in the direction of motion, the orbit size will increase. If the delta-V is applied in the direction opposite to the motion, the orbit size will decrease.

To make the maneuvers in-track, they are carried out at the vertices (i.e., perigee or apogee) of the elliptical orbit. Delta-V's applied at one vertex affect only the radius of the other. Thus, to raise the perigee radius, a velocity increment must be added at the apogee. Example 20.3 illustrates a Hohmann orbital transfer.

Noncoplanar Hohmann transfers involve delta-V's applied out of the plane of the satellite orbit. Instead of a simple arithmetic difference in speed, the velocities involved must be added vectorially using the law of the cosines. There are also other types of orbital transfers besides the Hohmann. These are beyond the scope of the introductory discussion given here.

There are many possible orbits around the earth. Among them are the *equatorial* orbit lying in the plane of the equator, *inclined* orbits in planes at various

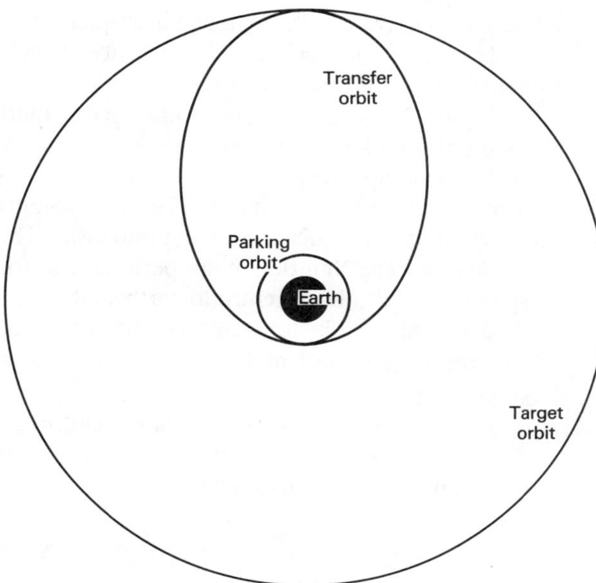

Figure 20.9 Coplanar Hohmann orbit transfer from the original parking orbit to an elliptical transfer orbit, which is tangent to both the parking orbit and the target orbit, and then to the target orbit.

angles to the equatorial plane, *polar* orbits in planes including the earth's poles, and *synchronous* orbits. A *synchronous* orbit is one in which the semimajor axis has the value that produces a period exactly equal to the time for the earth to rotate once on its axis. In the equatorial plane, this "stationary" satellite will appear to hover motionlessly over a point on the equator. The altitude for a synchronous circular orbit is 19,323 n. mi. above the equator. In an orbit inclined to the equatorial plane, the satellite will appear to describe a figure-eight curve of limited extent. The ground track of the figure-eight is called the *deadband* within which the satellite is always found. If the deadband is of negligible size, the satellite is said to be *geostationary*. Geostationary satellites are frequently used for communications.

The orbits we have discussed thus far are ideal orbits. An ideal orbit is one followed by a satellite on which the only force is the attraction of a spherical body. There are, however, other forces acting on a satellite. These forces may cause small deviations or perturbations in the orbits. Even though the effects are very small, a significant deviation from the desired orbit may occur over a long period of time. Among the causes of these perturbations are atmospheric drag, even if very small, the effects of the sun's electromagnetic radiation, which produces pressure on a satellite similar to wind pressure, the fact that the earth is not perfectly spherical, and gravitational forces of other bodies, such as the sun, moon, and planets.

The earth is slightly flattened at the poles and is approximately an ellipsoid of revolution. The deviation from spherical is small, the equator bulging only 11 n. mi. Nevertheless, the effect on certain orbits is detectable. In general, the higher the altitude of the orbit, the less the effect of the earth's shape.

The combined effects of the perturbing forces alter orbits enough so that for certain missions, for which precise orbits are required, periodic corrections must be made. These orbital adjustments are called *stationkeeping* maneuvers. While the adjustments may be both small and infrequent, they still require fuel to power the small on-board liquid-fuel rockets. The quantity of fuel carried will set a limit on the operational lifetimes of such satellites.

Figure 20.10 is a picture of the Space Shuttle, its huge droppable fuel tank, and its two solid rocket launching boosters. Most satellites are relatively small and designed to accomplish specialized functions such as communications, weather, navigation or scientific measurement. The Space Shuttle is an exception. It is designed to carry satellites and other equipment into orbit. The Space Shuttle is an airplane as well as a satellite. The Shuttle's wing permits it to follow a carefully controlled schedule of speed and altitude on return to earth so that the vehicle surface temperatures do not exceed the allowable temperatures for the structural materials. Thus, the Shuttle avoids the fiery end of most satellites that fall through the atmosphere.

Example 20.2

A satellite is to be boosted to an orbital altitude of 500 km. What is the orbital velocity and the orbital period?

Solution: From equation 20.15,

$$V_s = r_0\sqrt{\frac{g_0}{r_0 + h}}$$

r_0 is 6.378×10^6 m, $h = 500{,}000$ m, and $g_0 = 9.8$ m/s^2.

$$V_s = 6.378 \times 10^6\sqrt{\frac{9.8}{(6.378 \times 10^6) + (0.5 \times 10^6)}} = \underline{7613 \text{ m/s}}$$

The time for one revolution of the earth is

$$\tau = \frac{2\pi(r_0 + h)^{3/2}}{r_0\sqrt{g_0}} = \frac{2\pi[(6.378 \times 10^6) + (0.5 \times 10^6)]^{3/2}}{(6.378 \times 10^6)\sqrt{9.8}}$$

$$= \underline{5676.43 \text{ s}} = \underline{1.58 \text{ h}}$$

Example 20.3

A spacecraft has been launched into a circular parking orbit at an altitude of 300 n. mi. The final or target orbit, which lies in the same plane, is a circular orbit with an altitude of 7,500 n. mi. Determine the eccentricity and dimensional parameters of the appropriate Hohmann transfer orbit and the delta-V's required at each orbital transfer. What are the mass ratios, in terms of the ratio of fuel burned to initial weight, required to accomplish these maneuvers? Assume an effective rocket exhaust velocity of 3100 m/s.

Solution: The radius of the parking orbit is 300 n. mi. plus the radius of the earth. Noting that the radius of the earth is 20.92×10^6 ft,

$$h_{\text{park}} = 300 \text{ (n. mi.)} \times 6083 \text{ (ft/n. mi.)} = 1{,}824{,}900 \text{ ft}$$

$$r_{\text{park}} = 1{,}824{,}900 + 20{,}920{,}000 = \underline{22{,}744{,}900 \text{ ft}}$$

Similarly, the radius of the target orbit is

$$r_{\text{target}} = (7500 \times 6083) + 20{,}920{,}000 = \underline{66{,}542{,}500 \text{ ft}}$$

Figure 20.10 Space shuttle. (Courtesy of NASA.)

Referring to Figure 20.9, the transfer orbit will have a perigee radius equal to the radius of the parking orbit and an apogee radius equal to the radius of the target orbit. Since the sum of the radii for perigee and apogee equals $2a$, it follows that a for the transfer orbit is

$$a_{\text{transfer}} = \frac{r_p + r_a}{2} = \frac{22{,}744{,}900 + 66{,}542{,}500}{2} = \underline{44{,}643{,}700 \text{ ft}}$$

Thus the transfer orbit has a value of a of 44,643,700 ft, a perigee radius of 22,744,900 ft, and an apogee radius of 66,543,500 ft. From equation 20.23, the eccentricity is

$$e = 1 - \frac{r_p}{a} = 1 - \frac{22{,}744{,}900}{44{,}643{,}700} = \underline{0.49}$$

To inject the satellite from the parking orbit into the transfer orbit, the speed at the point of tangency of the orbits must be increased from the parking orbit speed to the transfer orbit speed (i.e., the perigee speed of the transfer orbit). The parking orbit speed is found from equation 20.15:

$$V_{\text{parking}} = r_0\sqrt{\frac{g_0}{r_0 + h}} = 20.92 \times 10^6\sqrt{\frac{32.17}{22{,}744{,}900}}$$
$$= 24{,}880 \text{ ft/s}$$

The transfer orbit perigee speed is found from equation 20.28:

$$V_p = \sqrt{\frac{g_0 r_0^2(1 + e)}{a(1 - e)}} = \sqrt{\frac{32.17(20.92 \times 10^6)^2(1 + 0.49)}{44{,}643{,}700(1 - 0.49)}}$$

$$= 30{,}354 \text{ ft/s}$$

Thus the first delta-V is

$$\Delta V = 30{,}354 - 24{,}880 = \underline{5474 \text{ ft/s}}$$

For the second maneuver from the transfer orbit to the target orbit, again refer to Figure 20.9, where it is seen that the radius of the circular target orbit equals the apogee radius of the elliptical transfer orbit. The speed of the target orbit is found from equation 20.15.

$$V_{\text{target}} = r_0\sqrt{\frac{g_0}{r_0 + h}} = 20.92 \times 10^6\sqrt{\frac{32.17}{66{,}542{,}500}}$$

$$= 15{,}546 \text{ ft/s}$$

The speed at the apogee of the transfer orbit can be found from equation 20.27. It is easier, however, to use equation 20.30. Then

$$V_a = V_p\frac{r_p}{r_a} = 30{,}354\frac{22{,}744{,}900}{66{,}542{,}500} = 10{,}375 \text{ ft/s}$$

The delta-V required to inject the satellite from the transfer orbit to the target orbit is the difference between the target speed and the apogee speed.

$$\Delta V = 15{,}546 - 10{,}375 = \underline{5171 \text{ ft/s}}$$

The mass ratio of the satellite (i.e., the ratio of the initial weight to the final weight) required to achieve the speed increases can be found from equation 20.3. Letting the weight ratio be R,

$$\log_e R = \log_e\left(\frac{\text{initial weight}}{\text{final weight}}\right) = \frac{\Delta V}{c_e}$$

$$R = e^{\Delta V/c_e}$$

For the first delta-V, with $c_e = 3100$ m/s or 10,171 ft/s,

$$R = e^{5474/10171} = e^{0.5382} = 1.713$$

The ratio of fuel to initial mass is $(R - 1)/R$ (the reader might like to prove this), so the fuel to initial mass ratio for the first orbit change is

$$\frac{1.713 - 1}{1.713} = \underline{0.416}$$

For the second delta-V, from the transfer to the target orbit, with $c_e = 3100$ m/s or 10,171 ft/s,

$$R = e^{5171/10171} = e^{0.5084} = 1.663$$

The ratio of fuel to initial mass for the second orbit change is

$$\frac{1.663 - 1}{1.663} = \underline{0.399}$$

Obviously, these major orbit changes consume major fuel quantities.

PROBLEMS

20.1. A satellite is found to complete one revolution around its earth orbit in 93 minutes. What is the satellite velocity? What is the orbital altitude?

20.2. A multistage rocket carrying an interplanetary spacecraft has been launched from the NASA space center in Florida. The spacecraft destination is Jupiter. The final stage of the rocket has been boosted to a speed of 8100 m/s at an altitude of 695 km. It is traveling perpendicular to the surface of the earth. The effective exhaust velocity of the final stage rocket is 3100 m/s and the burn time is 20 s. Determine the required mass ratio of the final stage rocket to avoid having the rocket fall back to earth (use an appropriate average value of g).

20.3. Consider a spacecraft in an elliptical orbit around the earth with a perigee altitude of 300 km and an apogee altitude of 1000 km. Determine the value of e, the orbit eccentricity. It is desired to enter a circular orbit at the apogee altitude. Assume that the effective rocket exhaust velocity is 3000 m/s. What is the magnitude of the speed increment required to accomplish the orbit circularization? How much fuel, expressed as a fraction of the spacecraft mass, is required to achieve this?

REFERENCES

20.1. Sutton, George P., and Ross, Donald M., *Rocket Propulsion Elements,* Wiley, New York, 1976.

20.2. Unpublished material by Andrew E. Turner, Angela M. Niles, Eugene L. Williams, and S. Rudolph Reyes.

Appendix A

CHARACTERISTICS OF THE STANDARD ATMOSPHERE

TABLE A.1 CHARACTERISTICS OF THE STANDARD ATMOSPHERE (SI UNITS)

Altitude, m	Temperature, T, K	Pressure, P, N/m^2	Density, ρ, kg/m^3	Speed of sound, m/s	Kinematic viscosity, m^2/s
0	288.16	1.01325^{+5}	1.2250	340.29	1.4607^{-5}
300	286.21	9.7773^{+4}	1.1901	339.14	1.4956
600	284.26	9.4322	1.1560	337.98	1.5316
900	282.31	9.0971	1.1226	336.82	1.5687
1,200	280.36	8.7718	1.0900	335.66	1.6069
1,500	278.41	8.4560	1.0581	334.49	1.6463
1,800	276.46	8.1494	1.0269	333.32	1.6869
2,100	274.51	7.8520	9.9649^{-1}	332.14	1.7289
2,400	272.57	7.5634	9.6673	330.96	1.7721
2,700	270.62	7.2835	9.3765	329.77	1.8167
3,000	268.67	7.0121	9.0926	328.58	1.8628
3,300	266.72	6.7489	8.8153	327.39	1.9104
3,600	264.77	6.4939	8.5445	326.19	1.9595
3,900	262.83	6.2467	8.2802	324.99	2.0102
4,200	260.88	6.0072	8.0222	323.78	2.0626
4,500	258.93	5.7752	7.7704	322.57	2.1167
4,800	256.98	5.5506	7.5247	321.36	2.1727

Altitude, m	Temperature, T, K	Pressure, p, N/m^2	Density, ρ, kg/m^3	Speed of sound, m/s	Kinematic viscosity, m^2/s
5,100	255.04	5.3331	7.2851	320.14	2.2305
5,400	253.09	5.1226	7.0513	318.91	2.2903
5,700	251.14	4.9188^{+4}	6.8234^{-1}	317.69	2.3522^{-5}
6,000	249.20	4.7217	6.6011	316.45	2.4161
6,300	247.25	4.5311	6.3845	315.21	2.4824
6,600	245.30	4.3468	6.1733	313.97	2.5509
6,900	243.36	4.1686^{+4}	5.9676^{-1}	312.72	2.6218^{-5}
7,200	241.41	3.9963	5.7671	311.47	2.6953
7,500	239.47	3.8299	5.5719	310.21	2.7714
7,800	237.52	3.6692	5.3818	308.95	2.8503
8,100	235.58	3.5140	5.1967	307.68	2.9320
8,400	233.63	3.3642	5.0165	306.41	3.0167
8,700	231.69	3.2196	4.8412	305.13	3.1046
9,000	229.74	3.0800	4.6706	303.85	3.1957
9,300	227.80	2.9455	4.5047	302.56	3.2903
9,600	225.85	2.8157	4.3433	301.27	3.3884
9,900	223.91	2.6906	4.1864	299.97	3.4903
10,200	221.97	2.5701	4.0339	298.66	3.5961
10,500	220.02	2.4540	3.8857	297.35	3.7060
10,800	218.08	2.3422	3.7417	296.03	3.8202
11,100	216.66	2.2346	3.5932	295.07	3.9564
11,400	216.66	2.1317	3.4277	295.07	4.1474
11,700	216.66	2.0335^{+4}	3.2699^{-1}	295.07	4.3475^{-5}
12,000	216.66	1.9399	3.1194	295.07	4.5574
12,300	216.66	1.8506	2.9758	295.07	4.7773
12,600	216.66	1.7654	2.8388	295.07	5.0078
12,900	216.66	1.6842	2.7081	295.07	5.2494
13,200	216.66	1.6067	2.5835	295.07	5.5026
13,500	216.66	1.5327	2.4646	295.07	5.7680
13,800	216.66	1.4622	2.3512	295.07	6.0462
14,100	216.66	1.3950	2.2430	295.07	6.3378
14,400	216.66	1.3308	2.1399	295.07	6.6434
14,700	216.66	1.2696	2.0414	295.07	6.9637
15,000	216.66	1.2112	1.9475	295.07	7.2995
15,300	216.66	1.1555	1.8580	295.07	7.6514
15,600	216.66	1.1023	1.7725	295.07	8.0202
15,900	216.66	1.0516	1.6910	295.07	8.4068
16,200	216.66	1.0033	1.6133	295.07	8.8119
16,500	215.66	9.5717^{+3}	1.5391	295.07	9.2366
16,800	216.66	9.1317	1.4683	295.07	9.6816
17,100	216.66	8.7119	1.4009	295.07	1.0148^{-4}
17,400	216.66	8.3115	1.3365	295.07	1.0637
17,700	216.66	7.9295^{+3}	1.2751^{-1}	295.07	1.1149^{-4}
18,000	216.66	7.5652	1.2165	295.07	1.1686
18,300	216.66	7.2175	1.1606	295.07	1.2249
18,600	216.66	6.8859	1.1072	295.07	1.2839
18,900	216.66	6.5696	1.0564	295.07	1.3457

TABLE A.2 CHARACTERISTICS OF THE STANDARD ATMOSPHERE (ENGLISH UNITS)

Altitude, ft	Temperature, T, °R	Pressure, p, lb/ft^2	Density ρ, lb s^2/ft^4 (slugs/ft^3)	Speed of sound, ft/s	Kinematic viscosity, ft^2/s
0	518.69	2116.2	2.3769^{-3}	1116.4	1.5723^{-4}
1,000	515.12	2040.9	2.3081	1112.6	1.6105
2,000	511.56	1967.7	2.2409	1108.7	1.6499
3,000	507.99	1896.7	2.1752	1104.9	1.6905
4,000	504.43	1827.7	2.1110	1101.0	1.7324
5,000	500.86	1760.9	2.0482	1097.1	1.7755
6,000	497.30	1696.0	1.9869^{-3}	1093.2	1.8201^{-4}
7,000	493.73	1633.1	1.9270	1089.3	1.8661
8,000	490.17	1572.1	1.8685	1085.3	1.9136
9,000	486.61	1512.9	1.8113	1081.4	1.9626
10,000	483.04	1455.6	1.7556	1077.4	2.0132
11,000	479.48	1400.0	1.7011^{-3}	1073.4	2.0655^{-4}
12,000	475.92	1346.2	1.6480	1069.4	2.1196
13,000	472.36	1294.1	1.5961	1065.4	2.1754
14,000	468.80	1243.6	1.5455	1061.4	2.2331
15,000	465.23	1194.8	1.4962	1057.4	2.2927
16,000	461.67	1147.5	1.4480^{-3}	1053.3	2.3541^{-4}
17,000	458.11	1101.7	1.4011	1049.2	2.4183
18,000	454.55	1057.5	1.3553	1045.1	2.4843
19,000	450.99	1014.7	1.3107	1041.0	2.5526
20,000	447.43	973.27	1.2673	1036.9	2.6234
21,000	443.87	933.26	1.2249^{-3}	1032.8	2.6966^{-4}
22,000	440.32	894.59	1.1836	1028.6	2.7724
23,000	436.76	857.24	1.1435	1024.5	2.8510
24,000	433.20	821.16	1.1043	1020.3	2.9324
25,000	429.64	786.33	1.0663	1016.1	3.0168
26,000	426.08	752.71	1.0292^{-3}	1011.9	3.1044^{-4}
27,000	422.53	720.26	9.9311^{-4}	1007.7	3.1951
28,000	418.97	688.96	9.5801	1003.4	3.2893
29,000	415.41	658.77	9.2387	999.13	3.3870
30,000	411.86	629.66	8.9068	994.85	3.4884
31,000	408.30	601.61	8.5841^{-4}	990.54	3.5937^{-4}
32,000	404.75	574.58	8.2704	986.22	3.7030
33,000	401.19	548.54	7.9656	981.88	3.8167
34,000	397.64	523.47	7.6696	977.52	3.9348
35,000	394.08	499.34	7.3820	973.14	4.0575
36,000	390.53	476.12	7.1028^{-4}	968.75	4.1852^{-4}
37,000	389.99	453.86	6.7800	968.08	4.3794
38,000	389.99	432.63	6.4629	968.08	4.5942
39,000	389.99	412.41	6.1608	968.08	4.8196
40,000	389.99	393.12	5.8727	968.08	5.0560
41,000	389.99	374.75	5.5982^{-4}	968.08	5.3039^{-4}
42,000	389.99	357.23	5.3365	968.08	5.5640
43,000	389.99	340.53	5.0871	968.08	5.8368
44,000	389.99	324.62	4.8493	968.08	6.1230
45,000	389.99	309.45	4.6227	968.08	6.4231
46,000	389.99	294.99	4.4067^{-4}	968.08	6.7380^{-4}
47,000	389.99	281.20	4.2008	968.08	7.0682
48,000	389.99	268.07	4.0045	968.08	7.4146
49,000	389.99	255.54	3.8175	968.08	7.7780

Altitude, ft	Temperature, T, °R	Pressure, p, lb/ft^2	Density ρ, lb s^2/ft^4 (slugs/ft^3)	Speed of sound, ft/s	Kinematic viscosity, ft^2/s
50,000	389.99	243.61	3.6391	968.08	8.1591
51,000	389.99	232.23	3.4692^{-4}	968.08	8.5588^{-4}
52,000	389.99	221.38	3.3072	968.08	8.9781
53,000	389.99	211.05	3.1527	968.08	9.4179
54,000	389.99	201.19	3.0055	968.08	9.8792
55,000	389.99	191.80	2.8652	968.08	1.0363^{-3}
56,000	389.99	182.84	2.7314^{-4}	968.08	1.0871^{-3}
57,000	389.99	174.31	2.6039	968.08	1.1403
58,000	389.99	166.17	2.4824	968.08	1.1961
59,000	389.99	158.42	2.3665	968.08	1.2547
60,000	389.99	151.03	2.2561	968.08	1.3161
61,000	389.99	143.98	2.1508^{-4}	968.08	1.3805^{-3}
62,000	389.99	137.26	2.0505	968.08	1.4481
63,000	389.99	130.86	1.9548	968.08	1.5189
64,000	389.99	124.75	1.8336	968.08	1.5932
65,000	389.99	118.93	1.7767	968.08	1.6712
66,000	389.99	113.39	1.6938^{-4}	968.08	1.7530^{-3}
67,000	389.99	108.10	1.6148	968.08	1.8387
68,000	389.99	102.06	1.5395	968.08	1.9236
69,000	389.99	98.253	1.4678	968.08	2.0230
70,000	389.99	93.672	1.3993	968.08	2.1219
71,000	389.99	89.305	1.3341^{-4}	968.08	2.2257^{-3}
72,000	389.99	85.142	1.2719	968.08	2.3345
73,000	389.99	81.174	1.2126	968.08	2.4486
74,000	389.99	77.390	1.1561	968.08	2.5653
75,000	389.99	73.784	1.1022	968.08	2.6938

Appendix B

DERIVATION OF THE COMPRESSIBLE FLUID BERNOULLI EQUATION

From equation 7.5,

$$\int_{p_T}^{p}\left(\frac{p}{p_T}\right)^{-1/\gamma}\frac{1}{\rho_T}\,dp + \int_0^V V\,dV = 0$$

so that

$$\frac{\gamma}{\gamma-1}\left[\frac{p^{1-1/\gamma}}{p_T^{-1/\gamma}}\frac{1}{\rho_T}\right]_{p_T}^{p} + \left[\frac{V^2}{2}\right]_0^V = \text{constant}$$

$$\frac{\gamma}{\gamma-1}\left[\frac{p^{(\gamma-1)/\gamma}}{p_T^{-1/\gamma}}\frac{1}{\rho_T} - \frac{p_T^{1-1/\gamma}}{p_T^{-1/\gamma}}\frac{1}{\rho_T}\right] + \frac{V^2}{2} = \text{constant}$$

$$\frac{\gamma}{\gamma-1}\left[\frac{p^{(\gamma-1)/\gamma}}{p_T^{-1/\gamma}}\frac{1(p_T)}{\rho_T(p_T)} - \frac{p_T}{\rho_T}\right] + \frac{V^2}{2} = \text{constant}$$

$$\frac{\gamma}{\gamma-1}\frac{p_T}{\rho_T}\left[\left(\frac{p}{p_T}\right)^{(\gamma-1)/\gamma} - 1\right] + \frac{V^2}{2} = \text{constant}$$

$$\frac{\gamma}{\gamma-1}\frac{p_T}{\rho_T}\left(\frac{p}{p_T}\right)^{(\gamma-1)/\gamma} + \frac{V^2}{2} = \text{constant} + \left(\frac{\gamma}{\gamma-1}\right)\frac{p_T}{\rho_T}$$

But when $V = 0$, $V^2 = 0$ and $p = p_T$.

$$\frac{\gamma}{\gamma - 1}\frac{p_T}{\rho_T}\left(\frac{1}{1}\right)^{(\gamma-1)/\gamma} + 0 = \text{constant} + \left(\frac{\gamma}{\gamma - 1}\right)\frac{p_T}{\rho_T}$$

Thus constant $= 0$ and

$$\frac{\gamma}{\gamma - 1}\frac{p_T}{\rho_T}\left(\frac{p}{p_T}\right)^{(\gamma-1)/\gamma} + \frac{V^2}{2} = \frac{\gamma}{\gamma - 1}\frac{p_T}{\rho_T} \qquad \text{compressible Bernoulli equation}$$

Appendix C

SUMMARY OF STATE AND ONE-DIMENSIONAL FLOW EQUATIONS

Equation of State

$$p = \rho RT \tag{5.1}$$

Incompressible Flow

$$V_1 S_1 = V_2 S_2 \qquad \text{continuity} \tag{6.2}$$

$$p_1 + \rho\frac{V_1^2}{2} = p_2 + \rho\frac{V_2^2}{2} \qquad \text{Bernoulli} \tag{6.7}$$

$$p_1 - p_2 = \frac{\rho}{2}(V_2^2 - V_1^2) = \frac{\rho V_2^2}{2}\left[1 - \left(\frac{V_1}{V_2}\right)^2\right] \tag{6.14}$$

$$V_0 = \sqrt{\frac{2(p_T - p_0)}{\rho}} \tag{6.11}$$

$$V_2 = \sqrt{\frac{2(p_1 - p_2)}{\rho[1 - (S_2/S_1)^2]}} \tag{6.16}$$

Compressible Flow

$$\rho_1 V_1 S_1 = \rho_2 V_2 S_2 \qquad \text{continuity} \tag{6.1}$$

$$\frac{p_2}{p_1} = \left(\frac{\rho_2}{\rho_1}\right)^\gamma \tag{7.1}$$

where γ is the ratio of the specific heats, $c_p/c_v = 1.4$ for air.

$$\frac{p_2}{p_1} = \left(\frac{T_2}{T_1}\right)^{\gamma/(\gamma-1)} \tag{7.2}$$

$$\frac{\rho_2}{\rho_1} = \left(\frac{T_2}{T_1}\right)^{1/(\gamma-1)} \tag{7.3}$$

$$\frac{\gamma}{\gamma - 1}\frac{p_T}{\rho_T}\left(\frac{p}{p_T}\right)^{(\gamma-1)/\gamma} + \frac{V^2}{2} = \frac{\gamma}{\gamma - 1}\frac{p_T}{\rho_T} \qquad \text{Bernoulli} \tag{7.6}$$

where $(\cdot)_T$ signifies stagnation ($V = 0$) conditions.

$$a^2 = \gamma RT, \qquad a = \sqrt{\gamma RT} \tag{7.14}$$

$$c_p T + \frac{V^2}{2} = c_p T_T \qquad \text{energy} \tag{7.15}$$

$$\frac{T_T}{T} = 1 + \frac{\gamma - 1}{2}M^2 \tag{7.18}$$

$$\frac{p_T}{p} = \left(1 + \frac{\gamma - 1}{2}M^2\right)^{\gamma/(\gamma-1)} \tag{7.19}$$

$$\frac{\rho_T}{\rho} = \left(1 + \frac{\gamma - 1}{2}M^2\right)^{1/(\gamma-1)} \tag{7.20}$$

$$M = \sqrt{\frac{2}{\gamma - 1}\left[\left(\frac{p_T}{p}\right)^{(\gamma-1)/\gamma} - 1\right]} \tag{7.21}$$

$$V_E = \sqrt{\frac{2}{\rho_s}(p_T - p)\frac{1}{1 + M^2/4 + M^4/40 + M^6/1600}} \tag{7.25}$$

$$\frac{dS}{S} = (M^2 - 1)\frac{dV}{V} \tag{7.40}$$

$$\left(\frac{S}{S^*}\right)^2 = \frac{1}{M^2}\left[\frac{2}{\gamma + 1}\left(1 + \frac{\gamma - 1}{2}M^2\right)\right]^{(\gamma+1)/(\gamma-1)} \tag{7.41}$$

where S = channel area at the point where the Mach number is M
S^* = throat area

INDEX

A

B

C

D

E

S